Third Edition

PCR
Technology

Current Innovations

Third Edition

PCR
Technology

Current Innovations

Edited by
Tania Nolan
Stephen A. Bustin

CRC Press
Taylor & Francis Group
Boca Raton London New York

CRC Press is an imprint of the
Taylor & Francis Group, an **informa** business

CRC Press
Taylor & Francis Group
6000 Broken Sound Parkway NW, Suite 300
Boca Raton, FL 33487-2742

First issued in paperback 2016

© 2013 by Taylor & Francis Group, LLC
CRC Press is an imprint of Taylor & Francis Group, an Informa business

No claim to original U.S. Government works

Version Date: 20130416

ISBN 13: 978-1-138-19858-6 (pbk)
ISBN 13: 978-1-4398-4805-0 (hbk)

Library of Congress Cataloging-in-Publication Data

PCR technology : current innovations. -- 3rd ed. / editors, Tania Nolan, Stephen A. Bustin.
    p. ; cm.
Includes bibliographical references and index.
Summary: "A technique used to amplify the number of copies of a specific region of DNA, the polymerase chain reaction (PCR) is at the forefront of the dramatic development of biochemistry. This text provides the tools for developing innovative approaches to using this leading technology. It includes theoretical considerations, discussions, and a selection of state-of-the-art techniques for mutation studies, clinical diagnosis, and the detection of food-borne pathogens. This edition also discusses the preparation of PCR experiments, includes examples of analytical PCR divided into qualitative and quantitative applications, and explores preparative methods that address DNA generation for further analysis and in vitro evolution"--Provided by publisher.
ISBN 978-1-4398-4805-0 (hardcover : alk. paper)
I. Nolan, Tania. II. Bustin, Stephen A., 1954-
[DNLM: 1. Polymerase Chain Reaction--methods. 2. Nucleic Acids--analysis. QU 450]

QP606.D46
572.8'636--dc23   2013010828

Visit the Taylor & Francis Web site at
http://www.taylorandfrancis.com

and the CRC Press Web site at
http://www.crcpress.com

Contents

PART I   Template Preparation

PART II   Reaction Components

PART III   Instruments

PART IV   Design, Optimization, QC, and Standardization

PART V   Data Analysis

PART VI   Applications

Preface

It is astonishing to consider how much innovation surrounds the PCR, a technique that is, superficially, as simple as you can get. But the concepts discussed in *PCR Technology: Current Innovations, Third Edition* demonstrate numerous innovative tweaks, adaptations, formulations, concepts, and applications. Indeed, it is its very simplicity that allows PCR to be continuously expanded and reinvigorated, placing this technology firmly at the top of molecular techniques and, one could argue, making it the most important scientific tool ever invented.

It is curious, then, that the origin of the PCR is somewhat controversial and that the first public report of its use (1985) is predated by a long way by its theoretical description (1971). Kleppe and Gobind Khorana,[1] the Nobel laureate, published a description of a technique that represented the basic principles of a method for nucleic acid replication: "The principles for extensive synthesis of the duplexed tRNA genes which emerge from the present work are the following:

1. The DNA duplex would be denatured to form single strands. This denaturation step would be carried out in the presence of a sufficiently large excess of the two appropriate primers.
2. Upon cooling, one would hope to obtain two structures, each containing the full length of the template strand appropriately complexed with the primer.
3. DNA polymerase will be added to complete the process of repair replication. Two molecules of the original duplex should result.
4. The whole cycle could be repeated, there being added every time a fresh dose of the enzyme."

The authors go on to posit that "it is, however, possible that upon cooling after denaturation of the DNA duplex, renaturation to form the original duplex would predominate over the template–primer complex formation. If this tendency could not be circumvented by adjusting the concentrations of the primers, clearly one would have to resort to the separation of the strands and then carry out repair replication. After every cycle of repair replication, the process of strand separation would have to be repeated." This passage has been quoted many times and extensively scrutinized by patent lawyers, since it appears to be a clear description of the forerunner to the process now recognized as PCR.

Of course, we now know that the vast excess of primers favors the template–primer complex formation and drives the PCR. We could speculate that had the authors carried out an empirical assessment of their idea, the PCR would have been invented 14 years earlier. However, the concept proposed by Kleppe and coworkers was simply way ahead of the available technology in 1971. First, oligonucleotide synthesis was an expensive process performed only by organic chemists, of whom Khorana's group were among the few. Synthetic oligonucleotides were not as readily available as in 1985 and have become a low-cost commodity item we do not even think about any longer. Second, the description preceded the invention of DNA sequencing by either Gilbert and Maxam[2] or Sanger[3,4] and so the target sequence required for primer design was not readily available. This is of course a crucial issue and sometimes forgotten by those who criticize Kary Mullis.

As a consequence, he was the first person to actually amplify DNA and so was awarded the Nobel Prize for the invention of the technique and acknowledged this pioneering work of Kleppe and Khorana: "He [Kleppe] almost had it. He saw the problems but didn't realise how fast things happen [in the PCR]."[5]

In his Nobel Prize acceptance speech, Mullis described the journey of discovery and development of PCR, summarizing: "With two oligonucleotides, DNA polymerase and the four nucleoside triphosphates I could make as much of a DNA sequence as I wanted and I could make it on a fragment of a specific size that I could distinguish easily."[6]

Originally, Mullis had assumed that when primers were added to denatured DNA, they would be extended, the extension products would then become unwound from their templates, be primed again, and the process of extension repeated. Unfortunately, these events do not simply occur by diffusion. The DNA must be heated to almost boiling after each round of synthesis, in order to denature the newly formed, double-stranded DNA. This inactivated the Klenow fragment of DNA polymerase I enzyme that was being used for synthesis and so more enzyme was required at the start of each cycle (as predicted by Kleppe, 1971). Arguably, the critical development leading to the universal adoption of the PCR technique was the concept of using the thermal stable DNA polymerase that could tolerate the high temperature of the repeated denaturation steps. The switch was made to the now commonly used *Taq* DNA polymerase that is extracted from the bacterium *Thermus aquaticus*,[7] which lives in thermal hot springs, and is resistant to permanent inactivation by exposure to high temperature.[8,9] This means that several rounds of amplification can be carried out in a closed reaction tube using the same batch of enzymes. In addition, using a thermally stable enzyme allowed for amplification of larger fragments and for the reaction to be performed at a higher temperature. The adoption of higher temperature reactions was sufficient to increase replication fidelity, reduce nonspecific product formation, and allow the products to be detected directly on ethidium bromide-stained, agarose gels.[10–12] An interesting aside is that the authors initially surmised that this enzyme might be useful as a reverse transcriptase, as it would be able to read through RNA secondary structure at elevated temperatures. They were not far wrong, since a related enzyme, *Tth* polymerase from *Thermus thermophiles*, does exactly that.

By 1989, the PCR technique was being used in all areas of modern biological sciences research, including clinical and diagnostic studies of detection of HIV in AIDS patients. The adoption of the PCR technique has been perhaps the major force driving the revolution in life sciences and associated fields we have all been privileged to witness.

The standard PCR is a deceptively simple process: low concentrations of template DNA, from a theoretical single copy to approximately 10^{11} copies, are combined with oligonucleotide primers in a reaction buffer containing variations on a basic composition consisting of ammonium sulfate, Tris, EDTA, BSA, β-mercaptoethanol, dNTPs, $MgCl_2$, KCl, NaCl, and DNA polymerase. Application of heat and a differential temperature profile, repeated 30–40 times, results in the amplification of the original target, which can then be visualized, purified, and used for a wide range of downstream manipulations.

In reality, there are numerous considerations for each of these components that transform the simple PCR concept into a series of steps that need to be carefully considered, optimized, and validated. The purification and handling of the template, prior to inclusion in the reaction, can have a profound effect on results. This is particularly important when a quantitative assessment is required. These factors are discussed in detail in Chapters 1 and 2. While the oligonucleotides may be considered a standard component of the reaction, these must be designed with care and the reaction conditions optimized to ensure that the results represent the true underlying biology, as described in Part IV. In addition, the manufacture and purification of the oligonucleotide can contribute to PCR behavior (Chapter 3) and so the appropriate synthesis and purification should be selected.

The absolute, optimum buffer composition is dependent upon the DNA polymerase used; different enzymes can affect PCR efficiency and therefore product yield.[13] In Chapter 4, Ernie Mueller presents a detailed discussion of the potential components for reaction buffers and in Chapter 8, new alternatives to the standard dNTPs are described.

As PCR reaches later cycles, it enters the plateau phase. This deviation from exponential amplification makes quantification inaccurate when based on estimates from end point product yield. Quantification using PCR requires that a modification to the basic PCR technique is made. This can involve digital PCR approaches or taking measurements earlier in the process, prior to the plateau phase. Higuchi et al.[14] recognized that the process of PCR could be tracked by including a fluorescent label into the reaction that could bind to the accumulating PCR product. As the PCR product increases, the fluorescent emission and therefore intensity of the signal also increase. In current real-time quantitative PCR (qPCR) technology, these signals are generated by inclusion of either fluorescent DNA-binding dyes or additional oligonucleotide probes. Fluorescent DNA-binding dyes, such as SYBR Green I, and related derivatives such as BEBO or BOXTO,[15,16] are included in the PCR buffer along with the DNA primers. As the target is amplified, the dye binds to the DNA product and adopts an alternative conformation. This conformational change results in an increase in fluorescent emission. Alternatively, a labeled primer can be included in the reaction, as described in Chapter 5. Additional specificity is achieved by including an additional oligo probe situated between the two primers. This oligo probe is labeled (Chapter 5) and, in most cases, also quenched. Various probe options are available but the most popular are the linear hydrolysis probe (also referred to as TaqMan probes),[17] Molecular Beacons,[18] Scorpions[19] (Chapter 6), and LightCycler® probes.[20] Linear hydrolysis probes are oligos with a fluorescent label on the 5′ end and a quencher molecule on the 3′ end. These are designed to have an annealing temperature above that of the primers and therefore, as the reaction is cooled from the melting temperature to the annealing temperature of the primers, the probe hybridizes to the target sequence. On further cooling, the primers hybridize and the new strand is elongated until the DNA polymerase reaches the 5′ of the probe. The probe is then cleaved by 5′–3′ exonuclease activity of the enzyme, releasing the fluorescent label. In this way, a fluorescent label should be released with each amplicon synthesized. In reality, between 4% and 47% of amplicons are detected using the linear probe system. Alternative probe systems may be used to increase detection sensitivity or specificity. Different probe methods provide different sensitivities of detection due to greater efficiency in separating the fluorescent label from the quencher.[21] For example, we have observed that incorporation of the locked nucleic acid (LNA™)-modified nucleotides (Chapter 7) residues into a linear hydrolysis probe can increase detection sensitivity by up to 10-fold.

Scorpion probes are a structured detection system. These combine the forward primer and detection probe into a single molecule, also holding the fluorophore and quencher in close proximity with a stem structure. Initially the primer region hybridizes and elongates from the single-stranded target. After melting away the template, the Scorpion opens and the probe region hybridizes to the target region, separating the label from the quencher. In Chapter 6, David Whitcombe provides a detailed description of how to design these molecules and also how to apply them to genotyping experiments.

The sensitivity of any assay is, in part, determined by the assay design. Much has been written about the design and optimization of qPCR assays.[22] In Chapters 12 through 14 guidelines are detailed that would enable any bioscientist to design and optimize their own PCR or qPCR assay regardless of the application. Chapter 15 contains a description of a case study that illustrated the dangers of ignoring best practice and working with poorly defined assays.

The replicative power of PCR is phenomenal. In theory, a single-template molecule should be replicated during each PCR cycle. Assuming absolute, perfect replication at each cycle, this would lead to 2^{40} or 1^{12} amplicon molecules after 40 rounds of amplification. By now, there could be something approaching a similar number of applications for the technique and a description of just a few of these is included in Part VI of this book. The sensitivity achieved using PCR allows infectious diseases such as HIV, TB, and malaria to be detected with tremendous sensitivity and treatment efficacy monitored; however, there is a requirement for this to be performed within the highly restrictive environment of the developing world. In Chapter 25, Clare Watt and Jim Huggett share their experience of setting up diagnostic services in more challenging situations.

In addition to genomic sequence analysis, it has long been believed that investigating specific mRNA sequences can be informative about the biology of the cell. Investigating gene quantity changes, for example, between normal tissues and diseased cells or looking for changes in gene expression in response to drug treatments is being used to understand how regulation of gene expression is a part of the complex system of control of cellular processes. Measuring mRNA requires an additional step in order to convert RNA to a DNA template that is suitable for PCR amplification.

This is carried out using a reverse transcriptase enzyme and extension from one or more oligo primers. Priming of the reverse transcription may be from a sequence-specific primer, from a series of random primers that hybridize along the length of the mRNA, or from a primer directed toward a tract of adenosines that is added to the 3′ end of most messenger RNA sequences, referred to as the poly A tail. After elongation from the primer, a double-stranded hybrid of RNA and DNA, called single-strand cDNA, is produced. This cDNA is then a suitable template for PCR and relative quantities of specific RNA templates are determined carefully in a process of semi-qPCR or qPCR (as described above).

The availability of vast amounts of sequence data from the various genome and transcriptome sequencing projects has also enabled the identification of small RNA species (ncRNA), including miRNA genetic markers that are characteristic for specific diseases. Using PCR to amplify these targets is particularly challenging because they are so small and do not have natural poly A tails. In Chapter 22, Castoldi et al. present a novel technique to quantify all RNA species in a sample, including miRNA and precursor molecules, thus enabling the components in a miRNA pathway to be quantified. These miRNA genetic markers can be tracked and used to monitor patient relapse, along with effectiveness of drug and transplant treatments (Chapter 18).

When launching into a project requiring PCR, it is important to ensure that instrumentation is appropriate and functioning correctly. PCR instruments have become notorious due to variability between instruments and within a single block. It is not unheard of to have a reaction that functions only in the central wells and not around the edges, due to lack of thermal uniformity (described in detail in Chapter 9). There is now a huge effort to develop and refine instrumentation. Drives to miniaturize (Chapter 11) and also to increase the uniformity and rate of reactions (Chapter 10) are described which give a wonderful insight into the world of engineering that addresses the biological need to push PCR further. Similarly, the fundamentals of the PCR technique are being adapted to quantify proteins using proximity ligation assay (Chapter 27) or to identify small sequence changes such as single-nucleotide polymorphisms (SNPs) using differences in the melting behavior of the PCR amplicon, referred to as high-resolution melting analysis (Chapter 28).

In order to protect against uncertainties due to reaction variability, it is important to include a series of controls alongside all samples for any experiment.

The choice of controls and whether to include them should always depend only upon the nature of the study. Factors such as the need to complete a paper/report/thesis/presentation, the cost of the experiment or the date of the next grant body's review meeting may be important to scheduling but they should not influence the experimental design. An inexpensive experimental design, run very quickly but without controls to verify that the findings are genuine, is not just completely worthless, but if published enters the peer-reviewed literature and becomes virtually impossible to erase. It amounts to a false economy of all resources to run inadequate experiments that could even result in retractions. Some controls must be obligatory and honest, original data should be presented for inspection, especially when the results of the study or analysis lead to life and death decisions. It is worth recognizing that the simple disregard for the data from controls in a study claiming a link between autism and the MMR (measles, mumps, rubella) vaccination lead to the distress of parents of children affected by the disorder and even to loss of life as a result of parents refusing to have their children vaccinated. The sobering story is told by Tania Nolan and Stephen Bustin in Chapter 15 as they outline the journey and justification for the MIQE (Minimum Information for Publication of Quantitative Real-Time PCR Experiments) guidelines.[23] It is easy for scientists who are engrossed in a research project and chained to a laboratory bench for 16 h/day to lose sight of their responsibility to the scientific community. While there is a constant and substantial pressure to publish, this should never be to the detriment of the best-quality experiments possible or without absolute consideration for the complete truth.

In conclusion, the work presented in this third edition of *PCR Technology: Current Innovations* demonstrates that there are still numerous new developments affecting the "simple" PCR and we would hazard a guess that when the fourth and fifth editions are published, the same will be true. We hope that readers of this book draw inspiration from the details presented by the numerous authors working in a wide range of fields and at the same time remember that while innovation is good, we must never lose sight of basic concepts of good science that can get overlooked in the rush to hop on the next bandwagon.

REFERENCES

1. Kleppe K, Ohtsuka E, Kleppe R et al. Studies on polynucleotides. XCVI. Repair replications of short synthetic DNAs as catalyzed by DNA polymerases. *J Mol Biol* 1971;56:341–361.
2. Gilbert W, Maxam A. The nucleotide sequence of the lac operator. *Proc Natl Acad Sci USA* 1973;70:3581–3584.
3. Sanger F, Nicklen S, Coulson AR. DNA sequencing with chain-terminating inhibitors. *Proc Natl Acad Sci USA* 1977;74:5463–5467.
4. Sanger F, Coulson AR. A rapid method for determining sequences in DNA by primed synthesis with DNA polymerase. *J Mol Biol* 1975;94:441–448.
5. Mullis KB, Francoise Ferre, Gibbs R. *The Polymerase Chain Reaction: A Textbook*. Birkhauser, Boston; 1994.
6. Mullis KB. *The Polymerase Chain Reaction (Nobel Prize Acceptance Speech)*. World Scientific Publishing, Singapore; 1993.
7. Chien A, Edgar DB, Trela JM. Deoxyribonucleic acid polymerase from the extreme thermophile *Thermus aquaticus*. *J Bacteriol* 1976;127:1550–1557.
8. Saiki RK, Gelfand DH, Stoffel S et al. Primer-directed enzymatic amplification of DNA with a thermo-stable DNA polymerase. *Science* 1988;239:487–491.
9. Lawyer FC, Stoffel S, Saiki RK et al. High-level expression, purification, and enzymatic characterization of full-length *Thermus aquaticus* DNA polymerase and a truncated form deficient in 5′ to 3′ exonuclease activity. *PCR Methods Appl* 1993;2:275–287.
10. Saiki RK, Scharf S, Faloona F et al. Enzymatic amplification of beta-globin genomic sequences and restriction site analysis for diagnosis of sickle cell anemia. *Science* 1985;230:1350–1354.
11. Mullis K, Faloona F, Scharf S et al. Specific enzymatic amplification of DNA *in vitro*: The polymerase chain reaction. *Cold Spring Harb Symp Quant Biol* 1986;51(Pt 1):263–273.
12. Mullis KB, Faloona FA. Specific synthesis of DNA *in vitro* via a polymerase-catalyzed chain reaction. *Methods Enzymol* 1987;155:335–350.
13. Wolffs P, Grage H, Hagberg O et al. Impact of DNA polymerases and their buffer systems on quantitative real-time PCR. *J Clin Microbiol* 2004;42:408–411.
14. Higuchi R, Fockler C, Dollinger G et al. Kinetic PCR analysis: Real-time monitoring of DNA amplification reactions. *Biotechnology (NY)* 1993;11:1026–1030.
15. Bengtsson M, Karlsson HJ, Westman G et al. A new minor groove binding asymmetric cyanine reporter dye for real-time PCR. *Nucleic Acids Res* 2003;31:e45.
16. Ahmad AI. BOXTO as a real-time thermal cycling reporter dye. *J Biosci* 2007;32:229–239.
17. Holland PM, Abramson RD, Watson R et al. Detection of specific polymerase chain reaction product by utilizing the 5′–3′ exonuclease activity of *Thermus aquaticus* DNA polymerase. *Proc Natl Acad Sci USA* 1991;88:7276–7280.
18. Tyagi S, Kramer FR. Molecular beacons: Probes that fluoresce upon hybridization. *Nat Biotechnol* 1996;14:303–308.
19. Whitcombe D, Theaker J, Guy SP et al. Detection of PCR products using self-probing amplicons and fluorescence. *Nat Biotechnol* 1999;17:804–807.
20. Wittwer CT, Herrmann MG, Moss AA et al. Continuous fluorescence monitoring of rapid cycle DNA amplification. *Biotechniques* 1997;22:130–138.
21. Wang L, Blasic JR, Jr., Holden MJ et al. Sensitivity comparison of real-time PCR probe designs on a model DNA plasmid. *Anal Biochem* 2005;344:257–265.
22. Nolan T, Hands RE, Bustin SA. Quantification of mRNA using real-time RT-PCR. *Nat Protoc* 2006;1:1559–1582.
23. Bustin SA, Benes V, Garson JA et al. The MIQE guidelines: Minimum information for publication of quantitative real-time PCR experiments. *Clin Chem* 2009;55:611–622.

Editors

Tania Nolan, BSc (Hons) PhD, earned a first class honors degree and was awarded the Excellence in Research Prize from undergraduate studies at Salford University and then a PhD in genetics from Manchester University, UK. She took an AstraZeneca Fellowship to study genetic regulation in breast cancer before moving to industry as an application scientist supporting the launch of a qPCR program. Dr. Nolan is an internationally recognized molecular biologist with a reputation for expertise in the field of mRNA quantification using RT-qPCR. She was a significant coauthor of the specialist textbook *The A–Z of Quantitative PCR* (ed. S.A. Bustin, IUL Press) along with numerous other book chapters and peer-reviewed papers. Among the publications are those describing the "SPUD" and "3′/5′mRNA integrity" assays. Dr. Nolan is an enthusiastic teacher and regularly organizes qPCR workshops. She has presented several plenary lectures and chaired sessions at major international qPCR meetings. Currently, she is a global manager, Technical and Application Support Team, Sigma Custom Products.

Stephen A. Bustin, BA(Mod) PhD FSB, obtained his PhD from Trinity College Dublin and has been working in the field of PCR (polymerase chain reaction) since 1987. He was professor of molecular science at Queen Mary University of London until 2012 and is currently professor of allied health and medicine at Anglia Ruskin University as well as visiting professor of molecular biology at the University of Middlesex. He acquired his first qPCR instrument in 1997 and has published numerous peer-reviewed papers that describe and use this technology. He wrote and edited the *A–Z of Quantitative PCR* (2004), universally acknowledged as the "qPCR Bible," edited *The PCR Revolution* (2011), and has written a series of *Definitive qPCR ebooks* (www.qPCRexpert.com). He led the international consortium that drew up the MIQE guidelines (2009) and is in constant demand as a speaker and teacher at international qPCR meetings and workshops. At the 2007 Autism trial at the Office of Special Masters of the U.S. Court of Federal Claims, he was an expert witness for the Department of Justice. Professor Bustin has extensive editorial involvements as editor-in-chief, *Gene Expression, International Journal of Molecular Sciences*, section editor, *Gene Expression, BMC Molecular Biology*, member of the editorial board of *Gene Regulation and Systems Biology* (Libertas Academica) and member of the editorial board of *Biomarkers in Medicine* (Future Science Group).

Contributors

Olga Adelfinskaya
TriLink Biotechnologies, Inc.
San Diego, California

Oto Akanji
Institute of Bioengineering
Queen Mary University of London
London, United Kingdom

Maria Galli de Amorim
Laboratory of Medical Genomics
AC Camargo Cancer Centre
São Paulo, Brazil

Ditte Andreasen
Exiqon A/S
Vedbaek, Denmark

Elena Hidalgo Ashrafi
TriLink Biotechnologies, Inc.
San Diego, California

Tzachi Bar
Labonnet
Ramat-Hasharon, Israel

Christiane Becker
Institute of Animal Nutrition
Liesel-Beckmann-Strasse
München, Germany

Vladimir Benes
European Molecular Biology Laboratory
Heidelberg, Germany

Einar S. Berg
Division for Infectious Disease Control
National Institute of Public Health
Oslo, Norway

Anders Bergkvist
Sigma-Aldrich Sweden AB
Stockholm, Sweden

Nick Burroughs
BJS Biotechnologies Ltd.
Perivale, Greenford
London, United Kingdom

Claire A. Bushell
LGC
Teddington, United Kingdom

Stephen A. Bustin
Postgraduate Medical Institute
Anglia Ruskin University
Chelmsford, United Kingdom

Mirco Castoldi
Clinic of Gastroenterology, Hepatology and
    Infectiology
Heinrich-Heine-University
Dusseldorf, Germany

Shiaw-Min Chen
Life Technologies, Inc.
Foster City, California

Tina Chowdhury
Institute of Bioengineering
Queen Mary University of London
London, United Kingdom

Paul Collier
European Molecular Biology Laboratory
Heidelberg, Germany

Theresa Creasey
Sigma-Aldrich
The Woodlands, Texas

Peter Damaschke
Department of Computer Science and
    Engineering
Chalmers University
Göteborg, Sweden

Philip Day
The Manchester Institute of
    Biotechnology
University of Manchester
Manchester, United Kingdom

Norha Deluge
Sigma-Aldrich
The Woodlands, Texas

Emmanuel Dias-Neto
Laboratory of Medical Genomics
AC Camargo Cancer Centre
and
Department and Institute of
    Psychiatry
Universidade de São Paulo
São Paulo, Brazil

Søren M. Echwald
Exiqon A/S
Vedbaek, Denmark

Malin Edling
NeuroSearch Sweden AB
Göteborg, Sweden

Giordana Feriotto
Department of Morphology
Ferrara University
Ferrara, Italy

Yannick Fillon
Cubist Pharmaceuticals
Lexington, Massachusetts

Simon Fredriksson
Olink Bioscience AB
Uppsala, Sweden

Roberto Gambari
Department of Life Sciences and
    Biotechnology
Ferrara University
Ferrara, Italy

Jorge A. Garcés
Genmark Dx
San Diego, California

Ray H. Gavin
Department of Biology
Brooklyn College–City University of
    New York
Brooklyn, New York

Michael Gotesman
Fish Medicine and Livestock Management
University of Veterinary Medicine
Vienna, Austria

Ashley R. Heath
Sigma-Aldrich
The Woodlands, Texas

Johannes Hedman
Department of Applied Microbiology
Lund University
Lund, Sweden

and

Department of Biology
Swedish National Laboratory of Forensic
    Science (SKL)
Linköping, Sweden

Tom Hendrikx
CYCLERtest BV
Landgraaf, the Netherlands

Jim Huggett
Nucleic Acid Metrology, Molecular
    and Cell Biology
LGC
Teddington, United Kingdom

Ehsan Karimiani
The Manchester Institute of Biotechnology
University of Manchester
Manchester, United Kingdom

Emmanouil Karteris
Department of Biosciences
Brunel University
Uxbridge, United Kingdom

Rickard Knutsson
Department of Bacteriology
National Veterinary Institute (SVA)
Uppsala, Sweden

Vlasta Korenkova
Institute of Biotechnology AS CR
Prague, Czech Republic

Kaarel Krjutškov
Estonian Genome Center
University of Tartu
Tartu, Estonia

Mikael Kubista
Institute of Biotechnology AS CR
Prague, Czech Republic

and

TATAA Biocenter
Göteborg, Sweden

Liu Xiao Kun
Temasek Life Sciences Laboratory
National University of Singapore
Singapore

Ulf Landegren
Department of Immunology, Genetics and
    Pathology/Molecular Medicine
Uppsala University
Uppsala, Sweden

Tony Le
TriLink Biotechnologies, Inc.
San Diego, California

Charlotta Löfström
National Food Institute
Technical University of Denmark
Søborg, Denmark

Maria Lövenklev
Department of Microbiology and Process
    Hygiene
Swedish Institute for Food and Biotechnology
    (SIK)
Göteborg, Sweden

Amelia Markey
School of Chemical Engineering and
    Analytical Science
Manchester Institute of Biotechnology
University of Manchester
Manchester, United Kingdom

Carlos Martinez
Technology Development and Engineering
Sigma Life Science–Custom Products
Sigma-Aldrich
The Woodlands, Texas

Andres Metspalu
Estonian Genome Center and Department of
    Biotechnology
Institute of Cell and Molecular biology
University of Tartu
Tartu, Estonia

Peter Mouritzen
Exiqon A/S
Vedbaek, Denmark

Ernest J. Mueller
Sigma-Aldrich Center for Biotechnology
St. Louis, Missouri

Azam Sheikh Muhammad
Department of Computer Science and
    Engineering
Chalmers University
Göteborg, Sweden

Gavin J. Nixon
LGC
Teddington, United Kingdom

Tania Nolan
Sigma-Aldrich Custom Products
Haverhill, United Kingdom

Xinghua Pan
Department of Genetics
Yale University School of Medicine
New Haven, Connecticut

Natasha Paul
TriLink Biotechnologies, Inc.
San Diego, California

Michael W. Pfaffl
Institute of Physiology
Center of Life and Food Sciences
    Weihenstephan
Technical University of Munich
Freising, Germany

Peter Rådström
Department of Applied Microbiology
Lund University
Lund, Sweden

Maido Remm
Department of Bioinformatics
University of Tartu
Tartu, Estonia

Irmgard Riedmaier
Institute of Physiology
Center of Life and Food Sciences
    Weihenstephan
Technical University of Munich
Freising, Germany

David W. Ruff
Fluidigm Corporation
South San Francisco, California

Vendula Rusnakova
Institute of Biotechnology AS CR
Prague, Czech Republic

Donald Salter
Institute of Genetics and
    Molecular Medicine
University of Edinburgh
Edinburgh, United Kingdom

Mark E. Shannon
Genomic Assays R&D
Life Technologies, Inc.
Foster City, California

Gregory L. Shipley
Shipley Consulting LLC
Missouri City, Texas

Sabrina Shore
TriLink Biotechnologies, Inc.
San Diego, California

Mary Span
CYCLERtest BV
Landgraaf, the Netherlands

Boel Svanberg
NeuroSearch Sweden AB
Göteborg, Sweden

David Svec
Institute of Biotechnology AS CR
Prague, Czech Republic

and

TATAA Biocenter
Göteborg, Sweden

Ales Tichopad
Charles University
Pilsen, Czech Republic

and

Biotechnology Institute
Academy of Science of the Czech Republic
Prague, Czech Republic

Alexander E. Urban
Department of Psychiatry and Behavioral
    Sciences
and
Center for Genomics and Personalized
    Medicine
Stanford University School of Medicine
Stanford, California

Marc Verblakt
CelsiusLabs
Landgraaf, the Netherlands

Clare Watt

Sherman M. Weissman
Department of Genetics
Yale University School of Medicine
New Haven, Connecticut

David Whitcombe
Clarity Now Limited
Northwich, United Kingdom

Selwyn A. Williams
Department of Biology
New York City College of Technology
Brooklyn, New York

Petra Wolffs
Department of Medical Microbiology
Maastricht University Medical Center
Maastricht, the Netherlands

Qin Xiang
Sigma-Aldrich
The Woodlands, Texas

Hong Yan
Temasek Life Sciences Laboratory
National University of Singapore
Singapore

Part I

Template Preparation

1 Pre-PCR Processing Strategies

Johannes Hedman, Maria Lövenklev, Petra Wolffs,
Charlotta Löfström, Rickard Knutsson, and Peter Rådström

CONTENTS

1.1 INTRODUCTION

Polymerase chain reaction (PCR) is a valuable tool for monitoring gene expression, quantifying food-borne pathogens, testing viral load, and also for forensic analysis and clinical diagnosis. Although PCR can be extremely effective with pure nucleic acids, its usefulness is limited, in part, by the presence of PCR-inhibitory substances originating from the samples or from the sample preparation, including the DNA extraction processes. These inhibitors can reduce or even block DNA amplification. Although many biological samples have been reported to inhibit PCR-based amplification, the biochemical and physical mechanisms and identities of many inhibitors remain unclear.

Pre-PCR processing is the strategy of combating PCR inhibition throughout the whole analysis chain, from sampling, via sample preparation, to the chemistry of the PCR.[1] In sampling, the amount of target cells/nucleic acids should be maximized and the amount of inhibitory compounds should be minimized. Sample preparation should remove or inactivate most inhibitors. The PCR chemistry, that is, the DNA polymerase with accompanying buffer and PCR facilitators, is vital for successful analysis. A customized PCR chemistry should serve to enable amplification, even in the presence of some inhibiting compounds.

The thermostable DNA polymerase is probably the most common target of PCR-inhibitory substances.[2] It is well known that the polymerization activity of DNA polymerases is affected by temperature, ionic strength, pH, buffer ions, sulfhydryl content, and other chemical agents.[3] Additionally, DNA polymerases from different sources exhibit different abilities to stay active in the presence of various disturbing compounds.[2] Recently, DNA analysis of complex crime scene samples was significantly improved by replacing the standard DNA polymerase Ampli*Taq* Gold with other more inhibitor-resistant DNA polymerases.[4] Another reason for the failure of PCR amplification is the status of the nucleic acids. For instance, any large or small molecule that binds to single- or double-stranded nucleic acids may alter the melting temperature.[5] This type of PCR inhibition can be caused by the presence of human immunoglobulin G, making the target nucleic acids unavailable for PCR.[6] Other causes of PCR inhibition are chelation of Mg^{2+} ions[7] or interference by cations such as Ca^{2+}, which compete with Mg^{2+} in binding to the DNA polymerase.[8] Lactoferrin inhibits PCR through its ability to release iron ions.[9] Generally, the PCR inhibitors may act through one or more of the following mechanisms: (i) inactivation of the thermostable DNA polymerase, (ii) degradation or capture of the nucleic acids, and (iii) interference with the real-time PCR quantitative (qPCR) detection system.

Humic acids have been found to quench the fluorescence from SYBR Green I dye in qPCR assays.[10] Recently, Half-Fraser media used for culture enrichment of *Listeria* have been demonstrated to emit background fluorescence that interfered with qPCR detection, possibly leading to false-negative results.[11] Additionally, nucleases from the sample may break down fluorescent probes. The complexity of qPCR has elevated the need for PCR-inhibitor inactivation or removal, and consequently several reports on the use of qPCR involve DNA extraction/purification and/or improved PCR chemistry, that is, pre-PCR processing.

1.2   SAMPLING

Sampling is the procedure of collecting or selecting representative samples from various organisms or environments. The choice of sampling technique depends on the purpose of the analysis, what to sample, the number of samples to be collected, the sample size, the type of contaminant, and the method of sample preparation. In many microbial samples such as foods, clinical specimens, and environmental samples, the target organisms are unevenly distributed, affecting the sampling procedure. Also, many microorganisms are attached to various surfaces where the critical step is to detach the organism to get a representative sample for PCR analysis. Swabbing techniques for various applications have been described in the literature, for example, to find fecal contaminations on beef carcasses,[12] to isolate *Escherichia coli* and *Enterobacteriaceae* on meat carcasses,[13–15] as well as for the recovery of *Listeria monocytogenes* from stainless steel,[16] saliva from human skin,[17] and human DNA from environmental samples.[18] As an alternative to swabbing surfaces, rinsing has been found to recover similar numbers of bacteria per unit area of poultry surfaces.[19] Sampling of surfaces is also used in combination with air sampling. Stetzenbach et al.[20] evaluated different swabbing techniques to recover bacteria from various indoor surfaces. Other swabbing techniques such as using sponge or macrofoam swabs have also been evaluated on various surfaces with different structures such as metal, glass, nylon cushion, and wood, for example, for the detection of *Erwinia herbicola* by qPCR[21] and for the recovery of *Bacillus anthracis* spores from steel.[22] It was concluded that the choice of the sampling method to obtain the best recovery of bacteria was dependent on the material and the surface structure.[21] In forensic analysis, tape lifting is a promising sampling technique, which has been successfully applied for sampling on human skin and clothing.[23,24]

Another approach is direct sampling, which involves securing the target organism/cells/nucleic acids directly from its surrounding matrix, for example, through excision of meat or clothing, or direct use of soil. These procedures usually give a good cell yield,[25] but also introduce high levels of PCR-inhibiting substances.[26,27] Excision can only be performed on a rather small area compared to swabbing. For soil, direct sampling is usually the only possible technique, making sample preparation and PCR chemistry customization very important.

1.3   SAMPLE PREPARATION

The objectives of sample preparation are (i) to reduce the size of the heterogeneous bulk sample to a small homogeneous PCR sample to ensure negligible variation between repeated analysis, (ii) to concentrate the target organism/cells/nucleic acids to a concentration within the practical operating range of the PCR system, and (iii) to remove or neutralize substances that may interfere with the PCR analysis. The type of the sample and the expected concentration of the analyte direct the prioritization of the above objectives and these, together with the needed time-to-results and cost aspects, will determine the choice of sample preparation method. Over the years, a large variety of methods have been developed, all of which affect the PCR analysis differently in terms of amplification efficiency, detection limit, and specificity. Sample preparation methods used for PCR can be divided into four major categories: (i) biochemical, (ii) culture enrichment, (iii) immunological, and (iv) physical methods. In recent years, combinations of these categories have increasingly been used (described in Section 1.3.5).

1.3.1 BIOCHEMICAL METHODS

The most traditional biochemical method is the extraction/purification of RNA or DNA with organic solvents, for example phenol–chloroform. The advantage of nucleic acid purification is that a homogeneous sample of high quality is obtained for reverse transcription and/or PCR amplification. In recent years, RNA and DNA extraction from a large variety of samples has been automated by using, for example, cationic magnetic beads or silica-based filters or suspensions to separate nucleic acids from sample matrices.[28,29] For clinical samples, such as blood, automated systems enable straightforward handling with few manual steps, but for more complex sample types, such as crime scene stains, several time-consuming pretreatment steps are generally needed prior to the automated procedure. In addition to the advances made on the extraction of total nucleic acids, new techniques are being developed to use differences in the biochemical composition of cell walls or membranes to selectively isolate DNA. By selective lysis of human blood cells and consequent destruction of human nucleic acids, followed by DNA extraction of the remaining bacterial cells and PCR, concentrations as low as 50 colony-forming units of *Staphylococcus aureus* per milliliter blood can be detected.[30] Another approach to accomplish a similar goal, that is, purification of bacterial DNA in a background of human DNA, was demonstrated by Horz et al.[31] In this study, total DNA was isolated after which bacterial DNA was purified by specific binding of nonmethylated CpG motifs, which are characteristic components of prokaryotic DNA. Biochemical methods can also be used for the separation of whole cells/particles from a complex sample. As an example, anion exchange filtration was used to separate norovirus particles from lettuce and fruits using charged nanoalumina filters instead of using filtration based on size.[32]

1.3.2 ENRICHMENT METHODS

Enrichment PCR involves the cultivation of the target microorganism prior to PCR. In spite of all the advances made in the area of biochemical, immunological, and physical sample preparation prior to molecular diagnostics testing, enrichment often remains essential for the detection of very low microbial loads. The aim of the enrichment culture is to provide detectable concentrations of viable target cells prior to PCR.[33] Furthermore, enrichment cultures prior to PCR analysis serve additional purposes, including the dilution of PCR-inhibitory substances present in the original sample matrix, dilution of dead target cells, and last but not least, the possibility of isolating the target cells for complementary microbial tests. Various enrichment PCR procedures have recently been used for the detection of bacteria such as pneumococci from nasopharyngeal samples,[34] *Campylobacter jejuni* from chicken,[35] or salmonellae from poultry fecal samples.[36] In addition, enrichment is also used prior to applying other molecular techniques such as microarrays, and diverse bacterial species were detected from positive blood cultures using a microarray-based assay.[37] The microarray system was developed to identify 50 bacterial species as well as the resistance marker mecA. Prior to microarray analysis, the samples were enriched by the traditional culture of blood cells, after which PCR-based amplification of *gyrB*, *parE*, and *mecA* from different species was performed. Using the new protocol, the identification of pathogens from blood cultures was achieved 18 h faster compared to conventional culture-based biochemical identification. For enrichment, selective and nonselective agar, as well as liquid enrichment media have been used, and the resulting specificity will depend on the characteristics of the medium and often varies with microorganism. Mafu et al.[38] used a single enrichment medium suitable for the enrichment of a variety of pathogenic microorganisms found on food contact surfaces. An additional challenge is that most media contain components that inhibit or interfere with PCR, such as blood, salts (e.g., $MgCl_2$ or bile salts), and/or components like malachite green oxalate.[11,39,40] For *Yersinia enterocolitica*, a PCR-compatible enrichment medium was developed, which removed the need for sample preparation.[41] This allowed simplified, integrated liquid handling; a closed-tube, qPCR analysis procedure; and the detection of small numbers of unlysed *Y. enterocolitica* in the presence of high concentrations of background flora.

1.3.3   Immunological Methods

Immunological sample preparation methods are mostly based on the use of magnetic beads coated with antibodies to separate target cells (human, bacterial, etc.) from their environment and often provide further concentration.[42] Because these methods are used to routinely purify intact cells containing all nucleic acids, immunomagnetic separation (IMS) methods are suited for both DNA and RNA detection.[43,44] A major advantage of these methods has proved to be that they are relatively easy to automate, for example, *E. coli* O157:H7 has been separated from poultry carcass rinse using automated IMS prior to microarray detection[45] and *Mycobacterium paratuberculosis* was separated from bovine milk using automated IMS prior to qPCR.[46] Special attention has to be paid to the fact that both the detection limit and the specificity of IMS-PCR protocols depend on the specificity of the antibodies and the binding capacity of the antibody-coated magnetic beads as well as on the characteristics of the PCR. The detection limit of an automated IMS-PCR method for the detection of *C. jejuni* from large-volume chicken skin rinses was limited by a relatively low recovery of magnetic beads from the system.[47] Furthermore, complex matrices can interfere with the interaction between antigen and antibodies. Finally, after immunocapturing, samples often require further processing such as lysis and washing or even nucleic acid purification prior to detection.[48]

1.3.4   Physical Methods

Aqueous two-phase systems,[49] buoyant density centrifugation,[50] flotation,[51] (ultra)centrifugation,[52] and filtration[53] are examples of methods that have been used as physical sample preparation methods prior to PCR. The success of these methods depends on the physical properties of the target cell, such as cell density and size. Knutsson et al.[54] have compared different physical sample preparation methods such as aqueous two-phase systems, buoyant density centrifugation, crude centrifugation, and dilution. In aqueous two-phase systems, PCR inhibitors and target cells were gently partitioned between two immiscible phases. For example, an aqueous two-phase system based on polyethylene glycol (PEG) 4000 and dextran 40 was used to detect *Y. enterocolitica* in pork and the enrichment medium.[55] The method is easy to perform, but phase separation may take 30–60 min. Buoyant density centrifugation has been shown to be a promising method if direct target detection is important. This treatment is used to isolate whole cells or particles with minimal processing.[51,56,57] Recently, Löfström et al.[58] developed a protocol that combines traditional buoyant density centrifugation using the flotation approach together with qPCR. The technique was used to extract and quantify living *Salmonella* cells in carcass gauze swabs from slaughterhouses and successfully minimized the influence of the carcass matrices containing PCR inhibitors derived from meat and blood, free nucleic acids, and microbial background flora. However, buoyant density-based methods have not been automated. Nonetheless, the main advantage of density centrifugation is the nondestructive extraction of intact target cells, thus improving detection limits.[54]

1.3.5   Combination of Methods

To obtain optimal detection limits, highest specificity, and shortest time-to-result, many sample preparation methods that have been published in recent years make use of a combination of the above methods. For example, the detection of very low concentrations of food-borne pathogens still generally requires enrichment. By combining it with other methods, such as IMS and/or DNA extraction followed by PCR detection, the time-to-result can be decreased.[48,59]

Another striking example demonstrated that by combining ultrafiltration with centrifugation and subsequent DNA purification using silica spin columns, concentrations as low as one echovirus, *Salmonella enterica* cell, or *Bacillus* spp. endospore could be detected per liter of water.[60] Although combining different methods can improve subsequent detection, it has to be borne in mind that such procedures often become prohibitively time-consuming and costly.

1.4 DNA POLYMERASES

A key component of PCR is the thermostable DNA polymerase, and any factor that compromises its enzymatic activity will generate a decreased formation of specific PCR products. The DNA polymerase can be degraded or denatured, or the enzymatic activity can be reduced by a wide variety of compounds that are present in the PCR samples.[2,61–63] A number of DNA polymerases are commercially available. Many of the currently available polymerases, including r*Tth* and *Tth* isolated from *Thermus thermophilus* and DyNazyme from *T. brockianus*, have been found by screening thermophilic organisms and phages for thermostable enzymes. Some of these enzymes, for example the *Taq* polymerase variants Ampli*Taq* Gold and Platinum *Taq*, have been improved by introducing hot-start properties, where the enzyme is activated by the heat in the first DNA denaturation step. The use of hot-start DNA polymerases has been shown to increase the product yield for highly degraded DNA from paraffin-embedded tissue[64] and from blood.[65,66]

Choosing the best-suited DNA polymerase is an important, but often overlooked, part of setting up a PCR system. The choice may be determined by several factors related to the application and influences the performance of a number of PCR-based applications such as restriction fragment length polymorphism (RFLP),[67] short tandem repeat (STR) analysis,[65] and randomly amplified polymorphic DNA (RAPD),[68,69] as well as peptide nucleic acid (PNA) clamp PCR,[70] PCR amplification of damaged DNA,[71] allele-specific PCR,[72] PCR mass spectrometry-based analyses,[73] bead-emulsion amplification,[74] reverse transcriptase PCR,[75] and qPCR.[76,77]

The various DNA polymerases differ in many features that are essential for PCR amplification such as thermostability,[70] extension rate,[78] processivity,[79,80] fidelity,[81,82] the ability to incorporate modified bases,[83,84] PCR-mediated recombination (chimera formation),[81] catalytic properties,[78] and termination of primer extension.[85] Furthermore, it has been shown that DNA polymerases from different sources may have different susceptibilities to PCR inhibitors.[2] For example, several *Taq* DNA polymerases have been shown to be susceptible to inhibition by various biological sample types, including clinical, environmental, forensic, and food samples, such as blood, cheese, feces, soil, and meat, as well as various ions.[2,62,86,87] Various studies have reported evaluations of the usefulness and characteristics of different DNA polymerases with respect to various PCR samples (Table 1.1). PCR detection was significantly improved by replacing *Taq* with *Tth* DNA polymerase for assays detecting *S. aureus* in bovine milk,[88] *Salmonella* in animal feed,[89] *Helicobacter hepaticus* in mice feces,[90] as well as influenza A in clinical specimens.[91] Furthermore, relief of inhibition from pig feces[92] and growth media[39] has been observed using the DNA polymerase r*Tth*. In several studies, the frequently used DNA polymerase Ampli*Taq* Gold has been shown to be inferior to alternative DNA polymerases in circumventing PCR inhibition from various sources.[86,87,93] For example, forensic DNA analysis of inhibited crime scene stains was significantly improved by replacing Ampli*Taq* Gold with one of the alternative DNA polymerases Bio-X-Act Short, Ex*Taq* Hot Start, or PicoMaxx High Fidelity.[4] Clearly, alternative polymerases can improve the performance of PCR and relieve the inhibition from biological samples.

Additionally, the properties of DNA polymerases may be improved by genetic modification using random mutagenesis or protein engineering (see Ref. [78] for a review). The introduction of specific mutations in both the wild-type *Taq* and its N-terminally truncated version Klen*Taq* enhanced resistance toward inhibitors in blood and soil.[93–95]

TABLE 1.1

Examples of Pre-PCR Processing Strategies

Sample Type	Inhibiting Substance[b]	Pre-PCR Processing[a]			
		Sample Preparation	DNA Polymerase	Facilitator[c]	Reference
Clinical					
Bile	–	Dilution, heating (98°C)	r*Tth*	Casein, formamide	[111]
Blood	–	Dilution, centrifugation, and washing with NaOH	*Taq* (Roche Diagnostics)	BSA	[6,9,109]
Blood	Hemoglobin, lactoferrin, immunoglobulin G	–	r*Tth*[d] (Applied Biosystems)	–	[2]
Feces	–	DNA extraction (Catrimox-14)	*Taq* (Takara Shuzo)	Gelatin	[112]
Feces	Bilirubin, urobilogens, bile salts, and humic materials	Filtration, DNA extraction (phenol–chloroform), Chelex-100	*Tth*[d] (Roche Diagnostics)	–	[90]
Feces (bovine)	Phytic acid	Stomaching, filtration, addition of NaOAc and phytase, centrifugation	*Taq* (Applied Biosystems)		[113]
Feces (porcine)	Polysaccharides, bilirubin, and bile salts	Enrichment and DNA extraction (PrepMan Kit [Applied Biosystems])	r*Tth*[d] (Applied Biosystems)	–	[92]
Middle ear effusion	–	DNA extraction (phenol–chloroform)	Ampli*Taq* Gold (Applied Biosystems)	–	[114]
Skeletal muscle tissue	Myoglobin	RNA extraction (modified acid guanidinium isothiocyanate phenol–chloroform method)	r*Tth*[d] (Applied Biosystems)	–	[87]
Swabs	–	RNA extraction (QIAamp Kit [Qiagen])	*Tth*[d] (Roche Diagnostics)	–	[91]
Tissue	Melanin	RNA extraction (phenol–chloroform)	Red Hot (Advanced Biotechnologies)	BSA	[115]
Tissue (paraffin-embedded)	Histological stains, preservatives	DNA extraction (QIA Quick Gel Extraction Kit [Qiagen])	Ampli*Taq* Gold (Applied Biosystems)	–	[64]
Vitreous fluids	–	–	*Tth*[d] (Promega Corporation), *Tfl*[d] (Promega Corporation)	Triton X-100	[116]
Urine	Urea, beta-HCG, crystals	DNA extraction (PEG)	Ampli*Taq* (Roche Diagnostics)	–	[117,118]
Environmental					
Gut content	–	DNA extraction (cetyltrimethylammonium bromide)	*Taq* (BioTherm)	BSA	[119]

TABLE 1.1 (continued)
Examples of Pre-PCR Processing Strategies

		Pre-PCR Processing[a]			
Sample Type	Inhibiting Substance[b]	Sample Preparation	DNA Polymerase	Facilitator[c]	Reference
Plant	Acidic polysaccharides	DNA extraction (cetyltrimethylammonium bromide)	Promega *Taq* (Promega Corporation)	Tween 20	[120]
Sludge	Polyphenols, humic acids, and heavy metals	Virus extraction, centrifugation with PEG, and RNA extraction (RNeasy Plant Mini Kit [Qiagen])	MuLV, Ampli*Taq* Gold (Applied Biosystems)	Polyvinyl-pyrrolidone (PVP), gp32	[108,121]
Soil	Humic materials	Enrichment and DNA extraction (QIAamp Tissue Kit [Qiagen])	Promega *Taq* (Promega Corporation)	–	[122]
Water	–	Filtration, enrichment, and DNA extraction (Isoquick DNA extraction [Ocra Research Ltd] or PrepMan Kit [Applied Biosystems])	Platinum *Taq* (Life Technologies)	–	[123,124]
Food					
Alfalfa seed	–	Enrichment and immunomagnetic separation	*Taq* (Roche Diagnostics)	Gelatin	[125,126]
Cold-smoked salmon	–	Silica column separation	*Taq* (Promega)	Tween 20	[127]
Meat	–	Enrichment and automated DNA extraction (Kingfisher, Thermo Labsystems, Helsinki, Finland)	*Tth* (Roche)	BSA, DMSO, glycerol	[59]
Meat (beef)	–	Enrichment and Percoll density centrifugation (flotation)	Ampli*Taq* (Applied Biosystems)	–	[57]
Meat (pork)	Collagen	DNA extraction (phenol–chloroform)	*Taq* (Roche Diagnostics)	MgCl$_2$	[7]
Milk	Proteases	DNA extraction (boiling, centrifugation), Chelex-100	*Tth*[d] (Roche Diagnostics)	–	[88,102]
Soft cheese	–	Aqueous two-phase systems	*Taq* (Roche Diagnostics)	–	[55]
Sugar solution	–	Filtration, cell lysis	*Taq* (Roche Diagnostics)	–	[128]
Forensic					
Bone	Calcium, collagen	DNA extraction (phenol, chloroform)	*Tth*[d] or *ExTaq HS*[d]	BSA	[86]

(*continued*)

TABLE 1.1   (continued)
Examples of Pre-PCR Processing Strategies

Sample Type	Inhibiting Substance[b]	Pre-PCR Processing[a]			Reference
		Sample Preparation	DNA Polymerase	Facilitator[c]	
Cigarette butts	–	DNA extraction (Chelex-100)	Bio-X-Act Short[d] (Bioline), Ex*Taq* HS[d] or PicoMaxx HF[d] (Stratagene)	BSA	[4]
Denim	Blue dye	DNA extraction (phenol, chloroform), Thiopropyl Sepharose 6B treatment	*Taq* (Perkin-Elmer)	BSA	[129]
Soil	Humic compounds	DNA extraction (phenol–chloroform) and washing with NaOH	*Taq*	BSA	[130]
Other					
Culture (bacteria)	–	Enrichment	r*Tth*[d] (Applied Biosystems)	Glycerol	[131]
Culture	–	Enrichment	*Tth*[d] (MBI Fermentas)	W1 detergent, BSA	[132]
DNA, RNA	Phenol	Chemical cell lysis	*Tth*[d] (Amersham)	–	[62]

[a] Pre-PCR processing aims at converting a complex biological sample containing the target microorganisms into PCR-amplifiable samples and includes sampling, sample preparation, DNA polymerases, and the use of facilitators.

[b] Inhibitor either suggested or identified in the reference.

[c] A facilitator is a compound that can improve the performance of PCR and which is added to the PCR reaction mix.

[d] An alternative DNA polymerase is used to overcome PCR inhibition.

1.5   PCR BUFFER COMPOSITION AND PCR FACILITATORS

The general PCR buffer composition is important for the activity of the DNA polymerase, with respect to both pH and cation content. It has been shown that PCR inhibition from components in blood samples can be relieved by elevating the pH of the buffer.[96] Additionally, increasing the concentration of Mg^{2+} ions may be a way of counteracting inhibition due to DNA intercalating dyes[97] or Ca^{2+} ions.[8]

Certain compounds, called amplification enhancers, amplification facilitators, or PCR facilitators, can improve the performance of PCR when added to the basic PCR master mix. They can be divided into five different groups: (i) proteins, (ii) organic solvents, (iii) nonionic detergents, (iv) biologically compatible solutes, and (v) polymers. With the commercial introduction of new DNA polymerases, a number of suppliers have added PCR facilitators to the accompanying buffers (a factor not always considered when the performances of different polymerases are compared).

PCR facilitators can affect amplification at different stages and under different conditions. They can increase or decrease the thermal stability of the DNA template. For example, organic solvents such as dimethyl sulfoxide (DMSO) and formamide destabilize nucleotide base pairs in solution.[98] Some amplification facilitators may affect the error rate of the DNA polymerase. Tween 20 reduces false terminations of the primer extension reaction.[99] Finally, facilitators can be used to relieve the amplification inhibition caused by complex biological samples. Many examples (Table 1.1) are known in this group, such as bovine serum albumin (BSA) and the single-stranded DNA-binding protein encoded by gene 32 of bacteriophage T4 (gp32). One of the proposed mechanisms behind the ability of BSA to relieve inhibition is that it binds to inhibitory compounds such as blood components, and phenolics.[100,101] Furthermore, proteins (BSA and gp32[102]) may act as a target for proteases, thereby protecting the polymerase or even the DNA itself (gp32[103]). gp32 also improves amplification by stabilizing ssDNA. PCR facilitators can be blended to take advantage of their different properties. Recently, the nonreducing detergent NP-40 was blended with the osmoprotecting sugar trehalose and L-carnitine, providing efficient amplification in the presence of inhibitors from blood and soil.[95]

1.6 CONCLUDING REMARKS

Pre-PCR processing includes all steps leading up to the PCR, that is, sampling, sample preparation, and PCR chemistry.[1] To fully benefit from the power of diagnostic PCR, the issue of inhibitors must be thoroughly addressed throughout the analytical process. When setting up a new assay, the first step of pre-PCR processing should be to identify the DNA polymerase buffer system most suitable for the analysis in question. Customizing the PCR chemistry has a great impact on the resistance to PCR inhibitors, since different DNA polymerases with their respective buffers and PCR facilitators have different abilities to withstand inhibiting substances from various sources.[2,4,77,100] Moreover, customizing the PCR chemistry does not affect the sample itself. Complex sample preparation, on the other hand, involves the risk of DNA loss. Recent research has shown that if an inhibitor-tolerant DNA polymerase and/or suitable PCR facilitators are chosen, it may be possible to add a biological sample directly to the PCR, without any sample preparation processes or cell lysis.[93–95,104] However, if the inhibitory effects are not eliminated through PCR chemistry customization, sampling and sample preparation must be improved.

The development of sampling techniques has not followed the quick development of PCR methods, and most sampling strategies used today, for example cotton swabbing, were developed for classical analysis methods,[105] and not for PCR. Considering the importance of sampling, more efforts need to be made to develop sampling strategies that are tailor-made and suitable for downstream PCR analysis.

The pretreatment of a complex biological sample is crucial, and for successful PCR, the following two requirements should be fulfilled[106]: (i) sufficient concentration of target nucleic acids and (ii) complete lack or low concentrations of PCR-inhibitory components in the sample. The aim of the sample preparation is to convert complex biological samples into homogeneous, PCR-amplifiable samples. Since most biological samples contain PCR inhibitors,[107] numerous pre-PCR processing protocols have been developed (Table 1.1). Many of these protocols combine sample preparation methods from different categories. A common strategy is to combine an enrichment method with a biochemical method[36,59,92,108] or with a physical sample preparation method.[109] In general, RNA/DNA extraction methods provide templates of high quality, but these methods are usually expensive and/or time-consuming. Physical methods are favorable as they do not influence the specificity of the PCR protocol in the way that the enrichment and immunological methods could. The reason for the many PCR protocols and pre-PCR processing methods employed is that the most suitable strategy depends on the specific nature of the sample and the purpose of the PCR analysis. For instance, various sample preparation methods have been developed to remove or reduce the effects of PCR inhibitors without knowing the identity of the PCR inhibitors or understanding the mechanisms of inhibition. Consequently, the conditions for DNA amplification must be optimized by the use of efficient pre-PCR processing.

Generally, several different pre-PCR processing strategies can be used, for example, (i) optimization of the DNA amplification conditions by the use of alternative DNA polymerases and/or amplification facilitators, (ii) optimization of the sample preparation method, and (iii) a combination of both strategies. The growing demand for rapid, robust, and user-friendly PCR protocols implies that research into pre-PCR processing is likely to expand in the future. In addition, we are gaining understanding of the mode of action of different PCR facilitators, as well as the mechanisms by which various factors negatively affect PCR. Recent work utilizing qPCR kinetics provided new insights into the manners of operation of the inhibitors humic acid, hematin, and melanin.[110] Studies such as this will make it easier to choose suitable modifications of the PCR chemistry to overcome the problem of PCR inhibition.

ACKNOWLEDGMENTS

This work was in part executed in the framework of the EU-project AniBioThreat (Grant Agreement: Home/2009/ISEC/AG/191) with the financial support from the Prevention of and Fight against Crime Programme of the European Union, European Commission—Directorate General Home Affairs. This article reflects the views only of the authors, and the European Commission cannot be held responsible for any use which may be made of the information contained herein. ML was financially supported by RISE, Research Institutes of Sweden. CL and PR wcrc financially supported in part by the European Union-funded Integrated Project BIOTRACER (contract FOOD-2006-CT-036272) under the 6th RTD Framework. PR was financially supported in part by the Swedish Research Council for Environment, Agricultural Sciences and Spatial Planning (FORMAS, 222-2007-373).

REFERENCES

1. Rådström, P., R. Knutsson, P. Wolffs, M. Lövenklev, and C. Löfström. 2004. Pre-PCR processing: Strategies to generate PCR-compatible samples. *Mol Biotechnol 26*:133–146.
2. Abu Al-Soud, W., and P. Rådström. 1998. Capacity of nine thermostable DNA polymerases to mediate DNA amplification in the presence of PCR-inhibiting samples. *Appl Environ Microbiol 64*:3748–3753.
3. Kornberg, A. 1980. *DNA Replication*. W.H. Freeman and Company, San Francisco, USA.
4. Hedman, J., A. Nordgaard, B. Rasmusson, R. Ansell, and P. Rådström. 2009. Improved forensic DNA analysis through the use of alternative DNA polymerases and statistical modeling of DNA profiles. *Biotechniques 47*:951–958.
5. Cantor, C.R., and P.R. Schimmel. 1980. *Biophysical Chemistry: The Behavior of Biological Macromolecules*. W.H. Freeman and Company, San Francisco, USA.
6. Abu Al-Soud, W., L.J. Jönsson, and P. Rådström. 2000. Identification and characterization of immuno-globulin G in blood as a major inhibitor of diagnostic PCR. *J Clin Microbiol 38*:345–350.
7. Kim, S., R.G. Labbe, and S. Ryu. 2000. Inhibitory effects of collagen on the PCR for detection of *Clostridium perfringens*. *Appl Environ Microbiol 66*:1213–1215.
8. Bickley, J., J.K. Short, D.G. McDowell, and H.C. Parkes. 1996. Polymerase chain reaction (PCR) detection of *Listeria monocytogenes* in diluted milk and reversal of PCR inhibition caused by calcium ions. *Lett Appl Microbiol 22*:153–158.
9. Abu Al-Soud, W., and P. Rådström. 2001. Purification and characterization of PCR-inhibitory components in blood cells. *J Clin Microbiol 39*:485–493.
10. Zipper, H., C. Buta, K. Lammle, H. Brunner, J. Bernhagen, and F. Vitzthum. 2003. Mechanisms underlying the impact of humic acids on DNA quantification by SYBR Green I and consequences for the analysis of soils and aquatic sediments. *Nucleic Acids Res 31*:e39.
11. Rossmanith, P., S. Fuchs, and M. Wagner. 2010. The fluorescence characteristics of enrichment media in the wavelength range of real-time PCR thermocycler optical path assignments. *Food Anal Methods 3*:219–224.
12. Dorsa, W.J., C.N. Cutter, and G.R. Siragusa. 1996. Evaluation of six sampling methods for recovery of bacteria from beef carcass surfaces. *Lett Appl Microbiol 22*:39–41.
13. Pearce, R.A., and D.J. Bolton. 2005. Excision vs sponge swabbing—A comparison of methods for the microbiological sampling of beef, pork and lamb carcasses. *J Appl Microbiol 98*:896–900.
14. Byrne, B., G. Dunne, J. Lyng, and D.J. Bolton. 2005. Microbiological carcass sampling methods to achieve compliance with 2001/471/EC and new hygiene regulations. *Res Microbiol 156*:104–106.
15. Lindblad, M. 2007. Microbiological sampling of swine carcasses: A comparison of data obtained by swabbing with medical gauze and data collected routinely by excision at Swedish abattoirs. *Int J Food Microbiol 118*:180–185.
16. Vorst, K.L., E.C. Todd, and E.T. Rysert. 2004. Improved quantitative recovery of *Listeria monocytogenes* from stainless steel surfaces using a one-ply composite tissue. *J Food Prot 67*:2212–2217.

17. Sweet, D., M. Lorente, J.A. Lorente, A. Valenzuela, and E. Villanueva. 1997. An improved method to recover saliva from human skin: The double swab technique. *J Forensic Sci 42*:320–322.

18. Toothman, M.H., K.M. Kester, J. Champagne, T.D. Cruz, W.S. Street, and B.L. Brown. 2008. Characterization of human DNA in environmental samples. *Forensic Sci Int 178*:7–15.

19. Gill, C.O., M. Badoni, L.F. Moza, S. Barbut, and M.W. Griffiths. 2005. Microbiological sampling of poultry carcass portions by excision, rinsing, or swabbing. *J Food Prot 68*:2718–2720.

20. Stetzenbach, L.D., M.P. Buttner, and P. Cruz. 2004. Detection and enumeration of airborne biocontaminants. *Curr Opin Biotechnol 15*:170–174.

21. Buttner, M.P., P. Cruz, L.D. Stetzenbach, and T. Cronin. 2007. Evaluation of two surface sampling methods for detection of *Erwinia herbicola* on a variety of materials by culture and quantitative PCR. *Appl Environ Microbiol 73*:3505–3510.

22. Hodges, L.R., L.J. Rose, A. Peterson, J. Noble-Wang, and M.J. Arduino. 2006. Evaluation of a macrofoam swab protocol for the recovery of *Bacillus anthracis* spores from a steel surface. *Appl Environ Microbiol 72*:4429–4430.

23. Li, R.C., and H.A. Harris. 2003. Using hydrophilic adhesive tape for collection of evidence for forensic DNA analysis. *J Forensic Sci 48*:1318–1321.

24. Gunnarsson, J., H. Eriksson, and R. Ansell. 2010. Success rate of a forensic tape-lift method for DNA recovery. *Problems Forensic Sci, 83*:243–254.

25. Snijders, J.M.A., M.H.W. Janssen, G.E. Gerats, and G.P. Corstiaensen. 1984. A comparative study of sampling techniques for monitoring carcass contamination. *Int J Food Microbiol 1*:229–236.

26. Tsai, Y.L., and B.H. Olson. 1992. Rapid method for separation of bacterial DNA from humic substances in sediments for polymerase chain reaction. *Appl Environ Microbiol 58*:2292–2295.

27. Tsai, Y.L., and B.H. Olson. 1992. Detection of low numbers of bacterial cells in soils and sediments by polymerase chain reaction. *Appl Environ Microbiol 58*:754–757.

28. Kessler, H.H., G. Muhlbaucr, E. Stelzl, E. Daghofer, B.I. Santner, and E. Marth. 2001. Fully automated nucleic acid extraction: MagNA Pure LC. *Clin Chem 47*:1124–1126.

29. Loens, K., K. Bergs, D. Ursi, H. Goossens, and M. Ieven. 2007. Evaluation of NucliSens easyMAG for automated nucleic acid extraction from various clinical specimens. *J Clin Microbiol 45*:421–425.

30. Hansen, W.L., C.A. Bruggeman, and P.F. Wolffs. 2009. Evaluation of new preanalysis sample treatment tools and DNA isolation protocols to improve bacterial pathogen detection in whole blood. *J Clin Microbiol 47*:2629–2631.

31. Horz, H.P., S. Scheer, M.E. Vianna, and G. Conrads. 2010. New methods for selective isolation of bacterial DNA from human clinical specimens. *Anaerobe 16*:47–53

32. Morales-Rayas, R., P.F. Wolffs, and M.W. Griffiths. 2009. Anion-exchange filtration and real-time PCR for the detection of a norovirus surrogate in food. *J Food Prot 72*:2178–2183.

33. Sharma, V.K., and S.A. Carlson. 2000. Simultaneous detection of *Salmonella* strains and *Escherichia coli* O157:H7 with fluorogenic PCR and single-enrichment-broth culture. *Appl Environ Microbiol 66*:5472–5476.

34. Carvalho, M.D., F.C. Pimenta, D. Jackson, A. Roundtree, Y. Ahmad, E.V. Millar, K.L. O'Brien, C.G. Whitney, A.L. Cohen, and B.W. Beall. 2010. Revisiting pneumococcal carriage using broth-enrichment and PCR techniques for enhanced detection of carriage and serotypes. *J Clin Microbiol 48*:1611–1618.

35. Rantsiou, K., C. Lamberti, and L. Cocolin. 2010. Survey of *Campylobacter jejuni* in retail chicken meat products by application of a quantitative PCR protocol. *Int J Food Microbiol 141*(Suppl. 1): S75–S79.

36. Löfström, C., F. Hansen, and J. Hoorfar. 2010. Validation of a 20-h real-time PCR method for screening of *Salmonella* in poultry faecal samples. *Vet Microbiol, 144*:511–514.

37. Tissari, P., A. Zumla, E. Tarkka, S. Mero, L. Savolainen, M. Vaara, A. Aittakorpi et al. 2010. Accurate and rapid identification of bacterial species from positive blood cultures with a DNA-based microarray platform: An observational study. *Lancet 375*:224–230.

38. Mafu, A.A., M. Pitre, and S. Sirois. 2009. Real-time PCR as a tool for detection of pathogenic bacteria on contaminated food contact surfaces by using a single enrichment medium. *J Food Prot 72*:1310–1314.

39. Knutsson, R., C. Löfström, H. Grage, J. Hoorfar, and P. Rådström. 2002. Modeling of 5′ nuclease real-time responses for optimization of a high-throughput enrichment PCR procedure for *Salmonella enterica*. *J Clin Microbiol 40*:52–60.

40. Hyeon, J.Y., I.G. Hwang, H.S. Kwak, C. Park, I.S. Choi, and K.H. Seo. 2010. Evaluation of PCR inhibitory effect of enrichment broths and comparison of DNA extraction methods for detection of *Salmonella enteritidis* using real-time PCR assay. *J Vet Sci 11*:143–149.

41. Knutsson, R., M. Fontanesi, H. Grage, and P. Rådström. 2002. Development of a PCR-compatible enrichment medium for *Yersinia enterocolitica*: Amplification precision and dynamic detection range during cultivation. *Int J Food Microbiol* 72:185–201.

42. Olsvik, O., T. Popovic, E. Skjerve, K.S. Cudjoe, E. Hornes, J. Ugelstad, and M. Uhlen. 1994. Magnetic separation techniques in diagnostic microbiology. *Clin Microbiol Rev* 7:43–54.

43. Park, S., B. Lee, I. Kim, I. Choi, K. Hong, Y. Ryu, J. Rhim, J. Shin, S.C. Park, H. Chung, and J. Chung. 2001. Immunobead RT-PCR versus regular RT-PCR amplification of CEA mRNA in peripheral blood. *J Cancer Res Clin Oncol* 127:489–494.

44. Mulholland, V. 2009. Immunocapture-PCR for plant virus detection. *Methods Mol Biol* 508:183–192.

45. Chandler, D.P., J. Brown, D.R. Call, S. Wunschel, J.W. Grate, D.A. Holman, L. Olson, M.S. Stottlemyre, and C.J. Bruckner-Lea. 2001. Automated immunomagnetic separation and microarray detection of *E. coli* O157:H7 from poultry carcass rinse. *Int J Food Microbiol* 70:143–154.

46. Metzger-Boddien, C., D. Khaschabi, M. Schonbauer, S. Boddien, T. Schlederer, and J. Kehle. 2006. Automated high-throughput immunomagnetic separation-PCR for detection of *Mycobacterium avium* subsp. *paratuberculosis* in bovine milk. *Int J Food Microbiol* 110:201–208.

47. Morales-Rayas, R., P.F. Wolffs, and M.W. Griffiths. 2008. Immunocapture and real-time PCR to detect *Campylobacter* spp. *J Food Prot* 71:2543–2547.

48. Mercanoglu Taban, B., U. Ben, and S.A. Aytac. 2009. Rapid detection of *Salmonella* in milk by combined immunomagnetic separation-polymerase chain reaction assay. *J Dairy Sci* 92:2382–2388.

49. Lantz, P.G., F. Tjerneld, E. Borch, B. Hahn-Hägerdal, and P. Rådström. 1994. Enhanced sensitivity in PCR detection of *Listeria monocytogenes* in soft cheese through use of an aqueous two-phase system as a sample preparation method. *Appl Environ Microbiol* 60:3416–3418.

50. Pertoft, H., K. Rubin, L. Kjellen, T.C. Laurent, and B. Klingeborn. 1977. The viability of cells grown or centrifuged in a new density gradient medium, Percoll. *Exp Cell Res* 110:449–457.

51. Pertoft, H. 2000. Fractionation of cells and subcellular particles with Percoll. *J Biochem Biophys Methods* 44:1–30.

52. Butot, S., T. Putallaz, and G. Sanchez. 2007. Procedure for rapid concentration and detection of enteric viruses from berries and vegetables. *Appl Environ Microbiol* 73:186–192.

53. Wolffs, P.F., K. Glencross, R. Thibaudeau, and M.W. Griffiths. 2006. Direct quantitation and detection of salmonellae in biological samples without enrichment, using two-step filtration and real-time PCR. *Appl Environ Microbiol* 72:3896–3900.

54. Knutsson, R., Y. Blixt, H. Grage, E. Borch, and P. Rådström. 2002. Evaluation of selective enrichment PCR procedures for *Yersinia enterocolitica*. *Int J Food Microbiol* 73:35–46.

55. Lantz, P.G., F. Tjerneld, B. Hahn-Hägerdal, and P. Rådström. 1996. Use of aqueous two-phase systems in sample preparation for polymerase chain reaction-based detection of microorganisms. *J Chromatogr B Biomed Appl* 680:165–170.

56. Kuji, N., T. Yoshii, T. Hamatani, T. Hanabusa, Y. Yoshimura, and S. Kato. 2008. Buoyant density and sedimentation dynamics of HIV-1 in two density-gradient media for semen processing. *Fertil Steril* 90:1983–1987.

57. Lindqvist, R. 1997. Preparation of PCR samples from food by a rapid and simple centrifugation technique evaluated by detection of *Escherichia coli* O157:H7. *Int J Food Microbiol* 37:73–82.

58. Löfström, C., J. Schelin, B. Norling, H. Vigre, J. Hoorfar, and P. Rådström. 2011. Culture-independent quantification of *Salmonella enterica* in carcass gauze swabs by flotation prior to real-time PCR. *Int J Food Microbiol* 145(Suppl. 1): S103–S109.

59. Josefsen, M.H., M. Krause, F. Hansen, and J. Hoorfar. 2007. Optimization of a 12-hour *Taq*Man PCR-based method for detection of *Salmonella* bacteria in meat. *Appl Environ Microbiol* 73:3040–3048.

60. Polaczyk, A.L., J. Narayanan, T.L. Cromeans, D. Hahn, J.M. Roberts, J.E. Amburgey, and V.R. Hill. 2008. Ultrafiltration-based techniques for rapid and simultaneous concentration of multiple microbe classes from 100-L tap water samples. *J Microbiol Methods* 73:92–99.

61. Akane, A., K. Matsubara, H. Nakamura, S. Takahashi, and K. Kimura. 1994. Identification of the heme compound copurified with deoxyribonucleic acid (DNA) from bloodstains, a major inhibitor of polymerase chain reaction (PCR) amplification. *J Forensic Sci* 39:362–372.

62. Katcher, H.L., and I. Schwartz. 1994. A distinctive property of *Tth* DNA polymerase: Enzymatic amplification in the presence of phenol. *Biotechniques* 16:84–92.

63. Rossen, L., P. Norskov, K. Holmstrom, and O.F. Rasmussen. 1992. Inhibition of PCR by components of food samples, microbial diagnostic assays and DNA-extraction solutions. *Int J Food Microbiol* 17:37–45.

64. Akalu, A., and J.K. Reichardt. 1999. A reliable PCR amplification method for microdissected tumor cells obtained from paraffin-embedded tissue. *Genet Anal 15*:229–233.

65. Moretti, T., B. Koons, and B. Budowle. 1998. Enhancement of PCR amplification yield and specificity using Ampli*Taq* Gold DNA polymerase. *Biotechniques 25*:716–722.

66. Kebelmann-Betzing, C., K. Seeger, S. Dragon, G. Schmitt, A. Moricke, T.A. Schild, G. Henze, and B. Beyermann. 1998. Advantages of a new *Taq* DNA polymerase in multiplex PCR and time-release PCR. *Biotechniques 24*:154–158.

67. Zsolnai, A., and L. Fesus. 1997. Enhancement of PCR-RFLP typing of bovine leukocyte adhesion deficiency. *Biotechniques 23*:380–382.

68. Löfström, C., J. Eriksson, A. Aspán, P. Häggblom, A. Gunnarsson, E. Borch, and P. Rådström. 2006. Improvement and validation of RAPD in combination with PFGE analysis of *Salmonella enterica* ssp. enterica serovar Senftenberg strains isolated from feed mills. *Vet Microbiol 114*:345–351.

69. Diakou, A., and C.I. Dovas. 2001. Optimization of random-amplified polymorphic DNA producing amplicons up to 8500 bp and revealing intraspecies polymorphism in *Leishmania infantum* isolates. *Anal Biochem 288*:195–200.

70. Gilje, B., R. Heikkila, S. Oltedal, K. Tjensvoll, and O. Nordgard. 2008. High-fidelity DNA polymerase enhances the sensitivity of a peptide nucleic acid clamp PCR assay for K-ras mutations. *J Mol Diagn 10*:325–331.

71. McDonald, J.P., A. Hall, D. Gasparutto, J. Cadet, J. Ballantyne, and R. Woodgate. 2006. Novel thermostable Y-family polymerases: Applications for the PCR amplification of damaged or ancient DNAs. *Nucleic Acids Res 34*:1102–1111.

72. Gale, J.M., and G.B. Tafoya. 2004. Evaluation of 15 polymerases and phosphorothioate primer modification for detection of UV-induced C:G to T:A mutations by allele-specific PCR. *Photochem Photobiol 79*:461–469.

73. Benson, L.M., A.P. Null, and D.C. Muddiman. 2003. Advantages of *Thermococcus kodakaraenis* (KOD) DNA polymerase for PCR-mass spectrometry based analyses. *J Am Soc Mass Spectrom 14*:601–604.

74. Tiemann-Boege, I., C. Curtis, D.N. Shinde, D.B. Goodman, S. Tavare, and N. Arnheim. 2009. Product length, dye choice, and detection chemistry in the bead-emulsion amplification of millions of single DNA molecules in parallel. *Anal Chem 81*:5770–5776.

75. Mullan, B., E. Kenny-Walsh, J.K. Collins, F. Shanahan, and L.J. Fanning. 2001. Inferred hepatitis C virus quasispecies diversity is influenced by choice of DNA polymerase in reverse transcriptase-polymerase chain reactions. *Anal Biochem 289*:137–146.

76. Kreuzer, K.A., A. Bohn, U. Lass, U.R. Peters, and C.A. Schmidt. 2000. Influence of DNA polymerases on quantitative PCR results using *Taq*Man probe format in the LightCycler instrument. *Mol Cell Probes 14*:57–60.

77. Wolffs, P., H. Grage, O. Hagberg, and P. Rådström. 2004. Impact of DNA polymerases and their buffer systems on quantitative real-time PCR. *J Clin Microbiol 42*:408–411.

78. Pavlov, A.R., N.V. Pavlova, S.A. Kozyavkin, and A.I. Slesarev. 2004. Recent developments in the optimization of thermostable DNA polymerases for efficient applications. *Trends Biotechnol 22*:253–260.

79. Davidson, J.F., R. Fox, D.D. Harris, S. Lyons-Abbott, and L.A. Loeb. 2003. Insertion of the T3 DNA polymerase thioredoxin binding domain enhances the processivity and fidelity of *Taq* DNA polymerase. *Nucleic Acids Res 31*:4702–4709.

80. Wang, Y., D.E. Prosen, L. Mei, J.C. Sullivan, M. Finney, and P.B. Vander Horn. 2004. A novel strategy to engineer DNA polymerases for enhanced processivity and improved performance *in vitro*. *Nucleic Acids Res 32*:1197–1207.

81. Lahr, D.J., and L.A. Katz. 2009. Reducing the impact of PCR-mediated recombination in molecular evolution and environmental studies using a new-generation high-fidelity DNA polymerase. *Biotechniques 47*:857–866.

82. Cline, J., J.C. Braman, and H.H. Hogrefe. 1996. PCR fidelity of *Pfu* DNA polymerase and other thermostable DNA polymerases. *Nucleic Acids Res 24*:3546–3551.

83. Strerath, M., C. Gloeckner, D. Liu, A. Schnur, and A. Marx. 2007. Directed DNA polymerase evolution: Effects of mutations in motif C on the mismatch-extension selectivity of thermus aquaticus DNA polymerase. *ChemBioChem 8*:395–401.

84. Sakaguchi, A.Y., M. Sedlak, J.M. Harris, and M.F. Sarosdy. 1995. Cautionary note on the use of dUMP-containing PCR primers with *Pfu* and Vent DNA polymerases. *Biotechniques 21*:368–370.

85. Modin, C., F.S. Pedersen, and M. Duch. 2000. Comparison of DNA polymerases for quantification of single nucleotide differences by primer extension assays. *Biotechniques 28*:48–51.

86. Eilert, K.D., and D.R. Foran. 2009. Polymerase resistance to polymerase chain reaction inhibitors in bone. *J Forensic Sci 54*:1001–1007.

87. Belec, L., J. Authier, M.C. Eliezer-Vanerot, C. Piedouillet, A.S. Mohamed, and R.K. Gherardi. 1998. Myoglobin as a polymerase chain reaction (PCR) inhibitor: A limitation for PCR from skeletal muscle tissue avoided by the use of *Thermus thermophilus* polymerase. *Muscle Nerve 21*:1064–1067.

88. Kim, C.H., M. Khan, D.E. Morin, W.L. Hurley, D.N. Tripathy, M. Kehrli, Jr., A.O. Oluoch, and I. Kakoma. 2001. Optimization of the PCR for detection of *Staphylococcus aureus nuc* gene in bovine milk. *J Dairy Sci 84*:74–83.

89. Löfström, C., R. Knutsson, C.E. Axelsson, and P. Rådström. 2004. Rapid and specific detection of *Salmonella* spp. in animal feed samples by PCR after culture enrichment. *Appl Environ Microbiol 70*:69–75.

90. Shames, B., J.G. Fox, F. Dewhirst, L. Yan, Z. Shen, and N.S. Taylor. 1995. Identification of widespread *Helicobacter hepaticus* infection in feces in commercial mouse colonies by culture and PCR assay. *J Clin Microbiol 33*:2968–2972.

91. Poddar, S.K., M.H. Sawyer, and J.D. Connor. 1998. Effect of inhibitors in clinical specimens on *Taq* and *Tth* DNA polymerase-based PCR amplification of influenza A virus. *J Med Microbiol 47*:1131–1135.

92. Dahlenborg, M., E. Borch, and P. Rådström. 2001. Development of a combined selection and enrichment PCR procedure for *Clostridium botulinum* types B, E, and F and its use to determine prevalence in fecal samples from slaughtered pigs. *Appl Environ Microbiol 67*:4781–4788.

93. Kermekchiev, M.B., L.I. Kirilova, E.E. Vail, and W.M. Barnes. 2009. Mutants of *Taq* DNA polymerase resistant to PCR inhibitors allow DNA amplification from whole blood and crude soil samples. *Nucleic Acids Res 37*:e40.

94. Kermekchiev, M.B., and W.M. Barnes. 2008. *Use of Whole Blood in PCR Reactions*. DNA Polymerase Technology, Inc., USA. US Patent 7462475.

95. Zhang, Z., M.B. Kermekchiev, and W.M. Barnes. 2010. Direct DNA amplification from crude clinical samples using a PCR enhancer cocktail and novel mutants of Taq. *J Mol Diagn 12*:152–161.

96. Nishimura, N., and T. Nakayama. 1999. *Process and Reagent for Amplifying Nucleic Acid Sequences*. Shimadzu Corporation, Kyoto, Japan. US Patent 5935825.

97. Nath, K., J.W. Sarosy, J. Hahn, and C.J. Di Como. 2000. Effects of ethidium bromide and SYBR Green I on different polymerase chain reaction systems. *J Biochem Biophys Methods 42*:15–29.

98. Lee, C.H., H. Mizusawa, and T. Kakefuda. 1981. Unwinding of double-stranded DNA helix by dehydration. *Proc Natl Acad Sci USA 78*:2838–2842.

99. Innis, M.A., K.B. Myambo, D.H. Gelfand, and M.A. Brow. 1988. DNA sequencing with *Thermus aquaticus* DNA polymerase and direct sequencing of polymerase chain reaction-amplified DNA. *Proc Natl Acad Sci USA 85*:9436–9440.

100. Abu Al-Soud, W., and P. Rådström. 2000. Effects of amplification facilitators on diagnostic PCR in the presence of blood, feces, and meat. *J Clin Microbiol 38*:4463–4470.

101. Kreader, C.A. 1996. Relief of amplification inhibition in PCR with bovine serum albumin or T4 gene 32 protein. *Appl Environ Microbiol 62*:1102–1106.

102. Powell, H.A., C.M. Gooding, S.D. Garrett, B.M. Lund, and R.A. McKee. 1994. Proteinase inhibition of the detection of *Listeria monocytogenes* in milk using the polymerase chain reaction. *Lett Appl Microbiol 18*:59–61.

103. Wu, J.R., and Y.C. Yeh. 1973. Requirement of a functional gene 32 product of bacteriophage T4 in UV, repair. *J Virol 12*:758–765.

104. Chang, C.W., D. Wang, and L.K. Henessy. 2012. *Method for Direct Amplification from Crude Nucleic Acid Samples*. Life Technologies Corporation, Carlsbad, CA, USA. US Patent 8173401.

105. Patterson, J.T. 1971. Microbiological assessment of surfaces. *J Food Technol 6*:63–72.

106. Lantz, P.G., W. Abu Al-Soud, R. Knutsson, B. Hahn-Hägerdal, and P. Rådström. 2000. Biotechnical use of polymerase chain reaction for microbiological analysis of biological samples. *Biotechnol Annu Rev 5*:87–130.

107. Wilson, I.G. 1997. Inhibition and facilitation of nucleic acid amplification. *Appl Environ Microbiol 63*:3741–3751.

108. Monpoeho, S., A. Dehee, B. Mignotte, L. Schwartzbrod, V. Marechal, J.C. Nicolas, S. Billaudel, and V. Ferre. 2000. Quantification of enterovirus RNA in sludge samples using single tube real-time RT-PCR. *Biotechniques 29*:88–93.

109. Abu Al-Soud, W., P.-G. Lantz, A. Bäckman, P. Olcén, and P. Rådström. 1998. A sample preparation method which facilitates detection of bacteria in blood cultures by the polymerase chain reaction. *J Microbiol Meth 32*:217–224.

110. Opel, K.L., D. Chung, and B.R. McCord. 2010. A study of PCR inhibition mechanisms using real time PCR. *J Forensic Sci* 55:25–33.

111. Abu Al-Soud, W., I.S. Ouis, D.Q. Li, A. Ljungh, and T. Wadström. 2005. Characterization of the PCR inhibitory effect of bile to optimize real-time PCR detection of *Helicobacter* species. *FEMS Immunol Med Microbiol* 44:177–182.

112. Uwatoko, K., M. Sunairi, A. Yamamoto, M. Nakajima, and K. Yamaura. 1996. Rapid and efficient method to eliminate substances inhibitory to the polymerase chain reaction from animal fecal samples. *Vet Microbiol* 52:73–79.

113. Thornton, C.G., and S. Passen. 2004. Inhibition of PCR amplification by phytic acid, and treatment of bovine fecal specimens with phytase to reduce inhibition. *J Microbiol Methods* 59:43–52.

114. Hendolin, P.H., L. Paulin, and J. Ylikoski. 2000. Clinically applicable multiplex PCR for four middle ear pathogens. *J Clin Microbiol* 38:125–132.

115. Eckhart, L., J. Bach, J. Ban, and E. Tschachler. 2000. Melanin binds reversibly to thermostable DNA polymerase and inhibits its activity. *Biochem Biophys Res Commun* 271:726–730.

116. Wiedbrauk, D.L., J.C. Werner, and A.M. Drevon. 1995. Inhibition of PCR by aqueous and vitreous fluids. *J Clin Microbiol* 33:2643–2646.

117. Behzadbehbahani, A., P.E. Klapper, P.J. Vallely, and G.M. Cleator. 1997. Detection of BK virus in urine by polymerase chain reaction: A comparison of DNA extraction methods. *J Virol Methods* 67:161–166.

118. Mahony, J., S. Chong, D. Jang, K. Luinstra, M. Faught, D. Dalby, J. Sellors, and M. Chernesky. 1998. Urine specimens from pregnant and nonpregnant women inhibitory to amplification of *Chlamydia trachomatis* nucleic acid by PCR, ligase chain reaction, and transcription-mediated amplification: Identification of urinary substances associated with inhibition and removal of inhibitory activity. *J Clin Microbiol* 36:3122–3126.

119. Juen, A., and M. Traugott. 2006. Amplification facilitators and multiplex PCR: Tools to overcome PCR-inhibition in DNA-gut-content analysis of soil-living invertebrates. *Soil Biol Biochem* 38:1872–1879.

120. Demeke, T., and R.P. Adams. 1992. The effects of plant polysaccharides and buffer additives on PCR. *Biotechniques* 12:332–334.

121. Koonjul, P.K., W.F. Brandt, J.M. Farrant, and G.G. Lindsey. 1999. Inclusion of polyvinylpyrrolidone in the polymerase chain reaction reverses the inhibitory effects of polyphenolic contamination of RNA. *Nucleic Acids Res* 27:915–916.

122. Campbell, G.R., J. Prosser, A. Glover, and K. Killham. 2001. Detection of *Escherichia coli* O157:H7 in soil and water using multiplex PCR. *J Appl Microbiol* 91:1004–1010.

123. Sails, A.D., A.J. Fox, F.J. Bolton, D.R. Wareing, D.L. Greenway, and R. Borrow. 2001. Development of a PCR ELISA assay for the identification of *Campylobacter jejuni* and *Campylobacter coli*. *Mol Cell Probes* 15:291–300.

124. Sails, A.D., F.J. Bolton, A.J. Fox, D.R. Wareing, and D.L. Greenway. 2002. Detection of *Campylobacter jejuni* and *Campylobacter coli* in environmental waters by PCR enzyme-linked immunosorbent assay. *Appl Environ Microbiol* 68:1319–1324.

125. Suslow, T.V., J. Wu, W.F. Fett, and L.J. Harris. 2002. Detection and elimination of *Salmonella mbandaka* from naturally contaminated alfalfa seed by treatment with heat or calcium hypochlorite. *J Food Prot* 65:452–458.

126. Soumet, C., G. Ermel, N. Rose, V. Rose, P. Drouin, G. Salvat, and P. Colin. 1999. Evaluation of a multiplex PCR assay for simultaneous identification of *Salmonella* sp., *Salmonella enteritidis* and *Salmonella typhimurium* from environmental swabs of poultry houses. *Lett Appl Microbiol* 28:113–117.

127. Simon, M.C., D.I. Gray, and N. Cook. 1996. DNA extraction and PCR methods for the detection of *Listeria monocytogenes* in cold-smoked salmon. *Appl Environ Microbiol* 62:822–824.

128. Lantz, P.G., R. Knutsson, Y. Blixt, W. Abu Al-Soud, E. Borch, and P. Rådström. 1998. Detection of pathogenic *Yersinia enterocolitica* in enrichment media and pork by a multiplex PCR: A study of sample preparation and PCR-inhibitory components. *Int J Food Microbiol* 45:93–105.

129. Shutler, G.G., P. Gagnon, G. Verret, H. Kalyn, S. Korkosh, E. Johnston, and J. Halverson. 1999. Removal of a PCR inhibitor and resolution of DNA STR types in mixed human-canine stains from a five year old case. *J Forensic Sci* 44:623–626.

130. Bourke, M.T., C.A. Scherczinger, C. Ladd, and H.C. Lee. 1999. NaOH treatment to neutralize inhibitors of *Taq* polymerase. *J Forensic Sci* 44:1046–1050.

131. Hoorfar, J., P. Ahrens, and P. Rådström. 2000. Automated 5′ nuclease PCR assay for identification of *Salmonella enterica*. *J Clin Microbiol* 38:3429–3435.

132. Laigret, F., J. Deaville, J.M. Bove, and J.M. Bradbury. 1996. Specific detection of *Mycoplasma iowae* using polymerase chain reaction. *Mol Cell Probes* 10:23–29.

2 Use of a Liposomal Internal Control Vehicle for Whole-Process Quality Assurance of Nucleic Acid Amplification

Einar S. Berg and Tania Nolan

CONTENTS

2.1   INTRODUCTION

2.1.1   Sources of False-Negative Results

When performing any scientific examination, a negative result is considered to be unreliable unless the chances of obtaining a false-negative result have been eliminated from the experimental system. It is important to minimize the likelihood of false-negative results since they lower the overall sensitivity, as well as the negative predictive value, of an analysis.[1] In a diagnostic setting, false-negative results may lead to inadequate treatment of the patients,[2,3] whereas in a public health setting, the surveillance and control of the spread of infectious disease becomes less efficient.[4] The principal sources of false-negative results in nucleic acid amplification assays are inhibition, human and technical operational errors, and nuclease degradation of the sample nucleic acid, especially RNA targets.

2.1.1.1   Inhibition

The presence of any inhibitory activity results in random errors, causing negative results in qualitative analysis[5] or underestimation of the amount of nucleic acid targets in a quantitative analysis.[6] Inhibitors may be introduced into a sample at any stage of the analytical process. The inhibitors can be derived from the specimen source matrix.[7–12] For example, nucleic acid prepared from soil might be accompanied with inhibitory humic substances.[13] The presence of heme from lysed erythrocytes in the nucleic acid isolated from whole-blood preparations is an inhibitor that can seriously disrupt genotyping experiments.[14,15] Inhibitors can be introduced during the preamplification, sample preparation process, such as traces of phenol from extraction reagents.[16–18] Similarly, when using sample preparation methods that include enzymatic lysis of the target cell/virus, the presence of inhibitors of these enzymes can result in impaired cell disruption. Reduced efficiency of the nucleic acid extraction process leads to loss of analysis target and lowered sensitivity of the assay. Accordingly, the presence of active proteases in this setting can be detrimental to the subsequent target sequence amplification and detection.[19]

2.1.1.2   Human and Technical Operational Errors

Several published reports have noted that inhibition is not the only source of false-negative results.[20–27] Random operational and technical errors occurring at any stage of the analysis have been regarded as plausible explanations.[5,11] Since amplification-based analyses require several steps, there are ample opportunities for error. Unfortunately, there are no adequate controls for poor laboratory practice although careful experimental design may reveal inadequate technical skills.

2.1.1.3   Presence of Nucleases

Contaminating RNases and DNases are a significant cause of false-negative results, although the presence of DNases is regarded as a relatively minor problem when compared to contamination with RNases, which are present almost everywhere. Active nucleases lead to sample nucleic acid degradation and removal of the target sequences.[28] Efficient inactivation/removal of RNases is of great importance. For instance, when analyzing an RNA message, the assay frequently contains only a limited number of copies of the mRNA. Loss of any of these copies would result in misinterpretation of the expression profile of the cell and the apparent ratio of expressed genes. In an infectious disease diagnostic setting, the samples may contain an extremely low copy number of target DNA or RNA and ultrasensitive assays for monitoring response to an antiviral treatment or by low-level viremia are accordingly very susceptible to nucleases. In comparison, when analyzing either relative gene copy number or the genotype of DNA targets, it is usual to use a higher input of nucleic acid. The multistep nature of the PCR and qPCR/RT-qPCR process makes it vulnerable to error from a variety of sources. To validate and improve the interpretation of data, it is critical to include the appropriate controls.

2.2 QUALITY ASSURANCE OF NUCLEIC ACID AMPLIFICATION ASSAYS

2.2.1 USE OF NAKED NUCLEIC ACID INTERNAL CONTROLS

To detect generic, rather than target-specific, amplification failures, a spiked nucleic acid control target can be included in the amplification reaction mixture. In contrast to the use of external controls where each sample must be tested in parallel, that is, with and without spiking of the control target, the internal controls enable surveillance of the amplification step within individual reactions. Internal controls may detect random errors that cannot be detected by external controls.[11,26,29] A positive internal control signal is analyzed during post-PCR result assessment and this is used to confirm that each individual reaction mixture was capable of amplifying a target of interest, if this had been present initially. Thus, the internal controls are used to validate target-negative amplification reactions. The most reliable detection of random errors is achieved when the internal control target is amplified with similar amplification efficiency to the analysis sequence of interest. The key is to avoid situations where there is preferential amplification of the internal controls and poor amplification of the targets of interest, for example, caused by partial inhibition.[30] Furthermore, the initial copy number of the internal control should be kept relatively low and/or balanced in favor of the target of interest.[11,26,31–35] In general, the internal control is used to identify the samples. This prompts for further treatment for inhibition removal and/or retesting such that the additional positive samples among the initial false-negative samples can be recognized. Accordingly, as the number of false-negatives is minimized, the analytical sensitivity of the testing is maximized.[26]

Conventional or naked internal controls consist of a known sequence of DNA or RNA dissolved in a suitable buffer such as TE buffer (10 mM Tris, 1 mM EDTA). Naked RNA internal controls can only be spiked into the sample at the lysis stage, pre-PCR, and after any RNases are inactivated. An internal control for RT-PCR analysis must be made of RNA to provide surveillance of the reverse transcriptase reaction prior to the DNA amplification. Subsequently, when the RNA control is converted into cDNA, it serves as an internal control also for the amplification step.[26,35,36] Note that target-specific reverse transcriptase- or amplification failures caused by faulty wild-type-specific primers that are not matching to a mutated target that could be accidentally present may not be identified by the use of internal controls, which are normally adapted to the wild-type target amplification.

While it is critical to control for unexpected errors during the reverse transcription and amplification phases of the experiment, the execution of the preamplification stages of the sample handling can also be crucial for the validity of the final results. Storage of the samples at an unfavorable temperature, such as during transportation to the analysis laboratory, or storage for an extended time period prior to specimen processing, can all lead to the deterioration of sample quality. This is a particular concern when attempting to analyze sequences in fragile organisms, such as hepatitis C virus (HCV), human immunodeficiency virus (HIV), and *Neisseria gonorrhoea*.[3,37–43] Spiking of a naked internal control directly into the sample at the sampling stage, for quality assurance of the whole process, is not advisable due to the presence of nucleases in the crude specimen matrix.[44,45] Furthermore, if the preamplification sample processing involves cell/virus-enrichment methodology by centrifugation, a naked internal control nucleic acid is not heavy enough to be collected with the sample nucleic acid entity. Therefore, naked internal controls cannot be used to report sample processing errors, such as unintentional removal of the cell pellet by careless aspiration of the supernatant.[46,47] The use of extensive nucleic acid purification methods to eliminate the inhibitory activity often involves numerous handling steps, potentially increasing the risk of operational errors.[48] Automation of the analysis procedure may reduce this risk but may not completely eliminate it.[49] The test failures can appear as demonstrated in Figure 2.1. Without analysis of internal controls, there is a risk that all of the 12 samples with analysis target signal below threshold are interpreted as true-negatives, especially if the risk for operational random errors is neglected. When the internal control signal is also considered, the strength of the negative test results is increased considerably. Accordingly, just two of the samples

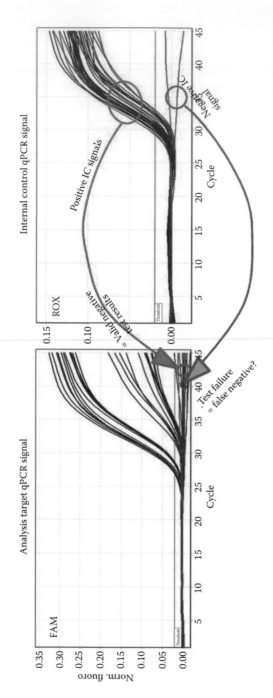

FIGURE 2.1 (**See color insert.**) Use of an internal control for quality assurance of a hepatitis C virus real-time RT-PCR assay. Red amplification curves: invalid HCV results disclosed by the negative internal control signal; green curves: valid negative HCV results confirmed by the positive internal control signal; blue curves: HCV positive samples.

in this experiment require further attention, that is, the two for which both the internal control and test assay present negative results. For all of the other samples with negative test results, the internal control data are positive, supporting a genuine negative test result.

Clearly, it is desirable for a chosen quality control system to be able to monitor the early stages of the analysis procedure, including sample handling and nucleic acid extraction.[44] However, naked internal controls cannot be used to monitor any cell-enrichment procedures. Moreover, they cannot monitor the efficacy of the lysis process or control that the integrity of target organism is preserved prior to analysis. Thus, whole-process quality assurance is not feasible when using naked internal controls. Monitoring of the entire process requires that the internal control is protected from destruction, which could be caused by the sample environment. Advantageously, the control should be contained within a tailored vehicle in the same way as the target nucleic acid is within the cell membrane/virus capsid. This is discussed in Section 2.2.2.

2.2.2 Whole-Process Quality Assurance by Use of Sheltered Internal Controls

A common approach for the detection of systemic error is to include weak positive samples, often called run-controls, among the unknown samples. These external controls allow systemic validation of the analysis on a day-to-day basis.[5] However, the detection of random errors among individual samples necessitates an approach whereby each sample has whole-process internal quality control surveillance.[6] Naked internal controls cannot be used. The active nucleases present in biological materials make enclosure and protection of the naked nucleic acid necessary before it can be spiked into the crude specimens.[43] Various protecting vehicles have been described for both RNA and DNA internal controls.[43,50,51] These can be divided into two groups, the endogenous and the exogenous internal control vehicles.

2.2.2.1 Endogenous Internal Controls

Sequences in the genome of the target cells/viruses can be employed as endogenous internal controls when they can be discriminated from the target of interest by targeted detection of mutations, polymorphisms, and so on. Alternatively, heterologous cells or viruses could provide the endogenous internal control target when they are reliably present in the host specimens.[52] The endogenous internal controls are purified alongside the host cell nucleic acid and therefore enable confirmation of the integrity of the nucleic acid in the sample, which might be an important feature for testing of possibly degraded or damaged nucleic acid due to certain sample treatments such as formalin fixation[53,54] or by testing of vulnerable RNA targets.[55–58] Thus, the endogenous internal controls provide quality assurance of the early stages of the analysis process. However, since these targets are different from the target under study, amplification of the internal control often requires another set of primers and testing in a multiplex PCR analysis setting. The alternative, a parallel and additional run to test the control assay in singleplex, corresponds to the use of external controls that may not enable efficient disclosure of random errors. The use of individual primer sets for the wild-type and internal control targets might lead to dissimilar amplification efficiencies with impaired ability to detect inhibition, decreased robustness, and lowered analytical sensitivity of the assay.[59,60] Another concern is that preferential amplification of the internal control may potentially generate false-negative, seemingly valid negative results for the target of interest. This scenario may occur when the copy number of the internal control is substantially higher than the copy number of the target of interest, for instance, as by the detection of low copy provirus in specimens prepared from white blood cells.[61]

2.2.2.2 Exogenous Internal Controls

Exogenous controls are required when a suitable endogenous control is not available or desirable. These consist of alien nucleic acid molecules encapsulated in a protective vehicle that is added to the samples. One option for the introduction of the exogenous control is through spiking part of the experimental sample with cells/viruses that are identical to those examined in the assay. This approach involves the use of an external control that requires subsequent parallel testing of spiked and nonspiked aliquots of the test specimens.[62] As such, the use of identical exogenous quality controls, that is, external controls, is less cost-effective than the use of internal controls that are different from the analysis targets.[63] The spiked external control does not enable the detection of random errors affecting the testing of just the nonspiked aliquot of the samples, which are the actual focus of the analysis. Detection of random errors that may occur anywhere in the analysis process requires internal controls with a form of cellular/virus-like protection. Four such categories of internal control vehicles of exogenous origin have been described below.

2.2.2.3   Internal Controls Retrieved from Unrelated Bacteria or Viruses

Whole-process quality assurance can be achieved by spiking the samples with nonrelated bacteria or viruses carrying a unique sequence that can be employed as an internal control target in the analysis. As with the use of endogenous internal controls, the detection of targets from different organisms often involves the use of several primer sets and a multiplex amplification.[64] Therefore, this approach has similar disadvantages that the amplification of the control may not reflect the inhibition of the amplification of the genuine target. If the internal control is at a significantly higher concentration, amplification of the target of interest could be suppressed. For these reasons, this approach is overall considered to be nonideal.

2.2.2.4   Internal Controls Retrieved from Related Bacteria or Viruses

Initial spiking of the samples with a few cells of a related organism or a few copies of a similar virus provides a more suitable internal control. Being similar and having the same primer sites, these internal controls are amplified with equal efficiency to the analysis targets. To avoid parallel testing of spiked and unspiked samples, that is, external control scenario, the PCR products from the target of interest and the internal control must be distinguishable. The internal control amplification target must therefore have a unique feature. Since these natural internal control vehicles are particle entities, they present the further advantage that they have similar physical and chemical properties to the organism that is the analysis target. This kind of internal control vehicle is suitable for the detection of bacterial 16S rRNA gene sequences[65–67] or other generic PCR assays.[68–70] This approach is preferable to the use of a spike derived from unrelated organisms.

2.2.2.5   Internal Controls Cloned into Genetically Modified Cells or Viruses

Versatile, exogenous control vehicles carrying suitable internal controls can be made with the use of recombinant DNA technology.[26,32,71–81] Genetically modified organisms such as *Escherichia coli* K12, M13-, λ-, and MS2 bacteriophages can serve as the internal control vehicles for the detection of bacteria, DNA viruses, and RNA viruses, respectively.[43,48,52,64,82–86] If the target sequence is originally present in the internal control organism, then it must be modified or removed prior to use. The most efficient quality control is achieved when the internal control vehicle has an encapsulation of a lipid membrane or protein envelope similar to the experimental target organism. This provides the internal control entity with similar physiochemical properties enabling security that experimental handling such as unfavorable storage will result in similar destruction, as well as provide the internal control vehicle with similar stability and disruption during the lysis process. Furthermore, similar density of the internal control entity might be of concern when the specimen preparation involves cell-enrichment methodology such as centrifugation. If the cell-enrichment protocol is based on immune-capture, the surface tags must be the same. In any case, the objective is to ensure similarly efficient co-separation and co-processing in the preamplification stages of the analysis. The spectrum of possible cloning hosts is fairly limited, so the above requirements may limit the use of genetically modified bacteria and viruses as internal control vehicles. In this context, it might be necessary to assess the storage stability of the internal control host in the given specimen matrix representative for the target organism environment.[87]

2.2.2.6   Internal Controls Inserted in Synthetic Vehicles Assembled *In Vitro*

Ideally, the internal control vehicle should have the same physical and chemical properties as the analysis target organism. In principle, the internal control nucleic acid can be encapsulated in any synthetic material that is capable of wrapping nucleic acid, for example, cationic polymers, cationic peptides, and other matrices such as polyacrylamide.[84,88,89] A closer alternative for the internal control can be entrapment in particles assembled from the natural building blocks constituting virus capsids, for example, prepared by disruption of viruses by chemical treatment.[90,91] These internal control vehicles are particularly suitable for the quality assurance of assays for the detection of viruses with a protein coat.[84,92] On the other hand, when the target organism has a lipid membrane, then the internal control vehicle should also have such an envelope. Liposomes can be regarded as the simplest cell membrane. These are capable of entrapping nucleic acids, thus resulting in nuclease protection.[93–95] In contrast to most viral capsids, the space available inside liposomes is very large. Therefore, there is no practical loading limitation with regard to the size or the copy number of internal control molecules to be entrapped in each liposomal particle.[94] DNA internal controls can be made by straightforward PCR and entrapped without the need for insertion into any cloning/plasmid vector. Similarly, RNA internal controls can be made by transcription of a PCR product that has an RNA polymerase promoter sequence inserted. Protocols for these are described below.[96] Furthermore, the liposome entities can be tailor-made to match the chemical and physical stability properties of the organism to be detected.[1,97,98] The incorporation of surface antigens on the liposome surface is feasible for the control of immuno-capture methodologies.[99,100] Thus, liposomes appear to be a versatile, cell-mimicking, internal control vehicle for DNA and/or RNA internal controls.[92]

2.2.3   Liposome-Based Internal Control Particle Technology

2.2.3.1   Light Internal Control Particles with Tailored Stability

In nucleic acid isolation procedures where the nuclease inactivation and lysis are done on the whole-specimen matrix, that is, without any enrichment of the analysis target entity, the functional requirements for the internal control vehicle are; that it must be nonpermeable to nucleases and it must have similar storage stability/lysis instability as the analysis target organism.[87] The liposomal internal control particles (IC particles) can be designed to meet these requirements. Basically, a high-stability tailoring is made by use of Tris buffer having a pH value near neutral, by inclusion of cholesterol for membrane sealing, α-tocopherol as an antioxidant, and a charged phospholipid for reduced liposome aggregation.[97] The details of this IC particle formulation can be found in Section 2.3. Actually, the quality assurance of the real-time PCR results demonstrated in Figure 2.1 was achieved by the use of IC particles synthesized according to this formulation.

2.2.3.2   Density-Modified IC Particles Tailored as Bacteria Cell Mimics

The commercial Roche Cobas Amplicor® CT assay has been used as a prototype test of a protocol that depends on low-speed centrifugation for the enrichment of the target cells in the early specimen preparation stages. As proof-of-principle, *Chlamydia trachomatis*-mimicking liposomal_IC particles were designed for this assay and synthesized by the "film-method."[93,94] The challenge for the design of the internal control vehicle was to be stable in urine specimens and under the initial mild detergent wash conditions, to have similar density as the bacterium elementary bodies enabling co-separation during the low-speed centrifugation and finally to dissolve and release the nucleic acid content by cell lysis. It was demonstrated that whole-process quality assurance was feasible when the liposomes served as internal control vehicle in the initial specimen wash and centrifugation stages prior to the lysis and PCR in the Roche analysis procedure.[50] As shown in Figure 2.2, it was an absolute necessity to increase the density of the IC particles by the loading of blue dextran polysaccharide into the core of the liposomes together with the internal control nucleic acid.

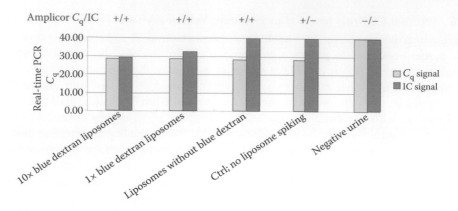

FIGURE 2.2 Whole-process quality assurance of the Cobas Amplicor® centrifugation-based *Chlamydia trachomatis* cell enrichment, specimen preparation method after initial spiking of an ordinary urine sample with blue dextran density-modified IC particles and qualitative signal detection in a Roche PCR and semi-quantitatively in a similar in-house real-time PCR.

2.2.3.3 Density-Modified IC Particles Tailored as Leukocyte Cell Mimics

Nucleic acid can be isolated with high yield from the leukocytes in whole blood. Commonly, these methods involve pre-lysis of the erythrocytes followed by enrichment of the leukocytes by low-speed centrifugation, cell wash, and lysis. The erythrocytes are removed as they contain hemoglobin, the heme moiety being PCR inhibitory. However, the subsequent cell washing process of repeated centrifugation and aspiration of the supernatant can easily lead to the removal of the leukocyte cell pellet, or part of it. Quality assurance of this critical step by the use of an internal control is thus advisable. Leukocyte mimicking IC particles designed for this purpose must be stable in blood and in red cell lysis buffer, must mimic the white cell deposition during the centrifugation and wash, and finally must dissolve in lysis buffer in a comparable manner to the white blood cells. Initial experiment showed that Ficoll, a neutral hydrophilic polysaccharide, could be loaded into the liposomes with higher concentration compared to negatively charged blue dextran. In addition, changing from the film liposome preparation method to the reversed-phase evaporation consistently resulted in liposomes with higher density.[101] Figure 2.3 shows the results from the testing of IC particles prepared with increased Ficoll after spiking in the blood collected in BD Vacutainer® CPT™ cell preparation tubes followed by direct separation of the plasma, leukocytes, and erythrocytes by density gradient centrifugation. Careful aspiration of small aliquots from the upper part of the plasma and from the leukocyte phases gave a more concentrated sample compared to the reference whole-blood matrix. The erythrocyte phase could not be collected with similar care since rough treatment was needed to remove the gel plug in the tube.

2.3 SYNTHESIS OF IC PARTICLES

2.3.1 Preparation of Naked Internal Control Nucleic Acid

Conveniently, the naked "ideal" DNA internal control nucleic acid can be prepared by the use of composite PCR primers that anneal to an alien internal control-specific spacer sequence.[71] This creates large amounts (cross-contamination risk!) of internal control nucleic acid having the assay primer sequences incorporated at the ends. Similarly, the use of the composite PCR primer approach can also put the T7 or SP6 RNA polymerase promoter sequence in the flank of a PCR product for the production of an RNA internal control by the use of a T7/SP6 *in vitro* transcription kit (Promega). The principles for the M13 phage-based synthesis of both DNA and RNA internal controls are given

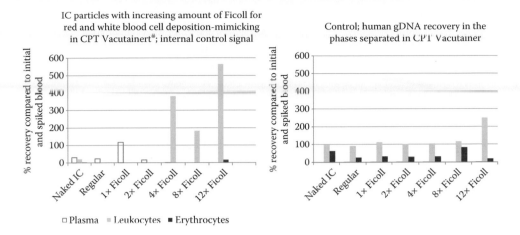

FIGURE 2.3 Co-separation of Ficoll density-modified IC particles with various blood cell types by centrifugation in BD Vacutainer® CPT™ cell preparation tubes followed by Roche's MagNa Pure LC total nucleic acid isolation and testing in internal control and human genomic DNA-specific singleplex real-time PCRs. Open bars: aliquots aspirated near the surface of the plasma phase Light gray bars: the leukocyte phase carefully harvested near the gel barrier. Black bars: aliquots from the red cell phase collected after removal of the gel barrier.

in Figure 2.4. An example of the use of composite primers for the synthesis of an HCV real-time RT-PCR RNA internal control assay is given below.

a. PCR generation of T7 RNA polymerase template (> RNA internal control size 99 bases):
   T7/HCV_up/M13_low-composite primer:
      5'-<u>taa tac gac tca cta tag gga ga</u>*CAT GGC GTT AGT ATG AGT GT*<u>g cgg ttt gcg tat</u>-3'
   HCV_low/M13_up-composite primer:
      5'-*GTT CCG CAG ACC ACT ATG* <u>ctg ttg ccc gtc t</u>-3'
b. HCV 5'UTR real-time PCR (product size 73 bp):
   HCV_up_primer:
      5'-*CAT GGC GTT AGT ATG AGT GT*-3'
   HCV_low_primer:
      5'-*GTT CCG CAG ACC ACT ATG*-3'
   HCV TaqMan probe:
      5'-(FAM)-CTC CAG GAC CCC CCC TCC-(BHQ-1)-3'
   Internal control_M13_TaqMan probe:
      5'-(JOE)-GTG AAA AGA AAA ACC ACC CTG GC-(BHQ-1)-3'

The best composite primers were selected after *in silico* examination of different combinations of T7-M13-target-of-interest sequences by the use of primer design software (http://www.humgen.nl/primer_design.html). The initial annealing and elongation of the composite primers in PCR enable them to function very effectively later as "big primers." In addition, the PCR is facilitated by the hot-start Taq polymerase as well as the PCR buffer IV given in Table 2.1, which gives a robust reaction with a wide range of possible optimum conditions for the magnesium concentration and annealing temperature and decreased primer–dimer formation. Therefore, it might be worthwhile to try the reaction mixture in Table 2.2 and a thermal cycling with annealing at 55°C for 1 min and 25–30 cycles protocol rather than an extensive PCR optimization strategy. The hybrid PCR product that shall serve as a DNA internal control may advantageously be purified to remove any primer–dimers prior to the entrapment in liposomes.

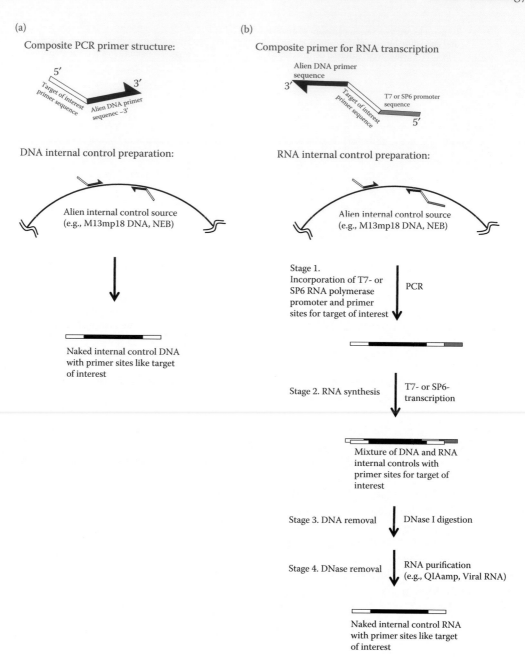

FIGURE 2.4 Principle of the generation of naked DNA internal controls (a) and RNA internal controls (b) by use of composite primers for adaption of the IC amplification to the wild-type target PCR primers.

With regard to the RNA internal controls, since the T7 (or SP6) RNA polymerase promoter is included in the internal control/T7 hybrid PCR product, the RNA polymerase will generate large amounts of internal control from an input of 10–20 ng DNA template. The use of an *in vitro* transcription kit, for example, RiboMax™ Large Scale RNA Production System (Promega Corporation), makes the DNA to RNA conversion process a straightforward task. After the transcription, the DNA template is removed by DNase digestion. Finally, the internal control RNA is purified and the DNase

TABLE 2.1
10× PCR Buffer IV (ABgene®)

Component	Final Concentration	Component Volume (μL)
Tris buffer, 1.0 M, pH 8.8	750 mM	750
Ammonium sulfate, 1.0 M	200 mM	200
Tween® 20 (Sigma-Aldrich)	1‰ (v/v)	1
Water, DNase-, RNase-free		478
Total volume		1000

Source: http://www.abgene.com/productDetails.asp?prodID = 16.

TABLE 2.2
1× PCR Reaction Mixture

Component	Final Concentration	Master Mix Volume (μL)
10× PCR buffer IV	1×	2.5
Magnesium chloride (25 mM)	3 mM	3.0
T7/HCV_forward/M13_reverse_composite primer (100 ng/μL)	150 nM	0.6
HCV_reverse/M13_forward_composite primer (100 ng/μL)	150 nM	0.3
dNTP mixture (10 mM of each nucleotide)	200 μM	0.8
Platinum® Taq DNA polymerase (5 U/μL, Invitrogen)	0.04 U/μL	0.2
M13mp18 RF I DNA (100 pg/μL, NEB)	4 pg/μL	1.0
Sterile deionized water		16.6
Total volume		25

is inactivated simultaneously in a common RNA extraction system, such as a QIAamp Viral RNA Mini Kit (Qiagen). To ensure input, and since the liposome entrapment efficiency may vary, a rough quantification of the internal control nucleic acid can be done at this stage of the process.

2.3.1.1 Preparation of Liposomal IC Particles

Several methods and various lipid formulations can be used for the preparation of liposomes.[89] The film method works well in the laboratory scale for the generation of liposomes loaded with nucleic acid by a freeze–thaw procedure.[87,88] The liposome formulation given in Table 2.3 is tailored to give a stable and protective vehicle for the internal control nucleic acid.[88] Initially, the liposome building blocks are dissolved in chloroform, and then deposited as a film on the tube wall as the solvent evaporates. When the film is rehydrated and exposed to vortex and ultrasonic agitation, liposomes are peeled off. The subsequent freeze–thaw causes the lipid bilayer to go through a phase-transition state where the nucleic acid can pass the liposome membrane. The rest of the nonencapsulated material should be removed by nuclease digestion.[46,88] A large amount of IC particles is produced in each batch. Thus, a considerable dilution of the liposome dispersion is necessary prior to final use.

2.3.2 Preparation of Stable Positively Charged IC Particles by Film Method

1. Dissolve the liposome components according to Table 2.3. A liquot 0.6 mL of the solution into 10 1.5-mL microtubes with O-ring screw cap. During the subsequent solvent evaporation, the tubes should be angled and occasionally turned for widest film deposition of the lipids (surplus tubes with dry film can be stored at –20°C for months).

TABLE 2.3

Formulation of Positively Charged Liposomes

Component	Amount, 10 vials	Proportion in Liposomes (%)
Palmitoyl-2-oleoyl-*sn*-glycero-3-phosphocholine, POPC (Lipoid GmbH, Ludwigshafen, Germany)	450 mg	82.5
Cholesterol (Sigma-Aldrich Chemie GmbH, Steinheim, Germany)	80 mg	15
Didodecyldimethylammonium bromide, DDAB (Fluka Chemie GmbH, Steinheim, Germany)	10 mg	2
α-Tocopherol (Sigma-Aldrich Chemie GmbH, Steinheim, Germany)	3 mg	0.5
Chloroform (Sigma-Aldrich Chemie GmbH, Steinheim, Germany)	6 mL[a]	
Each vial, aliquot volume	0.6 mL	

[a] Solvent to be evaporated by lipid film formation.

2. Add Tris buffer (500 μL, 10 mM, pH 7.5). Vortex the tube for 3 min, followed by sonication for 10 min in an ultrasonic bath at 4°C. Repeat the agitation until the lipid film has disappeared.
3. Add the internal control nucleic acid (100 μL, 10–100 ng) and freeze–thaw the sample 10 times, that is, freezing in liquid nitrogen and followed by thawing at room temperature for at least 15 min.
4. Removal of non-entrapped nucleic acid:
   i. DNA internal controls—DNase I digestion
       a. Combine the following:
           300 μL liposome/internal control dispersion
           5 mM $MgCl_2$
           0.1 mM DTT
           800–900 units DNase I
       b. Incubate at 37°C for 3 h followed by 75°C for 15 min
   ii. RNA internal controls—RNase I digestion
       Add 800–900 units of RNase I to 300 μL of liposome/internal control dispersion and incubate at 37°C for 2 h and then 75°C for 15 min.
5. Determination of the liposomal internal control load.
   Prior to use as a quality control tool, the amount of diluted IC particle dispersion to be spiked in the samples should be established to enable a suitable copy number input of internal control targets in the assay.[26] This can be done by practical functional testing by spiking of aliquots of a dilution series of the IC particles into a negative control sample matrix such as TE buffer, followed by extraction with the right nucleic acid isolation system, and signal determination, for example, by final real-time PCR analysis. The liposomal internal control nucleic acid should give the same signal as a naked internal control does when this is spiked into the sample's post lysis.

2.3.3 PREPARATION OF NEGATIVELY CHARGED IC PARTICLES WITH INCREASED DENSITY BY REVERSE-PHASE EVAPORATION METHOD

1. Dissolve the liposome components according to Table 2.4a and b. Aliquot 0.75 mL of the organic solution and 0.25 mL of the aqueous solution into 10 1.5-mL microtubes with O-ring screw cap. Vortex the tubes for 1 min, followed by sonication for 15 min in an ultrasonic bath at 4°C. During the subsequent organic solvent evaporation in a speedvac, care

TABLE 2.4

Formulation of Density-Modified IC Particles

a. Organic Phase

Component	Amount, 6 vials	Proportion in Liposomes (%)
Palmitoyl-2-oleoyl-*sn*-glycero-3-phosphocholine, POPC (Lipoid GmbH, Ludwigshafen, Germany)	150 mg	84.2
Cholesterol (Sigma-Aldrich Chemie GmbH, Steinheim, Germany)	25 mg	14
Sodium oleate (Lipoid GmbH, Ludwigshafen, Germany)	3.3 mg	1.6
α-Tocopherol (Sigma-Aldrich Chemie GmbH, Steinheim, Germany)	0.9 mg	0.2
Diethyl ether (Sigma-Aldrich Chemie GmbH, Steinheim, Germany)	5 mL	

b. Aqueous Phase

Liposome Name	Regular (µL)	1× Ficoll (µL)	2× Ficoll (µL)	4× Ficoll (µL	8× Ficoll (µL)	12× Ficoll (µL)
Internal control nucleic acid	10	10	10	10	10	10
Ficoll (5% w/v)[a] (Sigma-Aldrich Chemie GmbH, Steinheim, Germany)	0	20	40	80	160	240
TE buffer[a]	240	220	200	160	80	0

[a] Dissolved in TE buffer (10 mM Tris/1 mM EDTA, pH 7.5).

should be taken to avoid foam and bubbles. During the solvent removal process, a viscous gel is formed, which subsequently collapses into an aqueous suspension. Remove traces of the organic solvent by an additional 15 min in vacuum.

2. Vortex the tubes, followed by freeze–thaw, that is, freezing in liquid nitrogen and followed by thawing at room temperature for at least 15 min.
3. Remove the non-entrapped nucleic acid and subsequently determine the nucleic acid load as described above for the liposomes prepared by the film method.

2.4   CONCLUSION

A negative PCR result is unreliable unless internal controls are used to discern any false-negative results. Spiking with carefully designed, naked, internal control nucleic acids into the samples at the post lysis stage can provide partial quality assurance by the control of the amplification stages of the analysis, but not of the crucial specimen preparation steps. For the control to represent amplification of the desired target, the internal control and the wild-type targets should be amplified with comparable amplification efficiencies. Whole-process quality assurance can be achieved when the internal controls are entrapped in entities with comparable physical and chemical properties, thus enabling target cell/virus mimicking. The target cell/virus entity envelope stability, density, and surface antigen determinants as well as the specimen preparation protocol must be taken into account during the design of the suitable internal control vehicle. So far, the liposome-based IC particles technology seems to be the state of the art for the most versatile RNA and/or DNA internal control vehicle.

ACKNOWLEDGMENTS

This work was supported by Norwegian Research Council (Grant No 188010/I40) and IC Particles AS, Oslo, Norway.

REFERENCES

1. Black, C.M. 1997. Current methods of laboratory diagnosis of *Chlamydia trachomatis* infections. *Clin Microbiol Rev 10*:160–184.
2. Vernet, G. 2004. Molecular diagnostics in virology. *J Clin Virol 31*:239–247.
3. Yang, S., and R.E. Rothman. 2004. PCR-based diagnostics for infectious diseases: Uses, limitations, and future applications in acute-care settings. *Lancet Infect Dis 4*:337–348.
4. Quinn, T.C. 1994. Recent advances in diagnosis of sexually transmitted diseases. *Sex Transm Dis 21*:S19–S27.
5. Valentine-Thon, E. 2002. Quality control in nucleic acid testing—where do we stand? *J Clin Virol 25*:S13–S21.
6. Burkhart, C.A., M.D. Norris, and M. Haber. 2002. A simple method for the isolation of genomic DNA from mouse tail free of real-time PCR inhibitors. *J Biochem Biophys Methods 52*:145–149.
7. Abu Al-Soud, W., and P. Rådström. 1998. Capacity of nine thermostable DNA polymerases to mediate DNA amplification in the presence of PCR-inhibiting samples. *Appl Environ Microbiol 64*:3748–3753.
8. An, Q., J. Liu, W. O'Brien, G. Radcliffe, D. Buxton, S. Popoff, W. King et al. 1995. Comparison of characteristics of Q beta replicase-amplified assay with competitive PCR assay for *Chlamydia trachomatis*. *J Clin Microbiol 33*:58–63.
9. Mahony, J., S. Chong, D. Jang, K. Luinstra, M. Faught, D. Dalby, J. Sellors, and M. Chernesky. 1998. Urine specimens from pregnant and nonpregnant women inhibitory to amplification of *Chlamydia trachomatis* nucleic acid by PCR, ligase chain reaction, and transcription-mediated amplification: Identification of urinary substances associated with inhibition and removal of inhibitory activity. *J Clin Microbiol 36*:3122–3126.
10. Wiedbrauk, D.L., J.C. Werner, and A.M. Drevon. 1995. Inhibition of PCR by aqueous and vitreous fluids. *J Clin Microbiol 33*:2643–2646.
11. Burkardt, H.J. 2000. Standardization and quality control of PCR analyses. *Clin Chem Lab Med 38*:87–91.
12. Goessens, W.H.F., J.W. Mouton, W.I. van der Meijden, S. Deelen, T.H. van Rijsoort-Vos, N. Lemmens-den Toom, H.A. Verbrugh, and R.P. Verkooyen. 1997. Comparison of three commercially available amplification assays, AMP CT, LCx, and COBAS AMPLICOR, for detection of *Chlamydia trachomatis* in first-void urine. *J Clin Microbiol 35*:2628–2633.
13. Tsai, Y.L., and B.H. Olson. 1992. Rapid method for separation of bacterial DNA from humic substances in sediments for polymerase chain reaction. *Appl Environ Microbiol 58*:2292–2295.
14. Higuchi, R. 1989. *Simple and Rapid Preparation of Samples for PCR, PCR Technology: Principles and Applications for DNA Amplification*. Stockton Press, New York, USA.
15. Akane, A., K. Matsubara, H. Nakamura, S. Takahashi, and K. Kimura. 1994. Identification of the heme compound copurified with deoxyribonucleic acid (DNA) from bloodstains, a major inhibitor of polymerase chain reaction (PCR) amplification. *J Forensic Sci 39*:362–372.
16. de Lomas, J.G., F.J. Sunzeri, and M.P. Busch. 1992. False-negative results by polymerase chain reaction due to contamination by glove powder. *Transfusion 32*:83–85.
17. St Pierre, B., P. Neustock, U. Schramm, D. Wilhelm, H. Kirchner, and G. Bein. 1994. Seasonal breakdown of polymerase chain reaction. *Lancet 343*:673.
18. Rossen, L., P. Nørskov, K. Holmstrøm, and O.F. Rasmussen. 1992. Inhibition of PCR by components of food samples, microbial diagnostic assays and DNA-extraction solutions. *Int J Food Microbiol 17*:37–45.
19. Tan, S.C., and B.C. Yiap. 2009. DNA, RNA, and protein extraction: The past and the present. *J Biomed Biotechnol 2009*, article id: 574398, 10 pp.
20. Bassiri, M., P.A. Mårdh, and M. Domeika. 1997. Multiplex AMPLICOR PCR screening for *Chlamydia trachomatis* and *Neisseria gonorrhoeae* in women attending non-sexually transmitted disease clinics. The European Chlamydia Epidemiology Group. *J Clin Microbiol 35*:2556–2560.
21. Jungkind, D., S. Direnzo, K.G. Beavis, and N.S. Silverman. 1996. Evaluation of automated COBAS AMPLICOR PCR system for detection of several infectious agents and its impact on laboratory management. *J Clin Microbiol 34*:2778–2783.
22. Morré, S.A., I.G. van Valkengoed, R.M. Moes, A.J.P. Boeke, C.J.L.M. Meijer, and A.J.C. van den Brule. 1999. Determination of *Chlamydia trachomatis* prevalence in an asymptomatic screening population: Performances of the LCx and COBAS Amplicor tests with urine specimens. *J Clin Microbiol 37*:3092–3096.
23. Pasternack, R., P. Vuorinen, A. Kuukankorpi, T. Pitkajarvi, and A. Miettinen. 1996. Detection of *Chlamydia trachomatis* infections in women by Amplicor PCR: Comparison of diagnostic performance with urine and cervical specimens. *J Clin Microbiol 34*:995–998.

24. Pasternack, R., P. Vuorinen, and A. Miettinen. 1997. Evaluation of the Gen-Probe *Chlamydia trachomatis* transcription-mediated amplification assay with urine specimens from women. *J Clin Microbiol* 35:676–678.

25. Puolakkainen, M., E. Hiltunen-Back, T. Reunala, S. Suhonen, P. Lahteenmaki, M. Lehtinen, and J. Paavonen. 1998. Comparison of performances of two commercially available tests, a PCR assay and a ligase chain reaction test, in detection of urogenital *Chlamydia trachomatis* infection. *J Clin Microbiol* 36:1489–1493.

26. Rosenstraus, M., Z. Wang, S.Y. Chang, D. DeBonville, and J.P. Spadoro. 1998. An internal control for routine diagnostic PCR: Design, properties, and effect on clinical performance. *J Clin Microbiol* 36:191–197.

27. Vincelette, J., J. Schirm, M. Bogard, A.M. Bourgault, D.S. Luijt, A. Bianchi, P.C. Voorst Vader, A. Butcher, and M. Rosenstraus. 1999. Multicenter evaluation of the fully automated COBAS AMPLICOR PCR test for detection of *Chlamydia trachomatis* in urogenital specimens. *J Clin Microbiol* 37:74–80.

28. Ulrich, P.P., J.M. Romeo, L.J. Daniel, and G.N. Vyas. 1993. An improved method for the detection of hepatitis C virus RNA in plasma utilizing heminested primers and internal control RNA. *PCR Methods Appl* 2:241–249.

29. Wilson, I.G. 1997. Inhibition and facilitation of nucleic acid amplification. *Appl Environ Microbiol* 63:3741–3751.

30. Berg, E.S., G. Størvold, G. Ånestad, H. Moi, and K. Skaug. 2005. Reliability of the Amplicor™ internal control to disclose false-negative *Chlamydia trachomatis* PCR results. *J Microbiol Methods* 60:125–129.

31. Larzul, D., F. Guigue, J.J. Sninsky, D.H. Mack, C. Brechot, and J.L. Guesdon. 1988. Detection of hepatitis B virus sequences in serum by using *in vitro* enzymatic amplification. *J Virol Methods* 20:227–237.

32. McCulloch, R.K., C.S. Choong, and D.M. Hurley. 1995. An evaluation of competitor type and size for use in the determination of mRNA by competitive PCR. *PCR Methods Appl* 4:219–226.

33. McDowell, D.G., N.A. Burns, and H.C. Parkes. 1998. Localised sequence regions possessing high melting temperatures prevent the amplification of a DNA mimic in competitive PCR. *Nucleic Acids Res* 26:3340–3347.

34. Siebert, P.D., and J.W. Larrick. 1992. Competitive PCR. *Nature* 359:557–558.

35. Wang, A.M., M.V. Doyle, and D.F. Mark. 1989. Quantitation of mRNA by the polymerase chain reaction. *Proc Natl Acad Sci USA* 86:9717–9721.

36. Becker-Andre, M., and K. Hahlbrock. 1989. Absolute mRNA quantification using the polymerase chain reaction (PCR). A novel approach by a PCR aided transcript titration assay (PATTY). *Nucleic Acids Res* 17:9437–9446.

37. Chong, S., D. Jang, X. Song, J. Mahony, A. Petrich, P. Barriga, and M. Chernesky. 2003. Specimen processing and concentration of *Chlamydia trachomatis* added can influence false-negative rates in the LCx assay but not in the APTIMA Combo 2 assay when testing for inhibitors. *J Clin Microbiol* 41:778–782.

38. Land, S., S. Tabrizi, A. Gust, E. Johnson, S. Garland, and E.M. Dax. 2002. External quality assessment program for *Chlamydia trachomatis* diagnostic testing by nucleic acid amplification assays. *J Clin Microbiol* 40:2893–2896.

39. Cuypers, H.T., D. Bresters, I.N. Winkel, H.W. Reesink, A.J. Weiner, M. Houghton, C.L. van der Poel, and P.N. Lelie. 1992. Storage conditions of blood samples and primer selection affect the yield of cDNA polymerase chain reaction products of hepatitis C virus. *J Clin Microbiol* 30:3220–3224.

40. Damen, M., P. Sillekens, M. Sjerps, R. Melsert, I. Frantzen, H.W. Reesink, P.N. Lelie, and H.T. Cuypers. 1998. Stability of hepatitis C virus RNA during specimen handling and storage prior to NASBA amplification. *J Virol Methods* 72:175–184.

41. Grant, P.R., A. Kitchen, J.A. Barbara, P. Hewitt, C.M. Sims, J.A. Garson, and R.S. Tedder. 2000. Effects of handling and storage of blood on the stability of hepatitis C virus RNA: Implications for NAT testing in transfusion practice. *Vox Sang* 78:137–142.

42. Martin, D.H., C. Cammarata, B. Van Der Pol, R.B. Jones, T.C. Quinn, C.A. Gaydos, K. Crotchfelt et al. 2000. Multicenter evaluation of AMPLICOR and automated COBAS AMPLICOR CT/NG tests for *Neisseria gonorrhoeae*. *J Clin Microbiol* 38:3544–3549.

43. Pasloske, B.L., C.R. WalkerPeach, R.D. Obermoeller, M. Winkler, and D.B. DuBois. 1998. Armored RNA technology for production of ribonuclease-resistant viral RNA controls and standards. *J Clin Microbiol* 36:3590–3594.

44. Freeman, W.M., S.J. Walker, and K.E. Vrana. 1999. Quantitative RT-PCR: Pitfalls and potential. *Biotechniques* 26:112–22.

45. Kaucner, C., and T. Stinear. 1998. Sensitive and rapid detection of viable *Giardia cysts* and *Cryptosporidium parvum* oocysts in large-volume water samples with wound fiberglass cartridge filters and reverse transcription-PCR. *Appl Environ Microbiol* 64:1743–1749.

46. Taggart, E.W., C.L. Byington, D.R. Hillyard, J.E. Robison, and K.C. Carroll. 1998. Enhancement of the AMPLICOR enterovirus PCR test with a coprecipitant. *J Clin Microbiol 36*:3408–3409.

47. Hultman, T., S. Bergh, T. Moks, and M. Uhlen. 1991. Bidirectional solid-phase sequencing of *in vitro*-amplified plasmid DNA. *Biotechniques 10*:84–93.

48. Sambrook, J., E.F. Fritch, and T. Maniatis. 1989. *Molecular Cloning, A Laboratory Manual*. Cold Spring Harbor Laboratory Press, New York, USA.

49. Greenspoon, S.A., K.L. Sykes, J.D. Ban, A. Pollard, M. Baisden, M. Farr, N. Graham, B.L. Collins, M.M. Green, and C.C. Christenson. 2006. Automated PCR setup for forensic casework samples using the Normalization Wizard and PCR Setup robotic methods. *Forensic Sci Int 164*:240–248.

50. Berg, E.S., and K. Skaug. 2003. Liposome encapsulation of the internal control for whole process quality assurance of nucleic acid amplification-based assays. *J Microbiol Methods 55*:303–309.

51. WalkerPeach, C.R., M. Winkler, D.B. DuBois, and B.L. Pasloske. 1999. Ribonuclease-resistant RNA controls (Armored RNA) for reverse transcription-PCR, branched DNA, and genotyping assays for hepatitis C virus. *Clin Chem 45*:2079–2085.

52. Coutlee, F., M. de Ladurantaye, C. Tremblay, J. Vincelette, L. Labrecque, and M. Roger. 2000. An important proportion of genital samples submitted for *Chlamydia trachomatis* detection by PCR contain small amounts of cellular DNA as measured by beta-globin gene amplification. *J Clin Microbiol 38*:2512–2515.

53. Finch, J.L., R.M. Hope, and A. van Daal. 1996. Human sex determination using multiplex polymerase chain reaction (PCR). *Sci Justice 36*:93–95.

54. Shindo, M., T. Okuno, K. Arai, M. Matsumoto, M. Takeda, K. Kashima, M. Shimada, Y. Fujiwara, and Y. Sokawa. 1991. Detection of hepatitis B virus DNA in paraffin-embedded liver tissues in chronic hepatitis B or non-A, non-B, hepatitis using the polymerase chain reaction. *Hepatology 13*:167–171.

55. Bariana, H.S., A.L. Shannon, P.W.G. Chu, and P.M. Waterhouse. 1994. Detection of five seedborne legume viruses in one sensitive multiplex polymerase chain reaction test. *Phytopathology 84*:1201–1205.

56. Akhter, S., H. Liu, R. Prabhu, C. DeLucca, F. Bastian, R.F. Garry, M. Schwartz, S.N. Thung, and S. Dash. 2003. Epstein-Barr virus and human hepatocellular carcinoma. *Cancer Lett 192*:49–57.

57. Menzel, W., W. Jelkmann, and E. Maiss. 2002. Detection of four apple viruses by multiplex RT-PCR assays with coamplification of plant mRNA as internal control. *J Virol Methods 99*:81–92.

58. Svoboda-Newman, S.M., J.K. Greenson, T.P. Singleton, R. Sun, and T.S. Frank. 1997. Detection of hepatitis C by RT-PCR in formalin-fixed paraffin-embedded tissue from liver transplant patients. *Diagn Mol Pathol 6*:123–129.

59. Elnifro, E.M., A.M. Ashshi, R.J. Cooper, and P.E. Klapper. 2000. Multiplex PCR: Optimization and application in diagnostic virology. *Clin Microbiol Rev 13*:559–570.

60. Nelson, N.C., A.B. Cheikh, E. Matsuda, and M.M. Becker. 1996. Simultaneous detection of multiple nucleic acid targets in a homogeneous format. *Biochemistry 35*:8429–8438.

61. Mallet, F., C. Hebrard, J.M. Livrozet, O. Lees, F. Tron, J.L. Touraine, and B. Mandrand. 1995. Quantitation of human immunodeficiency virus type 1 DNA by two PCR procedures coupled with enzyme-linked oligosorbent assay. *J Clin Microbiol 33*:3201–3208.

62. Kolk, A.H., A.R. Schuitema, S. Kuijper, J. van Leeuwen, P.W. Hermans, J.D. van Embden, and R.A. Hartskeerl. 1992. Detection of *Mycobacterium tuberculosis* in clinical samples by using polymerase chain reaction and a nonradioactive detection system. *J Clin Microbiol 30*:2567–2575.

63. Kolk, A.H., G.T. Noordhoek, O. de Leeuw, S. Kuijper, and J.D. van Embden. 1994. *Mycobacterium smegmatis* strain for detection of *Mycobacterium tuberculosis* by PCR used as internal control for inhibition of amplification and for quantification of bacteria. *J Clin Microbiol 32*:1354–1356.

64. Cleland, A., P. Nettleton, L. Jarvis, and P. Simmonds. 1999. Use of bovine viral diarrhoea virus as an internal control for amplification of hepatitis C virus. *Vox Sang 76*:170–174.

65. Greisen, K., M. Loeffelholz, A. Purohit, and D. Leong. 1994. PCR primers and probes for the 16S rRNA gene of most species of pathogenic bacteria, including bacteria found in cerebrospinal fluid. *J Clin Microbiol 32*:335–351.

66. Klausegger, A., M. Hell, A. Berger, K. Zinober, S. Baier, N. Jones, W. Sperl, and B. Kofler. 1999. Gram type-specific broad-range PCR amplification for rapid detection of 62 pathogenic bacteria. *J Clin Microbiol 37*:464–466.

67. Relman, D.A. 1999. The search for unrecognized pathogens. *Science 284*:1308–1310.

68. Hoffmann, E., J. Stech, Y. Guan, R.G. Webster, and D.R. Perez. 2001. Universal primer set for the full-length amplification of all influenza A viruses. *Arch Virol 146*:2275–2289.

69. Kleter, B., L-.J. van Doorn, J. ter Schegget, L. Schrauwen, K. van Krimpen, M. Burger, B. ter Harmsel, and W. Quint. 1998. Novel short-fragment PCR assay for highly sensitive broad-spectrum detection of anogenital human papillomaviruses. *Am J Pathol 153*:1731–1739.

70. Miura, T., J. Sakuragi, M. Kawamura, M. Fukasawa, E.N. Moriyama, T. Gojobori, K. Ishikawa, J.A. Mingle, V.B. Nettey, and H. Akari. 1990. Establishment of a phylogenetic survey system for AIDS-related lentiviruses and demonstration of a new HIV-2 subgroup. *AIDS 4*:1257–1261.

71. Zimmermann, K., and J.W. Mannhalter. 1996. Technical aspects of quantitative competitive PCR. *Biotechniques 21*:268–272.

72. Griffiths, P.D., and V.C. Emery. 1991. Quantitative nucleic acid amplification. Patent Cooperation Treaty (PCT) publication number WO9302215. Priority date July 24, 1991.

73. Hahn, M., V. Dorsam, P. Friedhoff, A. Fritz, and A. Pingoud. 1995. Quantitative polymerase chain reaction with enzyme-linked immunosorbent assay detection of selectively digested amplified sample and control DNA. *Anal Biochem 229*:236–248.

74. Lock, M.J., P.D. Griffiths, and V.C. Emery. 1997. Development of a quantitative competitive polymerase chain reaction for human herpesvirus 8. *J Virol Methods 64*:19–26.

75. Nygren, M., M. Ronaghi, P. Nyren, J. Albert, and J. Lundeberg. 2001. Quantification of HIV-1 using multiple quantitative polymerase chain reaction standards and bioluminometric detection. *Anal Biochem 288*:28–38.

76. Rudi, K., O.M. Skulberg, F. Larsen, and K.S. Jakobsen. 1998. Quantification of toxic cyanobacteria in water by use of competitive PCR followed by sequence-specific labeling of oligonucleotide probes. *Appl Environ Microbiol 64*:2639–2643.

77. Bretagne, S., J.M. Costa, M. Vidaud, J. Tran, V. Nhieu, and J. Fleury-Feith. 1993. Detection of *Toxoplasma gondii* by competitive DNA amplification of bronchoalveolar lavage samples. *J Infect Dis 168*:1585–1588.

78. Cone, R.W., A.C. Hobson, and M.L. Huang. 1992. Coamplified positive control detects inhibition of polymerase chain reactions. *J Clin Microbiol 30*:3185–3189.

79. Gilliland, G., S. Perrin, K. Blanchard, and H.F. Bunn. 1990. Analysis of cytokine mRNA and DNA: Detection and quantitation by competitive polymerase chain reaction. *Proc Natl Acad Sci USA 87*:2725–2729.

80. Li, B., P.K. Sehajpal, A. Khanna, H. Vlassara, A. Cerami, K.H. Stenzel, and M. Suthanthiran. 1991. Differential regulation of transforming growth factor beta and interleukin 2 genes in human T cells: Demonstration by usage of novel competitor DNA constructs in the quantitative polymerase chain reaction. *J Exp Med 174*:1259–1262.

81. Uberla, K., C. Platzer, T. Diamantstein, and T. Blankenstein. 1991. Generation of competitor DNA fragments for quantitative PCR. *PCR Methods Appl 1*:136–139.

82. Natarajan, V., R.J. Plishka, E.W. Scott, H.C. Lane, and N.P. Salzman. 1994. An internally controlled virion PCR for the measurement of HIV-1 RNA in plasma. *PCR Methods Appl 3*:346–350.

83. Lai, K.K., L. Cook, S. Wendt, L. Corey, and K.R. Jerome. 2003. Evaluation of real-time PCR versus PCR with liquid-phase hybridization for detection of enterovirus RNA in cerebrospinal fluid. *J Clin Microbiol 41*:3133–3141.

84. Pasloske, B.L., D.B. DuBois, D. Brown, and M. Winkler. 1996. Ribonuclease-resistant RNA preparation and utilization. United states patent number US5939262. Priority date July 3, 1996.

85. Pear, W.S., G.P. Nolan, M.L. Scott, and D. Baltimore. 1993. Production of high-titer helper-free retroviruses by transient transfection. *Proc Natl Acad Sci USA 90*:8392–8396.

86. Verstrepen, W.A., S. Kuhn, M.M. Kockx, M.E. Van De Vyvere, and A.H. Mertens. 2001. Rapid detection of enterovirus RNA in cerebrospinal fluid specimens with a novel single-tube real-time reverse transcription-PCR assay. *J Clin Microbiol 39*:4093–4096.

87. WalkerPeach, C.R., and B.L. Pasloske. 2004. DNA bacteriophage as controls for clinical viral testing. *Clin Chem 50*:1970–1971.

88. Wagner, E., C. Plank, K. Zatloukal, M. Cotten, and M.L. Birnstiel. 1992. Influenza virus hemagglutinin HA-2 N-terminal fusogenic peptides augment gene transfer by transferrin–polylysine–DNA complexes: Toward a synthetic virus-like gene-transfer vehicle. *Proc Natl Acad Sci USA 89*:7934–7938.

89. Wolfert, M.A., P.R. Dash, O. Nazarova, D. Oupicky, L.W. Seymour, S. Smart, J. Strohalm, and K. Ulbrich. 1999. Polyelectrolyte vectors for gene delivery: Influence of cationic polymer on biophysical properties of complexes formed with DNA. *Bioconjug Chem 10*:993–1004.

90. Brady, J.N., J.D. Kendall, and R.A. Consigli. 1979. *In vitro* reassembly of infectious polyoma virions. *J Virol 32*:640–647.

91. Paintsil, J., M. Muller, M. Picken, L. Gissmann, and J. Zhou. 1998. Calcium is required in reassembly of bovine papillomavirus *in vitro*. *J Gen Virol 79*:1133–1141.

92. Berg, E.S., and K. Skaug. 2000. Use of nonviable particles comprising an internal control (IC) nucleic acid. New Zealand patent number NZ524881. Priority date August 30, 2000.

93. Bangham, A.D., M.M. Standish, and J.C. Watkins. 1965. Diffusion of univalent ions across the lamellae of swollen phospholipids. *J Mol Biol 13*:238–252.
94. Monnard, P.A., T. Oberholzer, and P. Luisi. 1997. Entrapment of nucleic acids in liposomes. *Biochim Biophys Acta 1329*:39–50.
95. Szoka, F., Jr., and D. Papahadjopoulos. 1980. Comparative properties and methods of preparation of lipid vesicles (liposomes). *Annu Rev Biophys Bioeng 9*:467–508.
96. Titlow, C.C., J.K. Andersen, J.A. Trofatter, and O. Breakfield. 1992. *In vitro* translation of proteins with terminal deletion by SP6 RNA polymerase-mediated transcription of PCR products. *PCR Methods Appl 2*:172–174.
97. Grit, M., and D.J. Crommelin. 1992. The effect of aging on the physical stability of liposome dispersions. *Chem Phys Lipids 62*:113–122.
98. Zuidam, N.J., H.K. Gouw, Y. Barenholz, and D.J. Crommelin. 1995. Physical (in)stability of liposomes upon chemical hydrolysis: The role of lysophospholipids and fatty acids. *Biochim Biophys Acta 1240*:101–110.
99. Olsvik, Ø., T. Popovic, E. Skjerve, K.S. Cudjoe, E. Hornes, J. Ugelstad, and M. Uhlén. 1994. Magnetic separation techniques in diagnostic microbiology. *Clin Microbiol Rev 7*:43–54.
100. Rongen, H.A., A. Bult, and W.P. van Bennekom. 2004. Liposomes and immunoassays. *J Immunol Methods 204*:105–133.
101. Szoka, F., Jr., and D. Papahadjopoulos. 1978. Procedure for preparation of liposomes with large internal aqueous space and high capture by reverse-phase evaporation. *Proc Natl Acad Sci USA 75*:4194–4198.

Part II

Reaction Components

3 Oligonucleotide Synthesis and Purification

Yannick Fillon, Theresa Creasey, Qin Xiang,
and Carlos Martinez

CONTENTS

3.1 INTRODUCTION

Oligonucleotide synthesis is the chemical synthesis of nucleic acids to assemble a defined sequence. In contrast to enzymatic synthesis, which in nature proceeds in a 5′ to 3′ direction, synthetic oligonucleotide synthesis predominantly takes place in the 3′ to 5′ orientation.

The process of oligonucleotide synthesis has evolved since the early 1950s, although the basic solid-phase synthesis methods that are currently used for the assembly of oligonucleotides have been in place for nearly 30 years. These methods rely primarily on nucleoside phosphoramidite (amidite) chemistry.[1–3] Recently, the field of oligonucleotide chemistry has broadened with the introduction of new monomers, activation chemistries, and solid supports. This has enabled the creation of novel oligonucleotide constructs that are suitable for many varied applications, such as construction of microarrays,[4,5] sequencing,[6–10] polymerase chain reaction (PCR),[11] and quantitative real-time PCR (qPCR).[12,13] This chapter contains a review of the synthesis and processing of oligonucleotides for optimal performance in PCR and qPCR assays.

3.2   NUCLEOSIDE PHOSPHORAMIDITES

Naturally occurring nucleosides are unsuitable substrates for use in synthetic oligonucleotide synthesis because they have relatively low reactivity and so their use would not allow the synthesis of oligonucleotides of sufficiently high yields or lengths to be useful in molecular biology applications. The rate of formation of the required internucleosidic bonds is greatly improved by using 3′-O-(N,N-diisopropyl phosphoramidite) nucleosidic derivatives (phosphoramidites).

A DNA primer used for PCR is constructed from individual phosphoramidites (dA, dC, dG, T), whereas RNA molecules are synthesized from protected ribonucleosides (A, C, G, U). Desired characteristics such as fluorescent labels and quenchers that are used for qPCR probes, LNA™, or other chemical motifs are introduced into the oligonucleotide by using modified phosphoramidites or postsynthetic reactive moieties such as N-hydroxysuccinimidyl esters. All the functional groups that are not involved in the synthesis process are protected to prevent side reactions occurring during the synthetic cycle (Figure 3.2).

Conventionally, the 5′ hydroxyl group is protected by an acid-labile 4,4′-dimethoxytrityl (DMT) group. Exocyclic amino functionalities on the purines and pyrimidines are, in general, protected to eliminate unwanted side reactions or to improve the solubility of the phosphoramidites in solvents such as acetronitrile or toluene. The only exceptions are thymidine and uracil, which do not bear an exocyclic amine group (Figure 3.1). To avoid the protecting groups from being removed when the acid-labile, DMT removal is carried out at the end of each cycle, the exocyclic amine protection used is only base labile.

There are two general categories of base-labile, exocyclic amine protection: those requiring *standard deprotection* protocol and those requiring *fast deprotection* protocol. In general, the

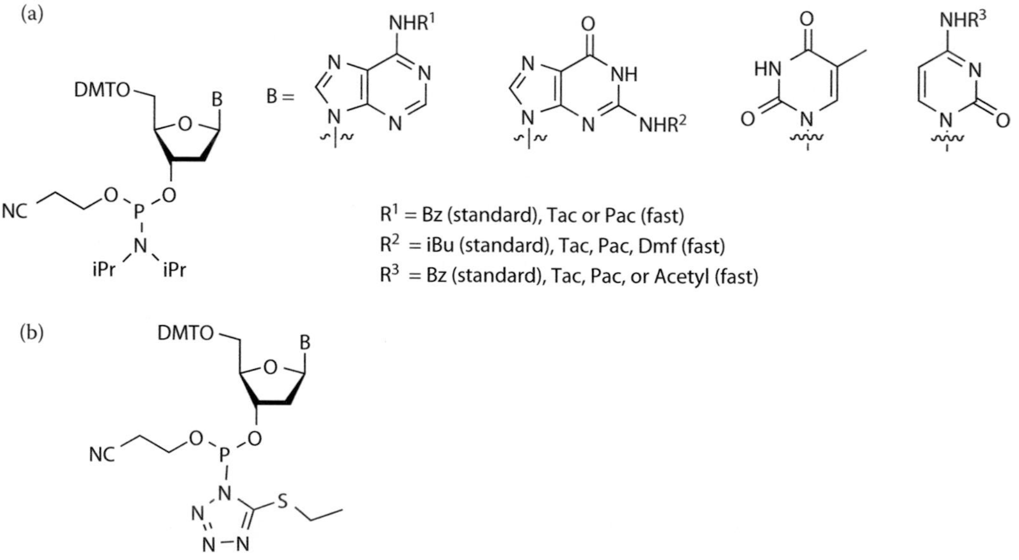

FIGURE 3.1   (a) Protecting groups on fast deprotecting amidites. Tac, *tert*-butylphenoxyacetyl; Pac, phenoxyacetyl; Dmf, dimethylformamidino protecting groups. (b) Activated phosphoramidite with 5-ethylthio-1H-tetrazole (ETT).

processing conditions required for oligonucleotides prepared with amidites that are suitable for fast deprotection protocols are milder than those required when working with standard deprotection protocol amidites.

The amidites supporting standard deprotection have the exocyclic amines protected with benzoyl (A, dA, C, and dC) and isobutyl (G, dG) moieties (Figure 3.1). An acetyl group may also be used to protect C and dC. For those where fast deprotection protocols are suitable, the exocyclic amines are manufactured with *tert*-butylphenoxyacetyl (Tac) (dG, dA, and dC) (Sigma-Aldrich) or phenoxyacetyl (Pac) (dG, dA, and dC) protecting groups.

Deprotection of oligonucleotides with standard protecting groups involves incubation under basic conditions (concentrated ammonia or a mixture of ammonium hydroxide and methylamine) and often requires elevated temperatures to achieve complete deprotection. Processing of oligonucleotides manufactured using fast deprotecting chemistries (Figure 3.1) requires shorter reaction times, lower temperatures, and milder conditions for deprotection. Selection of the optimal phosphoramidites depends on the desired manufacturing time (synthesis to final product) and the stability of modifications introduced during synthesis, as some modifications may be sensitive to the conditions required for the removal of the protecting groups.

Oligonucleotides without any modification are typically synthesized using standard phosphoramidites, as the components of the oligomer are robust and stable under the conditions needed for their processing. However, oligonucleotides containing fluorescent monomers and other labile modifications are synthesized using the fast deprotecting chemistry to minimize the decomposition of the modifier during processing. Degradation products will mostly affect the fluorescence output of the reporter fluorophore. In the case where the quencher is tetramethylrhodamine (TAMRA™), the same degradation risk exists, but in this case, it will also affect the quenching efficiency, resulting in more background fluorescence.

The formulation of amidites using high-quality, dry acetonitrile under moisture-free conditions (under an inert gas like argon or nitrogen) is important to ensure a high integrity of the amidite and for achieving maximum step-wise coupling efficiency. Amidites are packaged to accommodate the requirements of both commercial and custom synthesizers so that anhydrous conditions can be easily achieved.

3.3 SOLID-PHASE SYNTHESIS SUPPORT

The manufacture of oligonucleotides occurs on a solid support. Throughout the synthetic process, the oligonucleotide is covalently bound to the support via the 3′ hydroxyl or the 3′ moiety. Hence, the synthesis of an oligonucleotide also requires the selection of the appropriate, derivatized synthesis support. The main factors to consider when selecting a support are the cost of the support material, oligonucleotide length, and desired oligonucleotide yield. The most common solid supports are derivatized to initiate the synthesis at the 3′ end of the oligonucleotide such that synthesis occurs in a 3′ to 5′ direction. Supports are typically modified with a universal linker, a nucleotide (dA, dC, dG, T), or a modifier (e.g., amine, phosphate, biotin, or a fluorescent label) depending on the final application and the 3′ modification requirements of the oligonucleotide. Occasionally, there is a need for a 5′–3′ synthesis, in which case 5′ derivatized solid supports are used. This is the case when 2′, 3′-dideoxy nucleosides (ddA, ddC, ddG, ddT) need to be added to the oligonucleotide (such that there is no 3′-hydroxy available for synthesis) or in those cases where the modifier cannot withstand the synthesis conditions that are used during each coupling cycle.

Two of the most common supports are controlled pore glass (CPG) and macroporous polystyrene (MPPS).[14] CPG is economical and is available in a variety of pore sizes, providing flexibility for the oligonucleotide length and loading levels, which influences the final yield of the product. For oligonucleotide synthesis, CPG pore sizes range from 500 Å, which is used for the synthesis of oligonucleotides up to 50 bases, through to 2000 Å for the synthesis of oligonucleotides up to 150 bases. When using CPG to synthesize oligonucleotides shorter than 50 bases, steric and spatial hindrances are not a problem; however, the optimal pore size must be taken into consideration when producing oligonucleotides with lengths that are greater than 50 bases, since the increasing length and thus the size of the oligonucleotide will increase the steric hindrance and reduce coupling efficiencies. The quantity of the solid support and its loading capacity will determine the yield of the oligonucleotide. Loading capacity is a measure of the number of reactive sites per amount of support and ranges from 15 to 100 μmol/g. As an example, a reaction chamber that is loaded with 5 mg of a support with a loading capacity of 30 μmol/g will theoretically have 0.150 μmol (150 nmol) of reactive sites, whereas 5 mg of a support with a loading capacity of 80 μmol/g will theoretically have 0.400 μmol (400 nmol) of reactive sites. The yield of the crude material should approach this initial loading but will be affected by the coupling efficiency, cleaving, and purification conditions used to obtain the final purified product.

For 30–100 nmol of oligonucleotides of around 25–35 bases in length to be used in routine PCR research applications, CPG with a pore size of 500 Å and a loading capacity of 30 μmol/g is typically selected. The economics and availability of a wide assortment of standard and modified CPG supports make this the support of choice for the manufacturing of standard oligonucleotides.

An alternative solid support is MPPS, which is a highly cross-linked material produced from the polymerization of divinylbenzene, styrene, and 4-chloromethylstyrene. This results in a mechanically strong, free-flowing material that is easily controlled for particle size. Unlike CPG, MPPS supports are not porous and so the oligonucleotide is synthesized on the surface of the particle. Since the oligonucleotide is synthesized on the outer surface of the support, the physical restrictions caused by a porous material (like CPG) do not apply, thus allowing length flexibility. Therefore, the same solid support material can be used for the synthesis of oligonucleotides from 10 bases to greater than 150 bases in length. However, this nonporous nature of MPPS restricts the available surface area and therefore the loading capacity of the support. The loading capacity, synthesis column architecture, and resin swelling must all be taken into consideration when manufacturing oligonucleotides at larger scales on MPPS supports. The selection of this support over CPG will be driven by the instrument and scale of synthesis. For example, vendors of large-scale synthesizers have optimized the synthesis cycle for MPPS material.

Regardless of the support chosen, the fundamental processes of oligonucleotide synthesis are similar.

3.4 SYNTHESIS CYCLE

The synthesis methods used for the production of DNA- and RNA-based oligonucleotides require variations on four core steps. In addition, modified phosphoramidites can be incorporated in a similar manner to introduce none-nucleobase elements, such as amines and fluorescent dyes, into the oligonucleotides.

3.4.1 DETRITYLATION OR DEBLOCKING

The first step in the synthesis cycle requires the removal of the DMT 5′ hydroxyl protecting group from the oligonucleotide precursor that is bound to the support. DMT is removed by treatment with acid (trichloroacetic acid [TCA] or dichloroacetic acid [DCA] are commonly used) dissolved in dichloromethane or toluene. The removal of the DMT group liberates a trityl carbocation, which is orange in color and can provide a visual or quantitative assessment of the oligonucleotide synthesis process. Once the solid support has been treated with the acidic solution to remove the DMT protecting group and washed repeatedly, the synthesis of the oligonucleotide can commence with the coupling step by adding the first phosphoramidite monomer. The DMT group can be left on the oligonucleotide after the last cycle of coupling to aid in some purification processes, as discussed in Section 3.6.

3.4.2　Coupling

The process of coupling is the chain extension step during which an activated phosphoramidite is added to the oligonucleotide precursor on the solid support. A key component of the coupling reaction is the activation reagent, referred to as "the activator." The activator reacts with the phosphoramidites to create an "activated" species (Figure 3.1), which in turn reacts with the 5′ primary hydroxyl group of the growing oligonucleotide chain.

Various activation reagents are available, including 5-(bis-3,5-trifluoromethylphenyl)-1H-tetrazole Pharmadite (Activator 42), tetrazole, ethylthio tetrazole (ETT), and dicyanoimidazole (DCI). All these differ from each other in their acidity. The more acidic the activator, the faster the activation step, but the higher the risk of removing the DMT protecting group on the phosphoramidites, which may result in phosphoramidite self-reaction, that is, dimer formation. Dimerization of the phosphoramidite during the activation step may lead to unwanted oligonucleotide side products. This can result in faults in the final oligonucleotide sequence that are referred to as insertions and described as $n + 1$, $n + 2$, and so on, where n refers to the desired oligonucleotide length and the number refers to the number of additional bases added to the sequence. The proper selection and formulation of phosphoramidites and compatible activation chemistry contributes significantly to reducing unwanted side reactions, thereby producing high-quality oligonucleotides that can be further purified according to the requirements of their intended use.

3.4.3　Capping

The process of capping refers to the blocking of unextended chains from the coupling step. Although phosphoramidite coupling in solid-phase chemistry is an efficient reaction (>99% efficiency is routinely achieved), it is important that the approximately 1% of chains that are not coupled are blocked from further extension through an acetylation reaction, referred to as "capping." When using standard amidites, this is achieved by acetylation of the unreacted 5′ hydroxyl groups using acetic anhydride and 1-methylimidazole. When fast deprotecting amidites are used, it is recommended that the acetic anhydride solution be replaced by a Tac or Pac anhydride solution to avoid a transacetylation reaction at the N2 position on the dG amidite. The product of this reaction could be more difficult to remove under the mild deprotection conditions that are usually followed when the fast deprotecting chemistry is used.

Poor coupling followed by inefficient capping, detritylation, or both will lead to the production of oligonucleotide strands with base deletions, referred to as $n - 1$ side products. Furthermore, the detritylation reaction must be carefully fine-tuned to avoid the protonation of purine residues (dA and dG), which leads to the loss of the heterocyclic group (depurination) and formation of an abasic site.[15–17] Oligonucleotide fragmentation at abasic sites can occur during postsynthesis processing. Smaller oligonucleotide fragments are not always removed during purification and their remaining presence may lead to higher backgrounds in PCR reactions.

3.4.4　Oxidation: Formation of Stable Phosphotriester Bond

After coupling, a phosphite triester bond is formed. This is an unstable linkage under the oligonucleotide synthesis conditions. During the oxidation step, the oligonucleotide precursor on the support is treated with iodine and water in the presence of a weak base to oxidize the trivalent phosphite to a more stable pentavalent phosphate. This prevents the cleavage of this internucleotide bridge during the postsynthesis processing. Incomplete oxidation will produce a series of fragments that could carry over during the purification of the final product.

These four steps, in addition to intermediate wash steps, make up the synthesis cycle, as depicted in Figure 3.2. In summary, the oligonucleotide synthesis protocol is composed of repeated synthetic cycles, with the required, specified amidites being introduced at the coupling step during the

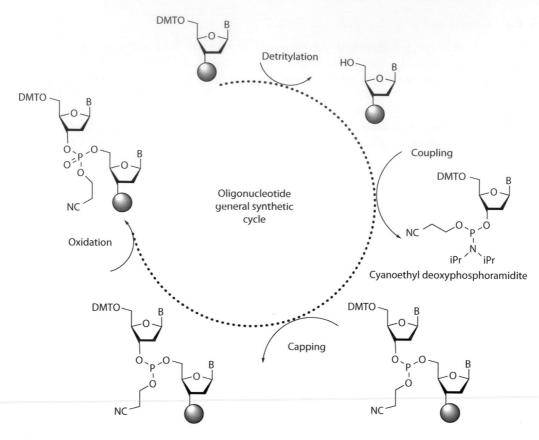

FIGURE 3.2 The general synthetic cycle of nucleic acids consists of four basic steps: Removal of the trityl protecting group, coupling of the incoming amidite, capping of unreacted hydroxyl groups, and oxidation of the phosphite into a phosphodiester.

protocol to result in an oligonucleotide of predefined length and sequence. Optimization of these cycles is critical to obtain high-quality oligonucleotides, which results in reliable PCR results.

In addition to the protection and synthetic chemistry, the postsynthetic process plays a key role in the quality of oligonucleotides used in PCR applications. Three general steps are carried out after the synthesis of the oligonucleotide is completed:

- Deprotection of the phosphate backbone
- Cleavage of the oligonucleotide from the support, including the elimination of the 3′-linker from syntheses on universal supports
- Deprotection of the exocyclic amines on the nucleobases (dA, dC, and dG)

In some instances, all of these processes are completed under the same conditions and in a single step, but for the most part, careful selection of individual reagents and conditions for each deprotection step is required to ensure that oligonucleotides are produced with the highest quality and maximum yields. As a result of the coupling step, the phosphate backbone is protected with a cyanoethyl group. Secondary amines (such as diethylamine) are used to eliminate this protecting group with the oligonucleotide still bound to the support. Deprotecting the phosphate backbone as a discreet processing step avoids the side reaction of the acrylonitrile adduct (cyanoethyl group) with other functional groups on the oligonucleotide, such as exocyclic amines or other primary amines. Such side reactions are undesirable because they have a direct effect on the hybridization efficiency of the primers (blocking of hydrogen bonding) or in the postsynthetic labeling of oligonucleotides with fluorescent molecules or other amine-reactive compounds. Cleaving the oligonucleotides from the support can be achieved through treatment with ammonium hydroxide, methylamine, or a mixture of both (AMA). Sufficient reaction times for the cleavage step are important to achieve maximum oligonucleotide yields. Finally, the complete deprotection of the nucleobase exocyclic amines must be carefully controlled to ensure that hydrogen bonding is not impaired, which may produce errors by the polymerase or even stop the PCR reaction. This is achieved with the same reagents as the cleaving step mentioned above, which produces the amide form of the protecting groups and the free exocyclic amine for each of the nucleobases.

For routine PCRs reactions, the vast majority of primers consist of dA, dC, dG, and T and require no further modifications. However, specialty primers or probes such as those used in qPCR are typically modified. The following section contains a focus on selected modifications that are used to enhance the performance of oligonucleotides in PCR and the implications of these requirements in the manufacturing processes.

3.5 OLIGONUCLEOTIDE MODIFICATIONS

There are countless modifications that can, potentially, be incorporated into oligonucleotides to create the unique molecules that are required for the varied applications of synthetic oligonucleotides. Some of these modification classifications include duplex stabilization molecules (MGB™, LNA, bridged modifiers, and propynes), labels (dyes, quenchers, biotin, and digoxigenin), backbone modifications (phosphonates, morpholinos, and thiophosphates), modifications for structural studies (isotope labels and halogen labels), and modifiers (spacers, aldehyde modifications, and dendrimers). The focus of this section is the common modifications that are used in the synthesis of probes that enhance the performance of primers and probes in PCR applications.

3.5.1 FLUORESCENTLY LABELED OLIGONUCLEOTIDES

The progress of a typical qPCR amplification is monitored by measuring the change in overall fluorescence over the course of the qPCR run. The reporting molecule can be incorporated either specifically bound to an oligonucleotide primer or probe or nonspecifically using a DNA binding dye, such as SYBR Green I® dye.

Where a probe is used for reporting, the change of fluorescence that is measured during qPCR is typically based on changes in fluorescence (or Förster) resonance energy transfer (FRET), which pairs an energy donor molecule with an acceptor molecule. The energy donor molecule, in its electronic excited state, transfers energy to a compatible acceptor molecule, that is located in close proximity (Förster radii), through nonradiative dipole–dipole coupling. An acceptor molecule is matched with a donor fluorophore when its excitation wavelength range is within the emission range of the donor. Acceptor dyes or molecules either fluoresce at their emission wavelength or dissipate the energy in the form of heat. Donor fluorophores are often referred to as the reporter and acceptors as the quencher.[18,19]

Probes for qPCR are generally labeled with either one or two dyes (reporter and quencher), usually incorporated at the 5′ and 3′ ends of the oligonucleotide. For instance, dual-labeled fluorescent hydrolysis probes (often referred to as TaqMan® probes), molecular beacons, and Scorpion™ probes are labeled with two dyes on the same oligonucleotide, whereas LightCycler® technology uses a two-dye and two-oligonucleotide system with the donor and acceptor dyes incorporated onto two separate oligonucleotides. The one-dye systems include Lux™ Primers and LightUp Probes® in which the fluorophore is quenched by the nucleobases on the probe oligonucleotide when it is unhybridized; however, upon hybridization to the target, the fluorescence signal significantly increases.[20,21]

In the synthesis of qPCR probes, the selection of the donor and acceptor molecules depends upon the instrumentation and assay design, that is, singleplex or multiplex. For example, the original format hydrolysis probe (Taqman) assays take advantage of the overlapping fluorescent emission range of FAM™ (fluorescein absorbance 480–600 nm) and the absorption range of TAMRA (tetramethylrhodamine—absorbance 530–570 nm). Using this combination, the energy from the excited FAM dye is transferred to TAMRA. When the two dyes are separated by cleavage of the 5′ reporter from the probe by the exonuclease activity of Taq polymerase during qPCR, the FRET effect will cease, resulting in detectable fluorescence of the FAM dye. However, the fluorescence and broad emission spectrum of the TAMRA dye can be detrimental to the signal-to-noise (S/N) ratio, as it overlaps with FAM's own emission spectrum. To address this issue, nonfluorescent dyes such as Dabcyl have been used. These nonfluorescent quenchers absorb the energy of the fluorophore and emit heat instead of light, which improves the overall S/N ratio. Dabcyl is optimal for quenching shorter-wavelength fluorophores, such as fluorescein. A series of more specific acceptor molecules, referred to as "dark quenchers," have been developed from the basic Dabcyl molecule. These quenchers are used to improve quenching efficiency when used with a broad range of donor dye wavelengths. Biosearch Technologies has developed a portfolio of dark quenchers referred to as "Black Hole Quencher™" or BHQ™, which display good stability, have a broad spectrum of absorption with low background fluorescence (improved S/N), and can be matched with a variety of fluorescent dyes (see Table 3.1). In addition, alternative dark quenchers include "Iowa Black™" FQ/RQ by IDT, and "QXL™ quenchers" from Kaneca. When designing probe sequences, it is important to note that fluorophores can be partially quenched by an adjacent guanine base and so it is generally recommended that the fluorophore is not placed next to the guanine base.[22] Nonetheless, this effect was successfully exploited in the design of "Lux" (light upon extension) primers where a fluorescein or JOE™ dye is quenched by a G base that is brought into close proximity via a hairpin conformation. A recent study used this particular effect for the design of a qPCR assay, in which a nonfluorescent probe consisting of a target-specific sequence at the 5′ end and a tailing sequence at the 3′ end was prepared. The tailing sequence hybridizes with an oligonucleotide labeled with a fluorescent dye. When the complex hybridized with a target during PCR, the fluorescent was quenched by a nearby guanine. The fluorescent quenching can be linked with the amount of target for quantification. The method generated reproducible results, comparable to hydrolysis probes with the added benefit of reducing the cost of the assay.[23] Several manufacturers have developed families of fluorescent dyes that cover the majority of the UV/VIS spectrum and can be easily incorporated into oligonucleotides during solid-phase synthesis (amidites) or postsynthesis in solution (NHS esters) using, for instance, an amine-modified probe. Examples of these dye families include ATTO (ATTO Tech/Sigma), ALEXA® (LTI), WellRED (Beckman Coulter), Cyanine® dyes (GE), and DyLight® dyes (Thermo) and all contain several different molecules per series.

For hydrolysis probes, the 3′ dye modification is most conveniently incorporated using a solid support preloaded with a quencher, for example, TAMRA-CPG or BHQ-CPG. The reporting fluorescent dye is introduced using an amidite-modified dye as shown in Figure 3.3 or as a post-solid-phase synthesis conjugation as shown in Figure 3.4. The dye selection will depend upon assay requirements and instrumentation.

TABLE 3.1
Common Quenchers and Fluorescent Dye Combinations

Quencher/Acceptor	Quenching Range (nm)	Possible Pairing Candidate
TAMRA		FAM
Dabcyl	390–510	FAM, JOE, HEX™, VIC®
BHQ-1	480–580	FAM, JOE, HEX, Cy3, TAMRA, ATTO 488, ALEXA 488
BHQ-2	550–650	ATTO 590, ALEXA 594, Texas Red®, Cy3.5
BHQ-3	620–730	ATTO 655, ALEXA 647, Cy5, Cy5.5

The composition of the oligonucleotide must be taken into account when processing oligonucleotides because some treatments can result in dye cleavage or degradation. For example, cyanine dyes are sensitive to extended treatment at higher temperature (~80°C) under basic conditions (aqueous ammonia); therefore, lower temperatures (~50°C for 120 min) are preferred and minimize degradation. This is an especially important consideration when working with dyes in the

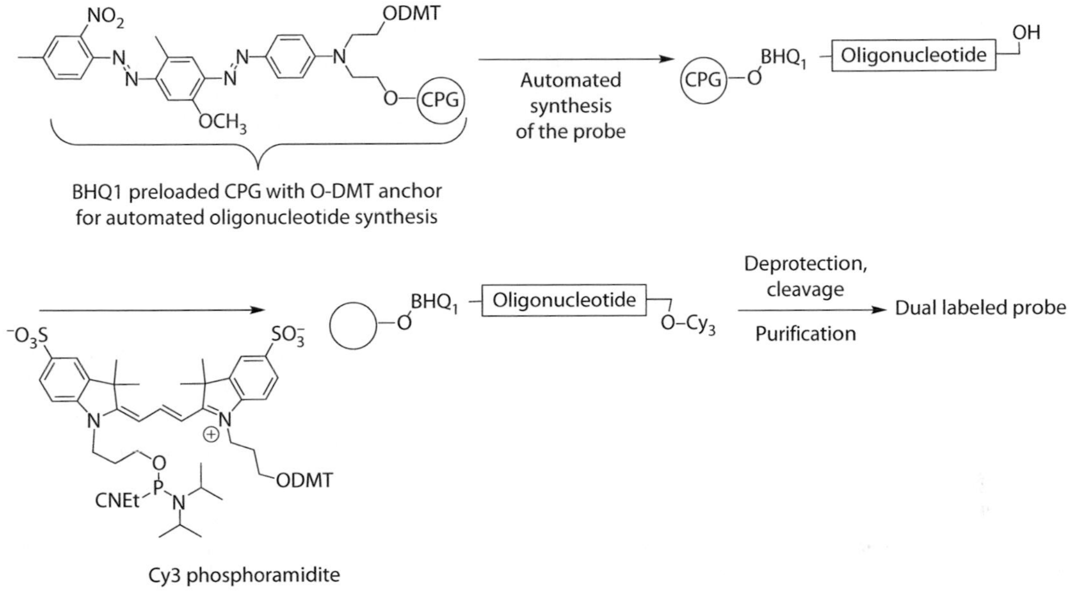

FIGURE 3.3   Example of the synthesis of a dual-labeled probe. The BHQ1 quencher is preloaded onto CPG support and the Cy3 amidite is coupled to the 5′ end of the oligo.

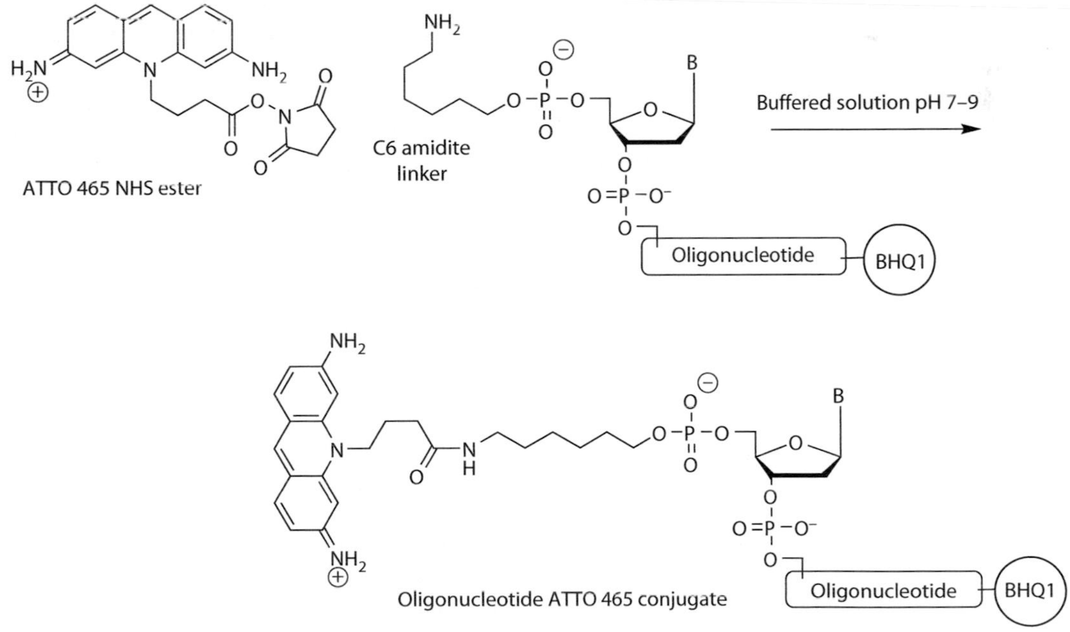

FIGURE 3.4   Aqueous conjugation in a buffer with a pH 7–9 of an ATTO dye to a 3′ BHQ1 labeled oligonucleotide.

higher absorption ranges (stability: Cy3 and Cy3.5 > Cy5 and Cy5.5 and Cy7). HEX (hexachloro-fluorescein) dye decomposes if the deprotection step is prolonged; therefore, higher temperatures (65°C) for a shorter time (less than 120 min) are required. To avoid degradation of TAMRA-labeled probes, t-butylamine is recommended as the base for cleaving and deprotecting because it is a hindered amine and therefore less nucleophilic.[24] The quality of the synthesis and cleavage of a dye-labeled oligonucleotide can easily be verified by mass spectrometry and the ratio of UV absorbance$_{260}$ and the maximum absorption wavelength of the fluorophore(s).

Some fluorophores are not available as amidites or are incompatible with processing conditions; therefore, these are conjugated to the oligonucleotides after cleavage from the solid support. To incorporate these specific modifications, the oligonucleotide can be synthesized to include a 5′ or 3′ nucleophilic linker, which is used in subsequent conjugation reactions (Figure 3.4). Common nucleophilic linkers include aliphatic amines (MMT-Amino C6, MMT-Amino C12, and TFA-Amino C6 from Sigma-Aldrich or Glen Research) and thiols (MMT—Thiol C6 and DMTO-C6—Thiol C6 or C3). The inclusion of an internal modifier can be achieved by using a modified base, such as an amino-modified thymine or cytosine (Glen Research, Chemgenes). The modified oligonucleotide can be conjugated to a dye in a similar manner as a 3′ or 5′ modification. The Scorpion probe is an example of an oligonucleotide with a dye conjugated to a modifier within the sequence. Following cleavage and deprotection, the MMT amino linker-modified oligonucleotides can be purified using reverse-phase cartridge methods to remove potential deletion sequences. A more stringent purification is typically not necessary at this step, especially for shorter oligonucleotides (less than 40 bases long). Caution must be taken to remove all traces of ammonia and tertiary amines prior to conjugation, as contamination will impact the final yield. The amino-modified oligonucleotides are then reacted with an excess of NHS-activated ester fluorophore (commercially available) in an aqueous buffer at an appropriate pH (~7.2–9.0) to ensure the nucleophilicity of the amine, minimize the hydrolysis of the NHS ester, and optimize yields.[25] Cleanup of the conjugate is achieved by using a size exclusion cartridge or by ethanol precipitation to remove the excess unreacted and hydrolyzed dye. The oligonucleotide–dye conjugate is then purified by reverse phase or ion exchange HPLC. Owing to the sensitivity and requirements for good S/N in qPCR assays, dye-modified oligonucleotides should be purified to eliminate free dye, $n - 1$, $n - x$, and $n + 1$ failure sequences that can lead to high background signals. Additionally, oligonucleotides can be conjugated using different chemistries such as the thiol and maleimide system, in which an oligonucleotide containing a C6 sulfhydryl linker to be modified at the 3′ or 5′ end is reacted with a maleimide-activated dye. Furthermore, using heterobifunctional linkers such as 6-maleimidohexanoic acid N-hydroxysuccinimide ester from Sigma-Aldrich can be a convenient way to conjugate an oligonucleotide to a unique modifier and/or surface.

While fluorophores and quenchers are the most common group of modifiers that are used in the detection and quantification of the qPCR, another group, namely structural modifiers, have been developed or adapted to the specific field of PCR to obtain greater sensitivity and specificity of the probes and primers. The incorporation of structural modifiers may be beneficial when designing assays with stringent requirements or challenging targets. The most commonly used structural modifiers are described below.

3.5.2 Structural Modifiers

3.5.2.1 Minor Grove Binder Probes

Kutyavin et al. demonstrated that the conjugation of a modified, trimeric organic molecule, derived from a naturally occurring antibiotic, to an oligonucleotide strongly stabilizes its hybridization to a complementary strand of RNA or DNA.[26–28] The tethered molecule was found to bind to the minor groove of a duplex with a high affinity for A–T-rich regions, leading to the use of the name minor groove binder (MGB) for this class of compound.

The higher melting temperature (T_m) resulting from the incorporation of an MGB into a probe sequence allows for the design of shorter oligonucleotides (10–15 bases) with the equivalent specificity of a longer probe with the same T_m. This makes the MGB modification applicable for qPCR applications, since it has a greater specificity for single or multiple base mismatches. The T_m increases when an MGB molecule flanks A/T-rich regions (up to ~20°C); however, the stabilization effect is lessened when the MGB molecule is adjacent to G/C-rich regions (~10°C).[28,29] MGB-modified probes are routinely used for qPCR and are comparable with standard designs such as hydrolysis probes.[30] The MGB molecule tethered to the 3′ end of the probe does not hinder the hydrolysis of the 5′ reporter by Taq DNA polymerase. Hydrolysis probes with 3′-MGB are typically synthesized using an MGB-preloaded synthesis support and standard solid-phase synthesis.[30] Probes with a 5′-MGB can be more challenging to synthesize and are generally more costly to manufacture. These probes utilize 5′-β-cyanoethyl phosphoramidites as opposed to the traditional 3′-β-cyanoethyl phosphoramidites, that is, oligonucleotides are synthesized starting from the 5′ end and proceed in a step-wise fashion to the 3′ end (Figure 3.5c). One example of 5′-MGB probe is MGB Eclipse™, which fluoresces upon hybridization and is not degraded during the qPCR assay due to the 5′ location of the MGB.[31] MGB as either an amidite or modified solid support is not commercially available. Quantitative real-time PCR probes incorporating MGB are available from Life Technologies Inc., Epoch Technologies (ELITech), and Thermo Fisher.

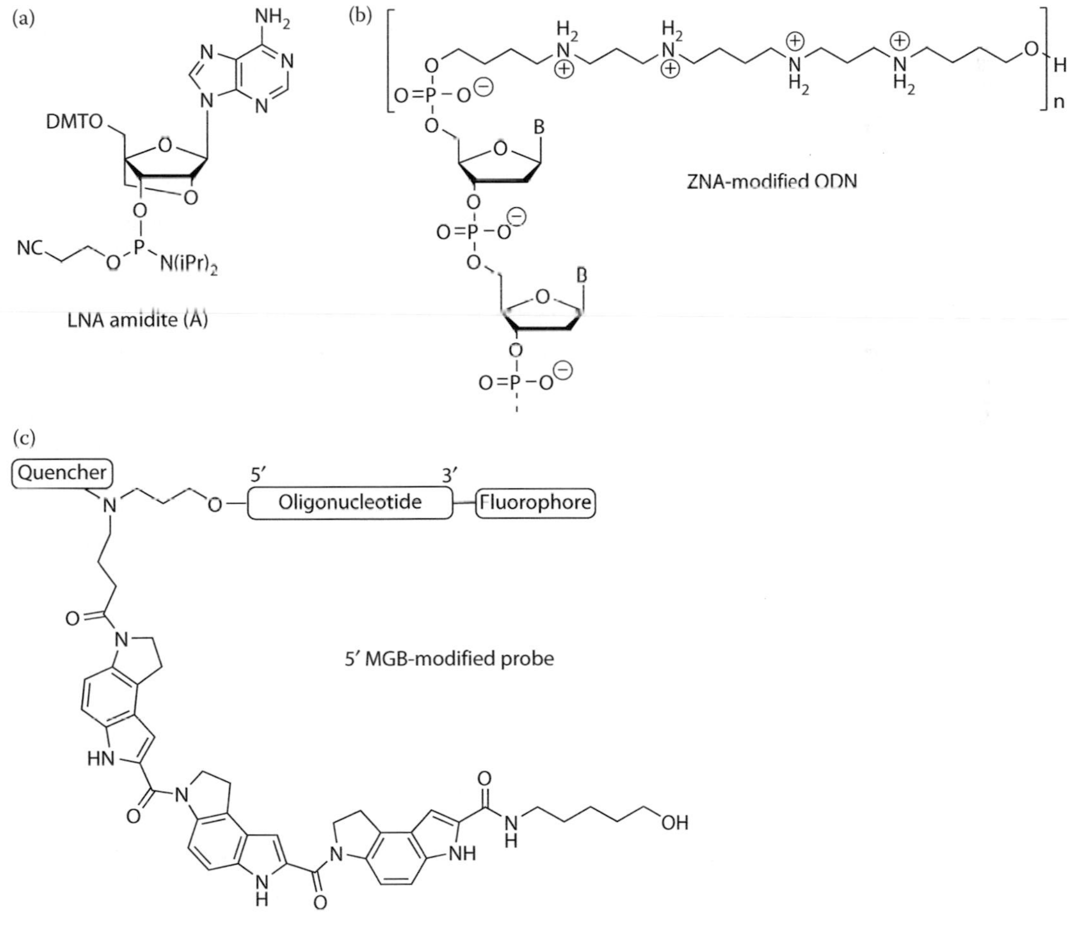

FIGURE 3.5 (a) LNA phosphoramidite. (b) 5′ ZNA modifier, several units can be attached to a primer. (c) Fluorescent probe with MGB modification.

3.5.2.2   Bridged Modifications: LNA™

Locked Nucleic Acids (LNA™) are a class of nucleic acid analogs that differ from standard ami-dites in that the furanose is bicyclic (Figure 3.5a). This modification is technically an RNA analog and utilizes a methylene bridge between the 2′-O and 4′-C positions to lock the conformation of the sugar ring. Oligonucleotides containing LNA residues along with DNA display improved hybridiza-tion potential and better mismatch sensitivity, as well as improved stability to endonucleases when compared to their natural DNA counterparts. An NMR study has shown that a change of confor-mation in the backbone of an LNA-modified oligonucleotide led to better base stacking, which improved the stability of the duplex.[32] LNA is commonly used in the field of qPCR for the enhance-ment of both primer and probe designs, including hydrolysis probes and molecular beacons.[33,34] The use of primers and probes modified with LNA, however, requires careful assay design to avoid mispriming and off-target extension.[35] LNA phosphoramidites are commercially available and are chemically robust, making them easy to incorporate into a standard oligonucleotide synthesis pro-tocol. LNA-modified probes are distributed by several oligonucleotide-manufacturing companies, including Sigma-Aldrich, under license from Exiqon.

3.5.2.3   Bridged Modifications: ENA®

Ethylene-bridged nucleic acids (ENA) is a bicyclic compound in which the 2′-O to 4′-C bridge is formed with two carbons rather than the single carbon as in LNA. The ring system locks the con-formation of the sugar in a 3′-endo position, yielding similar properties to that of the LNA-modified oligonucleotides, that is, increased resistance to nuclease and higher T_m when hybridizing with RNA and DNA.[36,37] ENA amidites for commercial supply are currently available only in Japan. Variations of ENA phosphoramidites are also reported, in which different chemical moieties are substituted on the bridge.

3.5.2.4   Zip Nucleic Acids or ZNA™.

A spermine (N,N'-bis(3-aminopropyl)butane-1,4-diamine)-based phosphoramidite coupled to an oli-gonucleotide has been reported to improve hybridization to a target sequence by folding over the strands in a zipper manner. These oligonucleotides are referred to as Zip Nucleic Acids, or ZNA™ oligonucleotides (Figure 3.5b) and were developed by Polyplus Transfection. ZNA properties have been reported to include a linear increase in T_m per spermine unit and faster annealing kinetics when compared to standard DNA or to LNA-modified oligonucleotides. These properties make ZNA oligo-nucleotides potential candidates for primers in fast cycling PCR.[38] Additionally, ZNA oligonucleotides were recently designed for use as hydrolysis probes in qPCR and multiplexing studies. In that study, the dual-labeled ZNA probes were reported to display lower background fluorescence when compared to identical DNA probes, while dual-labeled 17-mer ZNA probes were also shown to behave in a comparable manner to standard 22-mer DNA probes.[39] It should be noted that the protonation of a spermine unit changes the global charge of the oligonucleotide; that is, the number of spermines incor-porated into the sequence directly impacts the net charge of the oligonucleotide. Altering the charge of an oligonucleotide may have an impact on the manufacturing processes, particularly purification and the formulation of oligonucleotides, as the solubility in aqueous solutions may be reduced. Such char-acteristics should be taken into consideration when designing assays containing ZNA primers and/or probes. Polyplus Transfection distributes the spermine phosphoramidites, while a limited number of suppliers, including Sigma-Aldrich and Metabion, manufacture ZNA oligonucleotides.

3.5.2.5   L-DNA

L-DNA is the enantiomeric version of standard DNA (D-DNA); therefore, it has the same physi-cal properties as DNA but is unable to hybridize to natural DNA because L-DNA adopts a left-handed helix conformation.[40,41] The fact that L-DNA cannot hybridize with DNA was exploited in the design of an alternative form of molecular beacons.[42] In these, the stem was synthesized with L-deoxyphosphoramidites (Chemgenes) and the loop with standard phosphoramidites. This provided the molecular beacon with increased stability toward nucleases and prevented strand invasion that can typically cause false-positives and high background fluorescence.

3.5.2.6 Peptide Nucleic Acids

Peptide nucleic acids (PNAs) are also used in qPCR (light-up probes by Invitrogen)[43] as detection probes. PNA oligonucleotides are composed of a peptide-like backbone, with nucleobases. Enzymes, including nucleases or polymerases, do not recognize PNA oligonucleotides; therefore, they are very stable but can only be used as probes. The neutral, peptide-like backbone of PNA gives it a very high affinity for both DNA and RNA as static repulsion is minimal during hybridization.[44] The synthesis of such probes is different from the DNA oligonucleotide synthesis process and is more similar to peptide solid-phase chemistry, which uses "Fmoc" or "Boc" protecting group strategies. Similar to ZNA, PNA probes have limited solubility in aqueous systems and careful attention to the sequence design, modifications to improve solubility, and assay conditions is required. Some companies, such as AdvanDX, specialize in the production of PNA for diagnostic products and use them in assays designed for the rapid identification of bacterial infection and potential antibiotic resistance.

3.5.3 ALTERNATIVE MODIFICATIONS

The modifications described in the previous sections certainly do not represent an exhaustive list; other modifications are available, such as $2'$-O-methyl-$2'$-fluoro-nucleic acids, thiolated phosphate backbones, and unlocked nucleic acids (UNA). Each of these can be used to increase nuclease resistance of the oligonucleotide but do not typically increase affinity or sensitivity when used in PCR or qPCR assays.

All of the above modifications can potentially introduce impurities, which may have a negative impact on the PCR reaction. Many purification methods are available to address different requirements and to achieve the desired product specifications. In the following section, the most common purification methods for oligonucleotides that are used as primers and probes are described.

3.6 PURIFICATION

As described in Section 3.1, during oligonucleotide synthesis, each base is sequentially coupled to the growing chain in a $3'$ to $5'$ direction. For each coupling cycle, a small percentage of the oligonucleotide chains will not be extended, resulting in a mixture of full-length product (n) and truncated products ($n - 1, n - 2$, etc.). The criteria for the selection of the appropriate purification strategy will include the sensitivity of the application to $n - 1, n - 2, n - x$ truncations, length of the oligonucleotide, and quantity of the oligonucleotide required. For optimal performance in some applications, it is crucial that as much full-length (n) oligonucleotide be present. For others, the presence of shorter oligonucleotides ($n - 1, n - 2$, etc.) will not negatively impact the experimental results. Additional details on each of these purification techniques are summarized below.

Four of the most common purification strategies, in increasing order of stringency, are

- Desalt (DST)
- Reverse-phase cartridge (RPC)
- HPLC (reverse phase and ion exchange)
- PAGE (polyacrylamide gel electrophoresis)

3.6.1 DESALT PURIFICATION

Following oligonucleotide synthesis, the oligonucleotide is released from the solid support by incubation in basic solutions such as ammonium hydroxide. The desalting procedure removes residual by-products from the synthesis, cleavage, and deprotection processes, including side-chain protecting groups and organic salts. The desalting process does not remove truncation products ($n - 1$, $n - 2$, $n - x$). Desalting can be achieved by using size exclusion chromatography (i.e., sephadex), ethanol precipitation, or extraction (using butanol, ether, or ethyl acetate). For standard PCR, most primers are in the 20- to 30-base range and thus, the overwhelming abundance of full-length oligonucleotide outweighs any contributions from shorter products (>70–80%, depending on length). For oligonucleotides used as primers, desalting is a usually satisfactory purification method. The advantages of the desalting process are that it is economical, high yields are achieved, and oligonucleotides can be processed in parallel for rapid turnaround times. However, the overall purity is lower than that resulting from more stringent purification alternatives.

3.6.2   REVERSE-PHASE CARTRIDGE PURIFICATION

Separation on a reverse-phase cartridge offers the next level of purity. This separation is based on the hydrophobicity of 5′-dimethoxytrityl (DMT)-protected oligonucleotides. To distinguish between the full-length oligonucleotide and the shorter impurities, it is important that oligonucleotides be manufactured in the "trityl-on mode" (DMT-ON) as described in Section 3.1. The separation is based on the difference in the affinity of the full-length product, which contains a 5′-DMT group, and truncated sequences ($n - 1$, $n - 2$, $n - x$), which do not contain the DMT group. Many common 5′ modifications either contain trityl groups (biotin, phosphate, and primary amines) or are hydrophobic enough by themselves (such as cyanine dyes or WellRED dyes) and are compatible with reverse-phase cartridge purification. The optimal supports for reverse-phase cartridge purification include polystyrene-based supports (oligonucleotide $R3$) and silica-based supports such as C-18, C-5, and C-4 supports (e.g., Supelco, Phenomenex), which can be purchased in bulk or as individually packed columns. In general, after column loading, the full-length DMT-ON oligonucleotide is retained on the column and truncated sequences are washed off. The DMT group is then cleaved from the oligonucleotide while still on the cartridge, and the full-length product is eluted from the column in an aqueous solution, typically acetonitrile:water. The advantages of the reverse-phase purification are that it is economical, oligonucleotides can be processed in parallel for rapid turnaround times, and the purity levels achieved are generally >10% higher than for desalted oligonucleotides. Because a large portion of the impurities will be removed by this method, RPC is recommended for more stringent PCR-based experiments. Cartridge purification, however, is not a high-resolution technique. When the length of the oligonucleotides increases, the proportion of truncated sequences bearing the DMT tends to increase. These impurities will not be removed by RPC and thus for longer (>50 nucleotides) where a high proportion of full-length sequences are required, HPLC or PAGE purification is more suitable. This limitation is due to the lower effective resolution and due to the fact that the elution of impurities and full-length product does not occur in a controlled gradient as it does in HPLC.

3.6.3   HPLC (RP-HPLC AND IE-HPLC)

Reverse-phase HPLC purification (RP-HPLC) is one of the most versatile and widely utilized methods for the purification of oligonucleotides. Similar to cartridge purification, RP-HPLC is based upon the difference in affinity between DMT-ON full-length oligonucleotides and DMT-OFF truncations ($n - 1$, $n - 2$, $n - x$) or on the difference between dye labeled or unlabeled oligonucleotides. However, this method of purification offers the advantage of real-time monitoring of the separation via UV absorbance detection at 260 nm. In the case of oligonucleotides containing multiple modifications with distinct absorbance signatures (i.e., dual-labeled probes), modern detectors allow for the monitoring of multiple wavelengths and are capable of differentiating between multiple products based on the combined absorbance properties of the desired target, for example, an FAM-TAMRA-labeled oligonucleotide can be monitored at 260, 498, and 555 nm, therefore, monitoring both dyes and DNA simultaneously. In some cases, mass spectroscopy (MS) detectors can add a molecular weight dimension to the purification (also known as LC–MS for liquid chromatography–mass spectroscopy). A wide selection of columns, high column capacities, on-column detritylation procedures to reduce sample handling, and buffer systems that are volatile and achieve purity levels of >90% make this technique the preferred method for larger-scale and modified oligonucleotide purification.

If the oligonucleotide can be better separated from side products and truncations on the basis of the charge, then ion exchange HPLC (IE-HPLC) purification can be employed. This separation relies on the interaction of the negatively charged phosphate backbone of the oligonucleotide with the positively charged cations contained on the anion exchanger (typically Na^+).[45] Oligonucleotide purities of >90% can be achieved by ion exchange methods but only if the oligonucleotide length is limited to approximately 40 nucleotides. Column loading capacities are lower when compared with those of reverse phase and additional desalting steps are required to remove the salt buffers postpurification. The high salt content of the buffers also makes this technique less compatible with MS detectors. When used as a purification technique, desalting of the final product can be accomplished with the use of size exclusion columns or reverse-phase media.

The combination of these two techniques or tandem purification (typically IE-HPLC followed by RP-HPLC) can be used when a more stringent purification is needed for specific applications. Tandem HPLC purification is effectively used to purify dual-labeled probes and molecular beacons but is not limited to these products. Careful selection of the liquid chromatography technique will ultimately determine the success of the oligonucleotide in its intended application.

3.6.4 Polyacrylamide Gel Electrophoresis

The basis of PAGE separation is charge over molecular weight, leading to single base resolution. PAGE is suitable for oligonucleotides of any length.[46] In combination with UV shadowing, PAGE purification effectively separates full-length products from truncations; however, yields from PAGE are lower than for oligonucleotides purified using the previously described techniques but it produces a higher purity (>95%). Like any separation technique, the resolution from PAGE can be compromised if the gel is overloaded, that is, too much oligonucleotide added to the well; therefore, when compared with other methods, only small quantities of oligonucleotides can be purified by PAGE. The advantages of PAGE purification include parallel processing, effective purification of oligonucleotides with hairpin structures, and achievable purity levels of generally >95%. This technique is recommended when small quantities of highly purified product are required. PAGE purification is recommended for longer oligonucleotides (≥50 bases).

In summary, there are four major purification techniques for synthetic oligonucleotides. The selection of the purification technique will depend on the intended use of the oligonucleotide. Table 3.2 provides a summary of the recommended purification methods for different oligonucleotide types.

TABLE 3.2
Recommended Purification Methods by Oligo Type

Oligonucleotide Type	Purification Type			
	Desalted	RPC	HPLC	PAGE
PCR primers	X	X		
Single-labeled probes			X	
Dual-labeled probes			X	
Scorpions and synthetic templates			X	X

3.7   CONCLUSION

PCR and qPCR are the most commonly used techniques in molecular biology. Both require high-quality oligonucleotides with the possibility of a range of modifications. Careful selection of critical raw materials, the development of robust manufacturing processes, and a clear understanding of the relationship between product specifications (including purity requirements) and its final application are all essential for producing the quality of data necessary in today's highly scrutinized field of life sciences.

REFERENCES

1. Beaucage, S. L., Caruthers, M. H. 1981. Deoxynucleoside phosphoramidites—A new class of key intermediates for deoxypolynucleotide synthesis. *Tetrahedron Lett* 22(20), 1859–1862.
2. Sinha, N. D., Biernat, J., Köster, H. 1984. Polymer support oligonucleotide synthesis XVI: Synthesis of oligonucleotides using suitably protected deoxynucleoside-*N*-morpholinophosphoramidites on porous glass beads. *Nucleosides Nucleotides* 3(2), 157–171.
3. Caruthers, M. H. 1985. Gene synthesis machines: DNA chemistry and its uses. *Science* 230(4723), 281–285.
4. Hacia, J. G., Fan, J. B., Ryder, O., Jin, L. et al. 1999. Determination of ancestral alleles for human single-nucleotide polymorphisms using high-density oligonucleotide arrays. *Nat Genet* 22(2), 164–167.
5. Lausted, C., Dahl, T., Warren, C., King, K. et al. 2004. POSaM: A fast, flexible, open-source, inkjet oligonucleotide synthesizer and microarrayer. *Genome Biol* 5(8), R58.
6. Sanger, F., Coulson, A. R. 1975. A rapid method for determining sequences in DNA by primed synthesis with DNA polymerase. *J Mol Biol* 94(3), 441–448.
7. Sanger, F., Nicklen, S., Coulson, A. R. 1977. DNA sequencing with chain-terminating inhibitors. *Proc Natl Acad Sci USA* 74(12), 5463–5467.
8. Margulies, M., Egholm, M., Altman, W. E., Attiya, S. et al. 2005. Genome sequencing in microfabricated high-density picolitre reactors. *Nature* 437(7057), 376–380.
9. Schuster, S. C. 2008. Next-generation sequencing transforms today's biology. *Nat Methods* 5(1), 16–18.
10. Valouev, A., Ichikawa, J., Tonthat, T., Stuart, J. et al. 2008. A high-resolution, nucleosome position map of *C. elegans* reveals a lack of universal sequence-dictated positioning. *Genome Res* 18(7), 1051–1063.
11. Bartlett, J. M., Stirling, D. 2003. A short history of the polymerase chain reaction. *Methods Mol Biol* 226, 3–6.
12. VanGuilder, H. D., Vrana, K. E., Freeman, W. M. 2008. Twenty-five years of quantitative PCR for gene expression analysis. *Biotechniques* 44(5), 619–626.
13. Bustin, S. A., Benes, V., Garson, J. A., Hellemans, J. et al. 2009. The MIQE guidelines: Minimum information for publication of quantitative real-time PCR experiments. *Clin Chem* 55(4), 611–622.
14. Pon, R. T. 1993. Solid-phase supports for oligonucleotide synthesis. *Methods Mol Biol* 20, 465–496.
15. Adams, S. P., Kavka, K. S., Wykes, E. J., Holder, S. B. et al. 1983. Hindered dialkylamino nucleoside phosphite reagents in the synthesis of two DNA 51-mers. *J Am Chem Soc* 105(3), 661–663.
16. Efcavitch, J. W., Heiner, C. 1985. Depurination as a yield decreasing mechanism in oligodeoxynucleotide synthesis. *Nucleosides Nucleotides* 4(1), 267.
17. Septak, M. 1996. Kinetic studies on depurination and detritylation of CPG-bound intermediates during oligonucleotide synthesis. *Nucleic Acids Res* 24(15), 3053–3058.
18. Förster, T. 1948. Zwischenmolekulare energiewanderung und fluoreszenz. *Annal Physik* 437(1–2), 55–75.
19. Stryer, L., Haugland, R. P. 1967. Energy transfer: A spectroscopic ruler. *Proc Natl Acad Sci USA* 58(2), 719–726.
20. Svanvik, N., Nygren, J., Westman, G., Kubista, M. 2001. Free-probe fluorescence of light-up probes. *J Am Chem Soc* 123(5), 803–809.
21. Nazarenko, I., Lowe, B., Darfler, M., Ikonomi, P. et al. 2002. Multiplex quantitative PCR using self-quenched primers labeled with a single fluorophore. *Nucleic Acids Res* 30(9), e37.
22. Kurata, S., Kanagawa, T., Yamada, K., Torimura, M. et al. 2001. Fluorescent quenching-based quantitative detection of specific DNA/RNA using a BODIPY((R)) FL-labeled probe or primer. *Nucleic Acids Res* 29(6), E34.
23. Tani, H., Miyata, R., Ichikawa, K., Morishita, S. et al. 2009. Universal quenching probe system: Flexible, specific, and cost-effective real-time polymerase chain reaction method. *Anal Chem* 81(14), 5678–5685.

24. Mullah, B., Andrus, A. 1997. Automated synthesis of double dye-labeled oligonucleotides using tetra-methylrhodamine (TAMRA) solid supports. *Tetrahedron Lett* 38(33), 5751–5754.

25. Hermanson, G. T. 2008. *Bioconjugate Techniques*, 2nd Edition. Elsevier, London, UK, p. 238.

26. Hurley, L. H., Reynolds, V. L., Swenson, D. H., Petzold, G. L. et al. 1984. Reaction of the antitumor antibiotic CC-1065 with DNA: Structure of a DNA adduct with DNA sequence specificity. *Science* 226(4676), 843–844.

27. Sinyakov, A. N., Lokhov, S. G., Kutyavin, I. V., Gamper, H. B. et al. 1995. Exceptional and selective stabilization of A-T rich DNA.cntdot.DNA duplexes by *N*-methylpyrrole carboxamide peptides conjugated to oligodeoxynucleotides. *J Am Chem Soc* 117(17), 4995–4996.

28. Afonina, I., Zivarts, M., Kutyavin, I., Lukhtanov, E. et al. 1997. Efficient priming of PCR with short oligonucleotides conjugated to a minor groove binder. *Nucleic Acids Res* 25(13), 2657–2660.

29. Kutyavin, I. V., Lukhtanov, E. A., Gamper, H. B., Meyer, R. B. 1997. Oligonucleotides with conjugated dihydropyrroloindole tripeptides: Base composition and backbone effects on hybridization. *Nucleic Acids Res* 25(18), 3718–3723.

30. Kutyavin, I. V., Afonina, I. A., Mills, A., Gorn, V. V. et al. 2000. 3′-Minor groove binder-DNA probes increase sequence specificity at PCR extension temperatures. *Nucleic Acids Res* 28(2), 655–661.

31. Afonina, I. A., Reed, M. W., Lusby, E., Shishkina, I. G. et al. 2002. Minor groove binder-conjugated DNA probes for quantitative DNA detection by hybridization-triggered fluorescence. *Biotechniques* 32(4), 940–944, 946–949.

32. Vester, B., Wengel, J. 2004. LNA (locked nucleic acid): High-affinity targeting of complementary RNA and DNA. *Biochemistry* 43(42), 13233–13241.

33. Ugozzoli, L. A., Latorra, D., Puckett, R., Arar, K. et al. 2004. Real-time genotyping with oligonucleotide probes containing locked nucleic acids. *Anal Biochem* 324(1), 143–152.

34. Wang, L., Yang, C. J., Medley, C. D., Benner, S. A. et al. 2005. Locked nucleic acid molecular beacons. *J Am Chem Soc* 127(45), 15664–15665.

35. Levin, J. D., Fiala, D., Samala, M. F., Kahn, J. D. et al. 2006. Position-dependent effects of locked nucleic acid (LNA) on DNA sequencing and PCR primers. *Nucleic Acids Res* 34(20), e142.

36. Morita, K., Hasegawa, C., Kaneko, M., Tsutsumi, S. et al. 2002. 2′-O,4′-C-ethylene-bridged nucleic acids (ENA): Highly nuclease-resistant and thermodynamically stable oligonucleotides for antisense drug. *Bioorg Med Chem Lett* 12(1), 73–76.

37. Morita, K., Takagi, M., Hasegawa, C., Kaneko, M. et al. 2003. Synthesis and properties of 2′-O,4′-C-ethylene-bridged nucleic acids (ENA) as effective antisense oligonucleotides. *Bioorg Med Chem* 11(10), 2211–2226.

38. Moreau, V., Voirin, E., Paris, C., Kotera, M. et al. 2009. Zip nucleic acids: New high affinity oligonucleotides as potent primers for PCR and reverse transcription. *Nucleic Acids Res* 37(19), e130.

39. Paris, C., Moreau, V., Deglane, G., Voirin, E. et al. 2010. Zip nucleic acids are potent hydrolysis probes for quantitative PCR. *Nucleic Acids Res* 38(7), e95.

40. Urata, H., Shinohara, K., Ogura, E., Ueda, Y. et al. 1991. Mirror-image DNA. *J Am Chem Soc* 113(21), 8174–8175.

41. Urata, H., Ogura, E., Shinohara, K., Ueda, Y. et al. 1992. Synthesis and properties of mirror-image DNA. *Nucleic Acids Res* 20(13), 3325–3332.

42. Kim, Y., Yang, C. J., Tan, W. 2007. Superior structure stability and selectivity of hairpin nucleic acid probes with an L-DNA stem. *Nucleic Acids Res* 35(21), 7279–7287.

43. Petersen, K., Vogel, U., Rockenbauer, E., Nielsen, K. V. et al. 2004. Short PNA molecular beacons for real-time PCR allelic discrimination of single nucleotide polymorphisms. *Mol Cell Probes* 18(2), 117–122.

44. Nielsen, P. E., Egholm, M., Berg, R. H., Buchardt, O. 1991. Sequence-selective recognition of DNA by strand displacement with a thymine-substituted polyamide. *Science* 254(5037), 1497–1500.

45. S. Cohen, A., Vilenchik, M., Dudley, J. L., Gemborys, M. W. et al. 1993. High-performance liquid chromatography and capillary gel electrophoresis as applied to antisense DNA. *J Chromatogr A* 638(2), 293–301.

46. Srivatsa, G. S., Batt, M., Schuette, J., Carlson, R. H. et al. 1994. Quantitative capillary gel electrophoresis assay of phosphorothioate oligonucleotides in pharmaceutical formulations. *J Chromatogr A* 680(2), 469–477.

4 Augmenting PCR
Effects of Additives in the Polymerase Chain Reaction

Ernest J. Mueller

CONTENTS

4.1 INTRODUCTION

Since the invention of the polymerase chain reaction (PCR),[1,2] researchers have been busily applying this versatile technique to their individual purposes. Optimization of the reaction seems simple; a maximally active reaction only requires the proper conditions (correct pH, salts, temperature, and divalent cation concentrations) and a suitable substrate at saturating levels. Despite this, researchers have found it desirable to test many different additives to increase the efficiency of their amplification protocol. This chapter contains a review of the original literature to describe the effects and limitations imparted by PCR adjuncts.

4.2 POLYMERASE CHARACTERISTICS

Optimal reaction conditions for *Taq* DNA polymerase, the first and the most popular of the commercially available thermostable DNA polymerases, have been well established.[3] The enzyme has a broad pH optimum of pH 7–9,[4] but works best in Tris–HCl buffer.[5] Salts are also necessary, and either potassium or sodium chloride can be used in the reaction.[4] Divalent metal cations are absolutely required for enzyme activity, and the reaction is best supported with a magnesium concentration of 2–3 mM[5] although this is affected by nucleotide triphosphate concentrations.[7] When using *Taq* polymerase, high product yields are supported by the inclusion of 200 µM each dNTP, but this may be much higher than necessary; similar polI enzymes have reported K_m values near 1 µM[8] and Kunkel et al. measured decreases in fidelity when dNTP concentrations are increased from 1 to 100 µM.[9] Typical commercial reaction buffers contain, as a final concentration, 10 mM Tris–HCl, pH 8.3 (25°C), 50 mM KCl, 1.5–3.5 mM $MgCl_2$, 100–200 µM each dNTP, and 0.05% (v/v) each Tween-20 and IGEPAL NP40 detergent. The temperature optimum of *Taq* DNA polymerase lies between 75°C and 80°C.[5]

Other thermostable DNA polymerases have been used in PCR, and many of these enzymes amplify with higher fidelity than can be obtained by *Taq* polymerase.[9] Their buffers differ from *Taq*; for instance, the first commercial proofreading enzyme, *Pfu* polymerase, is used with an optimized reaction buffer containing 20 mM Tris–HCl, pH 8.8 (25°C), 10 mM KCl, 10 mM $(NH_4)_2SO_4$, 2 mM $MgSO_4$, 100–200 µM each dNTP, 0.1 mg/mL bovine serum albumin (BSA), and 0.1% (v/v) Triton X-100.[6] The major buffer differences are inclusion of (1) ammonium or magnesium sulfate and (2) detergents that vary between the trade named Tween, IGEPAL, and/or Triton classes. This observation apparently inspired researchers to investigate the effect of adding these constituents to reactions that were utilizing *Taq* DNA polymerase.

4.3   ADDITIVES TO ENHANCE ACTIVITY

Ammonium sulfate has been attributed to PCR enhancement in two separate publications. Olive et al.[10] observed increased PCR yield of cytomegalovirus DNA from urine using a buffer containing 17 mM $(NH_4)_2SO_4$, 67 mM Tris–HCl, pH 8.5, 7 mM $MgCl_2$, and 1.5 mM each dNTP. It is worth noting that this reaction composition is unusual, using very high levels of magnesium and dNTPs to generate short (79 bp) target. Nonetheless, the authors reported that their results correlated with alternative methods, including enzyme-linked immunosorbent assay (ELISA) and hybridization techniques. This unique observation was corroborated 5 years later,[11] the authors claiming that addition of ammonium sulfate substantially increased the success rate of their genomic fingerprinting assay. Data contrasting these two separate findings are shown in Figure 4.1; in this case, titrating potassium chloride and ammonium sulfate into the PCR gives the same profile of *Taq* activity, within the error of measurement. This previously unpublished data suggests that any enhancement of *Taq* polymerase by ammonium sulfate may be due to an effect on DNA rather than on the enzyme.

Detergents, on the other hand, have been demonstrated to be a requirement for DNA polymerase activity, and these compounds have always been among the original early buffer components in PCR application literature.[12,13] The addition of nonionic detergents was reported early on to enhance PCR yield,[14] and inclusion of a suitable detergent is absolutely required for *Taq* DNA polymerase thermostability, as shown in Figure 4.2 (Hernan, R. A. and Mueller, E. J.,

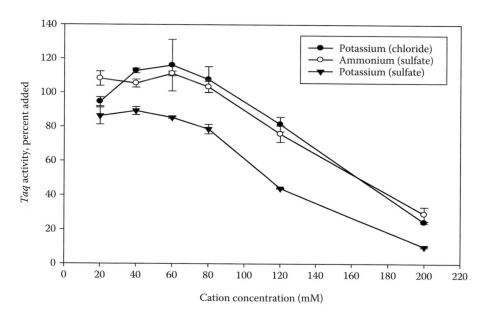

FIGURE 4.1   Cation concentration effect on *Taq* DNA polymerase activity. PCR assays using *Taq* DNA polymerase were performed essentially as described,[15] except that the substrate was activated calf thymus DNA and the base used in buffer titration was either potassium (filled symbols) or ammonium hydroxide (open symbols); this accounted for 20 mM of the monovalent cation in the mixture. Assay mixtures were supplemented with additional salt, as shown. Reactions were run in duplicate and data averaged.

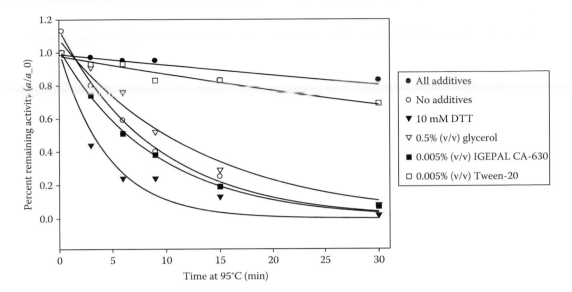

FIGURE 4.2   Effect of additives on *Taq* DNA polymerase stability. *Taq* DNA polymerase was formulated to contain 5 units/μL in 20 mM Tris–HCL, pH 8.3 (25°C), 100 mM KCl, 0.1 mM EDTA, 10% glycerol. It was assayed at 0.05 units/μL (1:100 dilution) final concentration. Each 100-μL reaction contained 20 mM Tris–HCl, pH 8.3 (25°C), 50 mM KCl, 1.5 mM $MgCl_2$, and 0.1% (v/v) glycerol in addition to the listed additives. Reactions were prepared on ice, overlayed with mineral oil, and placed in a heat block at 95°C. Samples were removed at 0, 3, 6, 9, 15, and 30 min and immediately assayed for activity as previously described.[15]

1997, Sigma-Aldrich, previously unpublished observations). The importance of detergents is also reflected in the patent literature,[16–18] and all commercial formulations of *Taq* DNA polymerase buffer contain these substances as stabilizers; the early enzyme formulations were listed to contain 0.5% (v/v) Tween-20 and 0.5% NP-40.

4.4   DNA MELTING ADJUNCTS

Assuming that the reaction is suitably formulated, PCR amplification should be achieved if a suitable template (substrate) is available. Much of the literature is thus focused on methods for effective amplification of less-than-perfect substrates, whether the cause lies with high-melting DNA or template that carries inhibiting contaminants (for example and further discussion, see Chapter 1).

Examples of poorly amplified templates have been reported throughout the 20 years of literature, and a variety of additives have been applied with the claim of melting DNA template and facilitating polymerase amplification. The first compound used in this manner was dimethyl sulfoxide (DMSO).[19,20] This discovery was followed by reports of other compounds known to affect DNA stability, including formamide,[21] glycerol,[22] betaine,[23–25] and tetramethyl ammonium chloride.[26] These and other additives seem to have different efficacy with different templates, and certainly each has a unique effective concentration range.

The group of DNA melting adjuncts was expanded and then extensively characterized by the work of Chakrabarti and Schutt. Their first published works focused on identifying new PCR-active amides,[27] sulfones,[28] and sulfoxides[29,30] by screening their effect on PCR yield. Thus, armed with a variety of newly discovered, variably effective PCR enhancers, Chakrabarti characterized the amplification effects of these compounds on a range of GC-containing templates. He also measured their separate effects on DNA melting, polymerase activity, and polymerase thermostability.[31] Chakrabarti showed that the effect of these additives was mainly due to the destabilization of the DNA duplex, but each adjunct also reduced the stability and activity of the polymerase to create a compound-specific window of effective concentration. Enhancement also appears to depend on template structure, and thus the literature evidence suggests that adjunct-mediated PCR enhancement is template- and adjunct-specific. In practical terms, this implies that each PCR enhancer must be optimized for each amplified DNA locus and the adjunct of choice.

4.5   ADDITIVES TO COMBAT SAMPLE-INTRODUCED INHIBITORS

The third application of PCR additives is to combat sample-introduced inhibitors. The phenomenon was described in the first decade of PCR literature and was initially counteracted by diluting samples past the inhibition threshold of the polymerase.[32] This tactic is not always possible, especially for clinical, environmental, and plant-derived nucleic acid samples.[33] These challenges have been reviewed (Ref. 34 and the references therein), with the more popular published adjuncts, including spermine,[35] polyvinylpyrrolidone,[36] BSA,[37,38] and a few others. Strategies to relieve inhibition depend on the inhibitor, and the lengthy listings of PCR inhibitors include many of the compounds used to enhance amplification efficiency as well as reagents required in the purification of nucleic acids.[39] In cases where inhibitors cannot be purified or diluted away, it appears that the solution, if possible at all, is empirical and as specific and difficult as the amplification of high-melting templates.

A new approach to PCR inhibition has recently been reported by Barnes and coworkers, pairing a polymerase engineered to tolerate a broad range of inhibitors[40] along with specific PCR enhancers.[41] Although this solution is relatively untested, it holds some promise in the analysis of clinical and environmental samples that are otherwise unable to be amplified.

4.6   SUMMARY

To summarize, adjuncts to the polymerase chain reaction have been used for three reasons—to increase polymerase activity, decrease template thermal stability, or mask the effects of polymerase inhibitors. For this first application, this brief review of the early literature and some supplementary data suggest that optimal polymerase activity is best achieved by commercially supplied buffers. Additives to destabilize high-melting templates can be effective, but have a narrow window of useful concentration that is template-dependent; extensive optimization is also required. Finally, using PCR adjuncts to neutralize inhibitors is a hit-and-miss process; there have been occasional effects reported with the use of adjuncts such as polyvinylpyrrolidone and BSA, but new polymerase mutants offer possible future uses when contaminating inhibitors cannot be purified or diluted away. It is hoped that this review and collection of literature sources will be useful for those researchers wishing to augment PCR.

REFERENCES

1. Mullis, K. B. 1987. Process for amplifying nucleic acid sequences, US4683202, Cetus Corporation, Emberville, CA, issued July 28.
2. Saiki, R. K., Scharf, S., Faloona, F., Mullis, K. B., Horn, G. T., Erlich, H. A., and Arnheim, N. 1985. Enzymatic amplification of beta-globin genomic sequences and restriction site analysis for diagnosis of sickle cell anemia. *Science* 230(4732): 1350–1354.
3. Kramer, M. F. and Coen, D. M. 2001. Enzymatic amplification of DNA by PCR: Standard procedures and optimization. *Curr. Protoc. Immunol.* May; Chapter 10:Unit 10.20.
4. Chien, A., Edgar, D. B., and Terla, J. M. 1976. Deoxyribonucleic acid polymerase from the extreme thermophile *Thermus aquaticus*. *J. Bacteriol.* 127(3): 1550–1557.
5. Frances, C., Lawyer, F. C., Stoffel, S., Saiki, R. K., Chang, S.-Y., Landre, P. A., Abrarnson, R. D., and Gelfand, R. D. 1993. High-level expression, purification, and enzymatic characterization of full-length *Thermus aquaticus* DNA polymerase and a truncated form deficient in 5′ to 3′ exonuclease activity. *Genome Res.* 2: 275–287.

6. Cline, J., Braman, J. C., and Hogrefe, H. H. 1996. PCR fidelity of Pfu DNA polymerase and other thermostable DNA polymerases. *Nucleic Acids Res.* 24(18): 3546–3551.

7. Roux, K. H. 1995. Optimization and troubleshooting in PCR. *Genome Res.* 4: S185–S194.

8. Astatke, M., Grindley, N. D. F., and Joyce, C. M. 1995. Deoxy triphosphate and pyrophosphate binding sites in the catalytically competent ternary complex for polymerase reaction catalysed by DNA polymerase I (Klenow fragment). *J. Biol. Chem.* 270(4): 1945–1954.

9. Eckert, K. A. and Kunkel, T. A. 1991. DNA polymerase fidelity and the polymerase chain reaction. *Genome Res.* 1: 17–24.

10. Olive, D. M., Simsek, M., and Al-Mufti, S. 1989. Polymerase chain reaction assay for detection of human cytomegalovirus. *J. Clin. Microbiol.* 27(6): 1238–1242.

11. Baransel, A., Dulger, H. E., and Tokdemir, M. 2004. DNA amplification fingerprinting using 10× polymerase chain reaction buffer with ammonium sulfate for human identification. *Saudi Med. J.* 25(6): 741–745.

12. Innis, M. A., Myambo, K. B., Gelfand, D. H., and Brow, M. A. D. 1988. DNA sequencing with *Thermus aquaticus* DNA polymerase and direct sequencing of polymerase chain reaction-amplified DNA. *Proc. Natl. Acad. Sci. USA* 85: 9436–9440.

13. Saiki, R. K., Scharf, S., Faloona, F., Mullis, K. B., Horn, G. T., Erhch, H. A., and Araheim, N. 1985. Enzymatic amplification of beta-globin genomic sequences and restriction site analysis for diagnosis of sickle cell anemia. *Science* 230: 1350–1354.

14. Bachmann, B., Lueke, W., and Hunsmann, G. 1990. Improvement of PCR amplified DNA sequencing with the aid of detergents. *Nucleic Acid Res.* 18(5): 1309.

15. Lawyer, F. C., Stoffel, S., Saiki, R. K., Myambo, K., Drummond, R., and Gelfand, D. H. 1989. Isolation, characterization and expression in *Escherichia coli* of the DNA polymerase gene from *Thermus aquaticus*. *J. Biol. Chem.* 264(11): 6427–6437.

16. Ward, B. W., Mueller, E. J., Copeland, J., and Vassar, D. 2008. Stabilized compositions of thermostable DNA polymerase and anionic or zwitterionic detergent. US20080145910, Sigma-Aldrich, St. Louis, MO, June 19.

17. Gelfand, D. H., Stoffel, S., and Saiki, R. K. 2000. Stabilized thermostable nucleic acid polymerase compositions containing non-ionic polymeric detergents, US 6,127,155, Roche Molecular Systems, Pleasanton, CA, October 3.

18. Fan, N., Loeffert, E., Erbacher, C., and Peters, L.-E. 2010. Polymerase stabilization by ionic detergents, WO2008107473, Qiagen, GMBH, Deutschland, April 22.

19. Winship, P. R. 1989. An improved method for directly sequencing PCR amplified material using dimethyl sulphoxide. *Nucleic Acids Res.* 17: 1266.

20. Hung, T., Mak, K., and Fong, K. 1990. A specificity enhancer for polymerase chain reaction, *Nucleic Acids Res.* 18(16): 4953.

21. Sarkar, G., Kapelner, S., and Sommer, S. S. 1990. Formamide can dramatically improve the specificity of PCR. *Nucleic Acids Res.* 18(24): 7465.

22. Varadaraj, K. and Skinner, D. M. 1994. Denaturants or cosolvents improve the specificity of PCR amplification of a G + C-rich DNA using genetically engineered DNA polymerases. *Gene* 140: 1–5.

23. Chamberlin, M. and Mytelka, D. 2000. Methods for the elimination of DNA sequencing artifacts, US6270962, University of California, August 7.

24. Li, W.-B., Jessee, J. A., Schuster, D., Xia, J., and Gebeyehu, G. 2004. Compositions and methods for enhanced synthesis of nucleic acid molecules, US6787305, Invitrogen Corporation, Carlsbad, CA, September 7.

25. Henke, W., Herdel, K., Jung, K., Schnorr, D., and Loening, S. A. 1997. Betaine improves the PCR amplification of GC-rich DNA sequences. *Nucleic Acids Res.* 25(19): 3957–3958.

26. Chevet, E., Lemaitre, G., and Katinka, M. D. 1995. Low concentrations of tetramethylammonium chloride increase yield and specificity of PCR. *Nucleic Acids Res.* 23(16): 3343–3344.

27. Chakrabarti, R. and Schutt, C. E. 2001. The enhancement of PCR amplification by low molecular weight amides. *Nucleic Acids Res.* 29(11): 2377–2381.

28. Chakrabarti, R. and Schutt, C. E. 2001. The enhancement of PCR amplification by low molecular-weight sulfones. *Gene* 274: 293–298.

29. Chakrabarti, R. and Schutt, C. E. 2002. Novel sulfoxides facilitate GC-rich template amplification. *Biotechniques* 32: 866–874.

30. Chakrabarti, R. and Schutt, C. E. 2007. Compositions and methods for enhancing polynucleotide amplification reactions. US 7,276,357, Trustees of Princeton University, Princeton, NJ, October 2.

31. Chakrabarti, R. 2003. Novel PCR-enhancing compounds and their modes of action, In *PCR Technology: Current Innovations*, 2nd edition. T. Weissensteiner, H. G. Griffin, and A. Griffin, eds., CRC Press, Boca Raton, FL, pp. 51–62.

32. Sarkar, G., Cassady, J. D., Pyeritz, R. E., Gilchrist, G. S., and Sommer, S. S. 1991. Isoleucine397 is changed to threonine in two females with hemophilia B. *Nucleic Acids Res.* 19(5): 1165.
33. Quintana, J., Segalés, J., Calsamiglia, M., and Domingo, M. 2006. Detection of porcine circovirus type 1 in commercial pig vaccines using polymerase chain reaction. *Veterinary J.* 171: 570–573.
34. López, M. M., Llop, P., Olmos, A., Marco-Noales, E., Cambra, M., and Bertolini, E. 2009. Are molecular tools solving the challenges posed by detection of plant pathogenic bacteria and viruses? *Curr. Issues Mol. Biol.* 11: 13–46.
35. Wan, C.-Y. and Wilkins, T. A. 1993. Spermidine facilitates PCR amplification of target DNA. *Genome Res.* 3: 208–210.
36. Koonjul, P. K., Brandt, W. F., Farrant, J. M., and Lindsey, G. G. 1999. Inclusion of polyvinylpyrrolidone in the polymerase chain reaction reverses the inhibitory effects of polyphenolic contamination of RNA. *Nucleic Acids Res.* 27(3): 915–916.
37. Lin, Z., Kondo, T., Minamino, T., Ohtsuji, M., Nishigami, J., Takayasu, T., Sun, R., and Ohshima, T. 1995. Sex determination by polymerase chain reaction on mummies discovered at Taklamakan desert in 1912. *Forensic Sci. Int.* 75(2–3): 197–205.
38. Kreader, C. A. 1996. Relief of amplification inhibition in PCR with bovine serum albumin or T4 gene 32 protein. *Appl. Environ. Microbiol.* 62(3): 1102–1106.
39. Charlotte, L., Bailey, L. B., and David, G. M. 2008. Factors affecting reliability and validity, In *Essentials of Nucleic Acid Analysis: A Robust Approach*, J. Keer and L. Birch, eds., Royal Society of Chemistry Publishing, Cambridge, UK.
40. Kermekchiev, M. B and Barnes, W. M. 2008. Use of whole blood in PCR reactions, US appl 20090170060, DNA Polymerase Technology, Inc., St Louis, MO, filed December 8.
41. Zhang, Z., Kermekchiev, M. B., and Barnes, W. M. 2010. Direct DNA amplification from crude clinical samples using a PCR enhancer cocktail and novel mutants of Taq. *J. Mol. Diagn.* 12(2): 152–161.

5 Real-Time Fluorescent PCR by Labeled Primer with a Single Fluorescent Molecule

Hong Yan and Liu Xiao Kun

CONTENTS

5.1 INTRODUCTION

5.1.1 BACKGROUND

Many formats of real-time, fluorescent polymerase chain reaction (PCR) detection have been developed during the past few years. They include DNA-binding dye-based methods,[1,2] hybridization probe methods,[3–5] and labeled primer methods.[6–8] DNA-binding dye-based methods are comparatively simple and cost-effective because the PCR product is detected through the binding of double-stranded DNA-specific dyes such as SYBR Green I. Free dyes in such methods, however, do not bind specifically to a target product because the fluorescence intensity depends on the total amount of double-stranded DNA, which can include target PCR products, nontarget PCR products, and primer–dimer products. Accordingly, real-time PCR using free dyes is not a suitable for multiplex detection. To overcome these drawbacks, many hybridization probe-based methods have been developed. These methods make use of oligonucleotide–dye conjugates (probes) that hybridize to an internal sequence of the amplified product. In one approach, an acceptor moiety will quench the fluorescence of a reporter moiety when they are put on the same probe. Separation of the reporter from the quencher by cleavage (e.g., hydrolysis probe or TaqMan assay) or secondary structure change (e.g., molecular beacon) leads to fluorescence emission. In another approach (e.g., Roche fluorescence

TABLE 5.1

Comparison of qPCR Signal Detection Systems

Method		Principle	Specification of Assay	Labeled Oligonucleotides/ Cost	Design or Optimization
Double-stranded DNA-binding dye	SYBR Green I	Dye binding to ds-DNA generates fluorescence	Low specificity due to nonspecific binding to nontarget PCR products and primer–dimers	Not applicable/low	Simple
Labeled primers	Amplifluor primer	Amplification leads to opening of scorpion tag of primer, resulting in release of fluorescence	Middle specificity due to two recognition sites	One primer labeled with two fluorescent group/ high	Middle
	Q-primer PCR	Hydrolysis of Q-primer releases fluorescence		One primer labeled with one fluorescent group/middle	Middle
Labeled probes	TaqMan probe	Hydrolysis of probe	High specificity due to three recognition sites	One probe labeled with two groups/ high	Complex
	Molecular beacon	Opening of molecular beacon			
	Dual hybridization probes	Hybridization of two probes to target sites will quench fluorescence	Highest specificity due to four recognition sites	Each of two probes labeled with one dye/high	More complex

resonance energy transfer [FRET]), two probes are labeled with either a reporter or acceptor dye, which hybridize to adjacent regions on target DNA. Probe binding puts the two dyes in close proximity, allowing energy transfer from the reporter to the acceptor dye. Such energy transfer excites fluorescence of the acceptor. In both approaches, progression of PCR can be monitored by measuring the increase of fluorescence resulting from the separation or joining of the reporter and the acceptor. The second hybridization probe-based approach provides a high level of specificity and sensitivity in signal detection. However, the presence of additional oligonucleotide hybridization probe(s) in addition to two PCR primers increases the complexity of the PCR system and limits the capability of multiplex detection. Moreover, labeling probe(s) with two dyes increases cost. To strike a better balance between specific detection and cost of operation, there have been efforts to develop alternative methods using labeled PCR primers to provide simpler alternatives,[6–8] but most of them still require labeling with two dyes for signal generation (see Table 5.1).

5.2 BODIPY® FL

BODIPY FL, developed by Molecular Probes (USA), is a 4,4-difluoro-5,7-dimethyl-4-bora-3a,4a-diaza-s-indacene-3-propionic acid. BODIPY FL can be attached to a nucleotide by a linker. Horn and coworkers[9] reported that the fluorescence from a probe modified with BODIPY FL was diminished after hybridization, and it was later found[10] that the quenching was caused by the interaction between BODIPY FL and a guanine base. The quenching was more dramatic when guanine is positioned opposite BODIPY FL. Compared with a few commercially available fluorescent dyes, BODIPY FL linked to an oligonucleotide was found to have the highest quenching rate (95%) by hybridization.[11] Based on the effective fluorescence quenching of BODIPY FL-modified primer, Kurata and coworkers[10] further developed a real-time quantitative PCR (qPCR) system using a PCR primer with BODIPY FL linked to a cytosine at its 5′ end. As the PCR primer is integrated into the product, the fluorescence of BODIPY FL was quenched. The quenching rate was proportional to the amount of target DNA. When compared to hybridization probe-based approaches, the above system needs only two primers in the reaction because one PCR primer has the additional function of being the probe as well. This simple design reduces the complexity of the PCR system. Using a guanine base instead of a special dye as the quencher also makes it a cost-effective system. This approach, however, relies on the decrease in total fluorescence for monitoring PCR progress.

The presence of high-level fluorescence background due to unused labeled primers will interfere with detecting the decrease of fluorescence. Moreover, calculating the decrease of fluorescence over background is not compatible with most commercial real-time PCR instruments since these all have software that is designed for detecting the increase in fluorescence.

5.3 PRINCIPLE OF Q-PRIMER qPCR

To overcome the limitation of detecting the decrease of fluorescence using the real-time PCR system described above, we have developed a novel real-time PCR system that uses a BODIPY FL labeled primer (Q-primer). As illustrated in Figure 5.1, a specific PCR primer (Q-primer)

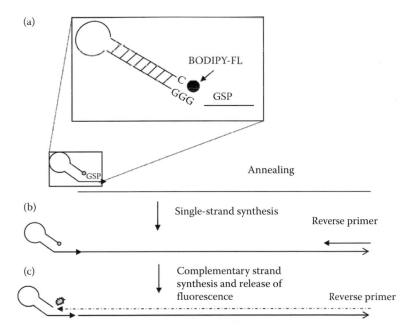

FIGURE 5.1 Mechanism of Q-priming PCR. (a) In an intact probe, a tight hairpin structure is formed to position BODIPY FL just opposite three guanidine bases that quench fluorescence from BODIPY FL. (b) From the second round of PCR, at the end of polymerization from a reverse primer, 5′ > 3′ exonuclease activity of *Taq* polymerase cleaves nucleotides attached to BODIPY FL, thereby releasing fluorescence. (c) With the progression of PCR, more BODIPY FL will be released and hence fluorescence increases. GSP, gene-specific primer.

consists of a gene-specific primer for PCR and a self-quenched probe (Q-probe) labeled with BODIPY FL at its 5′ end. The Q-probe forms a hairpin structure that causes the BODIPY FL to be placed opposite to three continuous guanidine bases at the complementary strand of the hairpin stem. As long as BODIPY FL remains attached to the Q-primer, the three guanidine bases will quench BODIPY FL effectively during all steps of PCR, except for the denaturing step when the hairpin is opened. At the beginning of the PCR, the PCR primer part of the molecule directs synthesis of the first strand. From the second round of PCR onward, an unlabeled reverse PCR primer directs synthesis of the complementary strand. When the *Taq* polymerase reaches the Q-probe, its 5′ > 3′ exonuclease activity will cleave the stem structure to release BODIPY FL and result in the emission of fluorescence. The quantity of free BODIPY FL is proportional to the quantity of a specific PCR product; hence, progression of PCR can be monitored by measuring the increase in fluorescence.

5.4 PROOF OF CONCEPT

To verify our postulated mechanism, described in Figure 5.1, a Q-primer (Q-primer 1) and a reverse primer (18s-rp1) were designed to amplify 18S rDNA in soybean genomic DNA (see Section 5.7). *Taq* polymerase (Roche, Germany) and Vent® (exo-) polymerase without 5′ > 3′ exonuclease activity (New England Biolabs, USA) were used in PCRs. During the progression of the PCR, an increase in fluorescence was detected for the reaction when using *Taq* polymerase but not for the reaction with Vent (exo-) polymerase (Figure 5.2a). However, when the PCR fragments were analyzed using agarose gel electrophoresis, the expected 219-bp PCR product was detected from both reactions (Figure 5.2b). From this result, it is clear that although both polymerases could amplify the target PCR product, the fluorescence increase was linked to the presence of 5′ > 3′ exonuclease activity of *Taq* polymerase. The PCR products were further analyzed through high-performance thin-layer chromatography (HPTLC). Exonuclease activity releases mostly the free form of BODIPY FL (Figure 5.2c). This further indicates that cleavage by 5′ > 3′ exonuclease activity is responsible for the release of BODIPY FL fluorescence.

5.5 APPLICATION IN qPCR

To evaluate the Q-priming system for use in real-time qPCR, we synthesized one artificial template (artificial template 1 shown in Table 5.2) that is derived from a sequence that is located in a conserved region of *Bacillus thuringiensis* Bt gene encoding Cry1A(c), an insect toxin. A Q-probe region was linked to the forward PCR primer (Q-primer 2) to amplify a 48-nt product together with an unlabeled reverse primer (Bt-rp1). Real-time PCRs (see Section 5.7) were conducted with the artificial template DNA at concentrations ranging from 10^0 to 10^7 copies and using three replicates per dilution (Figure 5.3a). Average threshold cycle (C_q) values were plotted against logarithms of starting copy numbers. There was a strict linear relationship between the C_q value and logarithm of starting copy numbers with a coefficient of 0.9945 (Figure 5.3b). Average amplification efficiency per cycle, however, was less than 90%. There is the need for further optimization to improve amplification efficiency.

We also noted that even a single copy of template was detectable in the experiment, suggesting a high level of sensitivity. The same Q-primer was also used to amplify PCR fragments as long as 400 bp with similar results. With the wide dynamic range of detection, we conclude that the Q-priming system is suitable for real-time quantification of target DNA.

5.6 APPLICATION IN GENOTYPING

Besides its simplicity and high sensitivity, the Q-primer real-time PCR was also found to be sensitive to single-nucleotide mismatches. Using synthesized templates and PCR primers with mismatches

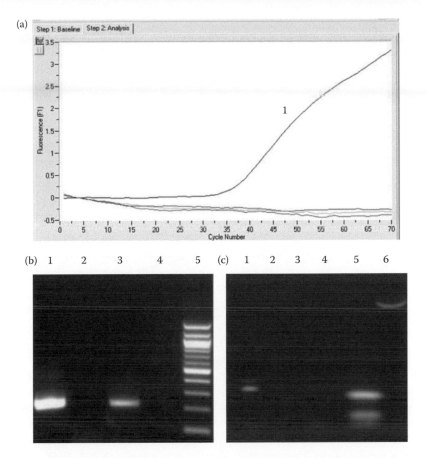

FIGURE 5.2 5′ > 3′ exonuclease activity of *Taq* releases fluorescence of BODIPY FL. Q-priming PCR was used to amplify soybean 18S rDNA. (1) By a DNA polymerase with 5′–3′ exonuclease activity (*Taq* polymerase); (2) negative control with *Taq* polymerase but no template DNA; (3) by a polymerase without 5′ > 3′ exonuclease activity (Vent exo-); (4) negative control with Vent (exo-) polymerase but no template. (a) Fluorescence intensity during progression of PCRs. (b) Agarose gel electrophoresis of PCR products. (5) 100-bp molecular size standards (New England Biolabs). (c) Fluorescence detection of PCR products by HPTLC. (5) Free dye marker; (6) 6-nt marker.

at different locations, we demonstrated that nucleotide mismatches at 3′ of Q-primer increases C_q values significantly but mismatches in the middle or 5′ end of primers have little effect (Figure 5.4).

This capability of discriminating single-nucleotide difference at extreme 3′ end of the primer can be exploited for detecting single-nucleotide polymorphism (SNP) and small deletions. Two Q-primers can be synthesized with each having its most 3′ end nucleotide(s) complementary to a

---

TABLE 5.2
Oligonucleotides Used in the Study

Target	Name	I. Sequence 5′–3′
18S rDNA	Q-primer 1	BODIPY FL-cctcgtcgccgcctgttcctaatacaataggaacag gcggcgacgagggatggggaatcttggacaatgg#
18S rDNA	18s-rp1	Ccgattcaccgcctacgt
Bt gene	Q-primer 2	BODIPY FL-cctcgtcgccatacaaggcgacgaggggagcgtgt ctggggtcctgattc#
Bt gene	Bt-rp1	cggatagggtaggttctggagtca
	Artificial template 1	tgagcgtgtctggggtcctgattcaggaactatgggaaacgccgctct

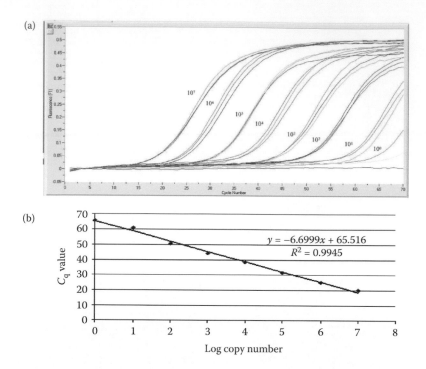

FIGURE 5.3   (a) Evaluation of Q-priming system for real-time qPCR amplification plot of a dilution series from 10^0 to 10^7 copies of an artificial template in steps of 10. Triplicate samples were used. (b) C_q as a function of the initial number of template copies (logarithmic scales) and linear regression of C_q versus log (template copies).

wild-type and a mutant allele. Together with one common reverse primer, the two Q-primers can be used in separate PCR reactions containing mutant and wild-type-specific probes to interrogate a sample of DNA. Samples from homozygous patients should generate lower C_q values from the mutant reaction than the wild-type reaction. By contrast, homozygous wild-type samples will give lower C_q value from the wild-type reaction. For heterozygotes, both mutant and wild-type alleles

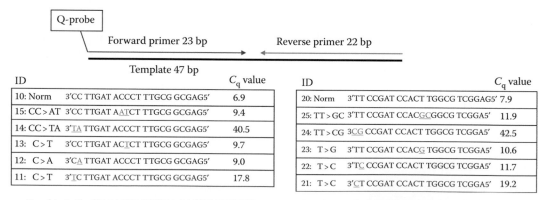

ID		C_q value
10: Norm	3′CC TTGAT ACCCT TTGCG GCGAG5′	6.9
15: CC > AT	3′CC TTGAT AATCT TTGCG GCGAG5′	9.4
14: CC > TA	3′TA TTGAT ACCCT TTGCG GCGAG5′	40.5
13: C > T	3′CC TTGAT ACTCT TTGCG GCGAG5′	9.7
12: C > A	3′CA TTGAT ACCCT TTGCG GCGAG5′	9.0
11: C > T	3′TC TTGAT ACCCT TTGCG GCGAG5′	17.8

Template 1: 5′... GG AACTA TGGGA AACGC CGCTC3′

ID		C_q value
20: Norm	3′TT CCGAT CCACT TGGCG TCGGAG5′	7.9
25: TT > GC	3′TT CCGAT CCACGCGGCG TCGGA5′	11.9
24: TT > CG	3CG CCGAT CCACT TGGCG TCGGA5′	42.5
23: T > G	3′TT CCGAT CCACG TGGCG TCGGA5′	10.6
22: T > C	3′TC CCGAT CCACT TGGCG TCGGA5′	11.7
21: T > C	3′CT CCGAT CCACT TGGCG TCGGA5′	19.2

Template 2: 5′... AA GGCTA GGTGA ACCGC AGCCT3′

FIGURE 5.4   Two 47 nt artificial templates with mismatches at different locations of the reverse primers. Compared to the perfect reverse primers, reverse primers with mismatches at 3′ end resulted in profound differences in C_q values. Two nucleotide mismatches at 3′ end resulted in higher C_q values than single mismatches. Mismatches in the middle and 5′ of reverse primers did not change C_q values significantly.

can amplify well, resulting in similar C_q values. A genotype can be decided based on the difference in C_q values (ΔC_q) between the two real-time PCRs.

To test the system, we selected three β-thalassemia mutations with high frequency in South east Asia. Based on the analysis of more than 3000 unrelated β-thalassemia patients, two point mutations [IVSI nt 5(G > C) (7.06%) and IVSII nt 654(C > T) (22.00%)], and one short deletion at Cd41/42(–TCTT) (26.09%) together account for >50% of mutations in Singapore (unpublished data). For each point mutation, we designed two Q-primers with the most 3′ nucleotide matching the wild type and the mutant sequence respectively. For the deletion mutant, two Q-primers were designed with 3′ differential sequences matching either the wild-type or the deletion mutant (Table 5.3).

For each sample, two reactions were set up to test the wild-type or mutant allele. After the PCR, the C_q values of the two reactions were collected and compared. For proof of concept, we tested a small number of heterozygote carriers and homozygous noncarriers (wild-type) that have been previously confirmed with the conventional reverse dot blot (RDB) analysis that has been used for routine screening and prenatal diagnosis. It was found that the observed differences in C_q values (ΔC_q) fall into two distinct groups: the group with small ΔC_q values being heterozygote carriers and the group of large ΔC_q values being homozygous noncarriers.

We defined the midpoint between two group averages as the threshold ΔC_q values for the three mutants as 8.25 for IVS2 nt 654(C > T); 4.5 for IVS1 nt 5(G > C); and 11 for CD41/42(–TCTT). Samples with ΔC_q values lower than the threshold value are heterozygote carriers and those with ΔC_q values higher than the threshold value are wild-type carrying no mutation in the target locus.

To further validate the high-throughput real-time PCR system on clinical samples, we tested 139 additional samples of unknown genotype, of which most were from carriers of one of the three β-thalassemia mutations as revealed by RDB. After real-time PCR, the genotype of each sample was deduced based on ΔC_q values in reference to respective threshold values. For each mutation, two distinct ΔC_q groups were observed corresponding to the wild-type and heterozygote mutant carriers. Application of a Student's t-test was used to analyze the data and this showed that ΔC_q values for these two groups were significantly different [Cd41/42(–TCTT), $P < 0.001$; IVSI nt 5(G > C),

TABLE 5.3
Oligonucleotides and Probes Used in β-Thalassemia Detection

Mutations	Primer Pair	Forward Primer (5′–3′)	Reverse Primer (5′–3′)
Cd41/42 (–TCTT)	Cd4142-rp1/ Cd4142-f1	BODIPY-FL CCTCGTCGCCATACAAGGCGACGAGG GATCCCCAAAGGACTCAACC	ACTCTCTCTGCCTATT GGTCTATTTT
	CD4142-10-R1/ Cd4142-f1	BODIPY-FL CCTCGTCGCCATACAAGGCGACGAGGGA TCCCCAAAGGACTCAAAG	
IVS1 nt 5(G > C)	IVS1-10-R1/ IVS1-f4	BODIPY-FL CCTCGTCGCCATACAAGGCGACGAGG CTCCTTAAACCTGTCTTGTAACCTTGATAC	TGAGGAGAAGTCT GCCGTTACTG
	IVS1-10-R2/ IVS1-f4	BODIPY-FL CCTCGTCGCCATACAAGGCGACGAGG CTCCTTAAACCTGTCTTGTAACCTTGATAG	
IVS1I nt 654(C > T)	IVS2-10-f1/ IVS-r4	BODIPY-FL CCTCGTCGCCATACAAGGCGACGAGG GTGATAATTTCTGGGTTAAGGT	ATAAAAGCAGAATGGTA GCTGGATTGTAG
	IVS2-10-f2/ IVS-r4	BODIPY-FL CCTCGTCGCCATACAAGGCGACGAGG GTGATAATTTCTGGGTTAAGGC	

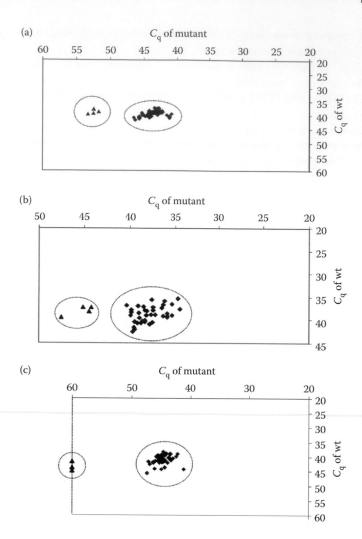

FIGURE 5.5 QPCR for detection of the three β-thalassemia mutations. Two Q-primers matching the mutation allele and the wild-type allele, respectively, were used for separate qPCR reactions for each sample. C_q values for the mutation allele and the wild-type allele are plotted for each sample. (a) IVSI nt 5(G > C), 39 heterozygote carriers and 4 wild-type. (b) IVSII nt 654(C > T), 44 heterozygote carriers and 4 wild-type. (c) Cd41/42(–TCTT), 45 heterozygote carriers and 3 wild-type. Diamonds and triangles represent heterozygote carriers and wild-types, respectively.

$P < 0.001$; and IVSII nt 654(C > T), $P < 0.001$]. Also, the results and predicted genotypes deduced from the real-time PCR system matched 100% with that previously derived using the RDB method (139 in 139) (Figure 5.5).

5.7 MATERIALS AND METHODS

5.7.1 MATERIALS

5.7.1.1 Instrument and Suppliers

Real-time PCR (qPCR) instrument: Roche LightCycler® was also used for normal qPCR; ABI7900 was used in high-throughput SNP detection. HPTLC equipment was from CAMAG company. HPTLC plate silica gel 60 F254 20 × 10 cm is from Merck (Germany). Reagent and suppliers

BODIPY FL, oligonucleotides, and hairpin fluorescence probe was commercially synthesized by Proligo (Sigma Aldrich) Singapore Pte Ltd. The *Taq* polymerase and dNTPs were purchased from Qiagen or Roche. *Taq* polymerase with 5 exo- was purchased from New England Biolabs.

5.7.2 METHODS

5.7.2.1 Sample Preparation

Blood DNA extraction was conducted from blood samples preserved in EDTA using PureGene Genomic DNA purification kit purchased from Qiagen. Plant DNA extraction was conducted by DNeasy plant mini kit Qiagen.

5.7.2.2 Real-Time qPCR Conditions

All qPCR reactions were carried out on a LightCycler® (Roche Molecular Biochemicals) in a 25-μL final volume. Each reaction contained 400 μM dNTPs, 0.2 μM Q-primer, and 0.4 μM reverse primer, 2.5U *Taq* polymerase (Roche), and 1 × *Taq* polymerase buffer, or 2.5U Vent (exo-) polymerase and buffer (New England Biolabs), 3.0 mM $MgCl_2$, and template DNA. Reaction conditions were an initial denaturation at 95°C for 2 min, followed by 70 cycles of 95°C for 15 s, annealing at 55°C for 15 s, and 72°C extension for 25 s, a single fluorescence measurement was made on channel 1 (Ex/Em = 470/530 nm) for each cycle at the extension step and data processed by LightCycler® Software (version 3, Roche) and cycle threshold (C_q) value calculated. PCR products were also analyzed on 2.0% agarose gel in 1 × TAE (Tris–acetate–EDTA) buffer.

5.7.2.3 High-Performance Thin-Layer Chromatography Analysis

HPTLC plate silica gel 60 F254 20 × 10 cm from Merck (Germany) was used as the stationary phase. Fifty microliters of qPCR products were collected by centrifugation and concentrated to 10 μL with Centricon (Eppendorf). Three microliter was sprayed on the silica gel plate as 6-mm bands (2 mm apart) by nitrogen gas with a Linomat 5 (CAMAG), 15 mm from lower edge of the plate. A 6 nt oligonucleotide labeled with BODIPY FL at 5′ end and free BODIPY FL (Sigma Aldrich) was used as a marker. Development was conducted in a 20 × 10 cm CAMAG twin trough chamber, saturated for 20 min with 10 mL of freshly prepared developing solvent (ammonium hydroxide:1-propanol = 7:11) per trough at a developing distance of 90 mm from the lower edge of the plate. The plate was dried in a fume hood and then photographed under UV light (366 nm) in a CAMAG Reprostar 3 chamber.

5.8 DISCUSSION

Like any other labeled primer qPCR systems, our system has a higher specificity when compared with dye-based methods because non-PCR products, such as primer–dimers, are not detectable. Because the Q-probe doubles as probe and PCR primer, this system is less complex than hybridization probe-based methods and can easily be included in a multiplex reaction system. The Q-probe has been designed with no significant sequence homology to any sequence in the GenBank database. The same Q-probe can easily be adapted to detect other target sequences by linking it to a different target-specific primer. This allows for easy transition from a qualitative PCR system to a highly specific quantitative PCR system. Compared with those systems using hybridization probes, this system might be less specific but would be advantageous because it is possible to amplify very short amplicons. We demonstrated that an amplicon as short as 48 bp could be amplified efficiently. Therefore, it is suitable to detect highly variable regions with only short consensus sequences, as demonstrated by our success in detecting a short consensus CryIA(b) region shared by the three commercial insect-resistant maize lines: Bt11, Event 176, and Mon810 (data not shown). Similarly, it has good potential in clinical application for the identification of microorganisms known to be highly variable with little short conserved regions such as *Mycobacterium tuberculosis*.

Another potential application could be the detection of transgenic elements in highly processed food products, from which only low-quality and short DNA fragments can be isolated. With our system, we were able to detect the presence of 18S rDNA in a highly degraded DNA from soybean sauce (data not shown).

We also successfully applied this system to discriminate SNPs. Its potential in diagnosis was validated by testing 139 clinical samples previously genotyped by RDB with 100% matching of results for three beta-thalassemia mutations. It takes about 2 h from setting up PCR to concluding genotyping, compared to 2 days for RDB. It is also much more time saving than other conventional methods such as allele-specific restriction enzyme analysis of PCR products, the amplification refractory mutation system, denaturing gradient gel electrophoresis and direct sequencing. All these methods require several hours, and sometimes days, for complete diagnosis.[12]

In comparison with another single vial qPCR amplifluor detection system,[13] only two oligonucleotides are required in our reaction, while five oligonucleotides are used in the amplifluor system. Thus, optimization will be more straightforward for our system.

Cost saving in reagents is another advantage of our system. First of all, reaction volume can be reduced to as little as 5 μL requiring only 1.25 pmol of Q-primer per reaction.

The cost for two Q-primers for each test can be as low as US 20 cents. After including costs of *Taq* polymerase, plate, and other reagents, we estimate the total consumable cost for each test can be less than one US dollar.

The system is also compatible with high-throughput analysis. With a qPCR instrument that can accommodate 384 reactions (e.g., ABI 7900), our method can simultaneously genotype 190 tests (each plate must include one positive and one negative control). With a 30-min PCR setup, a 90-min PCR amplification step, and a 10-min genotyping data analysis, one operator can process as many as 760 tests in up to four runs within a single day. Comparatively, the same number of tests with RDB requires as much as 1 week time of one operator.

As a homogeneous assay, the system can potentially be adapted in microfluid real-time platforms such as BioMark System (Fluidigm, CA, USA), which are able to reach 9216 reactions in a single run within 4 h.

ACKNOWLEDGMENT

This study was partially supported by a research grant from the Agri-Food and Veterinary Authority of Singapore.

REFERENCES

1. Higuchi R, Fockler C, Dollinger G, Watson R. 1993. Kinetic PCR analysis: Real-time monitoring of DNA amplification reactions. *Nat Biotechnol*, 11(9):1026–1030.
2. Wittwer CT, Herrmann MG, Moss AA, Rasmussen RP. 1997. Continuous fluorescence monitoring of rapid cycle DNA amplification. *BioTechniques*, 22(1):130–131, 134–138.
3. Heid CA, Stevens J, Livak KJ, Williams PM. 1996. Real time quantitative PCR. *Genome Res*, 6(10):986–994.
4. Holland PM, Abramson RD, Watson R, Gelfand DH. 1991. Detection of specific polymerase chain reaction product by utilizing the 5′–3′ exonuclease activity of *Thermus aquaticus* DNA polymerase. *Proc Natl Acad Sci USA*, 88(16):7276–7280.
5. Tyagi S, Kramer FR. 1996. Molecular beacons: Probes that fluoresce upon hybridization. *Nat Biotechnol*, 14(3):303–308.
6. Nazarenko IA, Bhatnagar SK, Hohman RJ. 1997. A closed tube format for amplification and detection of DNA based on energy transfer. *Nucleic Acids Res*, 25(12):2516–2521.
7. Whitcombe D, Theaker J, Guy SP, Brown T, Little S. 1999. Detection of PCR products using self-probing amplicons and fluorescence. *Nat Biotechnol*, 17(8):804–807.
8. Thelwell N, Millington S, Solinas A, Booth J, Brown T. 2000. Mode of action and application of Scorpion primers to mutation detection. *Nucleic Acids Res*, 28(19):3752–3761.

9. Horn T, Chang CA, Urdea MS. 1997. Chemical synthesis and characterization of branched oligodeoxy-ribonucleotides (bDNA) for use as signal amplifiers in nucleic acid quantification assays. *Nucleic Acids Res*, 25(23):4842–4849.

10. Kurata S, Kanagawa T, Yamada K, Torimura M, Yokomaku T, Kamagata Y, Kurane R. 2001. Fluorescent quenching-based quantitative detection of specific DNA/RNA using a BODIPY((R)) FL-labeled probe or primer. *Nucleic Acids Res*, 29(6):E34.

11. Torimura M, Kurata S, Yamada K, Yokomaku T, Kamagata Y, Kanagawa T, Kurane R. 2001. Fluorescence-quenching phenomenon by photoinduced electron transfer between a fluorescent dye and a nucleotide base. *Anal Sci*, 17(1):155–160.

12. Kanavakis E, Traeger-Synodinos J, Vrettou C, Maragoudaki E, Tzetis M, Kattamis C. 1997. Prenatal diagnosis of the thalassaemia syndromes by rapid DNA analytical methods. *Mol Hum Reprod*, 3(6):523–528.

13. Myakishev MV, Khripin Y, Hu S, Hamer DH. 2001. High-throughput SNP genotyping by allele-specific PCR with universal energy-transfer-labeled primers. *Genome Res*, 11(1):163–169.

6 Using Scorpion Primers for Genotyping

David Whitcombe

CONTENTS

6.1 INTRODUCTION

6.1.1 BACKGROUND AND AIMS

The development of quantitative real-time polymerase chain reaction (qPCR) has enabled significant contributions to many aspects of molecular analysis. This advance has been contingent on instrumentation and detection chemistries, that is, signaling systems that can generate amplification-dependent fluorescence, as well as instruments capable of performing the PCR cycling alongside the optical detection of emitted fluorescence. Numerous configurations of the optics have been used and a similar range of detection technologies has been brought into regular use.

Signaling systems can be divided into two main categories. First, "nonspecific" methods that are used to detect the presence of amplification but provide no definitive information as to the identity of the PCR product. Because the PCR tube remains closed throughout amplification and subsequent analysis, no amplicon size information is available, as there would be in a gel-based (post-PCR) analysis method. Methods in the "nonspecific detection" category include DNA-binding dyes (ethidium bromide, SYBR Green I, YO-PRO-1), which bind and fluoresce strongly in the presence of double-stranded (ds) DNA,[1] and amplifluor[2] in which a fluorophore/quencher pair is incorporated into a primer such that when the primer is incorporated into the dsDNA PCR product, the fluorophore and quencher become separated, permitting an increase in fluorescence that is proportional to the quantity of dsDNA in the reaction. A modification of the DNA-binding approach is to use a post reaction (high resolution), melting of the PCR products (described in Chapter 20 of this book), which can give useful information regarding the size and composition of the product. Second, there are the "specific detection" methods, which, generally speaking, are based around the concept of probing the PCR products with amplicon-specific sequences to unequivocally identify the PCR product. Such technologies include molecular beacons,[3] TaqMan/linear hydrolysis probes,[4–6] and the subject of this chapter, Scorpions.[7–9]

This chapter aims to

- Briefly describe the underlying concepts of Scorpions
- Explain the methodology for reaction design
- Illustrate the methodology by reference to genotyping reactions
- Discuss troubleshooting and possible pitfalls

6.1.2 OUTLINE OF THE SCORPIONS CONCEPT

Scorpions are molecules that integrate the functions of both primer and probe. There are two formats that have been used, the linear and the stem-loop configurations (see Figure 6.1).

The components of a Scorpion are

- A primer, which performs the usual function of binding template DNA and being extended by the *Taq* DNA polymerase.

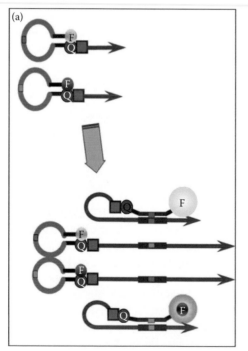 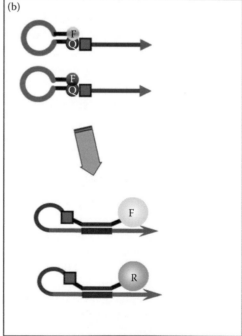

FIGURE 6.1   (See color insert.) Two formats for genotyping by Scorpions. (a) Allele-specific hybridization method in which the difference between the alleles is detected by the probe element. When the Scorpion is incorporated into a mismatched amplicon, the stem-loop of the primer construct is favored over the mismatched hybrid yielding improved discrimination. (b) Allele specific priming format. In this format, discrimination depends upon differential primer extension. The extended products are all then targets for the probe element.

- A blocking group, whose function is to "protect" the probe component from being copied into the amplicon. This is typically a large nonnucleotide group (such as hex-ethylene glycol [HEG]) that blocks template-dependent DNA synthesis. This nonnucleotide group is available as a monomer that can easily be incorporated into the primer synthesis.
- A signaling system that includes a fluorophore and quencher pair brought into close proximity by base pairing, in addition to a probe region that is complementary to a portion of the intended PCR amplicon. This element can be in either the stem-loop configuration (Figure 6.1a) or the linear, bimolecular format (Figure 6.1b).

The mechanism by which Scorpions generate signal is shown in Figure 6.2. After the primer element binds to its target and is extended in the usual way, the extended molecule now contains a region that is itself complementary to the probe region of the Scorpion molecule. *Important note*: the probe element is designed to bind the opposite strand to the primer portion because the probe binds the newly synthesized strand. Following a heat–cool cycle as part of the PCR, the probe region rapidly and efficiently "flips forward" and binds the complementary target on the same strand. The

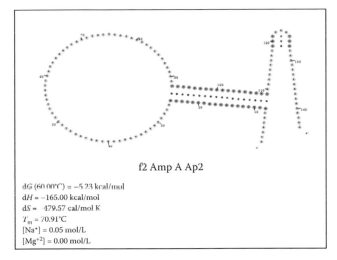

f2 Amp A Ap2

dG (60.00°C) = −5.23 kcal/mol
dH = −165.00 kcal/mol
dS = −479.57 cal/mol K
T_m = 70.91°C
[Na⁺] = 0.05 mol/L
[Mg⁺²] = 0.00 mol/L

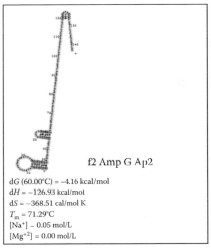

f2 Amp G Ap2

dG (60.00°C) = −4.16 kcal/mol
dH = −126.93 kcal/mol
dS = −368.51 cal/mol K
T_m = 71.29°C
[Na⁺] = 0.05 mol/L
[Mg⁺²] = 0.00 mol/L

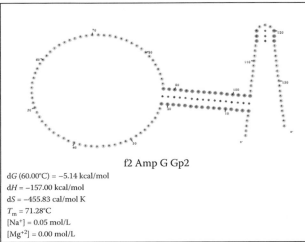

f2 Amp G Gp2

dG (60.00°C) = −5.14 kcal/mol
dH = −157.00 kcal/mol
dS = −455.83 cal/mol K
T_m = 71.28°C
[Na⁺] = 0.05 mol/L
[Mg⁺²] = 0.00 mol/L

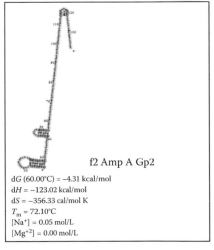

f2 Amp A Gp2

dG (60.00°C) = −4.31 kcal/mol
dH = −123.02 kcal/mol
dS = −356.33 cal/mol K
T_m = 72.10°C
[Na⁺] = 0.05 mol/L
[Mg⁺²] = 0.00 mol/L

FIGURE 6.2 Predicted folded amplicons for matched and mismatched Factor II probes. When the probe and amplicon match (left-hand panels), the probe binds efficiently, whereas in the mismatched versions (right-hand panels), the closed stem is more favored than the target binding.

strand is then copied as normal from the opposing primer that binds to the extended Scorpion molecule. The bound probe is displaced and reassumes the quenched configuration.

Some important features that derive from this unique mode of action are as follows:

- The unimolecular rearrangement is kinetically favored—the probe and its target are in close proximity on the same DNA strand. This means that the probe length can be shorter while maintaining a suitably high annealing temperature (T_m), which allows greater discrimination between matched and mismatched probes.
- The mechanism remains constant throughout the PCR and is not subject to significant competition from the complementary strand during the later stages of the PCR. This maintains the direct relationship between quantity of PCR product and fluorescent signal. In other systems based upon bimolecular probe binding, the signal per molecule decreases later in the PCR.
- The mechanism of signaling is identical to the mechanism of quenching (at least in the stem-loop format), which allows for accurate comparison of the thermodynamics of the two possible molecular structures of the extended Scorpion.
- The signal is at its maximum early in the annealing segment of a PCR cycle because the binding of the probe to the target is kinetically rapid and the probe has yet to be displaced by synthesis of the opposite strand.

6.1.3   GENOTYPING METHODS USING SCORPIONS

Scorpions may be used in two different configurations to permit genotyping: allele-specific hybridization and allele-specific priming (Figure 6.1).

6.1.3.1   Allele-Specific Hybridization

In this format, two different tail/probe regions are designed: one to each allele. The probes are designed to differentiate maximally between the two target sequences. This is analogous to using allele-specific TaqMan or molecular beacons. The specific and distinguishing features of Scorpions include:

- Shorter probes are possible because the unimolecular folding mechanism has the effect of increasing the kinetics of hybridization. Shorter probes mean that single mismatches are more easily discriminated.
- The stem of the basic Scorpion provides a thermodynamically viable alternative structure. In the case of a matched probe/target hybrid, this is of little significance, but in the situation where the probe and the target are mismatched, the alternative structure provides another level of discrimination.
- Because there is no probe cleavage, no free fluorophore accumulates and maximal signal is observed directly after a heat/cool step. To produce robust and clear discrimination, fluorescence measurement should be done at either 10°C below the T_m of the Scorpion's stem or more conveniently at an ambient temperature.

Genotyping can be achieved either by monitoring the two fluorophores during real-time PCR or, more typically, by endpoint analysis, post-PCR. In either case, the ratio of the two fluorophores is measured to determine the genotype of a sample.

6.1.3.2   Allele-Specific Priming

In this format, allele specificity comes from the differential priming observed when the 3′ base of a primer matches or mismatches its otherwise identical target. *Taq* polymerase can discriminate well between the two configurations and it is frequently possible to enhance this discrimination by introducing extra mismatches close to the terminal base (either adjacent or 2 away from the terminus [so-called −2 or −3 positions]). The assay is then configured so that each of the allele specific sequences is coupled to an amplicon-specific probe region. The probe sequence is the same for each variant under study, but a different fluorophore is used for each allele. Genotyping may be achieved in two different ways. First, endpoint fluorescence reads can be made as above and the ratio between the two fluorophores is used to determine the genotype. Second, real-time monitoring permits the user to establish the C_q determined using each fluorophore. An "early" C_q indicates a positive for the related allele and thus the genotype can be calculated by C_q comparison.

6.2 MATERIALS AND METHODS

6.2.1 Design of a Hybridization-Based Allele-Specific Genotyping Scorpion

6.2.1.1 Primers

Primers may be designed by any commonly used technique, typically using well-established software programs such as Primer3 (http://frodo.wi.mit.edu/primer3/). Use the energetics parameters prescribed by Santa Lucia et al.[10] and set the required T_m for 65°C (5°C above the actual annealing temperature of the PCR cycle). Aim for an amplicon size as small as possible (80–150 bp). There is a benefit in having the single nucleotide polymorphism (SNP) closer to one primer than the other (but at least 10 bases between the end of the primer and the SNP) to allow room for the probe.

6.2.1.2 Probe Design

Probe elements may be designed to be attached to either primer, but the smaller the "reach" (the distance from the probe to the SNP in the amplicon), the more efficient the probe and the shorter the amplicon may be. Shorter probes are more discriminatory since a single mismatch has a greater destabilizing effect. Allele discriminating probes can be designed by a number of methods and many software programs are available (including Primer3 discussed above) that incorporate allele discrimination-specific design options. As a "rule of thumb," the mismatch should be located close to the middle of the probe, which is, in general, the most destabilizing for the mismatched sequence. For probe designs, the following guidelines should be used:

- The probe region must not overlap with the primer sequence.
- The probe may be as much as 100 bases from the primer binding location (though 0–50 bases is much better).
- The probe should have no additional internal homology and no sequence homology with either primer (particularly the 3′-end of the primer that forms part of the Scorpion).
- A small amount of homology (one or two bases) at the 5′ and 3′ ends of the probe may be used as the start point for the stem.
- The stem must be 5–8 bases in length (including any endogeneous sequences from the previous point) and largely composed of Gs and Cs.
- The ΔG of the matched probe should be between −4 and −15 kcal/mol at PCR conditions (60°C, 2.5 mM Mg^{2+}, and 50 mM KCl); the T_m will typically be ~60°C.
- The ΔG of the stem should be −1.5 to −3.5 kcal/mol; a simple recommended sequence is fluorophore-CCGCGG-probe-CCGCGG-quencher-HEG-primer.
- For maximum discrimination, the mismatched amplicon should have a ΔG close to that of the stem.
- As with all fluorescent probe technologies, a G adjacent to the fluorophores is not recommended.

The stability of the probe on matched and mismatched targets can be calculated using the mfold DNA folding form[11]: http://mfold.rna.albany.edu/?q=mfold/DNA-Folding-Form.

To create the test sequence, paste the sequence of the Scorpion (probe, including the stem-loop, primer) to the amplicon sequences (this must be the same strand as the primer, not the complement!). Both the matched and mismatched amplicons can be tested in this way for both probe sequences (four possible combinations). Ensure that the parameters entered match the PCR conditions (60°C, 2.5 mM Mg, and 50 mM KCl).

6.2.2 Design of an Allele-Specific Priming Assay

6.2.2.1 Primers

Primers are designed to terminate at the allelic base. This means that there are typically two families of primers upon which to base the primer design. Primer T_ms are selected to be only 2.5°C above the intended annealing temperature. Take into account the buffer conditions when checking the calculated T_m in whichever software package is used. Allele-specific priming is greatly influenced by Mg^{2+} concentration; a lower concentration (1.5 mM) leads to more specific priming, but a concentration any lower than this may lead to poor PCR efficiency. A benefit of Scorpions is that they are able to tolerate a broad range of Mg concentrations, unlike TaqMan, which requires high Mg^{2+} for optimal probe cleavage and signal generation. Typical buffer conditions are

- 250 nM each primer
- 50 mM KCl
- 1.5–2.5 mM $MgCl_2$
- 200 nM each dNTP
- 1 U *Taq* polymerase

Typical cycling conditions are

- Heat activation of hot start enzyme where required
- 40 cycles of 95°C for 15 s; 60°C for 30 s
- Longer anneal/extend times may be needed for longer amplicons

6.2.2.1.1 Primer Survey

To obtain maximal discrimination between matched and mismatched primer/target combinations, the introduction of additional mismatches close to the 3′ terminus of the primers is recommended. This typically has the effect of destabilizing the mismatched complex more than the matched one. For situations in which the absolute optimal discrimination is required, it is possible to identify the "best" primers by performing a comprehensive survey of all the possible forward and reverse primers in combination with an appropriate opposing primer

- Primers have the correct T_m on matched template and are not prone to primer–dimers (PDs) with the opposing primer.
- Amplicons should be around 80–250 bp in size (the smaller the better—as long as there is adequate space for the probe).
- Seven primers per allele per direction are tested—one primer is a perfect match (no additional mismatch), three primers test the three possible mismatches at the penultimate base (–2 mismatches) and the other three test the three possible mismatches at the ante-penultimate base (–3 mismatches).
- Each potential primer is tested in real-time PCR with DNA samples that are homozygous for the two alleles in question (it may be necessary to construct a homozygote in a plasmid or in a "PCR cassette," if one homozygote is very rare).
- Detection is by DNA-binding dyes (a series of specific probes would be too expensive). We recommend the use of YO-PRO-1 (Molecular Probes) at 1 mM final concentration. We

have observed that other dyes (specifically SYBR-Green) may have a differential effect on the efficiency of different primers.

- A post-PCR melt curve should be incorporated into the amplification protocol to ensure that the product detected by this method is indeed amplicon-sized rather than the accumulation of PD.
- An ideal allele-specific primer amplifies efficiently on its matched template (similar to the "no additional mismatch" version) but does not amplify at all on the mismatch template. In reality, the matched template should give an "early C_q," while a mismatched template gives a "late C_q," typically at least 10 cycles later. This "window" of 10 cycles represents better than 1000-fold (2^{10}, assuming 100% PCR efficiency) discrimination between alleles. In practice, a window as small as five cycles can be used to produce a robust genotyping assay.
- It is possible that an otherwise "good" primer could be masked by the accumulation of PDs. If no other primers are suitable choices, it may be necessary to design an alternative opposing primer to remove this confounding factor.

6.2.2.1.2 Primer Look-Up Table

The above methodology is time-consuming and potentially costly, though for a high-value diagnostic product or any other application such as the detection of rare sequences in a mixed DNA population (e.g., emerging HIV variants), it may be money well spent.

An alternative approach is to use the tables produced by Fletcher et al. (manuscript in preparation), which provide a shortcut to primer design. The authors used a series of 64 synthetic templates to cover all possible 3′-terminal triplets, combined with the corresponding complementary primers. The extension efficiency of each combination was used to determine optimal mismatches for each possible allelic pair and the findings are summarized in Table 6.1. It is

TABLE 6.1
Recommended Additional Mismatches for All Template Triplets

Template (Note 3′–5′)	MM-2	MM-3	Note
AA-N	C/A	C/A (or A/A)	
AC-N	T/C or A/C	A/A or C/A	MM-3 favored
AG-N	G/G	A/A (or C/A)	MM-2 not for T-terminal primers
AT-N	C/T	A/A (or C/A)	MM-2 not for A-terminal primers
CA-N	C/A	A/C (or T/C)	
CC-N	A/C	A/C or T/C	MM-3 favored
CG-N	G/G	A/C (or T/C)	MM-2 not for T-terminal primers
CT-N	C/T	T/C (or A/C)	MM-2 not for T-terminal primers
GA-N	C/A	G/G or T/G	
GC-N	A/C	G/G or T/G	
GG-N	T/G	T/G or G/G	MM-3 favored strongly
GT-N	C/T	T/G or G/G	
TA-N	C/A	C/T (or G/T or T/T)	
TC-N	A/C or T/C	G/T (or C/T or T/T)	MM-3 favored
TG-N	G/G	Any mismatch	MM-2 not for T-terminal primers
TT-N	C/T	C/T (or G/T or T/T)	

Note: The recommendations refer to the primer sequence against the template sequence. For example, C/T refers to a C (primer) mismatched against a T (template). MM-2 is the mismatch to include at the −2 position and MM-3 the mismatch at the −3 position.

important to stress that these are the results from variants around a single primer sequence and future replications of this methodology might cause a revision of the guidelines. However, they are consistent with in-house knowledge resulting from the design of many hundreds of allele-specific primers.

6.2.2.2 Probe Design

The general guidelines for probe design are identical to those described in Section 6.1.3.1. In the case of allele-specific primers, though, only a single probe sequence is needed and there is no requirement to include any allele discrimination in the probe design.

6.3 RESULTS

6.3.1 Worked Example of Allele-Specific Hybridization

As an example, we have used the well-studied Factor II (prothrombin) mutation[12] for which amplification primers had previously been designed. Allele-specific probes were developed using a manual approach and tested using the *mfold* program.

To improve the discrimination, the length and strength of the stem sequences were adjusted to yield the following sequences:

Forward	TTGTGTTTCTAAAACTATGGTTCC
Reverse	AGTAGTATTACTGGCTCTTCCT
A-Probe	ACTCTCAGC**A**AGCCTCAA
G-probe	CTCTCAGC**G**AGCCTCA
A-Scorpion	f1-*ACCGCGC*ACTCTCAGC**A**AGCCTCAA*GCGC* *GGT*-que-heg-AGTAGTATTACTGGCTCTTCCT
G-Scorpion	f2-*CCGCGC*TCTCAGC**G**AGCCTCA*GCGCGG*-que-heg-AGTAGTATTACTGGCTCTTCCT

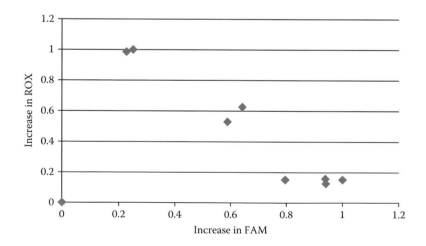

FIGURE 6.3 Factor II polymorphism genotyped using the allele-specific hybridization approach. Allele-specific Scorpion probes were used to genotype several samples of known genotype (the wild-type allele was labeled with FAM; the mutant with ROX). Clear genotype groups were easily distinguished.

Note that both the probe sequence and the stem for the A target are longer than the equivalents for the G variant. When incorporated into matched and mismatched amplicons, mfold predicts the structures in Figure 6.2.

When tested on a small group of samples of all three genotypes (cycling parameters as above), clear genotyping was observed (Figure 6.3).

6.3.2 WORKED EXAMPLE OF ALLELE-SPECIFIC AMPLIFICATION

The design constraints are much lower for the probe element of a Scorpion used in an allele-specific amplification approach. We used the UGT2B15*2 (a G > T change in codon 85) polymorphism[13] as our model system. The design process yielded the following sequences:

Forward T-primer	GAAGTTTATCCTACATCTTTAACTAAAAATT
Forward G-primer	GAAGTTTATCCTACATCTTTAACTAAAA**TT**G
Probe	*CCGCGCACCATATATCCATCTATCGAGCGCGG*
Reverse primer	CTTTACAGAGCTTGTTACTGTAGTCATAAT
T-Scorpion	FAM-CCGCGCACCATATATCCATCTATCGAGCGCGG-que-
	heg-GAAGTTTATCCTACATCTTTAACTAAAAATT
G-Scorpion	JOE-CCGCGCACCATATATCCATCTATCGAGCGCGG-que-
	heg-GAAGTTTATCCTACATCTTTAACTAAAA**TT**G

Note that the T-primer required no additional mismatch, probably due to the highly AT-rich sequence at the 3′-end. The G-primer was equally discriminatory with any of the possible −3 mismatches but the T on T mismatch was selected to keep the 3′-terminal region as AT-rich as possible. Note also that both Scorpions share the probe sequence and that a G/C base pair from the probe also participates in the stem.

Figure 6.4 shows the results of endpoint genotyping using this set of designs; all three genotypes are easily distinguished by simply plotting the two fluorescence values against each other.

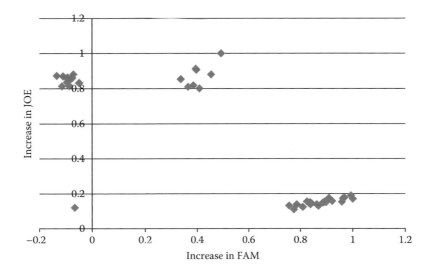

FIGURE 6.4 UGT2B15 genotyping using the allele-specific priming approach. Clear discrimination between all genotype groups was easily observed.

6.4 DISCUSSION

6.4.1 Hints

1. Real-time traces may show steady, linear increases even in the absence of template-specific amplification. Often, the instrument baseline setting software will eliminate this and no difficulty will ensue. The following should assist where a problem remains:
    a. This results from the stem being "too long." A long (eight or more base pairs) duplex can be a substrate for the exonuclease (TaqMan) function of the *Taq* polymerase. The stem should be designed to be no longer than six or seven base pairs.
    b. Occasionally, there are unexpected additional base pairs recruited from the probe itself—care should be taken to check for such possibilities at the design stage.
    c. It may be possible to substitute an exopolymerase such as KlenTaq,[14] which does not cleave the probe but still produces strong Scorpions signals.
2. For allele-specific hybridization assays, it may not always be possible to obtain the perfect hierarchy with regard to the mismatch target/closed stem. This does not mean the assay is not viable, merely that it may not be ideal.
    a. Ensure that the T_m (or ΔG) difference between matched and mismatched hybrids is maximized; likewise, the T_m of the stem (consistent with the other length constraints above).
    b. Consider reversing the probe direction (consistent with the amplicon constraints in Section 6.2.1.2).
    c. A postreaction melt curve will very often show the clear-cut genotype required.
3. For allele-specific PCR assays, it is sometimes the case that the different allele-specific primers give rise to substantially different amplification efficiencies (even if the discrimination is good in each case). This usually depends on unknown, context-specific factors, but may also reflect the GC content close to the 3′-terminus of the primer; higher GC often means greater priming efficiency.
    a. When selecting the mismatch for the two primers, attempt to ensure that both primers have the same number of Gs and Cs in the last three bases.
    b. Favor a −2 mismatch (more disruptive) for a primer that terminates with a G or a C and a −3 mismatch for an A or T terminal primer.
4. Scorpion signal strength is at its greatest immediately after the cool/anneal step of a PCR. Fluorescence should be read immediately after the PCR concludes or a heat/cool step 2 min at 95°C and cool rapidly to 25°C should be included immediately before reading the fluorescence.

6.4.2 Summary

Scorpions offer two distinct approaches to genotyping, each of which is simple to implement, robust, and reliable.

The allele-specific priming method does require a certain amount of prework to ensure that the primers are optimized with respect to their discrimination. Some general rules are presented to assist in the selection, though a comprehensive primer survey is recommended where absolutely optimal discrimination is required. For most genotyping applications, this is not necessary, but for the detection and quantitation of rare alleles in a mixed sample, it is essential.

REFERENCES

1. Higuchi R, Dollinger G, Walsh PS, Griffith R. Simultaneous amplification and detection of specific DNA sequences. *Biotechnology (NY)*. 1992;10(4):413–417.
2. Nazarenko IA, Bhatnagar SK, Hohman RJ. A closed tube format for amplification and detection of DNA based on energy transfer. *Nucleic Acids Research*. 1997;25(12):2516–2521.

3. Tyagi S, Kramer FR. Molecular beacons: Probes that fluoresce upon hybridization. *Nature Biotechnology.* 1996;14(3):303–308.
4. Heid CA, Stevens J, Livak KJ, Williams PM. Real time quantitative PCR. *Genome Research.* 1996;6(10):986–994.
5. Livak KJ. Allelic discrimination using fluorogenic probes and the 5′ nuclease assay. *Genetic Analysis: Biomolecular Engineering.* 1999;14(5–6):143–149.
6. Holland PM, Abramson RD, Watson R, Gelfand DH. Detection of specific polymerase chain reaction product by utilizing the 5′–3′ exonuclease activity of *Thermus aquaticus* DNA polymerase. *Proceedings of the National Academy of Sciences of the United States of America.* 1991;88(16):7276–7280.
7. Whitcombe D, Theaker J, Guy SP, Brown T, Little S. Detection of PCR products using self-probing amplicons and fluorescence. *Nature Biotechnology.* 1999;17(8):804–807.
8. Solinas A, Brown LJ, McKeen C et al. Duplex Scorpion primers in SNP analysis and FRET applications. *Nucleic Acids Research.* 2001;29(20):E96.
9. Thelwell N, Millington S, Solinas A, Booth J, Brown T. Mode of action and application of Scorpion primers to mutation detection. *Nucleic Acids Research.* 2000;28(19):3752–3761.
10. Santa Lucia J, Jr. A unified view of polymer, dumbbell, and oligonucleotide DNA nearest-neighbor thermodynamics. *Proceedings of the National Academy of Sciences of the United States of America.* 1998;95(4):1460–1465.
11. Zuker M. Mfold web server for nucleic acid folding and hybridization prediction. *Nucleic Acids Research.* 2003;31(13):3406–3415.
12. Poort SR, Rosendaal FR, Reitsma PH, Bertina RM. A common genetic variation in the 3′-untranslated region of the prothrombin gene is associated with elevated plasma prothrombin levels and an increase in venous thrombosis. *Blood.* 1996;88(10):3698–3703.
13. Guillemette C. Pharmacogenomics of human UDP-glucuronosyltransferase enzymes. *The Pharmacogenomics Journal.* 2003;3(3):136–158.
14. Barnes WM. The fidelity of *Taq* polymerase catalyzing PCR is improved by an N-terminal deletion. *Gene.* 1992;112(1):29–35.

7 LNA™
Adding New Functionality to PCR

Søren M. Echwald, Ditte Andreasen, and Peter Mouritzen

CONTENTS

7.1   INTRODUCTION

Real-time polymerase chain reaction (PCR) is the method of choice for many biomedical applications. Ever since its emergence more than a decade ago, there have been significant developments such as improvements to the assay design software, instruments, and detection chemistries. Still, the design of an efficient, sensitive and accurate quantitative PCR assay using standard DNA detection chemistry requires adherence to certain fundamental design rules that are dictated by the DNA composition of primers and probes. Melting temperature is determined by the primer and probe sequence and length, design of individual PCR primers and fluorescent probes have certain structural requirements, just as these designs must fit together under certain relative measures that are again restricted by the composition and sequence context of the desired target. This may, in many cases, complicate or even limit the practical applications of the technology, if, for example, the desired target sequences do not support a design according to guidelines or if the assay complexity prevents simultaneous detection of relevant biomarkers. This tends to be a particular challenge in diagnostic assay development where assay sensitivity and specificity are the key requirements but target sequences may be very limited due to similarity with nontarget sequences. The standard approaches for overcoming such limitations include moving the assay to another sequence, testing multiple suboptimal assays in search of functional assays, or changing cycling or assay chemistry conditions.[1] Overall, detection requirements outside the standard designs may require cumbersome optimization cycles and specialized settings that may work against the general preference of being able to run multiple assays under very similar conditions to facilitate high-throughput applications. In the following chapter, we will describe how the use of the RNA analog called locked nucleic acids (LNAs) can dramatically improve the design of qPCR assays through its ability to modify the thermal stability of any oligo while maintaining sequence recognition.

7.2   LNAs™ IN PCR

LNAs are nucleotide analogs that are conformationally locked in a C3-endo/N-type sugar conformation by an O2–C4_methylene linkage[2–6] (Figure 7.1) that leads to reduced conformational flexibility.[5,7] LNAs increase the thermal stability of oligonucleotides (by about 3–8°C per modified base in the oligo)[2,8,9] (Figure 7.1). This feature can be used in numerous ways to improve and facilitate PCR assay design. While DNA analogs have become a versatile tool for recognition of RNA and DNA targets and several analogs have been developed that are based on various backbone modifications,[10] LNAs remain one of the most useful modified backbones, as incorporation of a single LNA provides a substantial increase in duplex stability. At the same time, LNA amidites are compatible with the standard oligonucleotide synthesis and LNA/DNA bases can be freely interchanged in an oligo sequence by LNA spiking (see Table 7.1). Also the fact that LNA is well tolerated by enzymes relevant for molecular biology has made it increasingly popular.[11] Since the invention of LNA in 1997, the T_m of LNA-substituted DNA molecules have been widely studied and a highly predictive *in silico* T_m model, as well as several assay design-software packages are available to support the design of LNA/DNA chimeric molecules.[12–14] This has facilitated the application of LNA-incorporated nucleic acid probes and primers in the design of real-time PCR assays.

The ability to increase the thermal stability of probes and primers while maintaining the target recognition sequence and recognition ability is one of the key features of LNAs. As illustrated in Table 7.1, progressive substitution of DNA nucleotides with LNAs increases the melting temperature of the oligonucleotide (Table 7.1B,C) while maintaining the same recognition sequence, whereas the alternative strategy of increasing LNA substitutions allows shortening of the probe while maintaining the original T_m (Table 7.1D,E). Hydrolysis probes as short as 8-mers can be designed to have standard melting temperature in the 65–70°C range, thus greatly widening the design "window" of potentially suitable target sequences. In the following sections, we will address the advantages of using LNAs in PCR probes and primers as well as the use of PCR blocker probes, and some of the critical design parameters that are relevant when working with LNA. A selection of general LNA oligonucleotide design guidelines are listed in Box 7.1.

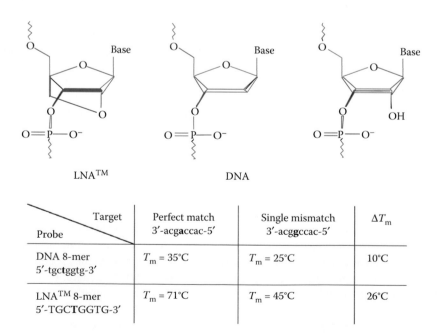

Probe \ Target	Perfect match 3′-acga**c**cac-5′	Single mismatch 3′-acg**g**ccac-5′	ΔT_m
DNA 8-mer 5′-tgc**t**ggtg-3′	$T_m = 35°C$	$T_m = 25°C$	10°C
LNA™ 8-mer 5′-TGC**T**GGTG-3′	$T_m = 71°C$	$T_m = 45°C$	26°C

FIGURE 7.1   Substitution of DNA with LNAs provides dramatically increased sensitivity and specificity in hybridization assays.

TABLE 7.1

T_m Effect of LNA™ Spiking

A.	tcgatcgattagctacgtacgta—23-mer DNA		T_m: 62°C
B.	tcgatcgatt**A**gctacgtacgta—23-mer DNA/LNA		T_m: 65°C
C.	tcgatc**G**att**A**gcta**C**gta**C**gta—23-mer DNA/LNA		T_m: 72°C
D.	− − tc**G**att**A**gcta**C**gtac—16-mer DNA/LNA		T_m: 62°C
E.	− − −c**GATTAGCT**− − − − − − 9-mer DNA/LNA		T_m: 62°C

Note: Replace DNA with LNA for higher T_m.
DNA, acgt; LNA, ACGT.

BOX 7.1   GENERAL LNA™ OLIGONUCLEOTIDE DESIGN GUIDELINES

1. Add LNAs at the site where specificity and discrimination are required (at the SNP position in allele-specific probes or at the 3′ end of allele-specific primers).
2. Avoid runs of more than three consecutive LNA bases, especially Gs. LNA stretches can have high affinity and the interaction between the nucleotides can cause the oligonucleotide to fold in on itself producing secondary structure, which will be detected as an unsuccessful synthesis.
3. Runs of three or more G bases are generally not recommended.
4. Avoid LNA: LNA interactions caused by self-complementarity or complementarity to other LNA-containing oligonucleotides in the assay. LNA binds very tightly to other LNA residues.
5. Each LNA base increases the T_m by approximately 2–4°C.
6. Blocks of LNA near the 3′ end of primers can prevent polymerase activity.
7. Keep the GC content between 30% and 60%.

7.2.1   LNA™ IN PCR PROBES

LNA modifications are widely applicable for various types of PCR probes, including both displacement- and hydrolysis-type probes. The key advantage of using LNAs in probes is the ability to increase their affinity and specificity toward their complementary DNA or RNA targets and the increased local specificity is made possible by the design of shorter probes.

Increased specificity at specific single-nucleotide polymorphism (SNP) sites can be achieved by spiking in LNA bases at, or very close to, a specific site of interrogation in a probe. The standard hydrolysis probes are often 20–25 bp in length, depending on sequence context, to maintain sufficient stability to bind to the target sequence before primer annealing and elongation, under PCR-cycling conditions. When such probes are employed to discriminate between SNP alleles, the difference in annealing between two allele-specific probes may be relatively small, given that the loss in T_m caused by a single mismatch, contributes on average only 1/25 of the combined T_m of the probe. The increased affinity resulting from inclusion of the LNA bases can increase the discriminatory ability of probes by adding to the ΔT_m between the match and mismatch probes. Consequently, the drop in T_m caused by a mismatch will equally be large, resulting in an increased specificity of the probe even in slightly reduced probe lengths.[15] A study by You et al.[16] investigated various LNA spiking options and demonstrated that placing a triplet of LNA residues centered on the mismatch was generally found to have the largest discriminatory power, with the exception of G–T mismatches where discrimination was found to be decreased. In cases where an existing assay needs optimization, adding in LNA at critical discriminatory positions in a probe can enhance specificity.[17] The high affinity of LNAs allows for the design of significantly shorter probes (as short as 8- or 9-mers) while still retaining sufficiently high melting temperature (65–70°C). Although the design needs to take into account the possible occurrence of such short sequences elsewhere in the target, short LNA probes are extremely specific since a single mismatch in an 8-mer will reduce T_m drastically and prevent binding. Short LNA probes are the core in Universal Probe Library (Roche).[18] Shorter LNA-containing probes have several potential advantages. First, the design flexibility is broadened, resulting in an increase in the availability of the potential probe targets. Several studies[19–23] have used short LNA probes as "catch-all" probes for multiple subtypes of viral and fungal targets. This approach can also reduce background fluorescence when employing such a single catch-all probe instead of several individual probes, each adding to the background. Better signal to noise is also seen as a result of increased quenching due to the shorter distance between the fluorophore and quencher and by less spurious binding of the probe during amplification. PCR efficiency has also been seen to increase[24–27] that is speculated to be a result of lower interference of the probe during amplification. Finally, better affinity in probes can increase sensitivity and requires less starting materials in the reaction.[24] Some general design criteria must be observed for LNA-containing probes, some of which are listed in Box 7.2.

When using hybridization probes and molecular beacons, the use of LNA leads to similar improvements to assay performance as described for hydrolysis probe style assays (as described above). Hybridization probe assays are particularly sensitive to secondary structure and interactions between the two probes, and the option to design shorter hybridization probes can facilitate the design. Molecular beacon probes containing LNAs have not been widely used but several studies indicate that increased affinity and lower background can be achieved by adding LNAs to the stem

---

BOX 7.2   DESIGN GUIDELINES FOR REAL-TIME qPCR PROBES

1. Typical T_m of dual-labeled probes: 65–70°C (i.e., slightly higher than the primer annealing temperature).
2. Optimal length of LNA-substituted dual-labeled probes: 14–18 nucleotides (note that these are 5–8 bases shorter than the corresponding DNA probes). Shorter, very efficient probes can be obtained by careful design.
3. Maintain T_m with LNA substitutions to match the T_m of the corresponding longer DNA probe.
4. Substitute every third base with LNAs in the central segment of the probe. Usually, 4–6 LNA substitutions are required to obtain a useful T_m.
5. Avoid stretches of more than 3 G DNA or LNA bases.
6. When detecting single-nucleotide mutations, select the probe sequence so that the mutation is located centrally in the probe. A single LNA at the SNP point, a triplet covering the mutation, or generally make a very short LNA probe is recommended. Positioning the SNP at the very 3′ or 5′ end or 1′-position from the ends may compromise discrimination.
7. Always check for possible secondary structures in the probe, especially caused by LNA:LNA interactions.
8. Position the dual-labeled probe as close as possible to the forward primer.
9. Avoid guanine (G) in the 5′-position next to the fluorophore. It has been shown that guanosine quenches the adjacent fluorophore.
10. Select the strand giving the lowest concentration of Gs in the probe.
11. Avoid longer stretches of identical nucleotides and especially Gs.
12. Keep the GC content between 30% and 60%.
13. Always allow at least one DNA in the 5′ end of the probe to allow exonuclease cleavage.

as well as the loop region.[28,29] In summary, inclusion of LNA in the design of real-time PCR probes greatly enhances the possibilities of utilizing a suitable template region.

7.2.2   LNA™ IN PCR PRIMERS

PCR primers have become easy to design because there are numerous software design packages and online design tools. As described for PCR probes, LNAs can increase the binding affinity and specificity when used in PCR primers. However, a DNA primer pair that is working perfectly may not always be enhanced further by incorporation of LNA. Instead, LNA should be considered as a supplementary tool in the design process, rescuing inferior assays, or for designing challenging assays. Adding LNA to PCR primers will increase T_m, thereby increasing the binding affinity or will allow for the shortening of primers that can increase the assay performance. In some instances (e.g., very AT-rich sequences), increasing the affinity of PCR primers will increase the overall efficiency of the PCR[30] and allows for the amplification of low-concentration targets.[31] In addition, when insufficient target sequence is available (e.g., shRNAs) or when multiple primers are combined in a single reaction to perform under similar conditions, LNAs can be used to adjust the T_m of the primers to match PCR-cycling conditions.[29] In other cases, shorter- but higher-affinity primers can prove more efficient in amplifying more degraded samples of lower purity[25,32,33] that are the key features in the fields of forensics or food testing.[34,35] Importantly, such primers may work less efficiently when compared to non-LNA-modified primers tested on highly purified DNA samples and hence it is important to test PCR design on the relevant sample DNA. In a more special application, LNAs were applied to "anchor" a PCR primer overlapping an AT-repeat sequence. This approach allowed the PCR assay to be used efficiently to distinguish between repeat lengths between 5 and $\geq$6 AT repeats.[36] For the above-mentioned applications, LNAs should be added to the 5′prime region of the primer where the modification will result in an increase in binding affinity but will not negatively affect specificity that is primarily determined by the 3′ end binding of the primer. Also, less than five LNA modifications should be added to a PCR primer, with interspersing DNA molecules to allow the polymerase to read through the primer with sufficient efficiency.[37] In an unusual study, Sun et al.[38] used LNA-spiked pentamer primers as universal primers to amplify genomic DNA. In this study, pentamer LNA primers with a T_m of 40–50°C showed high priming efficiency and unbiased amplification of genomic DNA from *Klebsiella pneumonia*. Another study has shown greatly increased specificity using LNA-spiked primers for discrimination of highly homologous, alternatively spliced isoforms.[39]

Allele-specific primers have been used for many applications including SNP typing and methylation analysis, and adding LNA modifications to the 3′ end of primers, primarily at the point of discrimination either at the last or penultimate nucleotide, can greatly enhance specificity.[40–42] Only a single LNA should be added in the 3′ end, since more LNAs can affect the ability of the *Taq* polymerase to recognize and elongate from the primer.[43] Finally, LNAs applied in reverse transcription (RT) primers in the preparatory steps of cDNA synthesis may also enhance target affinity and RT efficiency.[33,44] Some general design guidelines for LNA-containing primers are listed in Box 7.3.

7.2.3   LNA™ IN PCR-BLOCKING PROBES

Recently, significant interest has arisen in the detection of low-level mutations or minority alleles of mutated DNA from, for example, mosaic tumor samples or mixed infections.[45] These samples often contain a large excess of wild-type DNA and require the ability to detect minority alleles among 10^2 up to 10^5 wild-type alleles. To enhance amplification and detection of amplification of the desired allele, blocking or clamping probes have been used to selectively prevent amplification of the wild-type or undesired allele in the PCR.[45] Clamping probes can be designed to bind specifically to unwanted nontarget sequences and to prevent detection by interfering with the PCR primer sites,[46,47] with detection probe-binding sites,[47,48] or by inhibiting polymerase transcription of PCR

BOX 7.3   DESIGN GUIDELINES FOR PCR PRIMERS

1. LNAs should be introduced at the positions where specificity and discrimination is needed (e.g., 3′ end in allele-specific PCR and in the SNP position in allele-specific hybridization probes).
2. Avoid stretches of more than four LNA bases. LNA hybridizes very tightly when several consecutive residues are substituted with LNA bases.
3. Start by spiking LNA in the 5′ end of the primer (allow the 5′ end of the primer to anneal at high T_m avoiding random priming by unspecific annealing of the 3′ end).
4. Avoid LNA self-complementarity and complementarity to other LNA-containing oligonucleotides in the assay. LNA binds very tightly to other LNA residues.
5. The typical primer length of 18-mer should not contain more than eight LNA bases.
6. Each LNA base increases the T_m by approximately 2–4°C.
7. Do not use blocks of LNA near the 3′ end.
8. Keep the GC content between 30% and 60%.
9. Avoid stretches of more than 3 G DNA or LNA bases.
10. T_m of the primer pairs should be nearly equal.
11. Typical T_m of PCR primers for dual-labeled assays: 58–60°C.
12. For improvement of allele-specific PCR, a single LNA nucleotide should be placed in the terminal 3′ or the 3′-1 position. In both cases, the LNA base should correspond to the position of the polymorphism.

templates by binding between primers.[49,50] Early approaches have utilized PNA (peptide nucleic acid) probes for such applications; however, LNA modifications have several advantages over other technologies. First of all, the availability of accurate T_m prediction software for LNA/DNA chimeras allows for more precise fine tuning of the T_m of blocker probes, which is an important feature when designing allele-discriminating assays. This also facilitates designs that avoid fold-back structures or self-complementarities of the blocker probes. Finally, some protocols using PNA requires a four-step PCR, whereas LNA-based assays can run on the standard three-step (or two-step) PCR programs.[51,52] Hu et al.[52] describes the use of an LNA blocker probe that suppresses both the binding of detection probes to the wild-type alleles as well as inhibiting amplification of the wild-type allele. As yet there are only general rules that are available for designing of LNA blocker probes. Since PCR primer extension is dependent on the correct annealing of the 3′ end of the primer, primers and blocking primers should be designed to ensure high affinity at the discrimination site in the 3′ region. Prepens et al.[46] were able to suppress specific viral subtypes by more than 1000-fold by using LNA-blocking probes overlapping the PCR primer sites.[46] Probes should be adjusted in T_m to allow only the PCR blocker probe binding if there is a perfect match to avoid general PCR inhibition. Dominguez et al.[53] found that *Taq* polymerase was able to degrade an LNA probe and the use of an exonuclease-deficient polymerase was required to see an effect of the LNA probe. However, Sidon et al.[54] found no such effect of *Taq* polymerase on the blocking efficiency of LNA blockers. Generally, different *Taq* polymerases will have different degree of exonuclease activity that may be important when considering PCR-blocking probes. Also, the concentration of the probe may be important. After designing a PCR-blocking probe, Hu et al.[52] used a titration approach to assess the required amount of PCR blocker, where 0.005 mM LNA probes resulted in a marked decrease of the wild-type signal, while 0.01–0.02 mM LNA probe resulted in complete suppression of wild-type signal. Deckers et al.[55] used LNA blocking of the PCR to determine the differentiation of a total of 38-sequence subtypes, 10 more than when using the degenerate PCR alone. LNA-blocking probes should be synthesized with a 3′ phosphate or amino group to prevent extension of the probe/primer. For a general list of LNA design recommendations for PCR-blocking probes, see Box 7.4.

BOX 7.4   DESIGN GUIDELINES FOR LNA™ BLOCKER PROBES

1. Add 3′ phosphate or amino or other group to prevent extension by the polymerase.
2. Detection blocker probes: There are no empirical design rules but adjust T_m to allow the correct allele probe or the primer to bind.
3. Primer-blocking probes: Generally, primer extension is very dependent on the 3′ end; so, primers and blocking primers should be designed to position the discrimination site in the 3′ region.
4. PCR blockers: Design the probe to allow only the PCR blocker probe binding if there is a perfect match to avoid general PCR inhibition.

Clearly many publications describing the use of LNA-modified oligos in PCR show that LNA modifications provide the means to overcome difficult amplification situations. Well-functioning PCR designed with DNA may not benefit radically from the introduction of LNAs except that sensitivity and specificity may sometimes be improved to produce a better assay. Improvements may be seen both when LNA is applied in primers and in probes. Both FRET (fluorescence resonance energy transfer)-like detection probes and PCR-blocking probes may benefit from LNA that will help one to increase specificity of the PCR. In particular, with the PCR-blocking probes, there are many examples where incorporation of LNA seems to enable the detection of rare targets in a high background of molecules with close resemblance to the target. In the next section, we will focus on the description of special applications of LNA, where the high affinity of LNA is exploited to shorten primers or probes. In both cases, LNA is required for the applications; first, the specific and very sensitive detection of small target molecules and second, the development of a universal probe set for the detection of the mRNA transcriptome. The use of LNA in applied fields and diagnostic PCR will also be discussed.

7.3   SPECIAL APPLICATIONS OF LNA™ IN PCR

7.3.1   LNA™ IN MicroRNA Detection

MicroRNAs (miRNAs) are a novel class of regulatory molecules that are important for cell function and differentiation.[56] Quantification of miRNAs by qPCR represents a challenge because it is not possible to use conventional systems due to the short nature of the miRNAs (19–25 nt) and the high variability in the percentage GC nucleotide content. In addition, different miRNA families of homologous, closely related sequences exist within a given organism. The technical challenge of performing qPCR on the short miRNAs arises because it is impossible to design the required two PCR primers for a specific reaction within the available 19–25 nucleotides. In general, a single DNA primer is 18–24 nt long to achieve the optimal specificity and annealing temperature toward the target[57] and so a DNA primer is about the same length as an miRNA target. Various RT–qPCR techniques have been developed for miRNA. All these have the characteristic feature of the requirement for a nontarget-related nucleotide tag to be added at the 3′ end of the miRNA during the conversion of the miRNA into cDNA.[58,59] This creates a new primer-binding site on the extended cDNA together with the miRNA-specific primer-binding site. The added nucleotide tag can easily be designed to contain an optimal PCR priming sequence; the real challenge is the design of the miRNA-specific PCR primer. Here, the wide range of GC content in different miRNAs (ranging from as little as 16% in miR-548f to as much as 91% in miR-663) makes it virtually impossible to design the standard DNA PCR primers with similar annealing temperatures. Low-percentage GC content of the miRNA necessitates long PCR primers to reach a high enough annealing temperature. This may lead to a full-length microRNA primer and in some cases, even those do not reach the needed annealing temperature. For example, hsa-miR-1 has a low GC content and so (Figure 7.2) a DNA primer approach does not result in amplification, whereas an LNA enhancement gives good sensitivity for the same target. Even targets with medium GC percentage (exemplified by hsa-let-7a, hsa-miR-155, and hsa-miR-143 in Figure 7.2) may benefit from LNA enhancement with an approximately 100-fold increase in sensitivity. Furthermore, a full-length miRNA primer tends to be associated with severe PCR contamination problems. The reason for this is that the primer itself constitutes such a large portion of the miRNA sequence that it takes only a few reverse primer mispriming events to generate an amplicon that not only contains the correct miRNA target sequence but also has a size close to the correctly amplified miRNA. Consequently, misinterpretation of the results may occur. Conversely, high-percentage GC miRNA may lead to primers made almost entirely of GC. These will generally be unspecific, finding many alternative targets in the sample thus yielding undesired alternative amplicons.

Likewise, short-nucleotide stretches of palindromic sequence within the miRNA may be difficult to avoid. Primers that are designed to include such palindromic regions may result in amplification of unspecific primer–dimers. Also, primer–dimers may arise from the 3′ end complementarity of the primer pair.

Finally, the close resemblance of miRNA sequences belonging to the same family makes the discrimination of these difficult and requires considerable amounts of time to be spent on the assay design or optimization steps. The location of the mismatch within the primer sequence is important for the ability to discriminate; however, with LNA enhancement, it is possible to design primers with good discrimination regardless of the mismatch position within the miRNA sequence (Figure 7.3).

As described above, using LNA provides flexibility in the primer design for miRNA. The annealing temperature of the primers can be increased by exchanging DNA nucleotides with LNA within the primer. Consequently, the optimal annealing temperature of the primer can be maintained while

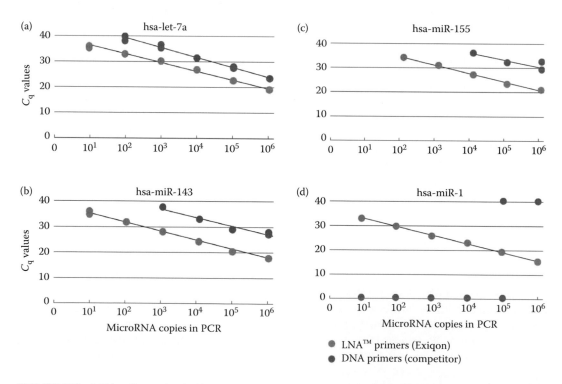

FIGURE 7.2 LNA-enhanced primers increase sensitivity by 100× compared to a DNA-based universal RT–qPCR system and enables the detection of low GC percentage of microRNA such as hsa-miR-1.

Target	Sequence	Assay	Percent detection
hsa-let-7c	UGAGGUAGUAGGUUGUAUGGUU	hsa-let-7c	100
hsa-let-7b	UGAGGUAGUAGGUUGU**G**UGGUU	hsa-let-7c	0.03
hsa-miR-99a	AACCCGUAGAUCCGAUCUUGUG	hsa-miR-99a	100
hsa-miR-100	AACCCGUAGAUCCGA**A**CUUGUG	hsa-miR-99a	0.00
hsa-miR-196b	UAGGUAGUUUCCUGUUGUUGGG	hsa-miR-196b	100
hsa-miR-196a	UAGGUAGUUUC**A**UGUUGUUGGG	hsa-miR-196b	0.01
hsa-miR-135b	UAUGGCUUUUCAUUCCUAUGUGA	hsa-miR-135b	100
hsa-miR-135a	UAUGGCUUUU**U**AUUCCUAUGUGA	hsa-miR-135b	0.03
hsa-let-7a	UGAGGUAGUAGGUUGUAUAGUU	hsa-let-7a	100
hsa-let-7e	UGAGGUAG**G**AGGUUGUAUAGUU	hsa-let-7a	0.00
hsa-miR-17	CAAAGUGCUUACAGUGCAGGUAG	hsa-miR-17	100
hsa-miR-106a	**A**AAAGUGCUUACAGUGCAGGUAG	hsa-miR-17	2.05

FIGURE 7.3 LNA-enhanced primers enable the discrimination between two targets with a single mismatch, regardless of the mismatch discrimination. The mismatch is shown in **bold**.

the nucleotide length of the primers is shortened, even for low GC percentage targets. Shortening of GC-rich primers may make rare AT bases comprise a larger proportion of the sequence, thus giving improved specificity. Designing shorter PCR primers spiked with LNA gives a higher degree of design freedom to avoid palindromic sequence or primers that are complementary to each other. Shortening of primers also allows for the design of a primer set where both primers are specific to the miRNA in question thus maximizing specificity and avoiding the risk of misinterpreting the detection of miRNA with primer–dimer products generated from full-length miRNA primers. Finally, LNA spiking increases the ability of primers to discriminate between sequences with close resemblance thus facilitating discrimination of nearly homologous miRNAs within the same family.[44] The current commercially available LNA-based miRNA–qPCR system uses SYBR Green I dye for detection and this facilitates a post-PCR quality control of the PCR itself through conducting an analysis of the amplicon dissociation curve. This allows for checking specificity of the amplification reaction by verifying that only one amplicon of the correct size is amplified as well as the absence of primer–dimers. However, coamplification of very closely related miRNA families may still be difficult to discriminate since they can differ only in one or two nucleotides. The PCR quality control facilitated by the use of a double-strand DNA-binding dye, here, SYBR Green I dye, is clearly one of the advantages of dye-based detection systems over probe-based systems. This high level of sensitivity becomes very useful when working with samples such as body fluids (e.g., blood serum and plasma) that have a very low content of nucleic acids including miRNA.

7.3.2 UNIVERSAL PROBE LIBRARY

The Universal Probe Library is based on a library of only 165 short oligonucleotides, 8–9 nucleotides in length. These are hydrolysis probes containing LNA, labeled with a quencher at the 3′ end and a fluorophore at the 5′ end.[18,60] The library of probes is sufficient for detection of the entire transcriptomes of several organisms including human, mouse, and rat. A subset of just 90 of these probes will allow the design of a probe-based real-time PCR assay for 98% of all human transcripts. The high-transcriptome coverage is obtained by designing the probes to be complementary to carefully selected 8- or 9-mer motifs that are highly prevalent in the transcriptomes of the higher organisms. Each probe binds, on average, to over 7000 transcripts in the human transcriptome while each transcript is detected by approximately 16 probes. The specificity of the assays that can be designed with the probes is provided by the probe and primer pair in combination. The entire assay design process is conducted live with a free web-based software[61] that identifies the probes that are matching a given transcript followed by the design of the required PCR primers for each of these. Given the many matching probes per transcript, assays can be designed that will allow the detection of transcripts with different splice forms at the same time (common assay) or the distinction between several splice forms with a probe that binds only to one or the other transcript (differentiating assay). Only one specific transcript is detected in each PCR assay as defined by the specific set of PCR primers. The design of functional, short-hydrolysis probes that are only 8–9 nucleotides long is obviously enabled by including LNA in the oligonucleotide to adjust the Tm to a temperature that allows the probes to bind to their target sites despite the short length. Therefore, the probes are compatible with the standard PCR conditions and detection formats. The probes are used together with normal *Taq* polymerases with 5′–3′ hydrolysis activity in a conventional two-step PCR-cycling protocol with annealing for 20 s at 60°C followed by denaturation at 95°C. The Universal Probe Library-based qPCR assays perform well in both rotor cyclers and plate-based cyclers (96–1536 wells) as well as on more high-throughput microfluidic platforms such as the BioMark Fluidigm platform.[62]

7.3.3   LNA™ in Applied Fields and Diagnostic PCR

LNA has been applied to a wide variety of diagnostic and industrial applications including viral, bacterial, and other infectious diseases, food safety, forensics, and other applications. A nonexhaustive list of applications is listed in Table 7.2 along with references to published papers. Among the applications listed, many of the infectious disease assays utilize the capability of LNA-modified oligonucleotides to be used to either increase specificity toward single-nucleotide variants with important diagnostic value or to assign multiplex or catch-all abilities to assays to cover whole strains or species. Such assays are typically restricted by limited availability of relevant target sequences and LNAs are used to adjust the assay to work with the available target. In the field of oncology, more emphasis is laid on the ability of LNA to block majority alleles to increase the detection of somatic minority alleles, with importance for prognosis or treatment selection. The current list is not exhaustive and represents applications where the use of LNA is highlighted. However, it is likely that the use of LNA in diagnostic applications will increase in the future and will lead to LNA being used as the standard ingredient in PCR-based diagnostics. One such use could be within applications of digital PCR[63] where specificity in the detection step is important.

7.4   CONCLUDING REMARKS

The specificity and simplicity of PCR, together with the increase in new sequence information generated from genome- and transcriptome-sequencing projects, has made it the preferred analytical method for most biomedical research. The emergence of next-generation sequence technologies is bound to increase the requirement for analysis of nucleic acids in both research and translational settings. Despite this, the basic PCR assay design characteristics for primers and probes have remained relatively unchanged for many years, while most improvements to the technology have been implemented through enhanced polymerases, reagents, sample preparation, and instruments. However, at the same time, the emergence of more clinical applications of PCR with requirements for very specific analysis has increased the need for tools to fine-tune assays to accommodate different needs just as requirements for robustness and the ability to multiplex assays under similar conditions to allow high-throughput analysis are in high demand.

As one of the few technologies, LNA has shown the capability to enhance the characteristics of the PCR by modifying the basic ingredients of the reaction: primers and probes. As presented, there are already many specific key improvements to PCR provided by LNA such as the ability to make

TABLE 7.2
LNAs™ in Applied Diagnostics and Applied Fields

Field	Application	Reference
Bacteria	Detection of thermotolerant Campylobacter (*Campylobacter jejuni, Campylobacter coli*, and *Campylobacter lari*)	25
Bacteria	Sample preparation for bacteria in blood	27
Bacteria	Multiplex PCR assay for identification of USA300 community-associated methicillin-resistant *Staphylococcus aureus*	36
Bacteria	Quantification and discrimination of *Chlamydia pneumonia*	64
Bacteria	Prevalence of ParC mutations in *Streptococcus*	65
Bacteria	Detection of *Mycobacterium avium* subspecies paratuberculosis in bovine feces	66
Bacteria	Detection and differentiation of *Clostridium chauvoei* and *Clostridium septicum* in clostridial myonecrosis	67
Bacteria	Detection of *Bacillus anthracis* strains	68
Bacteria	Quantification of 16S rDNA by real-time PCR	69
Cancer	MPL (W515K or W515 L) mutation in chronic myeloproliferative disorders	70
Cancer	EGFR mutations in advanced nonsmall-cell lung cancer	71
Cancer	EGFR mutations in nonsmall-cell lung cancer	72
Cancer	Detection of 1100delCCHEK2 germ-line mutation in patients with hematologic malignancies	73
Cancer	Promoter methylation analysis of O6-methylguanine-DNA methyltransferase in glioblastoma	40
Cancer	EGFR mutations in nonsmall-cell lung cancer	74
Cancer	Ki-ras gene mutations in patients with sporadic colorectal carcinomas	48
Cancer	Mutations in the Nucleophosmin gene in acute myeloid leukemia	75
Cancer	Sensitive detection of the JAK2V617F somatic single-base sequence variant	54
Cancer	Detection of K-ras mutations in azoxymethane-induced aberrant crypt foci in mice	52
Cancer	Detection of single-nucleotide minority mutations from clinical specimens	53
Cardiac	SOD3 Arg(213) Gly(R213G) SNP detection	76
Forensics	Increased sensitivity and performance with LNA primers	34
Forensics	Increased trace DNA amplification success with LNA primers	35
Fungi	Detection and identification of *Candida* species in blood	19
GMO	Detection of MON 810 GM maize (also known as YieldGard®)	77
GMO	Detection of MON 810 GM maize (also known as YieldGard)	78
GMO	Risks associated with recombination in transgenic plants expressing viral sequences	33
Parasite	Detection of benzimidazole resistance alleles in *Haemonchus contortus*	79
Parasite	Diagnosis of *Acanthamoeba keratitis*	80
Parasites	Detection of minority Pfcrt K76T mutant population of *Plasmodium falciparum*	51
Veterinary	Detection of the mutation responsible for progressive rod–cone degeneration in *Labrador Retriever* dogs	81
Virus	Detection of Clade 1 and 2 H5N1 Influenza A virus	22
Virus	Broadly targeted multiprobe qPCR for detection of coronaviruses	21
Virus	Discrimination between mild and severe *Citrus tristeza* virus isolates	82
Virus	Detection of Norovirus Genogroup 1	24
Virus	Screening for a key oseltamivir resistance H275Y-substitution in H1N1	83
Virus	Diagnosis of *Newcastle disease*	20
Virus	Quantitative detection and genotyping of HPV-prognostic risk group	23
Virus	Detection of African and South American yellow fever virus strains	84
Virus	Detection of novel Influenza H1N1	85
Virus	Monitoring HCV–RNA viral load	86
Virus	Analysis of hepatitis B virus mutants	87
Virus	Type-specific identification of herpes simplex	88

continued

TABLE 7.2 (continued)
LNAs™ in Applied Diagnostics and Applied Fields

Field	Application	Reference
Virus	Typing (A–B) and subtyping (H1–H3–H5) of Influenza A virus	89
Virus	Detection of emerging Rhinoviruses and Enteroviruses	90
Virus	Comprehensive detection of human Rhinoviruses	29
Virus	Detection of antiviral resistance in hepatitis B virus clinical isolates	91
Virus	Discovery of herpesviruses in multi-infected primates	46
Virus	Combined UL33–UL55 genotyping in human cytomegalovirus	55
Virus	Hepatitis B virus 1762T-1764A double mutation in hepatocellular carcinomas	49
Virus	K65R resistance mutation in patients infected with subtype C human immunodeficiency virus type 1	92

very short probes and primers. However, since in many cases, a good PCR assay can be designed using the standard DNA-based reagents; LNA should be considered as a "problem-solver" option whenever the standard PCR fails or when exceptional requirements are needed. Typically, PCR assays are designed to fit into an existing setup including reagents, instruments, or even cycling conditions that facilitate routine and high-throughput analysis, where changes to such infrastructure can be costly or time-consuming. In that context, LNA can be considered as a toolbox that will allow PCR assays to be designed to fit the assay conditions and requirements, rather than having to modify the existing procedures or instrumentation to fit a new assay. In other cases, LNA modifications can provide novel and unique characteristics to PCR assays, which allow for completely new uses. Although this chapter contains a number of options and guidelines for using LNA, there are likely to be many other applications for LNA in PCR. Consequently, we believe LNA represents an untapped potential for improving the assay design and for expanding the applications of PCR.

ACKNOWLEDGMENT

Part of this work was supported by the European Union Sixth Framework Programme Integrated Project SIROCCO (Grant LSHG-CT-2006-037900).

REFERENCES

1. Bustin SA. 2004. *A–Z of quantitative PCR* (ed. S.A. Bustin), International University Line (IUL) La Jolla, CA, USA (ISBN 0-9636817-8-8).
2. Koshkin AA, Singh SK, Nielsen P, Rajwanshi VK, Kumar R, Meldgaard M, Olsen CE, and Wengel J. 1998. LNA (locked nucelic acid): Synthesis of the adenine, cytosine, guanine, 5-methylcytosine, thymine and uracil bicyclonucleoside monomers, oligomerisation, and unprecedented nucleic acid recognition. *Tetrahedron* 54: 3607–3630.
3. Obika S, Nanbu D, Hari Y, Andoh J-I, Morio K-I, Doi T, and Imanishi T. 1998. Stability and structural features of the duplexes containing nucleoside analogs with a fixed *N*-type conformation. 2′-O, 4′-C methylene ribonucleosides. *Tetrahedron Lett.* 39: 5401–5404.
4. Wengel J, Koshkin A, Singh SK, Nielsen P, Meldgaard M et al. 1999. LNA (locked nucleic acid). *Nucleosides & Nucleotides* 18: 1365–1370.
5. Singh SK, Nielsen P, Koshkin AA, and Wengel J. 1998. LNA (locked nucleic acids): Synthesis and high-affinity nucleic acid recognition. *Chem Commun.* 4: 455–456.
6. Nielsen KE, Singh SK, Wengel J, and Jacobsen JP. 2000. Solution structure of an LNA hybridized to DNA: NMR study of the d(CT(L)GCT(L)T(L)CT(L)GC):D(GCAGAAGCAG) duplex containing four locked nucleotides. *Bioconjug Chem.* 11: 228–238.
7. Petersen M, Nielsen CB, Nielsen KE, Jensen GA, Bondensgaard K et al. 2000. The conformations of locked nucleic acids (LNAs). *J Mol Recogn.* 13: 44–53.

8. Braasch DA and Corey DR. 2001. Locked nucleic acid (LNA): Fine-tuning the recognition of DNA and RNA. *Chem Biol*. 8(1): 1–7.

9. Vester B and Wengel J. 2004. LNA (locked nucleic acid): High-affinity targeting of complementary RNA and DNA. *Biochemistry* 26;43(42): 13233–13241.

10. Freier SM and Altmann KH. 1997. The ups and downs of nucleic acid duplex stability: Structure – stability studies on chemically modified DNA. RNA duplexes. *Nucleic Acids Res*. 25(22): 4429–4443.

11. Di Giusto DA and King GC. 2004. Strong positional preference in the interaction of LNA oligonucleotides with DNA polymerase and proofreading exonuclease activities: Implications for genotyping assays. *Nucleic Acids Res*. 32(3): e32.

12. Tolstrup N, Nielsen PS, Kolberg JG, Frankel AM, Vissing H, and Kauppinen S. 2003. OligoDesign: Optimal design of LNA (locked nucleic acid) oligonucleotide capture probes for gene expression profiling. *Nucleic Acids Res*. 1;31(13): 3758–3762.

13. McTigue PM, Peterson RJ, and Kahn JD. 2004, Sequence-dependent thermodynamic parameters for locked nucleic acid (LNA)–DNA duplex formation. *Biochemistry* 43: 5388–5405.3

14. Tolstrup N, Kolberg JG, Tøstesen E, Salomon J, Willenbrock H, Vissing H, Kauppinen S, Sørensen N, Mouritzen P. LNA™ Oligo Tools and Design Guidelines. Exiqon, 2012. www.exiqon.com/oligo-tools.

15. Letertre C, Perelle S, Dilasser F, Arar K, and Fach P. 2003. Evaluation of the performance of LNA and MGB probes in 5′-nuclease PCR assays. *Mol Cell Probes*. 17(6): 307–311.

16. You Y, Moreira BG, Behlke MA, and Owczarzy R. 2006. Design of LNA probes that improve mismatch discrimination. *Nucleic Acids Res*. 34 (8): e60.

17. Ugozzoli LA, Latorra D, Puckett R, Arar K, and Hamby K. 2004. Real-time genotyping with oligonucleotide probes containing locked nucleic acids. *Anal Biochem*. 324(1): 143–152.

18. Mouritzen P, Nielsen PS, Jacobsen N, Noerholm M, Lomholt C, Pfundheller HM, Ramsing NB, Kauppinen S, and Tolstrup N. 2004. The ProbeLibrary™: Expression profiling 99% of all human genes using only 90 dual-labeled real-time PCR Probes. *Biotechniques* 37: 492–495.

19. Innings Å, Ullberg M, Johansson A, Rubin CJ, Noreus N, Isaksson M, and Herrmann B. 2007. Multiplex real-time PCR targeting the RNase P RNA gene for detection and identification of *Candida* species in blood. *J Clin Microbiol*. 45(3): 874–880.

20. Cattoli G, De Battisti C, Marciano S, Ormelli S, Monne I, Terregino C, and Capua I. 2009. False-negative results of a validated real-time PCR protocol for diagnosis of Newcastle disease due to genetic variability of the matrix Gene. *J Clin Microbiol*. 47(11): 3791–3792.

21. Muradrasoli S, Mohameda N, Hornyákb A, Fohlmand J, Olsene B, Belákb S, and Blomberg J. 2009. Broadly targeted multiprobe QPCR for detection of coronaviruses: Coronavirus is common among mallard ducks (*Anas platyrhynchos*). *J Virol Methods* 159: 277–287.

22. Thanh TT, Pawestri HA, Ngoc NM, Hien VM, Syahrial H et al. 2010. A real-time RT-PCR for detection of Clade 1 and 2 H5N1 influenza: A virus using locked nucleic acid (LNA) TaqMan probes. *Virol J*. 22(7): 46.

23. Leo E, Venturoli S, Cricca M, Musiani M, and Zerbini M. 2009. High-throughput two-step LNA real time PCR assay for the quantitative detection and genotyping of HPV prognostic-risk groups. *J Clin Virol*. 45(4): 304–310.

24. Dreier J, Störmer M, Mäde D, Burkhardt S, and Kleesiek K. 2006. Enhanced reverse transcription-PCR assay for detection of norovirus genogroup I. *J Clin Microbiol*. 44(8): 2714–2720.

25. Josefsen MH, Löfström C, Sommer HM, and Hoorfar J. 2009. Diagnostic PCR: Comparative sensitivity of four probe chemistries. *Mol Cell Probes*. 23(3–4): 201–203.

26. Costa JM, Ernault P, Olivi M, Gaillon T, and Arar K. 2004. Chimeric LNA/DNA probes as a detection system for real-time PCR. *Clin Biochem*. 37(10): 930–932.

27. Störmer M, Kleesiek K, and Dreier J. 2007. High-volume extraction of nucleic acids by magnetic bead technology for ultrasensitive detection of bacteria in blood components. *Clin Chem*. 53(1): 104–110.

28. Kim Y, Sohn D, and Tan W. 2008. Molecular beacons in biomedical detection and clinical diagnosis. *Int J Clin Exp Pathol*. 1(2): 105–116.

29. Lu X, Holloway B, Dare RK, Kuypers J, Yagi S, Williams JV, Hall CB, and Erdman DD. 2008. Real-time reverse transcription-PCR assay for comprehensive detection of human rhinoviruses. *J Clin Microbiol*. 46(2): 533–559.

30. Levin JD, Dean F, Samala MF, Kahn JD, and Peterson RJ. 2006. Position-dependent effects of locked nucleic acid (LNA) on DNA sequencing and PCR primers. *Nav Res.*, 34: 20.

31. Malgoyre A, Banzet S, Mouret C, Bigard AX, and Peinnequin A. 2007. Quantification of low-expressed mRNA using 5' LNA-containing real-time PCR primers. *Biochem Biophys Res Commun*. 2;354(1): 246–252.

32. Ballantyne KN, van Oorschot RA, and Mitchell RJ. 2008. Locked nucleic acids in PCR primers increase sensitivity and performance. *Genomics*. 91(3): 301–305.

33. Turturo C, Friscina A, Gaubert S, Jacquemond M, Thompson JR, and Tepfer M. 2008. Evaluation of potential risks associated with recombination in transgenic plants expressing viral sequences. *J Gen Virol*. 89(1): 327–335.

34. Ballantyne KN, van Oorschot RA, and Mitchell RJ. 2010. Increased amplification success from forensic samples with locked nucleic acids. *Forensic Sci Int Genet*. 5(4): 276–280.

35. Ballantyne KN, Oorschot RAH, and Mitchell RJ. 2008. Locked nucleic acids: Increased trace DNA amplification. Success with improved primers. *Forensic Sci Int: Genet Suppl Ser*. 1(2008): 4–6.

36. Bonnstetter KK, Wolter DJ, Tenover FC, McDougal LK, and Goering RV. 2007. Rapid multiplex PCR assay for identification of USA300 community-associated methicillin-resistant *Staphylococcus aureus* isolates. *J Clin Microbiol*. 45(1): 141–146.

37. Latorra D, Arar K, and Hurley JM. 2003. Design considerations and effects of LNA in PCR primers. *Mol Cell Probes*. 17(5): 253–259.

38. Sun Z, Chen Z, Hou X, Li S, Zhu H, Qian J, Lu D, and Liu W. 2008. Locked nucleic acid pentamers as universal PCR primers for genomic DNA amplification. *PLoS One*. 3(11): e3701.

39. Wan G and Too H-P. 2006. Discrimination of alternative spliced isoforms by real-time PCR using locked nucleic acid (LNA) substituted primer. *Mol Eng Biol Chem Syst*. (MEBCS), http://hdl.handle.net/1721.1/30385

40. Morandi L, Franceschi E, de Biase D, Marucci G, Tosoni A, Ermani M, Pession A, Tallini G, and Brandes A. 2010. Promoter methylation analysis of O6-methylguanine-DNA methyltransferase in glioblastoma: Detection by locked nucleic acid based quantitative PCR using an imprinted gene (SNURF) as a reference. *BMC Cancer*. 18(10): 48.

41. Gustafson KS. 2008. Locked nucleic acids can enhance the analytical performance of quantitative methylation-specific polymerase chain reaction. *J Mol Diagn*. 10(1): 33–42.

42. Nakitandwe J, Trognitz F, and Trognitz B. 2007. Reliable allele detection using SNP-based PCR primers containing locked nucleic acid: Application in genetic mapping. *Plant Methods*. 7(3): 2.

43. Latorra D, Campbell K, Wolter A, and Hurley JM. 2003. Enhanced allele-specific PCR discrimination in SNP genotyping using 3′ locked nucleic acid (LNA) primers. *Hum Mutat*. 22(1): 79–85.

44. Andreasen D, Fog JU, Biggs W, Salomon J, Dahslveen IK, Baker A, and Mouritzen P. 2010. Improved microRNA quantification in total RNA from clinical samples. *Methods*. 50: S6–S9.

45. Milbury CA, Li J, and Makrigiorgos GM. 2009. PCR-based methods for the enrichment of minority alleles and mutations. *Clin Chem*. 55(4): 632–640.

46. Prepens S, Kreuzer KA, Leendertz F, Nitsche A, and Ehlers B. 2007. Discovery of herpesviruses in multi-infected primates using locked nucleic acids (LNA) and a bigenic PCR approach. *Virol J*. 6(4): 84.

47. Oldenburg RP, Liu MS, and Kolodney MS. 2008. Selective amplification of rare mutations using locked nucleic acid oligonucleotides that competitively inhibit primer binding to wild-type DNA. *J Invest Dermatol*. 128(2): 398–402.

48. Beránek M, Jandík P, Šácha M, Rajman M, Sákra L, Štumr F, Soudková E, Živný P, and Havlíček K. 2006. LNA clamped PCR: A specific method for detection of Ki-ras gene mutations in patients with sporadic colorectal carcinomas Klin. *Biochem Metab.*, 14(4–35): 217–220.

49. Ren XD, Lin SY, Wang X, Zhou T, Block TM, and Su YH. 2009. Rapid and sensitive detection of hepatitis B virus 1762T/1764A double mutation from hepatocellular carcinomas using LNA-mediated PCR clamping and hybridization probes. *J Virol Methods*. 158(1–2): 24–29.

50. Hummelshoj L, Ryder LP, Madsen HO, and Poulsen LK. 2005. Locked nucleic acid inhibits amplification of contaminating DNA in real-time PCR. *Biotechniques*. 38(4): 605–610.

51. Senescau A, Berry A, Benoit-Vical F, Landt O, Fabre R, Lelièvre J, Cassaing S, and Magnaval JF. 2005. Use of a locked-nucleic-acid oligomer in the clamped-probe assay for detection of a minority Pfcrt K76T mutant population of *Plasmodium falciparum*. *J Clin Microbiol*. 43(7): 3304–3308.

52. Hu Y, Le Leu RK, and Young GP. 2009. Detection of K-ras mutations in azoxymethane-induced aberrant crypt foci in mice using LNA-mediated real-time PCR clamping and mutant-specific probes. *Mutat Res*. 677(1–2): 27–32.

53. Dominguez PL and Kolodney MS. 2005. Wild-type blocking polymerase chain reaction for detection of single nucleotide minority mutations from clinical specimens. *Oncogene*. 24(45): 6830–6834. Erratum in: *Oncogene*. 2006 Jan 26;25(4): 656.

54. Sidon P, Heimann P, Lambert F, Dessars B, Robin V, and El Housni H. 2006. Combined locked nucleic acid and molecular beacon technologies for sensitive detection of the JAK2V617F somatic single-base sequence variant. *Clin Chem*. 52(7): 1436–1438.

55. Deckers M, Hofmann J, Kreuzer KA, Reinhard H, Edubio A, Hengel H, Voigt S, and Ehlers B. 2009. High genotypic diversity and a novel variant of human cytomegalovirus revealed by combined UL33/UL55 genotyping with broad-range PCR. *Virol J.* 26(6): 210.

56. Kim N, Han J, and Siomi MC. 2009. Biogenesis of small RNAs in animals. *Nat Rev Mol Cell Biol.* 10: 126–139.

57. Dieffenbach CW, Lowe TM, and Dveksler DS. 1993. General concepts for PCR primer design. *Genome Res.* 3: S30–S37.

58. Raymond CK, Roberts BS., Garrett-Engele P, Lim LP, and Johnson JM. 2005. Simple, quantitative primer-extension PCR assay for direct monitoring of microRNAs and short-interfering RNAs. *RNA.* 11: 1737–1744.

59. Chen C, Ridzon DA, Broomer AJ, Zhou Z, Lee DH et al. 2005. Real-time quantification of microRNAs by stem-loop RT PCR. *Nucleic Acids Res.* 33: e179.

60. Mouritzen P, Noerholm M, Nielsen PS, Jacobsen N, Lomholt C, Pfundheller HM, and Tolstrup N. 2005. ProbeLibrary: A new method for faster design and execution of quantitative real-time PCR. *Nat Methods* 2(2205): 313–317.

61. Universal ProbeLibrary Assay Design Center. Roche Applied Science, 2012. http://www.roche-applied-science.com/sis/rtpcr/upl/index.jsp?id=UP030000.

62. Livak K. BioMark Dynamic Arrays for Single-Cell Gene Expression Analysis, Poster 1045, Stem Cells Conference Europe, 2008. http://www.fluidigm.com/home/fluidigm/Posters/FLDM_BMK_MRKT00097.pdf

63. Vogelstein B and Kintzler KW. 1999. Digital PCR PNHS 96: 9236–9241.

64. Rupp J, Solbach W, and Gieffers J. 2006. Single-nucleotide-polymorphism-specific PCR for quantification and discrimination of *Chlamydia pneumoniae* genotypes by use of a "locked". *Nucleic Acid Appl Environ Microbiol.* 72(5): 3785–3787.

65. Decousser J-W, Methlouthi I, Pina P, Collignon A, Allouch P, and on behalf of the ColBVH Study Group. 2006. New real-time PCR assay using locked nucleic acid probes to assess prevalence of ParC mutations in fluoroquinolone-susceptible *Streptococcus pneumoniae* isolates from France. *Antimicrob Agents Chemother.* 50(4): 1594–1598.

66. Schönenbrücher H, Abdulmawjood A, Failing K, and Bülte M. 2008. New triplex real-time PCR assay for detection of mycobacterium avium subsp. *Paratuberculosis* in bovine feces. *Appl Environ Microbiol.* 74(9): 2751–2758.

67. Halm A, Wagner M, Köfer J, and Hein I. 2010. Novel real-time PCR assay for simultaneous detection and differentiation of *Clostridium chauvoei* and *Clostridium septicum* in clostridial myonecrosis. *J Clin Microbiol.* 48(4): 1093–1098.

68. Wattiau P, Klee SR, Fretin D, Hessche MV, Ménart M, Franz T, Chasseur C, Butaye P, and Imberechts H. 2008. Occurrence and genetic diversity of *Bacillus anthracis* strains isolated in an active wool-cleaning factory. *Appl Environ Microbiol.* 74(13): 4005–4011.

69. Goldenberg O, Landt O, Schumann RR, Göbel UB, and Hamann L. 2005. Use of locked nucleic acid oligonucleotides as hybridization/FRET probes for quantification of 16S rDNA by real-time PCR. *Biotechniques.* 38(1): 29–30, 32.

70. Pancrazzi A, Guglielmelli P, Ponziani V, Bergamaschi G, Bosi A, Barosi G, and Vannucchi AM. 2008. A sensitive detection method for MPLW515 L or MPLW515K mutation in chronic myeloproliferative disorders with locked nucleic acid-modified probes and real-time polymerase chain reaction. *J Mol Diagn.* 10(5): 435–441.

71. Maemondo M, Inoue A, Kobayashi K, Sugawara S, Oizumi S et al. North-East Japan Study Group. 2010. Gefitinib or chemotherapy for non-small-cell lung cancer with mutated EGFR. *N Engl J Med.* 24,362(25): 2380–2388.

72. Inoue A, Kobayashi K, Usui K, Maemondo M, Okinaga S et al. North East Japan Gefitinib Study Group. 2009. First-line gefitinib for patients with advanced non-small-cell lung cancer harboring epidermal growth factor receptor mutations without indication for chemotherapy. *J Clin Oncol.* 20,27(9): 1394–1400.

73. Collado M, Landt O, Barragán E, Lass U, Cervera J, Sanz MA, and Bolufer P. 2004. Locked nucleic acid-enhanced detection of 1100delC*CHEK2 germ-line mutation in Spanish patients with hematologic malignancies. *Clin Chem.* 50(11): 2201–2204.

74. Tanaka T, Nagai Y, Miyazawa H, Koyama N, Matsuoka S et al. 2007. Reliability of the peptide nucleic acid-locked nucleic acid polymerase chain reaction clamp-based test for epidermal growth factor receptor mutations integrated into the clinical practice for non-small cell lung cancers. *Cancer Sci.* 98(2): 246–252.

75. Laughlin TS, Becker MW, Liesveld JL, Mulford DA, Abboud CN, Brown P, and Rothberg PG. 2008. Rapid method for detection of mutations in the *nucleophosmin* gene in acute myeloid leukemia. *J Mol Diagn*. 10(4): 338–345.

76. Brugè F, Littarru GP, Silvestrini L, Mancuso T, and Tiano L. 2009. A novel real time PCR strategy to detect SOD3 SNP using LNA probes. *Mutat Res*. 2,669(1–2): 80–84.

77. Buh Gasparic M, Cankar K, Zel J, and Gruden K. 2008. Comparison of different real-time PCR chemistries and their suitability for detection and quantification of genetically modified organisms. *BMC Biotechnol*. 6(8): 26.

78. Salvi S, D'Orso F, and Morelli G. 2008. Detection and quantification of genetically modified organisms using very short, locked nucleic acid TaqMan probes. *J Agric Food Chem*. 25,56(12): 4320–4327.

79. Walsh TK, Donnan AA, Jackson F, Skuce P, and Wolstenholme AJ. 2007. Detection and measurement of benzimidazole resistance alleles in *Haemonchus contortus* using real-time PCR with locked nucleic acid TaqMan probes. *Vet Parasitol*. 31,144(3–4): 304–312.

80. Thompson PP, Kowalski RP, Shanks RM, and Gordon YJ .2008. Validation of real-time PCR for laboratory diagnosis of *Acanthamoeba keratitis*. *J Clin Microbiol*. 46(10): 3232–3236.

81. Gentilini F, Rovesti GL, and Turba ME. 2009. Real-time detection of the mutation responsible for progressive rod-cone degeneration in *Labrador Retriever* dogs using locked nucleic acid TaqMan probes. *J Vet Diagn Invest*. 21(5): 689–692.

82. Ruiz-Ruiz S, Moreno P, Guerri J, and Ambrós S. 2009. Discrimination between mild and severe *Citrus tristeza* virus isolates with a rapid and highly specific real-time reverse transcription-polymerase chain reaction method using TaqMan LNA probes. *Phytopathology*. 99(3): 307–315.

83. van der Vries E, Jonges M, Herfst S, Maaskant J, Van der Linden A et al. 2010. Evaluation of a rapid molecular algorithm for detection of pandemic *Influenza A* (H1N1) 2009 virus and screening for a key oseltamivir resistance (H275Y) substitution in neuraminidase. *J Clin Virol*. 47(2010): 34–37.

84. Weidmann M, Faye O, Faye O, Kranaster R, Marx A, Nunes MR, Vasconcelos PF, Hufert FT, and Sall AA. 2010. Improved LNA probe-based assay for the detection of African and South American yellow fever virus strains. *J Clin Virol*. 48(3): 187–192.

85. Wenzel JJ, Walch H, Bollwein M, Niller HH, Ankenbauer W et al. 2009. Library of prefabricated locked nucleic acid hydrolysis probes facilitates rapid development of reverse-transcription quantitative real-time PCR assays for detection of novel influenza A/H1N1/09 virus. *Clin Chem*. 55(12): 2218–2222.

86. Morandi L, Ferrari D, Lombardo C, Pession A, and Tallini G. 2007. Monitoring HCV RNA viral load by locked nucleic acid molecular beacons real time PCR. *J Virol Methods*. 140(1–2): 148–154.

87. Sun Z, Zhou L, Zeng H, Chen Z, and Zhu H. 2007. Multiplex locked nucleic acid probes for analysis of hepatitis B virus mutants using real-time PCR. *Genomics*. 89(1): 151–159.

88. Meylan S, Robert D, Estrade C, Grimbuehler V, Péter O, Meylan PR, and Sahli R. 2008. Real-time PCR for type-specific identification of herpes simplex in clinical samples: Evaluation of type-specific results in the context of CNS diseases. *J Clin Virol*. 41(2): 87–91.

89. Suwannakarn K, Payungporn S, Chieochansin T, Samransamruajkit R, Amonsin A et al. 2008. Typing (A/B) and subtyping (H1/H3/H5) of influenza A viruses by multiplex real-time RT-PCR assays. *J Virol Methods*. 152(1–2): 25–31.

90. Tapparel C, Cordey S, Van Belle S, Turin L, Lee WM, Regamey N, Meylan P, Mühlemann K, Gobbini F, and Kaiser L. 2009. New molecular detection tools adapted to emerging rhinoviruses and enteroviruses. *J Clin Microbiol*. 47(6): 1742–1749.

91. Fang J, Wichroski MJ, Levine SM, Baldick CJ, Mazzucco CE et al. 2009. Ultrasensitive genotypic detection of antiviral resistance in hepatitis B virus clinical isolates. *Antimicrob Agents Chemother*. 53(7): 2762–2772.

92. Toni TA, Brenner BG, Asahchop EL, Ntemgwa M, Moisi D, and Wainberg MA. 2010. Development of an allele-specific PCR for detection of the K65R resistance mutation in patients infected with subtype C human immunodeficiency virus type 1. *Antimicrob Agents Chemother*. 54(2): 907–911.

8 Modified dNTPs
A Toolbox for Use in PCR

*Natasha Paul, Olga Adelfinskaya, Elena Hidalgo Ashrafi,
Tony Le, and Sabrina Shore*

CONTENTS

8.1 INTRODUCTION

Polymerase chain reaction (PCR) is a widely practiced technique that is utilized in a number of applications in medical diagnostics, forensic DNA analysis, genome sequencing, and basic research. Using PCR, a defined nucleic acid sequence of interest is amplified using a thermostable DNA polymerase in combination with a pair of primers and 2′-deoxyribonucleoside 5′-triphosphates (dNTPs).[1] The naturally occurring dNTPs are the most commonly used DNA polymerase substrates, where an equimolar mixture of the 5′-triphosphates of 2′-deoxyadenosine (dATP), 2′-deoxycytidine (dCTP), 2′-deoxyguanosine (dGTP), and 2′-deoxythymidine (dTTP) is routinely employed in PCR protocols. While the natural dNTPs are best suited for routine PCR, the inclusion of modified dNTPs into a PCR protocol can provide a wide range of beneficial effects. As depicted in Figure 8.1a, there are three main components to a dNTP—the 2′-deoxyribose sugar, the nucleobase, and the triphosphate,

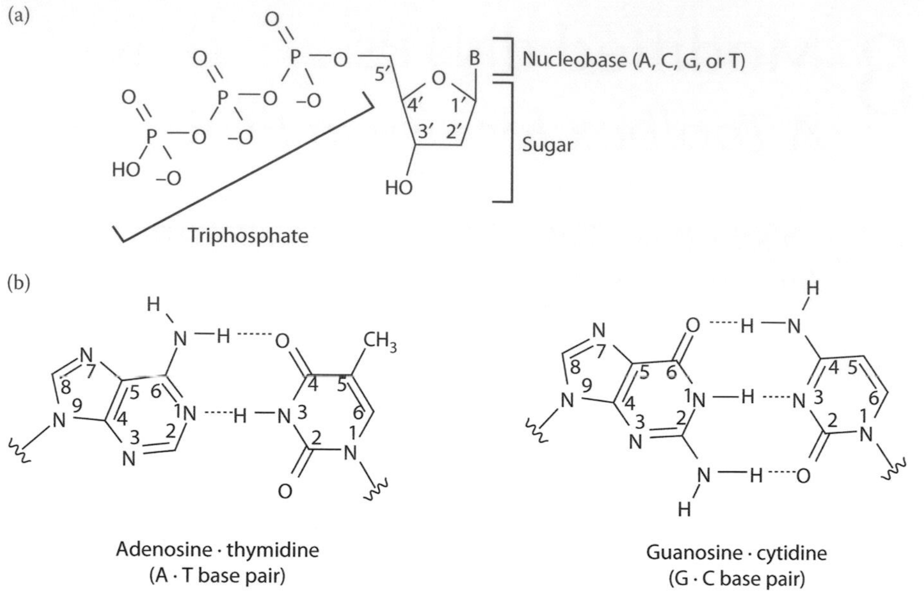

(a)

(b)

Adenosine · thymidine
(A · T base pair)

Guanosine · cytidine
(G · C base pair)

FIGURE 8.1   General structure of (a) a dNTP and (b) Watson–Crick base pairs, where A · T base pairs have two hydrogen bonds and G · C pairs have three hydrogen bonds.

each of which can be modified. Although there are a multitude of possibilities for chemical modification within the dNTP framework, for practical use in PCR, it is critical that the dNTP modifications do not significantly reduce the efficiency of nucleotide incorporation and recognition by DNA polymerase.

Herein, we will describe four classes of modified dNTPs as unique sets of tools that are amenable for use in PCR. The first class of dNTP analogs can be applied to improve the specificity of PCR. This class includes dNTP analogs that are applicable to Hot Start amplification schemes and dNTP analogs that modulate Watson–Crick hydrogen bonding and secondary structure formation. The second class of dNTP analogs can be used to introduce mutations during random mutagenesis PCR protocols. The use of mutagenic dNTP analogs increases the error rate of PCR relative to natural dNTPs. As applied to protein evolution schemes, the ability to introduce random amino acid substitutions into a gene of interest allows for identification of beneficial mutations, especially when the protein structure is unknown. The third class of dNTP analogs allows for expansion of the genetic code from the two traditional Watson–Crick base pairs (A · T and G · C; Figure 8.1b)[2] to three or more orthogonal base pairs. Although the identification of a novel base pair that can replicate with only its complementary counterpart can be quite challenging, if successful, it has wide applicability in nucleic acid detection and in the translation of proteins with nonstandard amino acids. The fourth and final class of dNTP analogs can be used to incorporate a variety of functionalities into the resultant amplicon. In this section, we will focus on the direct and indirect modification approaches that are used for the introduction of moieties, such as fluorophores and affinity tags, into a PCR amplicon. The dynamic nature of PCR continues to pave the way for advancements in the scientific field, allowing for different levels of control and many new discoveries in nucleic acid amplification.

8.1.1   Modified dNTPs to Improve the Specificity of PCR

The ability of PCR to amplify a nucleic acid sequence with high specificity and accuracy makes it one of the most powerful and versatile molecular biology techniques. When designing a PCR assay, specific target amplification can be compromised by a number of factors including nonspecific primer extension and sequence composition effects. Nonspecific primer extension, which commonly occurs at lower, less stringent temperatures, can result in primer–dimer formation and mis-priming amplification products which can, in turn, impact the yield of the desired amplicon.[3] Sequence composition also plays a role in the determination of the efficiency of amplification of a given target. Localized high GC or AT content may also preclude amplification of a target under the standard PCR protocols. There are several approaches, including routinely used Hot Start activation technologies, which have been developed to help overcome each of these challenges. Herein, we discuss two types of modified dNTPs that can be applied to improve the stringency of PCR, heat-labile dNTPs, and permanently modified dNTPs.

Heat-labile dNTPs contain a chemical modification group that is selected to block DNA polymerase activity. The start of PCR is controlled by heat-mediated removal of the thermolabile modification group to produce the corresponding unmodified dNTP that is a natural DNA polymerase substrate (Figure 8.2a). The premise of a heat-labile or Hot Start dNTP is analogous to the many Hot Start technologies that are available commercially. Although each Hot Start activation approach blocks DNA polymerase activity in different ways, all involve the release of a blocking moiety at elevated temperatures and are shown to improve the specificity of PCR by suppressing mis-priming and primer–dimer formation. There are two described approaches for the generation of heat-labile dNTPs with application to Hot Start activation schemes in PCR. Bonner[4] describes thermolabile glyoxal modifications to the exocyclic amino groups of the guanine nucleobase as a Hot Start mechanism. Similarly, Koukhareva and Lebedev[5] describe several 3′-thermolabile sugar modification groups, including tetrahydrofuranyl, for use in Hot Start PCR.[6] In each case, incubation at elevated temperatures releases the modification group to produce the corresponding unmodified dNTP. However, modification of the sugar is a more versatile Hot Start method than glyoxal modification, as the modification is not limited to a specific nucleobase, thereby reducing the influence of sequence contexts. The use of modified dNTPs as a Hot Start method is a fairly recent development, with versatility for use with a variety of standard DNA polymerases.

The second type of dNTP modification is one that results in a permanent alteration, where the presence of the modification may influence the secondary structure formation or Watson–Crick base pairing stability within a PCR amplicon. While temperature control methods such as Hot Start can be used in most cases, they may not always improve PCR performance. When the target of interest has a high GC composition, the tendency for intramolecular secondary structure formation within the template sequence can present a unique set of challenges. In addition to strong Watson–Crick hydrogen pairing within a G · C pair, Hoogsteen bond formation is also prominent in these regions and is responsible for the formation of strong secondary structures, such as G quadruplexes. As a result, PCR amplification can result in a little-to-no amplicon yield with a significant amount of mis-priming artifacts. Secondary structure disruption is critical for successful amplification of GC-rich regions, with the addition of 7-deaza-dGTP providing significant benefit. By replacement of the N7 nitrogen of dGTP with a *C–H* (Figure 8.2c), one of the hydrogen bonds involved in Hoogsteen pair formation is disrupted while maintaining the Watson–Crick base pairing face of the nucleobase. Therefore, regions of high secondary structure are destabilized to improve PCR amplification. Also, 7-deaza-dGTP can be used in conjunction with altered thermal cycling protocols and PCR enhancers, such as betaine and dimethyl sulfoxide (DMSO), to allow for targets with as high as 83% GC content to be successfully amplified.[7,8]

Alternatively, the use of permanently modified dNTPs can influence the strength of the A · T and G · C base pairs in a PCR amplicon. In standard Watson–Crick base pairs, A · T pairs contain two hydrogen bonds, while G · C pairs contain three hydrogen bonds (Figure 8.1b). Watson–Crick base pairing strength can be modulated either by altering the number of hydrogen bonds in a base pair or by introducing substituents that influence duplex stability. One interesting approach to reducing the strength of GC-rich targets is to employ dNTP analogs that inverse the natural hydrogen-bonding rule in DNA duplexes.[9] By substitution of 2′-deoxyinosine 5′-triphosphate (dITP) for dGTP and 2-amino-dATP (dDTP) for dATP in PCR, the base pair between inosine and cytidine is reduced in

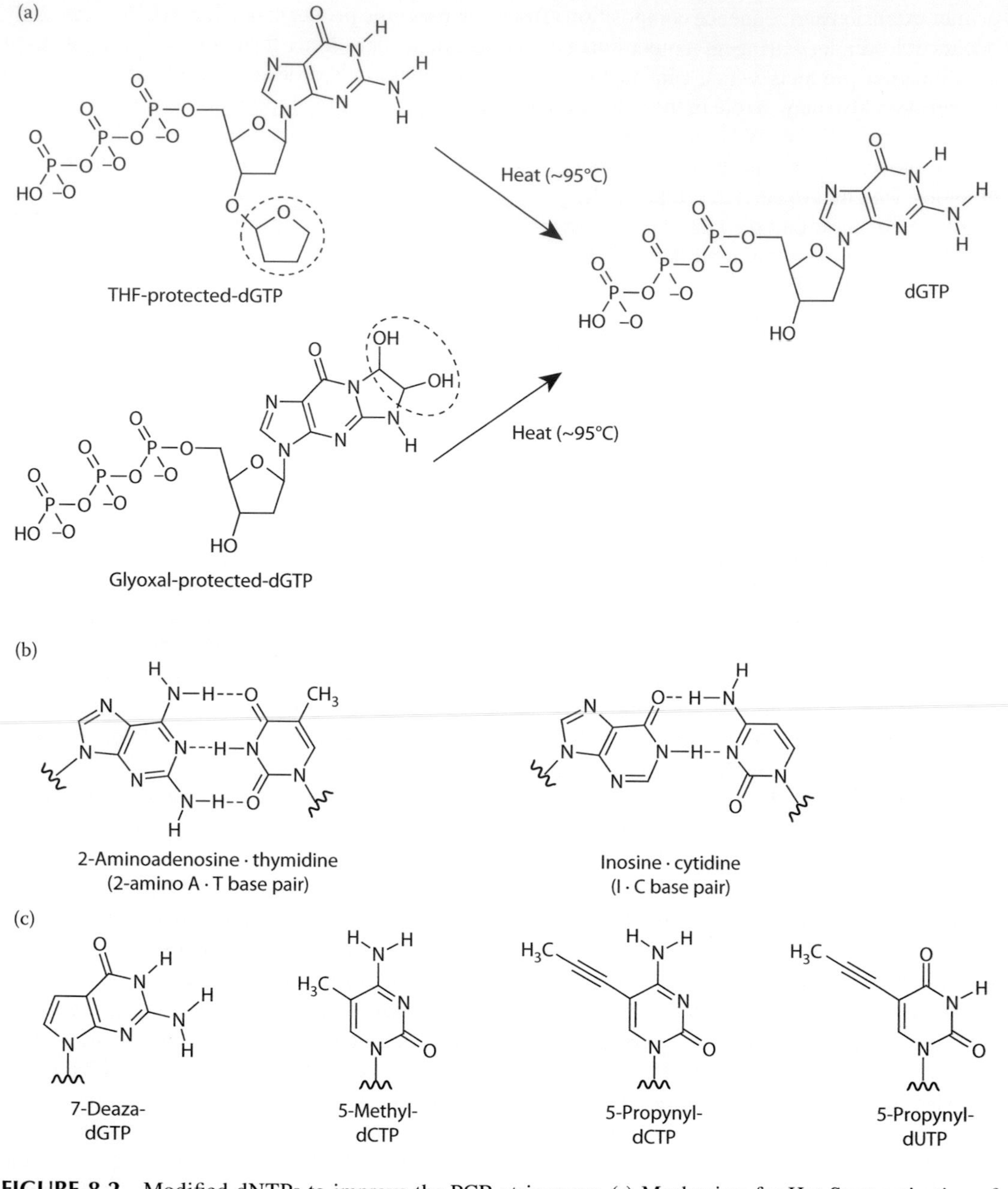

(a)

THF-protected-dGTP

Heat (~95°C)

dGTP

Glyoxal-protected-dGTP

Heat (~95°C)

(b)

2-Aminoadenosine · thymidine
(2-amino A · T base pair)

Inosine · cytidine
(I · C base pair)

(c)

7-Deaza-
dGTP

5-Methyl-
dCTP

5-Propynyl-
dCTP

5-Propynyl-
dUTP

FIGURE 8.2 Modified dNTPs to improve the PCR stringency. (a) Mechanism for Hot Start activation of heat-labile dNTPs. Both 3′-THF-modified dNTPs and glyoxylated dGTP contain modification groups that can be removed by a heat activation step to generate the corresponding unmodified dNTP. Although THF can be introduced onto all four dNTPs, only 3′-THF-dGTP is shown for simplicity. (b) Base pairs containing inverted hydrogen-bonding strengths. (c) Modified dNTPs that improve specificity by reducing secondary structure and increasing the strength of a Watson–Crick base pair. Only the nucleobase portion of the dNTP is shown.

strength to contain only two hydrogen bonds, while the base pair between 2-amino-adenosine and thymidine is increased in strength to contain three hydrogen bonds (Figure 8.2b). Concurrent substitution of dITP and dDTP for dGTP and dATP, respectively, in combination with careful control of denaturation conditions, allows for selective enrichment of GC-rich targets that were not amplifiable by PCR under standard conditions.[9] In another approach, modified dNTPs that stabilize DNA duplexes through base stacking interactions can provide additional benefits, such as the ability to use shorter primers and probes in real-time PCR without the need for costly modified primers and/ or probes.[10] For AT-rich regions, incorporation of dDTP and/or the duplex-stabilizing nucleotide, 5-propynyl-dUTP (5-Pr-dUTP), strengthens DNA duplexes by increasing the number of hydrogen bonds in the "A · T" base pair and/or by increasing the base stacking, respectively.[10] Similarly, dCTP analogs containing major groove substituents such as 5-methyl-dCTP or 5-propynyl-dCTP can easily be substituted to stabilize duplex interactions in GC-rich regions. By strategic modulation of the base pair stability of a DNA duplex by the careful selection of the dNTP analogs employed, PCR performance can be influenced significantly.

8.1.2 Modified dNTPs for Introduction of Mutations during PCR

The paradigm for high-fidelity DNA replication relies on specific DNA polymerase recognition and incorporation of a Watson–Crick complementary dNTP opposite a template nucleobase. Although DNA polymerase misincorporation is generally undesirable, modified dNTPs that directly degenerate DNA polymerase recognition can be used in random mutagenesis procedures in PCR. The concept of degeneracy describes a nucleobase analog with indiscriminate DNA polymerase recognition properties. Unlike the natural nucleobases that are replicated only with high specificity with their corresponding Watson–Crick complement, degenerate nucleobases are nonspecific and can be incorporated opposite two or more natural nucleobases within a DNA template. In turn, when degenerate nucleobases are recognized by DNA polymerases within a template, these analogs can code for incorporation of two or more of the natural nucleotides. Degenerate nucleobases are also known as "convertides" because of the duality of incorporation and replication of the nucleobase specificity.[11] As is the case for a wide variety of modified dNTPs, the effective use of convertible dNTPs in random mutagenesis protocols requires efficient DNA polymerase incorporation and recognition, with comparable performance to natural dNTPs.[11] With all the criteria for successful use of a degenerate dNTP analog in PCR, the identification of a modified dNTP that has the correct balance between indiscriminate DNA polymerase incorporation and recognition with successful replication can be challenging. Nonetheless, nucleobases that demonstrate several-fold degeneracy have found use in several important applications such as ligation and mutagenic PCR.

Two mechanisms that influence the convertible hydrogen-bonding patterns evident for degenerate nucleobases are tautomeric equilibrium and conformational changes. Both natural and modified nucleobases undergo rapid keto ↔ enol and amino ↔ imino tautomeric changes, with natural nucleobases existing in their keto and amino forms 99.99% of the time.[2] For mutagenesis, a shift in the tautomeric equilibrium toward the under represented isomer can cause a change in the base pairing capabilities of the nucleobase.[2] Notable nucleobase modifications that shift the tautomeric equilibrium include purines with substitution of the O6 oxygen with sulfur[2] and cytidines with introduction of an electronegative moiety such as hydroxyl, methoxy, or semicarbazide at the N^4-amino group.[12] Unlike tautomerization, which involves hydrogen migration, conformational changes occur as a result of rotation about a single bond, without the need for bond breakage. One commonly described conformational change in DNA is the rotation of the nucleobase about the glycosidic bond. When the nucleobase is in the *anti* conformation, it presents its Watson–Crick base pairing face to the opposite strand; when the nucleobase is in the *syn* conformation, nonstandard base pairs, such as purine–purine base pairs, are possible.[2] There is a variety of degenerate dNTPs amenable for use in random mutagenesis PCR. In the following section, we will focus on a description of the convertible nucleotides dPTP and 8-*oxo*-dGTP (Figure 8.3a).

The behavior of two well-characterized nucleoside analogs, dP and 8-*oxo*-dG (6-(2-deoxy-β-D-ribofuranosyl)-3,4-dihydro-8*H*-pyrimido-[4,5-C][1,2]oxazin-7-one and 8-oxo-2′-deoxyguanosine, respectively), was studied in PCR and their inclusion into the protocol was found to introduce a wide spectrum of mutations.[13,14] The dP nucleoside was designed as a conformationally stable, bicyclic analog of N^4-methoxy-2′-deoxycytidine that exists in both the imino and amino tautomeric forms

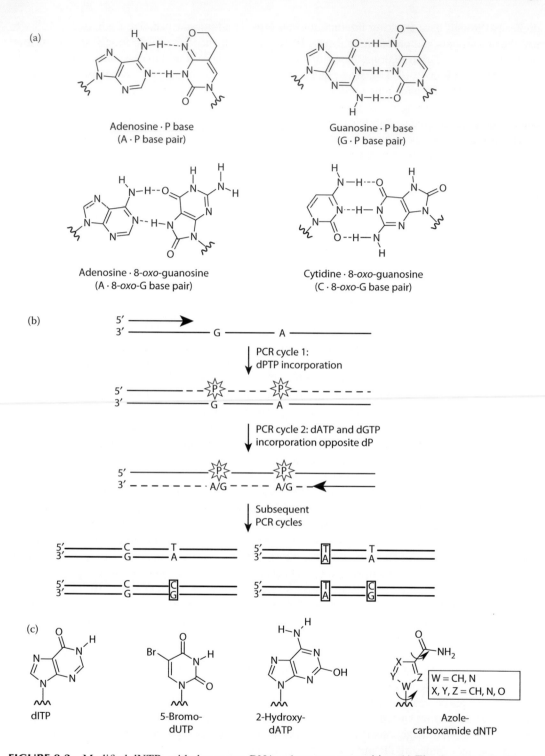

FIGURE 8.3 Modified dNTPs with degenerate DNA polymerase recognition. (a) The degenerate hydrogen bonding of P base with adenosine and guanosine results from tautomerism, while the hydrogen bonding of 8-*oxo*-guanosine with adenosine and cytidine results from conformation changes. (b) Hypothetical scheme for the generation of $G \cdot C \rightarrow A \cdot T$ and $A \cdot T \rightarrow G \cdot C$ transition mutations by the use of dPTP in PCR. The generated mutations are indicated by a rectangular box. (c) Other modified dNTPs with degenerate DNA polymerase recognition. Only the nucleobase portion of the dNTP is shown.

(5% and 95% of the time, respectively), allowing for formation of stable base pairs with adenosine and guanosine (Figure 8.3a). In Figure 8.3b, the concept of a convertible nucleobase is demonstrated, as dPTP can be incorporated opposite guanosine and can be coded for dATP incorporation. Similarly, dPTP can be incorporated opposite adenosine and can be coded for dGTP incorporation. The result of several PCR thermal cycles of incorporation and recognition of the dP nucleobase is a series of transition mutations: $A \cdot T \rightarrow G \cdot C$ and $G \cdot C \rightarrow A \cdot T$. The second mutagenic analog, 8-*oxo*-2′-deoxyguanosine, does not display tautomeric ambiguity but imparts degeneracy in base pairing through changes in the conformation about the glycosidic bond (syn $\leftrightarrow$ anti). When in the *anti* conformation, normal Watson–Crick base pairing between 8-*oxo*-guanosine and cytidine can occur, while in the *syn* conformation, 8-*oxo*-guanosine can form a Hoogsteen base pair with adenosine (Figure 8.3a). While dPTP could be substituted for dTTP or dCTP in PCR without significant influences on yield, complete substitution of 8-*oxo*-dGTP for dGTP was not possible.[14] Therefore, Brown and coworkers performed random PCR mutagenesis using dPTP and/or 8-*oxo*-dGTP in combination with the four natural nucleotides dATP, dCTP, dGTP, and dTTP.[14]

Brown's random mutagenesis PCR experiments demonstrated that when dPTP or 8-*oxo*-dGTP was added in an equimolar ratio to a mixture of natural nucleotides, the number of mutations was directly dependent on the number of PCR cycles, thus suggesting the means of controlling the extent of mutagenesis. When dPTP was used in PCR, both types of transition mutation are seen ($A \cdot T \rightarrow G \cdot C$ and $G \cdot C \rightarrow A \cdot T$ in approximately 5:1 ratio),[14] where the tendency for the $A \cdot T \rightarrow G \cdot C$ transition mutation is because the tautomeric equilibrium of dP is predominantly in the amino form. When 8-*oxo*-dGTP was included in the reaction, two predominant types of transversion mutations were identified ($A \cdot T \rightarrow C \cdot G$ and $T \cdot A \rightarrow G \cdot C$ in approximately 2:3 ratio), and although the number of mutations increased proportionally with the number of PCR cycles, the rate of mutation accumulation was slower than for reactions containing dPTP.[13,14] When both dPTP and 8-*oxo*-dGTP were employed, 6%, 11%, and 19% mutagenesis rates were obtained after 10, 20, and 30 cycles, respectively. Interestingly, this combination of mutagenic bases did not produce the same mutation levels as when dPTP and 8-*oxo*-dGTP were used in PCR alone. Although both transition and transversion mutations were produced, the $A \cdot T \rightarrow G \cdot C$ transition mutation was predominant, due to more efficient dPTP incorporation. As suggested by Brown and colleagues, this bias can possibly be solved by adjusting the concentrations of dPTP and 8-*oxo*-dGTP.

It is also feasible to broaden the mutational spectrum in PCR by using alternate mutagenic analogs (Figure 8.3c). Although dITP can be substituted for dGTP to reduce the number of hydrogen bonds in a $G \cdot C$ pair (Figure 8.2b),[9] it can also be used in PCR in combination with manganese chloride and a low concentration of dGTP to produce transition mutations ($A \cdot T \rightarrow G \cdot C$ and $G \cdot C \rightarrow A \cdot T$ in approximately 2:1 ratio) and a low level of transversion mutations.[15] 5-Bromo-dUTP (5-Br-dUTP) is another notable analog that causes both types of transition mutation to occur when used in PCR.[16] When 2-hydroxy-dATP (2-OH-dATP) is added to an equal mixture of all four dNTPs in PCR, a low level of mutations is accumulated, with transition mutations ($A \cdot T \rightarrow G \cdot C$) and transversion mutations ($A \cdot T \rightarrow C \cdot G$ and $G \cdot C \rightarrow T \cdot A$) being the major events.[17] The azole carboxamides are a class of nucleobases that undergo conformational rotation about the glycosidic and carboxamide bonds to present base pairing ambiguity.[18–20] Although the use of this class of analogs in PCR has yet to be described, notable members of this group result in two- to three-fold degeneracy in their incorporation and recognition. Overall, the use of degenerate dNTPs for random mutagenesis is a powerful tool. While there are many candidate analogs that show promise, none is completely degenerate, presenting a unique set of chemical and biochemical challenges for the years to come.

8.1.3 NOVEL BASE PAIRS WHICH CAN BE REPLICATED DURING PCR

The design of new, unnatural base pairs that can be introduced into DNA has raised interest among scientists due to their potential use in expanding the genetic alphabet. For an unnatural base pair $(X \cdot Y)$ to function in DNA replication, it must be orthogonal to the existing base pairs ($A \cdot T$ and

G · C), meaning that the X · Y pair should replicate with high specificity without formation of "mispairs" with the natural nucleobases in DNA. In the context of PCR, frequent mispairs of an unnatural nucleobase (X or Y) with any of the natural nucleobases (A, C, G, or T) can be detrimental, as the genetic code has the potential to be steadily altered over the course of several cycles of amplification. Another important requirement for the use of orthogonal base pairs in PCR is their ability to be replicated with efficiency similar to the natural base pairs, without termination of replication after nucleotide incorporation (dXTP or dYTP).[21,22] Herein, we will discuss several successful unnatural base pairs that have been designed and utilized in PCR, with focus on the classic isoG · isoC base pair designed by Benner and coworkers.[23]

Several different laboratories have designed new, orthogonal base pairs using tactics that include varying the hydrogen-bonding patterns within the basic purine and pyrimidine frameworks (Figure 8.4a). One of the earliest designs of an altered hydrogen-bonding pattern was the isocytosine · isoguanine (isoG · isoC) base pair described by the Benner laboratory.[23] Although this base pair is unique and differs from the natural base pairs, there are problems with specificity, as isodGTP can be incorporated across from thymidine when the O_2 oxygen is in the enol tautomeric form.[24] Interestingly, when isoguanosine is in the enol form, it is equivalent to the 2-hydroxy-adenosine analog (Figure 8.3c) that was described in the previous section for use in random mutagenesis schemes.[25] The degeneracy of isoguanosine can be overcome by the use of a slightly modified version of dTTP, 2-thio-dTTP, allowing for successful use in PCR.[22,26,27] Many unnatural base pairs with varied hydrogen bonding have since been described, with noteworthy analogs including X · K (xanthosine · 2,4-diaminopyrimidine) and Z · P (6-amino-5-nitro-3-(1′-β-2′-deoxyribofuranosyl)-2(1*H*)-pyridone · 2-amino-8-(1′-β-D-2′-deoxyribofuranosyl)-imidazo[1,2-*a*]-1,3,5-triazin-4(8*H*)-one) as shown in Figure 8.4a.[22,28,29]

The concept that defined that hydrogen bonding was required to form a base pair was challenged by the Kool laboratory, which designed a 4-methyl-benzimidazole · difluorotoluene (Z · F) pair (Figure 8.4a) that relied on shape complementarity and hydrophobic interactions, rather than hydrogen bonding.[30] Although this unnatural base pair was not orthogonal and was rather a variation on an A · T base pair, it expanded the understanding of a base pair to include forces other than hydrogen bonding. Subsequently, the PICS · PICS and 7AI · 7AI hydrophobic base pairs (Figure 8.4a) were developed by Romesberg et al.[31] While these analogs could be incorporated opposite one another with high specificity, further DNA polymerase extension was inhibited. These findings led Romesberg to evaluate several further generations of hydrophobic nucleobase analogs, utilizing large-scale screening approaches and DNA polymerase evolution to identify a successful orthogonal base pair.[32] Their most promising unnatural base pair, described to date, is the 5SICS · NaM pair (Figure 8.4a), which can be efficiently and specifically replicated in PCR.[33] Hirao and coworkers have developed another notable unnatural base pair, which is based on specific shape-based complementarity and is named the Ds · Px (7-(2-thienyl)-imidazo[4,5-b]pyridine · 2-nitro-4-propynylpyrrole) pair (Figure 8.4a).[34] The Ds · Px pair could be utilized in PCR, using Deep Vent (exo+) DNA polymerase, with >99.9% fidelity per cycle. These proof-of-principle studies demonstrate the feasibility of replicating nonnatural base pairs in PCR and pave the way for more advanced applications.

There are two major applications for the use of orthogonal base pairs: Nucleic acid detection and the translation of proteins containing nonstandard amino acids. One notable approach to the detection of PCR products in qPCR has been to incorporate one of the unnatural nucleobases (X) near the 5′-end of a PCR primer. Once the PCR primer is successfully extended to make a complementary copy of the nucleic acid target, then the resultant strand can serve as a template where the complementary unnatural nucleobase (dYTP) can be incorporated opposite X (Figure 8.4b). Variations in this basic approach have applied an isoC-modified primer containing a fluorophore and a isodGTP analog containing a Dabcyl quencher for use in qPCR detection and include the commercialized Plexor (Promega) and Multicode (Eragen) technologies.[35,36] Another notable variation with applications in multiplexed, nested PCR schemes employs dP-containing PCR primers, with dZTP incorporation (Figure 8.4a) allowing for successful replication.[29] The ability to amplify a sequence containing an

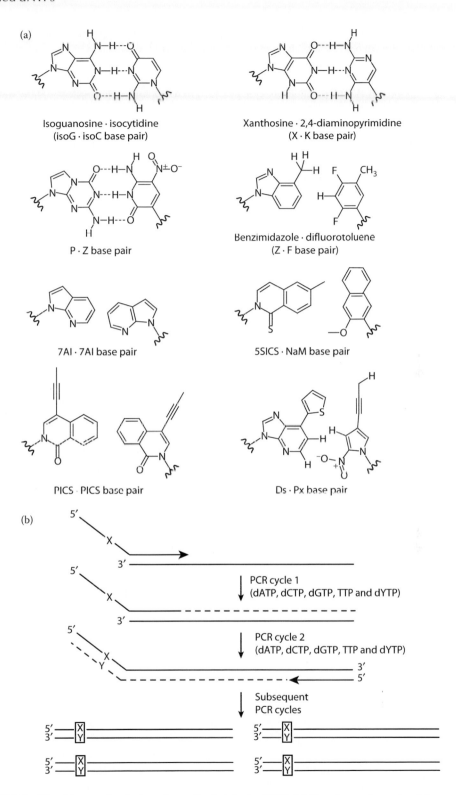

FIGURE 8.4 Novel base pairs that can be replicated during PCR. (a) Structures of unnatural base pairs that can be recognized by DNA polymerases in PCR. (b) Scheme for the introduction of replicable orthogonal base pair (X · Y) by the use of a modified PCR primer. The unnatural base pair is indicated by a rectangular box.

orthogonal base pair using PCR allows for subsequent *in vitro* transcription of the resultant PCR product and provides an RNA template containing an unnatural base for use in translation.

While the process of transcription and translation has yet to be applied *in vivo*, it is worth noting that PCR amplification and *in vitro* transcription using the 5SICS · NaM unnatural base pair have recently been achieved.[33,37] In these studies, a 134-bp DNA duplex with a 5SICS · NaM unnatural base pair was successfully amplified using PCR containing the dNTP analogs of dA, dC, dG, dT, d5SICS, and dNaM in combination with Deep Vent DNA polymerase.[33] The fidelity for replication of the 5SICS · NaM unnatural base pair in PCR was found to be greater than 99%. In further studies, 35 nucleotide transcription templates were prepared that contained a single d5SICS or dNaM modification. These sequences were used as templates for *in vitro* transcription using T7 RNA polymerase in the presence of rNTP versions of A, C, G, U, 5SICS, and NaM.[37] Results indicate that there was selective incorporation of the NaM analog when using templates containing d5SICS, and selective incorporation of the rNTP analog of 5SICS with dNaM-containing templates. Furthermore, from measurement of the kinetics of replication, it is clear that there is only a modest (~20-fold) decrease in the rate of transcript formation relative to reactions containing only natural rNTPs. The successful and selective replication of the 5SICS · NaM unnatural base pair by both DNA and RNA polymerases can be used in a number of *in vitro* applications that take advantage of an expanded genetic alphabet. By increasing the number of nucleobases in RNA from four to six, unnatural base codons are generated that have the potential for use in translation schemes that direct the incorporation of unnatural amino acids to produce proteins with novel functions.[38]

8.1.4   MODIFIED dNTPs FOR INTRODUCTION OF FUNCTIONALITIES DURING PCR

A convenient way of introducing functional motifs into a sequence of interest is via modification of the DNA with chemically altered dNTPs. This approach has wide applicability to the fields of molecular biology and nanotechnology. Applications that employ DNA modification are numerous and include fluorescent *in situ* hybridization (FISH), gene-expression quantification, single-nucleotide polymorphism (SNP) analysis, DNA sequencing, and molecular diagnostics.[39,40] While chemical modifications can be introduced into sequences via solid-phase oligonucleotide synthesis, the preparation of long sequences with functional tags can often be challenging. Therefore, the enzymatic incorporation of modified dNTPs provides a viable alternative to introduce functionality. The commonly used techniques for the enzymatic modification of nucleic acids of unknown samples include nick translation and random labeling.[39,40] In contrast, PCR amplification in the presence of modified dNTPs allows for the modification of a specific sequence of duplex DNA.[41] The incorporation of functionality during PCR can be as easy and convenient as including the corresponding modified dNTPs in the reaction. However, the modified dNTPs must be good DNA polymerase substrates to achieve high-density functionalization without sacrificing amplicon yield.[39,40] The common considerations for efficient PCR modification include DNA polymerase choice, the extent of substitution of the modified dNTP for its natural counterpart, and the chemical structure of the modified dNTP.[40–42]

While *Taq* DNA polymerase is routinely used in PCR, DNA polymerases from archaeal sources show significant promise for improved incorporation of modified dNTPs. Although modified dNTPs are generally incorporated less efficiently than their natural counterparts, the use of Vent$_R$™ (exo-) DNA polymerase allows for efficient incorporation of both biotin- and fluorophore-modified dNTPs.[43] Other noteworthy DNA polymerases that provide robust amplicon formation in the presence of modified dNTPs include *Tgo* DNA polymerase[44] and KOD Dash DNA polymerase,[45] which allow for complete substitution of rhodamine-green-*X*-dUTP and fluorescein-modified dUTP, respectively, for dTTP. These findings were corroborated by a recent PCR study in which a variety of C5-modified dCTP and dUTP analogs were evaluated and it was found that *Pfu*, *Vent* (exo-), and *Pwo* DNA polymerases provided the best performance.[41] As all these studies demonstrated some differences in the preferences of DNA polymerases for modified substrates, it is advisable to screen a variety of DNA polymerases to determine the optimal assay performance.

Efficient recognition of modified dNTPs by DNA polymerases is influenced by characteristics such as the size and structure of the functional group, the charge of the analog, the position of attachment, and the type of nucleobase. The most efficiently incorporated functionalized dNTPs contain modifications to the major groove of the nucleobase, the 5-position of pyrimidines, and the 7-position of purines.[40] In addition to the site of modification, the flexibility and length of the linker arm between the nucleobase and the functional group is also important. dNTPs with more flexible linker arms, such as alkane and *cis*-alkenyl, are less suitable DNA polymerase substrates than analogs with more rigid linker arms, such as trans-alkenyl and alkynyl.[40,43] As discussed in Section 8.1.1, alkynyl linkers provide an added benefit of improving the stability of the resultant DNA duplex.[10,40] The length of the linker arm also plays a role in the incorporation of modified dNTPs. When considering the design of biotinylated-dNTP analogs, shorter linker arms (i.e., biotin-4-dUTP) are better DNA substrates; however, the association between biotin and streptavidin improves with linker arm length (i.e., biotin-11-dUTP or biotin-16-dUTP), necessitating a compromise in linker arm length to achieve efficient incorporation and detection.[35,46]

The introduction of functional groups into a DNA sequence by PCR can be achieved by direct labeling (Figure 8.5a) and indirect labeling (Figure 8.5b). When using direct labeling, modified dNTPs containing the desired functionality serve as DNA polymerase substrates, allowing for incorporation of the functional group during PCR. Indirect labeling first employs a dNTP derivative with a reactive moiety that is introduced enzymatically during PCR. After PCR, a secondary-labeling step is performed to introduce functional groups via chemical coupling or conjugation to the reactive moiety (Figures 8.5b,c). While direct labeling approaches are desirable due to their inherent simplicity, indirect labeling approaches are used when the dNTP analogs containing the desired functionality are poor DNA polymerase substrates.[39] One traditional chemical coupling approach to indirect labeling is to employ dNTP derivatives containing an amino group (such as aminoallyl and aminopropynyl) as the reactive moiety (Figure 8.5d). The amino group is then conjugated to an *N*-hydroxysuccinimide (NHS) ester of the desired functionality to form an amide linkage (Figure 8.5c), where the commonly used functionalities include dyes.[39] Secondary labeling can suffer from low-conjugation efficiencies, making quantitative modification difficult to achieve.[40] Although the use of chemical labeling reagents necessitates clean up and purification to remove any unconjugated NHS ester,[40] many manufacturers offer spin columns as integral components of kits for this purpose. A more recently described approach named "Click chemistry"[47] also shows promise for use in indirect labeling schemes because of the efficient and reliable reaction of azido and alkynyl moieties in the presence of copper (*I*).[39] For successful use in secondary labeling schemes, the dNTP is commonly modified with an alkynyl moiety and reacted after PCR with the azido-modified functionalities (Figures 8.5c,d).[39]

The use of modified dNTPs provides a powerful means for the incorporation of different functionalities into nucleic acids during PCR. For optimal performance in PCR, the modified dNTP needs to be incorporated efficiently, necessitating decisions to be made in the labeling approach (direct vs. indirect), the choice of DNA polymerase, linker type and length, and the extent of substitution for the natural counterpart. While common functionalities that are introduced include fluorophores, affinity handles such as biotin, and reactive groups for indirect labeling, the possibilities are endless and may include magnetic properties, electrical conductivity, or functionalities with strong antibody affinity.[39,48]

8.2 PROTOCOLS

8.2.1 BASIC PCR PROTOCOL

8.2.1.1 Equipment

Pipettes capable of dispensing volumes from <1 to 200 µL, benchtop centrifuge, thermal cycler, electrophoresis equipment, and ultraviolet (UV) transilluminator.

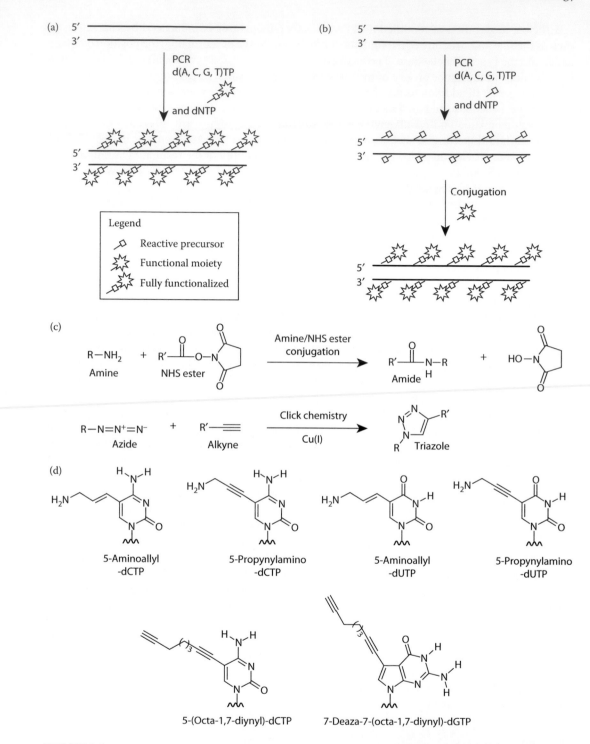

FIGURE 8.5 Introduction of functionality using PCR. (a) Direct labeling scheme for introduction of functionality. (b) Indirect labeling scheme for introduction of functionality. (c) Chemical coupling or conjugation schemes that are traditionally used in indirect labeling schemes. (d) Modified dNTPs that are routinely used in amine/NHS ester-mediated and "Click chemistry"-mediated indirect labeling schemes.

TABLE 8.1
Reaction Components for a Basic PCR Protocol

	Component	Final Concentration (in a 25-µL Reaction)	Master Mix Volume (µL) for One Reaction
i	PCR buffer (10×)	1×	2.5
ii	Magnesium chloride (50 mM)	2.5 mM	1.25
iii	Forward/reverse primer	50–500 nM	Variable
iv	dNTP mixture (10 mM of each nucleotide)	0.2 mM	0.5
v	*Taq* DNA polymerase (5 U/µL)	0.05 U/µL	0.25
	Sterile deionized water	Up to 20 µL	Up to 20
	Total volume (µL)	20	20

8.2.1.2 Supplies

Sterile filter pipette tips, sterile 1.5-mL screw-top microcentrifuge tubes, thin-walled 200-µL PCR tubes, dNTP mix (10 mM each of dATP, dCTP, dGTP, and dTTP), primer pair (cartridge purified), *Taq* DNA polymerase (5 U/µL), and supplied buffers (10× PCR buffer and 50 mM MgCl$_2$; Invitrogen), high-performance liquid chromatography (HPLC) water, and DNA template sample (variable source).
Method:

1. With the exception of the DNA polymerase, thaw all reaction components at room temperature or on ice, vortex to mix, centrifuge briefly, and store on ice.
2. Prepare a Master Mix containing all components except for the DNA template sample. Add each of the components as shown in Table 8.1 (multiply amounts by the number of reactions needed) into a 1.5-mL microcentrifuge tube, on ice.
3. Mix the Master Mix by pipetting up and down, pulse spinning if necessary to collect the sample at the bottom of the tube.
4. Aliquot 20 µL of the Master Mix into the appropriate number of 200-µL thin-walled PCR tubes, and add 5 µL of the DNA template sample to reach a final reaction volume of 25 µL.
5. Pulse spin the PCR tubes and place it into a thermal cycler with a heated lid.
6. Perform the following thermal cycling protocol:
   a. 95°C for 2–10 min
   b. [95°C for 30 s, 48–60°C for 30 s, and 72°C for 0.5–2 min] 25–50 cycles
   c. 72°C for 10 min
7. Analyze an aliquot of the completed reaction by agarose gel electrophoresis, with visualization on a transilluminator.

8.2.2 PCR Protocol I for Improved Specificity Using Hot Start dNTPs

The following alternate protocol provides a procedure for the use of Hot Start or 3′-THF dNTPs as described in Section 8.1.1 (also known as CleanAmp™ dNTPs) in PCR.

8.2.2.1 Additional Supplies

CleanAmp™ dNTP Mix [10 mM each of CleanAmp™ dATP, dCTP, dGTP, and dTTP (TriLink BioTechnologies, Inc, Cat. No. N-9501 or N-9506)].
Alternate method:

1. Follow the above basic PCR protocol in Section 8.2.1 with the following alterations:
2. Replace component **iv** from Table 8.1 with the CleanAmp™ dNTP Mix and add 1.0 µL per reaction, for a final concentration of 0.4 mM.

Important notes:

1. CleanAmp™ dNTPs should be thawed on ice or at room temperature. Do not heat the solution to thaw as it may result in loss of the CleanAmp™ modification.
2. In addition to *Taq*, CleanAmp™ dNTPs can be used with a number of DNA polymerases, allowing for conversion of a reaction into the corresponding Hot Start version.
3. Do not allow the CleanAmp™ dNTPs to stay at room temperature for more than 6 h.
4. For more challenging templates, the CleanAmp™ dNTP concentration can be increased in 0.2 mM increments, adding an additional 1 mM of magnesium chloride for each 0.2 mM increment.
5. For more information on CleanAmp™, dNTPs, and for use in other types of protocols, see http://www.trilinkbiotcch.com/products/cleanamp/dntps.asp.

8.2.3   PCR PROTOCOL II FOR IMPROVED SPECIFICITY USING 7-dEAZA-dGTP FOR AMPLIFICATION OF GC-RICH TARGETS

The use of 7-deaza-dGTP enables amplification of sequences with high GC content. There are many alternate protocols to consider when using 7-deaza-dGTP. Some require additives such as betaine or DMSO and others require altered thermal cycling protocols. The following is an example of a protocol employing 7-deaza-dGTP with altered PCR thermal cycling parameters called "Slowdown PCR."[8] This protocol accompanies Section 8.1.1.

8.2.3.1   Additional Supplies

7-Deaza-dGTP (available from numerous sources including TriLink Biotechnologies, Inc [Cat. No. N-1044], Jena Bioscience [Cat. No. NU-1179], and Roche Applied Science [Cat. No. 10 988 537 001]).

Alternate method: Follow the above basic PCR protocol in Section 8.2.1 with the following alterations:

1. Include magnesium chloride (component **ii** from Table 8.1) to a 1.5 mM final concentration, adding 0.75 µL per reaction.
2. Replace component **iv** from Table 8.1 with the following three components:
   a. Add 0.5 µL of a 10 mM solution of d(A, C, T)TP mix to reach a final concentration of 0.2 mM
   b. Add 0.125 µL of a 10 mM solution of dGTP to reach a final concentration of 0.05 mM
   c. Add 0.375 µL of a 10 mM solution of 7-deaza-dGTP to reach a final concentration of 0.15 mM
3. Replace the thermal cycling protocol from step 6 with the following touchdown PCR protocol:
   a. 95°C for 5 min
   b. [95°C for 30 s, 70–53°C for 30 s, and 72°C for 40 s] 48 cycles, decreasing the annealing temperature by 1°C over the range of 70–53°C.
   c. [95°C for 30 s, 58°C for 30 s, and 72°C for 40 s] 15 cycles

Important notes:

1. Set the heating ramp rate to 2.5°C/s and the cooling ramp rate for reaching the annealing temperature to 1.5°C/s.
2. dGTP should be in a 1:3 ratio with 7-deaza-dGTP.[49]
3. Although *Taq* DNA polymerase was successful in the Slowdown PCR technique, other polymerases may not work with this procedure.

4. Interestingly, we have found that the use of a CleanAmp™ version of 7-deaza-dGTP in conjunction with the other CleanAmp™ dNTPs (CleanAmp™ 7-deaza-dGTP Mix [TriLink Biotechnologies: Cat. No. N-9504]) was able to amplify sequences with up to 80% GC-rich content under similar thermal cycling conditions as described in Protocol B, with the exception of the need to use a shorter, 1-s annealing step.

8.2.4 MUTAGENIC PCR PROTOCOL

Random mutations can be introduced by including the mutagenic bases dPTP and 8-*oxo*-dGTP into a slightly modified PCR protocol to achieve both transition and transversion mutations. The number of mutations (rate of mutagenesis) can be easily controlled by varying the number of PCR cycles (5–30 cycles).[14] The rate of mutagenesis is dependent on the number of PCR cycles and when using *Taq* DNA polymerase, a 6%, 11%, and 19% mutagenesis rate is achievable after 10, 20, and 30 cycles, respectively, when both dPTP and 8-*oxo*-dGTP are used. The following protocol accompanies Section 8.1.2.

8.2.4.1 Additional Supplies

Mutagenic nucleotides dPTP and 8-*oxo*-dGTP can be purchased individually (from TriLink Biotechnologies [Cat. No. N-2037 and N-1066, respectively] and Jena Bioscience [Cat. No. NU-1119 and NU-1117, respectively]), as a dNTP mutagenesis mix (TriLink Biotechnologies [Cat. No. N-2501]), or as a dNTP mutagenesis kit (Jena Bioscience [Cat. No. PP-101]).

Alternate method: Follow the basic PCR protocol in Section 8.2.1 with the following alterations:

1. Reduce the amount of magnesium chloride (component **ii** from Table 8.1) to a 2.0 mM final concentration, adding only 1.0 µL per reaction.
2. Employ the forward/reverse primer (component **iii** from Table 8.1) at 500 nM concentration.
3. Use the 10 mM dNTP mixture (component **iv** from Table 8.1) at 0.5 mM final concentration, adding 1.25 µL per reaction.
4. Include both 8-*oxo*-dGTP and dPTP at a 0.5 mM concentration, adding 1.25 µL of each nucleotide per reaction from 10 mM stock solutions.
5. Add 10 fmol of DNA template to each reaction.
6. Replace the thermal cycling protocol from step 6 with the following protocol: (92°C for 1 min, 55°C for 1.5 min, and 72°C for 5 min) for 10–30 cycles, depending on the desired mutagenesis rate.
7. Prior to cloning and any subsequent steps, perform a second PCR following the same protocol as in steps 1–7 with the following exceptions:
   a. Omit 8-*oxo*-dGTP and dPTP from the setup (step 5).
   b. Utilize 1 µL of the product from steps 1–7 as the template.
   c. Thermal cycle as in step 7, performing 20–30 cycles.

Important notes:

1. To eliminate nonnatural nucleobases from the target product, the PCR product should be subjected to another PCR protocol containing only natural nucleotides (see step 7).
2. If desired, further fine tuning of the rate of mutagenesis and/or mutational bias could potentially be achieved through modulating the amounts (concentration and ratio) of mutagenic nucleotides.[14]

8.2.5 PCR PROTOCOL WITH NOVEL BASE PAIRS

One of the most straightforward ways to replicate an unnatural base pair (X · Y) in PCR is to introduce one of the analogs (X) into a PCR primer and to include dYTP into your PCR protocol

(Figure 8.4b). As described in Section 8.1.3, this basic approach has been described for replication of DNA containing isoC · isoG and P · Z base pairs.[29,35,36] Herein, we will describe a PCR protocol using isoC and isoG.

8.2.5.1   Additional Supplies

Oligonucleotides containing isodC (prepared using the corresponding phosphoramidite [available from Glen Research: 5-Me-isodC-CE phosphoramidite Cat. No.10-1067]) can be synthesized by several companies including TriLink Biotechnologies, Integrated DNA Technologies, and Eurogentec. IsodGTP is available from Chemgenes Corporation (Cat. No. NTP-4396).

Alternate method: Follow the above basic PCR protocol in Section 8.2.1 with the following alterations:

1. Substitute one of the primers (component **iii** from Table 8.1) with an isodC-containing primer, and use both primers at a final concentration of 10–400 nM.
2. Include isodGTP at 0.02–0.1 mM final concentration.
3. The amount of *Taq* DNA polymerase (component **v** from Table 8.1) can be varied from 1.25 to 5 units per reaction.

Important notes:

1. To improve the specificity of amplification of isodC and isodG in PCR, dTTP can be substituted with 2-thio-dTTP to avoid mismatch formation between isodG and dT.
2. Variations in this basic approach have applied a fluorophore-containing iso-dC-modified primer and a isodGTP analog containing a Dabcyl quencher for use in real-time detection and include the commercialized Plexor (Promega) and Multicode (Eragen) technologies.[35,36]
3. Another variation of this approach can employ a synthetic template that includes one or more of the unnatural nucleotides (*X* or *Y*) to replicate a six-base alphabet.

8.2.6   Protocol for Introduction of Functional Groups during PCR

In this protocol, for introduction of functional groups, a general strategy for the use of modified dNTPs in PCR (Section 8.1.4) will be described. Although complete substitution of a natural dNTP for its modified counterpart is possible in PCR, there are many cases where complete substitution can cause partial or complete inhibition of amplicon formation. To compensate for the lower efficiency of incorporation of the modified dNTP relative to its natural counterpart, the modified dNTP should only partially be substituted for its natural counterpart. In this protocol, a strategy to find the optimal percentage of modified dNTP substitution will be described. This example uses biotin-16-aminoallyl-dUTP in a direct labeling approach.[50]

8.2.6.1   Additional Supplies

Deoxynucleotide Solution Set (New England Biolabs, Cat No. N0446S), biotin-16-aminoallyl-2′-deoxyuridine-5′-triphosphate (TriLink Biotechnologies, Inc., Cat No. N-5001, Roche Applied Science, Cat No. 11 093 070 910).

Alternate method: Follow the above basic PCR protocol in Section 8.2.1 with the following alterations:

1. Instead of the equimolar dNTP mix (component **iv** in Table 8.1) used in the protocol in Section 8.2.1, prepare five dNTP mixes in which the percent of biotin-16-AA-dUTP is titrated from 100% to 0% relative to dTTP:
   a. Mix 1 (0% substitution); the concentration of dATP, dCTP, dGTP, and dTTP are each at 10 mM.

b. Mix 2 (25% substitution); the concentration of dATP, dCTP, and dGTP are each at 10 mM, the concentration of dTTP is at 7.5 mM, and the concentration of biotin-16-AA-dUTP is at 2.5 mM.

c. Mix 3 (50% substitution); the concentration of dATP, dCTP, and dGTP are each at 10 mM, the concentration of dTTP is at 5.0 mM, and the concentration of biotin-16-AA-dUTP is at 5.0 mM.

d. Mix 4 (75% substitution); the concentration of dATP, dCTP, and dGTP are each at 10 mM, the concentration of dTTP is at 2.5 mM, and the concentration of biotin-16-AA-dUTP is at 7.5 mM.

e. Mix 5 (100% substitution); the concentration of dATP, dCTP, and dGTP are each at 10 mM, and the concentration of biotin-16-AA-dUTP is at 10 mM.

2. For each dNTP mix prepared in step 2, prepare a Master Mix and set up PCR experiments as described in steps 1–7 from 1a, substituting the dNTP mixes (1–5) from above for the dNTP mixture (**iv** in Table 8.1).

3. After visualizing the gel on a transilluminator, use densitometry to determine the percentage of substitution that will allow for efficient amplicon formation without sacrificing amplicon yield.

Important notes:

1. 100% biotin-16-AA-dUTP substitution for dTTP causes complete reaction inhibition.[50]

2. As the percentage of biotin-16-AA-dUTP increases, a decrease in the mobility of the amplicon is evident by agarose gel analysis.

3. Biotinylated dNTPs can be incorporated by both reverse transcriptases and DNA polymerases.

4. **Protocol IIF** can be applied to indirect labeling schemes. When aminoallyl- or propynyl-amino-modified dNTPs are used, the resultant PCR amplicon can be conjugated with the desired NHS ester. Refer to the manufacturers of NHS esters for efficient conjugation and purification procedures.

8.3 CONCLUDING REMARKS

The use of modified dNTPs in PCR provides a great deal of flexibility in molecular biology and chemical biology. Although work with modified dNTPs is steadily expanding, many applications that can benefit from the use of modified dNTPs have potentials that have yet to be fully realized. The application of specificity analogs can improve the performance of PCR, allowing for enhanced sequence detection, with the potential for improved performance in cloning and genotyping applications. Degenerate nucleotide analogs can not only be used in protein evolution schemes, but they can also be used in introducing a controlled rate of mutations into a sequence of interest for use in *in vitro* evolution and aptamer selection. The novel base pairs are not only of interest from a basic science point of view, but they also offer orthogonal, replicatable base pairs with a variety of uses in nucleic acid detection and protein translation studies. The ability to introduce functionality into a nucleic acid target by slightly modifying the basic structure of the nucleobase of the dNTP allows a controlled rate of modification, simply by varying the extent of substitution for its natural counterpart. This allows a means of introducing chemical tags that allow for advanced functionalization of a DNA duplex. These tags can be used for detection or for introduction of enhanced properties in applications such as nanotechnology and aptamer selection schemes. While the possibilities for the use of modified dNTPs in PCR are seemingly endless, it is interesting to think about what may be achievable if applied to other nucleic acid replication formats such as *in vitro* transcription, next-generation sequencing, and nucleic acid detection.

ACKNOWLEDGMENTS

We thank R. Hogrefe and M. Crane for helpful suggestions and for critical reading of the manuscript.

REFERENCES

1. Saiki, R.K., Gelfand, D.H., Stoffel, S., Scharf, S.J., Higuchi, R., Horn, G.T., Mullis, K.B., and Erlich, H.A. 1988. Primer-directed enzymatic amplification of DNA with a thermostable DNA polymerase. *Science*, **239**, 487–491.
2. Saenger, W. 1984. *Principles of Nucleic Acid Structure*. Springer, Berlin.
3. Chou, Q., Russell, M., Birch, D.E., Raymond, J., and Bloch, W. 1992. Prevention of pre-PCR mis-priming and primer dimerization improves low-copy-number amplifications. *Nucleic Acids Res*, **20**, 1717–1723.
4. Bonner, A.G. 2003. Reversible chemical modification of nucleic acids and improved method for nucleic acid hybridization. U.S. Patent No. US2003162199 A1.
5. Koukhareva, I. and Lebedev, A. 2009. 3′-Protected 2′-deoxynucleoside 5′-triphosphates as a tool for heat-triggered activation of polymerase chain reaction. *Anal Chem*, **81**, 4955–4962.
6. Le, T. and Paul, N. 2009. Improved PCR flexibility with Hot Start dNTPs. *Biotechniques*, **81**, 880–881.
7. Frey, U.H., Bachmann, H.S., Peters, J., and Siffert, W. 2008. PCR-amplification of GC-rich regions: Slowdown PCR. *Nat Protoc*, **3**, 1312–1317.
8. Musso, M., Bocciardi, R., Parodi, S., Ravazzolo, R., and Ceccherini, I. 2006. Betaine, dimethyl sulfoxide, and 7-deaza-dGTP, a powerful mixture for amplification of GC-rich DNA sequences. *J Mol Diagn*, **8**, 544–550.
9. Suspene, R., Renard, M., Henry, M., Guetard, D., Puyraimond-Zemmour, D., Billecocq, A., Bouloy, M., Tangy, F., Vartanian, J.P., and Wain-Hobson, S. 2008. Inversing the natural hydrogen bonding rule to selectively amplify GC-rich ADAR-edited RNAs. *Nucleic Acids Res*, **36**, e72.
10. Kutyavin, I.V. 2008. Use of base-modified duplex-stabilizing deoxynucleoside 5′-triphosphates to enhance the hybridization properties of primers and probes in detection polymerase chain reaction. *Biochemistry*, **47**, 13666–13673.
11. Bergstrom, D.E. 2009. Unnatural nucleosides with unusual base pairing properties. *Curr Protoc Nucleic Acid Chem*, **37**, 1.4.1–1.4.32.
12. Suzuki, T., Moriyama, K., Otsuka, C., Loakes, D., and Negishi, K. 2006. Template properties of mutagenic cytosine analogues in reverse transcription. *Nucleic Acids Res*, **34**, 6438–6449.
13. Zaccolo, M. and Gherardi, E. 1999. The effect of high-frequency random mutagenesis on *in vitro* protein evolution: A study on TEM-1 beta-lactamase. *J Mol Biol*, **285**, 775–783.
14. Zaccolo, M., Williams, D.M., Brown, D.M., and Gherardi, E. 1996. An approach to random mutagenesis of DNA using mixtures of triphosphate derivatives of nucleoside analogues. *J Mol Biol*, **255**, 589–603.
15. Spee, J.H., de Vos, W.M., and Kuipers, O.P. 1993. Efficient random mutagenesis method with adjustable mutation frequency by use of PCR and dITP. *Nucleic Acids Res*, **21**, 777–778.
16. Ma, X., Ke, T., Mao, P., Jin, X., Ma, L., and He, G. 2008. The mutagenic properties of BrdUTP in a random mutagenesis process. *Mol Biol Rep*, **35**, 663–667.
17. Kamiya, H., Ito, M., and Harashima, H. 2004. Induction of transition and transversion mutations during random mutagenesis PCR by the addition of 2-hydroxy-dATP. *Biol Pharm Bull*, **27**, 621–623.
18. Adelfinskaya, O., Nashine, V.C., Bergstrom, D.E., and Davisson, V.J. 2005. Efficient primer strand extension beyond oxadiazole carboxamide nucleobases. *J Am Chem Soc*, **127**, 16000–16001.
19. Hoops, G.C., Zhang, P., Johnson, W.T., Paul, N., Bergstrom, D.E., and Davisson, V.J. 1997. Template directed incorporation of nucleotide mixtures using azole-nucleobase analogs. *Nucleic Acids Res*, **25**, 4866–4871.
20. Paul, N., Nashine, V.C., Hoops, G., Zhang, P., Zhou, J., Bergstrom, D.E., and Davisson, V.J. 2003. DNA polymerase template interactions probed by degenerate isosteric nucleobase analogs. *Chem Biol*, **10**, 815–825.
21. Herdewijn, P. 2008. *Modified Nucleosides: In Biochemistry, Biotechnology, and Medicine. Part I, 3.3.2.* Wiley-VCH, Weinheim.
22. Hirao, I. 2006. Unnatural base pair systems for DNA/RNA-based biotechnology. *Curr Opin Chem Biol*, **10**, 622–627.

23. Switzer, C., Moroney, S.E., and Benner, S.A. 1989. Enzymatic incorporation of a new base pair into DNA and RNA. *J Am Chem Soc*, **111**, 8322–8323.

24. Maciejewska, A.M., Lichota, K.D., and Kusmierek, J.T. 2003. Neighbouring bases in template influence base-pairing of isoguanine. *Biochem J*, **369**, 611–618.

25. Kamiya, H., Suzuki, A., Kawai, K., Kasai, H., and Harashima, H. 2007. Effects of 8-hydroxy-GTP and 2-hydroxy-ATP on *in vitro* transcription. *Free Radic Biol Med*, **43**, 837–843.

26. Johnson, S.C., Sherrill, C.B., Marshall, D.J., Moser, M.J., and Prudent, J.R. 2004. A third base pair for the polymerase chain reaction: Inserting isoC and isoG. *Nucleic Acids Res*, **32**, 1937–1941.

27. Sismour, A.M. and Benner, S.A. 2005. The use of thymidine analogs to improve the replication of an extra DNA base pair: A synthetic biological system. *Nucleic Acids Res*, **33**, 5640–5646.

28. Lutz, M.J., Held, H.A., Hottiger, M., Hubscher, U., and Benner, S.A. 1996. Differential discrimination of DNA polymerase for variants of the non-standard nucleobase pair between xanthosine and 2,4-diamino-pyrimidine, two components of an expanded genetic alphabet. *Nucleic Acids Res*, **24**, 1308–1313.

29. Yang, Z., Chen, F., Chamberlin, S.G., and Benner, S.A. 2010. Expanded genetic alphabets in the polymerase chain reaction. Angewandte Chemie Int Ed, **49**, 177–180.

30. Morales, J.C. and Kool, E.T. 2000. Functional hydrogen-bonding map of the minor groove binding tracks of six DNA polymerases. *Biochemistry*, **39**, 12979–12988.

31. McMinn, D.L., Ogawa, A.K., Wu, Y., Liu, J., Schultz, P.G., and Romesberg, F.E. 1999. Efforts toward expansion of the genetic alphabet: DNA polymerase recognition of a highly stable, self-pairing hydrophobic base. *J Am Chem Soc*, **121**, 11585–11586.

32. Leconte, A.M., Chen, L., and Romesberg, F.E. 2005. Polymerase evolution: Efforts toward expansion of the genetic code. *J Am Chem Soc*, **127**, 12470–12471.

33. Malyshev, D.A., Seo, Y.J., Ordoukhanian, P., and Romesberg, F.E. 2009. PCR with an expanded genetic alphabet. *J Am Chem Soc*, **131**, 14620–14621.

34. Kimoto, M., Kawai, R., Mitsui, T., Yokoyama, S., and Hirao, I. 2009. An unnatural base pair system for efficient PCR amplification and functionalization of DNA molecules. *Nucleic Acids Res*, **37**, e14.

35. Moser, M.J., Marshall, D.J., Grenier, J.K., Kieffer, C.D., Killeen, A.A., Ptacin, J.L., Richmond, C.S. et al. 2003. Exploiting the enzymatic recognition of an unnatural base pair to develop a universal genetic analysis system. *Clin Chem*, **49**, 407–414.

36. Nolte, F.S., Marshall, D.J., Rasberry, C., Schievelbein, S., Banks, G.G., Storch, G.A., Arens, M.Q., Buller, R.S., and Prudent, J.R. 2007. MultiCode-PLx system for multiplexed detection of seventeen respiratory viruses. *J Clin Microbiol*, **45**, 2779–2786.

37. Seo, Y.J., Matsuda, S., and Romesberg, F.E. 2009. Transcription of an expanded genetic alphabet. *J Am Chem Soc*, **131**, 5046–5047.

38. Wang, L. and Schultz, P.G. 2002. Expanding the genetic code. *Chem Commun (Camb)*, **7**, 1–11.

39. Gierlich, J., Gutsmiedl, K., Gramlich, P.M., Schmidt, A., Burley, G.A., and Carell, T. 2007. Synthesis of highly modified DNA by a combination of PCR with alkyne-bearing triphosphates and click chemistry. *Chemistry*, **13**, 9486–9494.

40. McGall, G.H. 2005. In Vaghefi, M. (ed.), *Nucleoside Triphosphates and Their Analogs: Chemistry, Biotechnology, and Biological Applications*. 1st Edition, Chapter 10. CRC Press, Boca Raton, FL, pp. 270–316.

41. Kuwahara, M., Nagashima, J., Hasegawa, M., Tamura, T., Kitagata, R., Hanawa, K., Hososhima, S., Kasamatsu, T., Ozaki, H., and Sawai, H. 2006. Systematic characterization of 2′-deoxynucleoside-5′-triphosphate analogs as substrates for DNA polymerases by polymerase chain reaction and kinetic studies on enzymatic production of modified DNA. *Nucleic Acids Res*, **34**, 5383–5394.

42. Giller, G., Tasara, T., Angerer, B., Muhlegger, K., Amacker, M., and Winter, H. 2003. Incorporation of reporter molecule-labeled nucleotides by DNA polymerases. I. Chemical synthesis of various reporter group-labeled 2′-deoxyribonucleoside-5′-triphosphates. *Nucleic Acids Res*, **31**, 2630–2635.

43. Tasara, T., Angerer, B., Damond, M., Winter, H., Dorhofer, S., Hubscher, U., and Amacker, M. 2003. Incorporation of reporter molecule-labeled nucleotides by DNA polymerases. II. High-density labeling of natural DNA. *Nucleic Acids Res*, **31**, 2636–2646.

44. Foldes-Papp, Z., Angerer, B., Ankenbauer, W., and Rigler, R. 2001. Fluorescent high-density labeling of DNA: Error-free substitution for a normal nucleotide. *J Biotechnol*, **86**, 237–253.

45. Obayashi, T., Masud, M.M., Ozaki, A.N., Ozaki, H., Kuwahara, M., and Sawai, H. 2002. Enzymatic synthesis of labeled DNA by PCR using new fluorescent thymidine nucleotide analogue and superthermophilic KOD–DNA polymerase. *Bioorg Med Chem Lett*, **12**, 1167–1170.

46. Rashtchian, A. and Mackey, J. 1987. *Labeling and Detection of Nucleic Acids*. Springer-Verlag, Berlin.

47. Kolb, H.C., Finn, M.G., and Sharpless, K.B. 2001. Click chemistry: Diverse chemical function from a few good reactions. *Angew Chem Int Ed*, **40**, 2004–2021.
48. Ogino, H., Fujii, M., Satou, W., Suzuki, T., Michishita, E., and Ayusawa, D. 2002. Binding of 5-bromo-uracil-containing S/MAR DNA to the nuclear matrix. *DNA Res*, **9**, 25–29.
49. McConlogue, L., Brow, M.A., and Innis, M.A. 1988. Structure-independent DNA amplification by PCR using 7-deaza-2′-deoxyguanosine. *Nucleic Acids Res*, **16**, 9869.
50. Yee, J. and Paul, N. 2010. PCR incorporation of modified dNTPs: The substrate properties of biotinylated dNTPs. *Biotechniques*, **48**, 333–334.

Part III

Instruments

9 Thermocycler Calibration and Analytical Assay Validation

Mary Span, Marc Verblakt, and Tom Hendrikx

CONTENTS

9.1 INTRODUCTION

Since the invention of polymerase chain reaction (PCR) in 1983,[1] much research has been done on the impact of the different variables within the pre-PCR and PCR process on the final PCR result. This research has mainly focused on the impact of (1) sampling methods,[2] (2) RNA and DNA isolation and purification procedures,[3,4] (3) quality and purity of RNA, cDNA, and DNA, and ways to determine this,[5,6] (4) primer and probe design strategies, (5) primer and probe quality and purity,[7] (6) RT- and PCR-priming strategies,[6] (7) detection chemistries,[6,8–10] (8) types of RT and PCR enzymes,[11,12] (9) PCR buffer compositions,[12] and (10) PCR enhancers and inhibitors.[13] Yet, not much research has been done on the impact of thermocyclers on the PCR result. Thermocyclers are considered to be constants rather than variables, although most researchers are familiar with the effect that particular PCRs function well on certain thermocyclers and fail, or generate different results, on others.[14–16]

With the more recent use of PCR in diagnostics, the call for quality control (QC) is increasing in accredited- and quality-aware laboratories. An increasing number of laboratories either elect, or are required, to obtain an ISO 17025[17] or ISO 15189[18] accreditation to guarantee the quality of the results generated. At the same time, the research community's call for biologically meaningful conclusions is increasing in parallel. In 2009, a group of leading qPCR scientists published the MIQE guidelines,[19] which assist qPCR users in designing a robust qPCR experiment that leads to trustworthy and biologically meaningful results that can be reproduced in any other laboratory. As a result of this call for QC, the number of thermocycler calibrations and assay–thermocycler validations has strongly increased. Through these calibrations, it has become clear that there is a substantial variation not only between different brands of thermocyclers, but also between models of the same brand, between individual serial numbers of one model, and also within one thermocycler. This variation can have a substantial impact on the outcome of PCRs or qPCRs. The effects move on a sliding scale from slightly less-efficient PCRs to complete failure.

By knowing the thermal performance of a thermocycler, it is possible to control the effects of thermocycler variability. Guidelines are offered as to how thermocyclers can be aligned and programmed to mimic each other, to use the complete thermocycler capacity of a laboratory. Furthermore, guidelines are offered on how calibration results can be used for validation purposes when working under accreditation.

9.2   TERMS AND DEFINITIONS

For terms and definitions, see Table 9.1 and Figure 9.1.

TABLE 9.1
Terms and Definitions

Calibration	The total set of operations that establish, under specified conditions, the relationship between values of quantities indicated by a measuring instrument or a measuring system or values represented by a material measure, and the corresponding values realized by an international traceable reference standard
	Traceability is guaranteed by the use of international traceable reference standards and also by the calculated uncertainty of a calibration and performance of the calibration by technically competent calibration engineers
	Calibration does not include adjustment
Adjustment	Adjustment of an indicated value of the instrument within given specifications or tolerances. Adjustments are always accompanied by an "as-found" or "as-received" calibration certificate and an "as-left" or "after adjustment" calibration certificate
Validation	Confirmation by examination and provision of objective evidence that the particular requirements for a specific intended use are met. In other words, do certain assays in combination with certain equipment generate the required results?
Verification	Quality control, via a defined procedure, to check whether a system or test still meets the specifications
Plateau phase/hold phase	Phase of the PCR process at which temperature is kept steady
Ramp phase	Phase of the PCR process at which temperature is changing toward the next plateau phase
Set temperature	The temperature that was programmed to be reached
Accuracy	Difference between average reaction block temperature and set temperature at a defined time point
Uniformity/spread	Difference in temperature between hottest and coldest wells in the reaction block at a defined time point
Overshoot	Overshooting of temperature, above the set temperature when ramping up
Undershoot	Overshooting of temperature, below the set temperature when ramping down Note: an undershoot is defined as an overshoot going down
Ramp rate	Speed of heating (heat rate) or cooling (cool rate) while cycling up or down
Hold time/plateau time	Time duration of the plateau phase

Note: For more detailed metrology definitions, refer to the *International Vocabulary of Metrology.*[20]

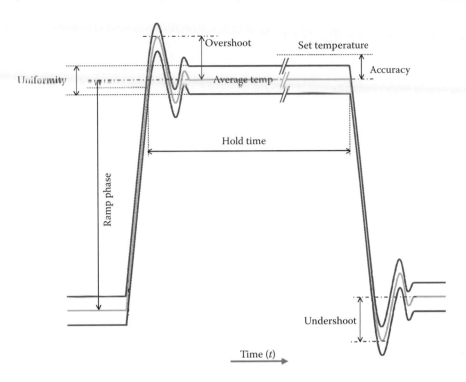

FIGURE 9.1 Graphical explanation of thermocycler temperature parameters.

9.3 THERMOCYCLER TECHNICAL DESIGN

9.3.1 VARIOUS TECHNOLOGIES

Since the invention of PCR, many different types and models of thermocyclers have been designed and manufactured. Owing to continuous development, a wide variety of heating and cooling techniques and temperature-control mechanisms have been used in thermocyclers in the past decades.

In the early days of PCR, a typical thermocycler would use a heater combined with a liquid-compression cooling system. With the need to design faster thermocyclers, Peltier-based thermocyclers were introduced. The latest developments permit completing a PCR under 30 min, and use either miniaturized blocks, ceramic heaters, or heated air. However, the Peltier-based thermocycler remains the most common type of thermocycler. The principle of a Peltier element is that it either heats or cools depending on how the electrical current is applied to the element. As a result, these elements can, very quickly, alternate from heating to cooling and the reverse. Most Peltier-based thermocyclers are designed according to a "sandwich" construction. The reaction block that is visible to the user is the top of the construction. Underneath the block, one or more temperature sensors are positioned such that they monitor the reaction block temperature and provide input to the control mechanism that regulates the heat generated by the heater and Peltier element. During the heating phase, the heat of the heater and the Peltier element is transferred to the reaction block and then to the reaction tubes and the reagents inside. During the cooling phase, the Peltier element gets cold on one side and transfers this cold to the reaction block; on the other side, the Peltier element generates heat, which is transferred to the heat sink and then ventilated to the environment via the fan.

9.3.2 SPEED VERSUS UNIFORMITY

Thermocyclers are designed to function as instruments that can hold a defined steady temperature for a defined time, then change temperature, and then again hold a defined steady temperature during a defined time. This requirement, as simple as it may seem, is the largest challenge when designing and constructing an accurate and uniform thermocycler. To design a fast thermocycler, the mass that needs to be heated must be reduced to a minimum to allow fast energy transfer during heating and cooling. However, to design an instrument that can hold its temperature in a defined, steady, and uniform way, a high mass is required. Thermocyclers with massive blocks are much more uniform, and also relatively slow. Thermocyclers with low mass blocks are fast, but due to their speed, they are also generally less controlled and therefore these blocks may easily overshoot the set temperature by several degrees before falling back to the set temperature. Depending on the control mechanism and the number of sensors that monitor the reaction block temperature, these fast thermocyclers can not only show high overshoots, but can also show highly nonuniform overshoots and plateau phases. This is especially the case in thermocyclers that are controlled only by a single sensor. For those thermocyclers that can be programmed in "standard" or "fast" mode, the two different modes can lead to substantially different thermal profiles that not only differ in ramp rate, but also in height and duration of the overshoot and uniformity of both overshoot and plateau phase. Summarizing, in block-based thermocyclers, high ramping speeds can be associated with high degrees of nonuniformity, especially during ramping, and also with high and poorly controlled overshoots.

9.4 THERMOCYCLER VARIABILITY AND PRACTICAL CONSEQUENCES

9.4.1 INTER- AND INTRA-THERMOCYCLER VARIABILITY

The thermal performance of a thermocycler is dependent on a number of variables. The main variables are the block "sandwich" construction, the different types and qualities of components used, the differences in technical design, the number of sensors, and the temperature-control mechanism. Most PCR users know from experience that some PCRs provide good results on certain thermocyclers, but fail on others. When these thermocyclers are of different brands, this is accepted and even considered as common sense as they are perceived to be different. However, different models of the same brand are often expected to function similarly while it is considered that different serial numbers of the same model should perform as identical copies. As shown in the figures below, thermocyclers show a substantial variation, not only between brands (Figure 9.2), but also between models (Figures 9.2 and 9.3) of the same brand, between individual serial numbers of the same model and brand (Figure 9.3), and even within one thermocycler (Figure 9.4). Each thermocyler has a unique thermal "fingerprint."

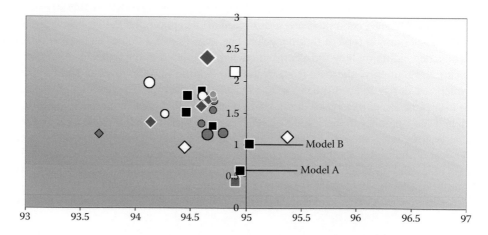

FIGURE 9.2 Average temperature accuracy and temperature uniformity of different models and brands of thermocyclers at 30 s, 95°C (symbols of the same color represent thermocyclers of the same brand, size of the symbols represents spread within the subpopulation).

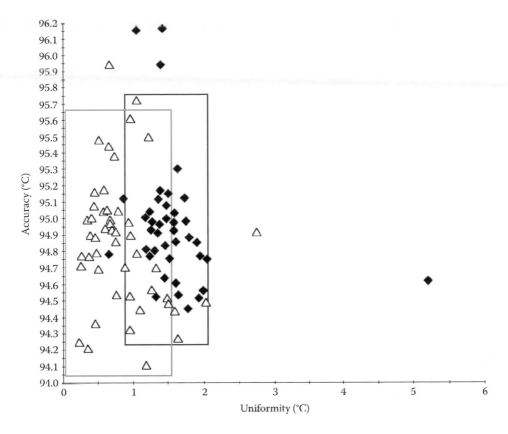

FIGURE 9.3 Individual temperature accuracy and nonuniformity at 30 s, 95°C of thermocyclers of model A (triangles) and model B (diamonds) of the same brand (triangles and diamonds represent individual serial numbers).

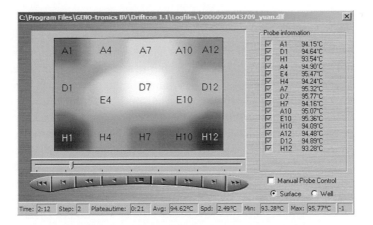

FIGURE 9.4 Temperature nonuniformity within individual thermocycler blocks.

9.4.2 CONSEQUENCES OF THERMOCYCLER VARIABILITY

The inter-thermocycler variability described above is the reason underlying the observation that PCRs function on certain thermocyclers and fail on others, or lead to different results on different thermocyclers.[14–16] The intra-thermocycler variability is the reason why certain wells generate false-negatives while in other wells, positive results are generated, during the same run, on the same thermocycler.[21–23] But thermocycler variability can cause more than just false-negatives. The effects move on a sliding scale from slightly less-efficient PCRs, which still give a result, to complete failure. The first category is often not noticed. In the case of expression profiling via qPCR, this category can lead to incorrect conclusions of gene up- or downregulation. Less-efficient PCRs with lower yields can, for example in the case of minimal residual disease in leukemia, lead to lower cancer cell counts and a clinical decision to not administer a second chemotherapy, although in reality, it is required. False-negatives are extremely risky, in general, as they can lead to incorrect conclusions and results, treatment, or categorization. They can lead to patients being diagnosed as healthy, while in reality, they might be infected with a life-threatening virus.

9.5   EFFECT OF THERMOCYCLER VARIABILITY ON A PCR OR qPCR

Since PCR is a dynamic process with ramping phases of different rates and plateau phases of different temperatures, several types of variability can occur. The most common types of variability are deviating plateau temperatures, high overshoots/undershoots and nonuniform plateau temperatures, ramping phases. The effects of these types of variability are dependent on the phase of the PCR.

The purpose of the denaturation phase is to denature the double-stranded DNA, to obtain two single strands to which the primers can bind. Denaturation is typically performed at 94–95°C. However, at the same time, the *Taq* polymerase is inactivated by the high temperature required for the denaturation. This inactivation is mainly temperature-, but is also time-dependent. At 95°C, the half-life time of *Taq* polymerase is 40 min, whereas at 98°C, the half-life time is reduced to 5 min. Therefore, incubating a PCR at high temperatures for long periods, or overshooting the target temperature by several degrees, can quickly lead to negative results, due to inactivation of the enzyme before detectable levels of amplicon are generated. Most false-negative PCR results are caused by problems in the denaturation phase, especially by high overshoots or high plateau temperatures.[21,22] If the temperature of the denaturation phase is too low, no denaturation will occur and therefore also no amplification. The minimum denaturation temperature is mainly dependent on the length and GC content of the DNA target and secondary structures. Typically, denaturation functions over a temperature range of about 2–3°C at 30-s denaturation time.

The purpose of the annealing phase is to allow the DNA primers that initiate the elongation to bind to the correct target sequence. This binding should preferably be as specific as possible. The specificity is highly temperature- and salt concentration-, but is hardly time-dependent. Annealing is typically performed between 45°C and 70°C. As temperature optimization of a PCR is typically only done for the annealing phase and not for the other phases, the annealing temperature is perceived as the most critical temperature step to allow a PCR to succeed. The annealing temperature is indeed critical for the specificity of the PCR,[23] but is much less critical to produce a result. Annealing functions over a much wider temperature range than denaturation. Typically annealing functions over a temperature range of about 5–7°C, albeit in varying degrees of specificity. Therefore, the risk of missing a result, due to a false-negative, is significantly higher when inappropriate temperatures are used in the denaturation phase. Thermal performance problems at the annealing phase typically express themselves as nonspecificity or lower yields than expected and are therefore more easily identified.

The purpose of the elongation phase is to synthesize the new strand of DNA, complementary to the template strand. The polymerization rate of *Taq* polymerase is temperature-dependent (2000 bases/min at 72°C), but elongation functions over a range of 55–85°C. To obtain highly efficient PCRs, many users design PCR products to be smaller than 200 bp, which means that elongations only take a few seconds, often much shorter than the protocol times programmed. Elongation is generally insensitive to temperature variability at the level that occurs in thermocyclers although nonuniformity of the thermocycler could lead to different polymerization rates. However, in general, excess times of elongation are used and therefore slower rates are compensated by longer times.

Although the post-qPCR analysis meltcurve step is not part of the actual qPCR, it is also influenced by thermal variability. The goal of a meltcurve is to identify whether the correct amplicon has been amplified and determine the presence of any nonspecific products or primer–dimers in the reaction. In the case of high-resolution melting (HRM), the goal is genotyping or mutation scanning.[24] Temperature nonuniformity during ramping can lead to shifts in meltcurves. What seem to be amplicons of different lengths and/or sequences, due to their differing T_m (melting temperature), are in reality identical amplicons melted at different times. The meltcurve plots the first negative derivative of the relative fluorescence versus the temperature, but in reality, relative fluorescence is plotted versus time instead of temperature. If a thermocycler is nonuniform, different wells will reach the T_m at different times and therefore a shift in time, depicted as temperature, occurs.

9.6 THERMOCYCLER CALIBRATION METHODS

Currently, only a few thermocycler calibration methods exist that are metrologically correct.

From a molecular biologist perspective, the best way to calibrate a thermocycler is by mimicking the PCR process. In other words, put tubes in all wells of the block, fill them with reaction mix, and put sensors in, close the tube lids, close the thermocyclers heated lid, and run the PCR protocol that is normally used in the laboratory. This would come as close as possible to the real temperature inside a particular tube, filled with a particular reaction mix, during a particular PCR protocol, in a particular instrument. The phrasing of the previous description already illustrates the major shortcoming of the proposed method; in that it cannot be used in a standardized way. For each combination of the tube, mix, protocol, and instrument, this "calibration" should be repeated, requiring a tremendous workload, generating results that cannot be compared to each other or to any kind of standard, or to specifications of the thermocycler manufacturer. Furthermore, this way of "calibrating" would introduce many variables that are uncontrolled and therefore add a large component to the measurement uncertainty, ending up with uncertainties well over 2°C. Practically, this means that when the uncertainty is, for example, 2°C, while measuring a defined 96°C, the thermocycler's actual temperature could be anywhere between 94°C and 98°C. As described in Section 9.4, this difference in temperature does have tremendous effects on the inactivation of the *Taq* polymerase. Therefore, this level of uncertainty is not acceptable for thermocycler calibration, although mimicking the PCR process may seem a simple and attractive choice.

From a clinical chemist's perspective, the best way to calibrate a thermocycler is by using a standardized method for measurement that is representative of the PCR process. This standardized method allows one calibration for all the different types of PCRs and also allows comparison to a standard. For a calibration to be representative of the PCR, it needs to take into account both the dynamic and the static part of the PCR process, as certain problems cannot be diagnosed by just checking the static part of a PCR.[21,22] Furthermore, a number of different temperatures, preferably close to denaturation, annealing, and elongation temperatures, should be evaluated since their effects can vary with temperature (see Section 9.4).

From a metrologist's perspective, the best way to calibrate a thermocycler is by measuring in a defined and traceable manner, by qualified personnel, under defined environmental conditions, with the lowest achievable and known measurement uncertainty, excluding as many noncontrolled variables as possible. In metrology, calibration is defined as the total set of operations that establish, under specified conditions, the relationship between values of quantities indicated by a measuring instrument or measuring system or values represented by a material measure, and the corresponding values realized by an international traceable reference standard. Traceability is guaranteed by the use of international traceable reference standards and the expression of the calibration values in SI units, and also by the calculated uncertainty of a calibration and the fact that only technically competent and qualified calibration engineers perform the calibrations. Any uncertainty should be as minimal as technically possible. Only if the requirements above are met is a calibration considered to be fully traceable to the International Temperature Standard (ITS-90). To be compliant to the ISO 10725 standard,[17] additional requirements need to be met.

Any laboratory running PCR-based tests that is accredited under ISO 17025 or ISO 15189, or any CE-IVD PCR kit manufacturer that is certified under ISO 13485, should, according to these international standards, calibrate their thermocyclers in a traceable way and should therefore conform to ISO 17025.

In summary, for a calibration to be certain, compliant to ISO 17025 and representative of the PCR process, the calibration should meet the following criteria:

1. Measure all thermal characteristics that influence the outcome of a PCR, including accuracy, uniformity, overshoot/undershoot, heat/cool rate, and hold time.
2. Be performed in a dynamic manner, as PCR is a dynamic and not a static process.
3. Be performed simultaneously with multichannels, to exclude any time effects between well performance.
4. Be performed under controlled environmental conditions.
5. Be performed by qualified and trained personal.
6. Be traceable to the ITS-90 and express calibration values in SI units with a calculated uncertainty.

To be able to meet these criteria, a typical thermocycler calibration system will be a physical, sensor-based, multichannel system that measures dynamically, with a frequency of more than once per second, and that can be calibrated so that it is traceable to the ITS-90 via a temperature reference standard. To minimize the measurement uncertainty, the influence of poorly controlled variables such as reaction tubes and thermocycler-heated lids are excluded. By calibrating with traceably calibrated equipment, under defined environmental conditions, with qualified personal, directly in the block, targeting a statistically relevant number of wells, it is possible to obtain measurement uncertainties as low as 0.1°C for a multichannel system. This allows making a certainty statement about the measured value and allows comparison to specifications, either laboratory defined or manufacturer defined.

The following methods are either uncertain, not ISO 17025 compliant, or not representative of the PCR process due to the reasons discussed above. Any method based on PCR assays, qPCR assays, thermochromic liquid crystals, single-channel thermometers (analog and digital), and multichannel static measurements will not show all the characteristics that determine the thermal profile of a thermocycler and therefore these methods are not representative of the PCR process. Any method based on in-tube measurements is accompanied by significant uncertainty arising from contact errors between tubes and the reaction block, and from errors due to nonuniformity in the wall thickness and heat conductivity of the tubes. The results are therefore associated with high overall uncertainty, disallowing certainty statements about the thermocyclers' performance. Any method based on measuring all the wells of the reaction block simultaneously will require the addition of a substantial mass to the block and will lead to biased dynamic values since ramp rates and overshoots can be strongly influenced by the addition of mass leading to incorrect results.

9.7 EVALUATING CALIBRATION RESULTS

Calibration results are often evaluated by categorizing a thermocycler to be "in" or "out" of manufacturer specifications. However, evaluating the thermal performance of a thermocycler goes far beyond categorizing a thermocycler to be "in" or "out" of specifications. A thermocycler can be well within manufacturer specifications for accuracy and uniformity at 95°C and, at the same time, it shows severe problems at other temperatures. Therefore, the complete thermal profile of a thermocycler should be evaluated, including accuracy, uniformity, ramp rates, overshoots, and hold times.

The most crucial parameters to evaluate for a calibration certificate are the thermocycler uniformity during ramping, overshoot and plateau, and the height and length of the overshoots. High nonuniformities and high and long overshoots typically cause variation in PCR results due to different levels of inactivation of the *Taq* polymerase in different wells.

The thermocycler performance guidelines in Table 9.2 allow for an objective evaluation of the thermocyclers performance and can be used as a checklist. The guidance criteria in Tables 9.3 and 9.4 can be used to categorize thermocyclers. These criteria are based on uniformity and over-shoots/undershoots as these characteristics of a thermocycler cannot, in most cases, be directly modified by programming the thermocycler differently. The average temperature (accuracy) can be

---

TABLE 9.2
Thermocycler Performance Guidelines

Item Calibration Certificate	Comment	Performance Criterium to Check
Uniformity during plateau	Large nonuniformities during plateau phases are an indication of suboptimal thermal control of the block. Fast and one-sensor thermocyclers typically show larger nonuniformities during plateau phases than standard and multisensor thermocyclers Typically the uniformity improves as the set temperature gets closer to the environmental temperature or with increasing hold times	Check that no large nonuniformities are present during plateaus See Table 9.3 for categorization Check if plateau uniformity improves with decreasing set temperature
Uniformity during overshoot	Large nonuniformities during overshoots are an indication of suboptimal thermal control of the block. Fast thermocyclers typically show larger nonuniformities during overshoots than standard thermocyclers Typically the overshoot uniformity improves as the set temperature gets closer to the environmental temperature	Check that no large nonuniformities are present during overshoots See Table 9.3 for categorization Check if overshoot uniformity improves with decreasing set temperature
Uniformity during ramping	Large nonuniformities during ramping are an indication of suboptimal thermal control of the block. Fast thermocyclers typically show larger nonuniformities during ramping than standard thermocyclers Typically the ramping uniformity improves when the ramping speed is reduced Furthermore, the ramping uniformity improves when the difference in temperature between two plateaus decreases, that is, the ramping nonuniformity while heating from 30°C to 95°C is larger than while heating from 50°C to 70°C	Check that no large nonuniformities are present during ramping See Table 9.3 for categorization Check if ramping uniformity improves with decreasing delta temperature
Average temperature per time point (accuracy)	The average temperature shows less deviation from the set temperature with increasing hold times	Check whether average temperature approaches the set temperature with increasing hold times
Temperature per channel per time point	Values at 30 s are most representative as the thermocycler has completed overshoot or approach of the plateau and has had some time to reach equilibrium Minimum/maximum temperatures that strongly deviate from the average temperature are an indication for cold/hot spots	Check for values strongly deviating from the set temperature In the case of extreme cold/hot spots, avoid using the wells concerned

(continued)

TABLE 9.2   (continued)
Thermocycler Performance Guidelines

Item Calibration Certificate	Comment	Performance Criterium to Check
Overshoot (average/ maximum)	High and long overshoots at 95°C lead to increased *Taq* polymerase inactivation	Check that overshoots do not exceed average plateau temperatures by more than 5°C for all temperatures
	If the overshoot at 95°C is highly nonuniform, the *Taq* polymerase inactivation can vary substantially per well and can even lead to positive results in certain reactions and can lead to false-negatives in others	Check that overshoot does not last longer than 10 s for all temperatures
		Check whether all channels go through overshoot for all temperatures
	High and long overshoots at 50°C and 60°C can cause mispriming during annealing	Check nonuniformity during overshoot for all temperatures
	If the overshoot at 50°C and 60°C is highly non uniform, mispriming can happen in certain wells and not in others, leading to nonspecific results in certain wells and not in others	See Table 9.4 for categorization
Average ramp rate	The average ramp rate is determined between 10% and 90% of the ramp	Check whether the thermocycler is in or out of specifications
	The ramp rate is strongly dependent on the brand and model of the thermocycler	Check whether the thermocycler is capable of cooling down to the lowest programmed temperature, within a reasonable time
	Typically cooling rates are slower than heating rates, especially in air-driven thermocyclers	
	Thermocyclers with fast ramp rates typically show higher overshoots, especially in air-driven thermocyclers	
	Slow ramping down can be an indication of cooling problems either caused by the thermocycler or the environment	
	Be aware that thermocyclers that cool by convection cannot reach temperatures lower than 10–15°C above ambient temperature	
Maximum ramp rate	The maximum ramp rate represents the point during the ramp at which the thermocycler heats or cools the fastest	Check whether the thermocycler is in or out of specifications
	Fast thermocyclers typically show larger nonuniformities during ramping than standard thermocyclers	Check whether ramping uniformity improves with decreasing delta temperature
	Furthermore, the ramping uniformity improves when the difference in temperature between the two plateaus decreases, that is, the ramping nonuniformity while heating from 30°C to 95°C is larger than while heating from 50°C to 70°C	
Hold time	The hold time is the duration of the plateau phase	Check whether the hold time corresponds to the protocol of the instrument
Curve morphology	The curve morphology of the temperature graph is dependent on the brand and model of the thermocycler, either a curve in which the plateau temperatures are approached or a curve with overshoots after which the plateau temperatures are achieved can be seen	Check whether thermocyclers of the same brand and model show similar curve morphologies

TABLE 9.2 (continued)
Thermocycler Performance Guidelines

Item Calibration Certificate	Comment	Performance Criterium to Check
Divergation/ convergation/ oscillation of channels at plateau phase	Divergation and oscillation of channels at plateau phase is an indication for suboptimal thermal control of the reaction block and can typically be found with blocks monitored by one or few sensors Divergation and oscillation during the plateau phase at 50°C and 60°C can cause mispriming during annealing Divergation and oscillation during the plateau phase at 95°C can cause variation in *Taq* polymerase inactivation and therefore can cause variation in yields Convergation typically occurs when a thermocycler is left for a longer time at the plateau phase	Check whether divergation/ convergation/oscillation of channels at the plateau phase occurs
Manufacturer specifications	Manufacturer specifications are specifications as provided by the thermocycler manufacturer These specifications are stated without any measurement uncertainty The manufacturer specifications allow categorizing a thermocycler to be in or out of manufacturer specifications	Check whether the thermocycler is specified at 90°C or 95°C as thermocyclers specified at 90°C or 95°C are typically more uniform than those specified at 70°C or 50°C Check whether the thermocycler is in or out of specifications for accuracy, uniformity, heat, and cool rate

TABLE 9.3
Guidance Criteria for Uniformity

Phase	Excellent Cycler	Good Cycler	Moderate Cycler	Poor Cycler
Ramping (°C)	<3	3–4	5–8	>8
Overshoot (°C)	<2	<2	2–3	>3
Plateau (95°C, 30 s) (°C)	<0.6	0.6–1	1–2	>2

TABLE 9.4
Guidance Criteria for Overshoots at 95°C—Duration of Maximum Overshoot

Max Overshoot at 95°C (°C)	Good Cycler (s)	Moderate Cycler (s)	Poor Cycler (s)
105	<0.3	0.3–0.5	>0.5
102	<0.7	0.7–1	>1
100	<2	1–4	>4
97.5	<7	7–10	>10
96.5	<12	12–15	>15

adjusted by modifying the set temperature and therefore is a less crucial criterium for categorizing a thermocycler.

9.8 MODIFYING THERMOCYCLER PERFORMANCE

Objective evaluation of thermocycler performance leads to the conclusion that each thermocycler has a unique thermal fingerprint. No thermocycler is an identical copy of another thermocycler, even when comparing cyclers with different serial numbers of the same model and brand. Currently, most cyclers can only be calibrated and not adjusted, neither by the user, nor by the manufacturer. However, by modifying the programmed protocol, any user can modify the performance. The programmed protocols will differ by the thermocycler, but the resulting thermal profiles will be identical or close to identical. The parameters that can be modified by programming are accuracy, ramp rate, overshoot, and hold time. Uniformity can be indirectly influenced by modifying other parameters, but cannot be modified directly and is therefore the most difficult parameter in a thermocycler to adjust and to control.

The average plateau temperatures (accuracies) can be modified by adapting the set temperature. Thermocyclers with average temperatures below/above the required temperature can be programmed at higher/lower set temperatures so that the required temperature will be achieved. Take the accuracy of 30 s at plateau and increase/decrease by the difference between average and set temperature to obtain the required temperature. For example, if a thermocycler reaches 94.5°C when it is programmed at 95°C, change the set temperature to 95.5°C to allow the thermocycler to reach 95°C. Verify the result of the adapted programming by a calibration and check whether the correct result has been achieved. If not, fine tune the correction required. Register the modified program and resulting temperatures in the lab journal.

Ramp rates can be modified by adapting the heat and cool rates. When the maximum ramp rate is the default setting, be aware that the ramp rate can only be reduced and not increased. So, if two thermocyclers should function alike, adapt to the thermocycler with the lowest ramp rate or a fixed ramp rate. Register the modified programs and resulting ramp rates in the lab journal. In general, reducing the ramp rate will result in an improved uniformity during ramping and overshooting. Furthermore, the height of the overshoot will be reduced.

Thermocyclers have a thermal profile (curve morphology) that is inherent to the model and brand. Certain thermocyclers have a thermal profile with overshoots, others without overshoot. The advantage of an overshoot at denaturation is that plateau times can be drastically reduced as the reaction mix heats up much faster. The disadvantage is that, in the case of high and poorly controlled overshoots, fast inactivation of the *Taq* polymerase occurs as the reaction mix also goes through the overshoot. Depending on the requirement, overshoots can be programmed in or out:

- If the requirement is fast cycling and fast results, it is recommended to leave overshoots at denaturation or program them in
- If the requirement is minimal *Taq* polymerase inactivation, maximum number of cycles, and sensitive detection, it is recommended to leave out overshoots at denaturation, minimize them, or program them out

Overshoots at denaturation can be programmed out by adding a short plateau at a slightly lower temperature to the program. For example, 30 s 95°C, 30 s 62°C, and 30 s 72°C can be modified into 1 s 90°C, 30 s 95°C, 30 s 62°C, and 30 s 72°C. In this way, the thermocycler ramps up with maximum speed to 90°C, overshoots it, continues to 95°C in a slower and more controlled way, and therefore hardly overshoots 95°C, saving the *Taq* polymerase activity. Depending on the brand and model of the thermocycler, this requires finetuning. Programming overshoots in is also possible. For example, 30 s 95°C, 30 s 62°C, and 30 s 72°C can be modified into 1 s 98°C, 30 s 95°C, 30 s 62°C, and 30 s 72°C. In this way, the plateau times of the denaturation phase can be reduced, reducing the total run time. Certain thermocyclers allow for programming of the reaction volume. In these thermocyclers, the overshoot can also be reduced/increased by programming a volume that is lower/higher than the volume used. The effect is not linear and should be verified by a calibration.

The only parameter that, unfortunately, cannot be modified is the uniformity of a thermocycler. Uniformity can only be indirectly influenced by modifying the ramp rate or by adding mass to the reaction block and loading it evenly. It is therefore the most difficult parameter in a thermocycler to modify and to control.

The effect of modifying the ramp rate depends on the design and construction of the thermocycler, mainly on the number of Peltier elements, the number of sensors, and the control mechanism. Slowing down the ramping rate typically leads to lower overshoots and improved uniformity during both the ramping phase and the plateau phase. The same phenomenon can, in certain thermocyclers, also be obtained by filling all unused wells with tubes containing water (using the same volume as the samples). In this way, more mass is added to the reaction block and the mass is distributed evenly, slowing the thermocycler down, forcing it to heat evenly, and therefore improving its uniformity. In well-regulated thermocyclers, adding mass has little effect on the uniformity. It will mainly increase the height of the overshoot. Uneven loading can lead to deterioration of the thermocycler uniformity over time. The left side of the thermocycler blocks is often more worn out than the right side due to left-to-right loading when using single tubes or strips. The best thermal performance of a block is obtained if it is loaded evenly. In other words, distribute the tubes or strips evenly over the block, starting from the middle of the block.

Adapted thermocycler protocols can be used to bring the thermocycler's performance back to the required performance. However, adapted protocols can also be used to align several thermocyclers. This can be done between different thermocyclers of the same model, and also between models and even between brands. It is straightforward to put in place and once all thermal characteristics are known, it allows a laboratory to use its full thermocycler capacity, in the sense that each assay can be run on each thermocycler.

Be aware that the uniformity cannot be directly adapted and that ramp rates can only be decreased and not increased. Therefore, initial validation should always be done on the least uniform thermocycler.

Protocol for aligning thermocyclers:

1. Calibrate all thermocyclers to be aligned.
2. Select the thermocycler with the lowest ramp rate.
3. Adjust the ramp rates of all thermocyclers to the slowest thermocycler.
4. Adjust the accuracy either to an absolute temperature or to a selected thermocycler (take 30-s plateau values from the calibration data).
5. Program overshoots in or out to obtain the desired thermal profile (take the overshoot value from 95°C step).
6. Verify the effects of adapting the protocol by a second calibration.
7. Finetune if necessary and verify again.
8. During assay validation (Section 9.8), validate the assay on the thermocycler with the highest nonuniformity. If the reaction produces a positive result in all 96 wells, it will also function on all aligned thermocyclers and does not require additional validation on these instruments.

9.9   ASSAY VALIDATION

9.9.1   INTRODUCTION

Both the ISO 17025 and ISO 15189 standard and also many other regulations require (q)PCR assay validation in addition to thermocycler calibration. The principle is to prove that the thermocycler is suitable to run a particular assay, as being within specifications is not a guarantee that a certain kit will function on a particular thermocycler, even when initially validated by the manufacturer on that particular model and brand. As all thermocyclers are more or less nonuniform, a validation done in just one row or column of the reaction block is not sufficient as it might not cover the most extreme positions, the hot and the cold spots of the block. An assay might function in the validation test wells and might not function in others as they are too cold to denature or too hot, leading to *Taq* polymerase inactivation before a detectable result is generated. Therefore, intra-thermocycler variability should be taken into account during the validation process, not only by the laboratory, but also by the kit manufacturer.

The methods below describe several ways to perform assay validation, taking intra-thermocycler variability into account. The most suitable method of those described below will depend on the equipment available, the workload involved, and the type of laboratory.

9.9.2 Hot–Cold Spot Validation Method

In the hot–cold spot validation method, the hottest and the coldest positions in the reaction block are determined via a calibration. Two positive controls and two negative controls are used. The two positive controls are positioned on the hottest spot and on the coldest spot. The two negative controls are positioned on the second hottest and second coldest spot. If the (q)PCR assay functions correctly on the temperature extremes, it will also do so at all temperatures in between and hereby the evidence is provided that the assay will function in all wells.

Protocol for hot–cold spot validation:

1. Calibrate the thermocycler (include a 95°C step).
2. Take the highest and second highest temperature and the lowest and second lowest temperature at 10 s into the 95°C step.
3. Position the positive controls in the hottest and coldest well and position the negative controls in the second hottest and second coldest well.
4. Run PCR protocol as usual.
5. Check whether the positive and negative controls give the correct result.
6. If yes, from a temperature perspective, the assay will give reliable results in all other wells. If no, avoid the hottest/coldest wells and position the positive and negative controls on the second hottest/coldest and third hottest/coldest well and start from step 4 or decrease/increase the set temperature thermocycler closer to the target temperature and start from step 1.

The advantage of this method is that it can be put into place very easily and quickly on any thermocycler and therefore is the recommended method for laboratories running many different assays over time, such as research laboratories.

The disadvantages are that controls can end up positioned in the middle of sample series or that the hot and cold spots can move through the reaction block over time.

9.9.3 Thermal Boundary Validation Method

The thermal boundary method requires a determination of the temperature extremes at which the (q)PCR assay still gives the correct result. The minimum and maximum denaturation, annealing, and elongation temperatures are determined and then a thermocycler is qualified as being within these thermal boundaries or not. If the thermocycler, including its full nonuniformity, lies within the thermal boundaries, the cycler is qualified as suitable. The advantage of this method is that it is very exact and allows laboratory technicians to position controls wherever they like. It also allows kit manufacturers to specify the thermal boundaries of a kit, instead of protocols for particular models and brands of thermocyclers on which the kit has been validated by the manufacturer. The thermal boundary method is a solution for CE-IVD kits that do not function correctly in the hands of end users and do not achieve the sensitivity and reproducibility claimed, although used exactly according to the protocol, on the thermocycler on which the kit has been validated. The cause for this is, in many cases, due to limited validation by the manufacturer and not taking thermocycler variability into account (see Section 9.9.5).

Laboratories that use CE-IVD kits can also determine these thermal boundaries themselves. In this way, kits can be used on alternative thermocyclers than the ones on which the kit was originally validated. According to the ISO 17025 and ISO 15189 standard, this is called "use of a standardized method outside its intended scope." The thermal boundaries method allows the use of the laboratory's full thermocycler capacity and allows universal thermocycler use. The disadvantage of this method is that it is initially labor-intensive as the thermal boundaries need to be determined for the denaturation, annealing, and elongation phase. Sometimes, it is also necessary to define the ramp rates and the height and length of the overshoots.

But once defined, it can be used for a long time. This method is, therefore, recommended for PCR and qPCR kit manufacturers and diagnostic laboratories that repeat the same test frequently over a long period of time.

Protocol for thermal boundaries validation:

1. Calibrate the thermocycler both in gradient and nongradient mode. Performance in the gradient mode can differ from performance in the nongradient mode, the real temperatures that the thermocycler reaches, in both modes needs to be determined.
2. If two different thermocyclers are used, and when the thermocycler that is normally used does not have a gradient, check the relationship between both cyclers and align them or take the difference into account in the calculations as made below.
3. Set up 2.5-mL mastermix of the selected (q)PCR assay, pipet 25 µL in each well of a 96-well plate.
4. Run PCR with a gradient at the denaturation temperature (90–100°C).
5. Determine the lowest (D_L) and highest (D_H) denaturation temperature at which the PCR still gives a result, based on the calibration data, not based on the thermocycler program.
6. Determine the optimal denaturation temperature ($D_O = (D_L + D_H)/2$).
7. Run PCR with D_O as denaturation temperature and gradient at annealing temperature (±5°C theoretical annealing temperature).
8. Determine the lowest (A_L) and highest (D_H) annealing temperature at which the PCR still gives a result, based on the calibration data, not based on the thermocycler program.
9. Determine the optimal annealing temperature ($A_O = A_L + A_H/2$).
10. Elongation analysis is less critical, but can also be done.
11. Optimal protocol: D_O, A_O, and E_O.
12. Thermal boundaries: $D_L–D_H$, $A_L–A_H$, and $E_L–E_H$.
13. Calibrate the unknown cycler, check whether the thermocycler lies with its full uniformity (accuracy ± uniformity) within $D_L–D_H$, $A_L–A_H$, and $E_L–E_H$.
14. If the thermocycler lies within defined boundaries, the cycler can be used without any adjustment.
15. If the thermocycler lies completely or partially outside defined boundaries, either adjust the protocol to bring the thermocycler within boundaries (see Section 9.7) or qualify as not suitable.
16. Optional: Use the recommended + and – control positions (based on calibration) to check for drifting over time.
17. Optional: Define the ramp rates and the height and length of overshoots. In cases of a defined ramp rate, the total run time can be used as daily QC to monitor the thermocycler performance and to check that no modifications to the protocol have been made.

9.9.4 Approximated Thermal Boundary Validation Method

When using the approximated thermal boundary validation method, the thermal boundaries are not determined exactly, but are approximated by programming the thermocycler a few degrees off-plateau. The resulting approximated temperature boundaries are more narrow than when applying the thermal boundary validation method, but in other respects, the method is comparable. If the thermocycler, including its full nonuniformity, lies within the approximated thermal boundaries, the cycler is qualified as suitable. The advantage of this method is that it is still specific, although less exact than the thermal boundary method, and still allows laboratory technicians to position controls wherever they like. It also allows kit manufacturers to specify the approximated thermal boundaries of a kit, instead of requiring protocols for particular models and brands of thermocyclers on which the kit has been validated by the manufacturer.

This method is also suitable for laboratories that do not have a gradient thermocycler in their instrument portfolio. The disadvantage of this method is that it is initially labor-intensive as the approximated thermal boundaries need to be defined for the denaturation, annealing, and elongation phase. Sometimes, it is necessary to verify the effect of the ramp rates and the overshoots. But once defined, it can be used for a long time. This method is, therefore, recommended for PCR and qPCR kit manufacturers and diagnostic laboratories that repeat the same test frequently over a long period of time.

The workload of this initial validation using approximated thermal boundaries can be reduced by designing all assays to function at the same denaturation, annealing, and elongation temperature. In other words, in the design phase, the thermal boundaries can be defined and only then, a verification needs to be performed.

Protocol for approximated thermal boundary validation:

1. Calibrate the designated thermocycler to determine the real temperatures that the thermocycler reaches.
2. Set up 2.5-mL mastermix of the selected (q)PCR assay, pipet 25 μL in each well of a 96-well plate.
3. Run PCR with denaturation temperature at 1°C or 2°C below and above the normally used temperature.
4. Check whether the PCR still gives a result at the lowest programmed denaturation temperature (D_{AL}) and at the highest programmed denaturation temperature (D_{AH}), based on the calibration data, not based on the thermocycler program.
5. Run PCR with an annealing temperature at 1°C or 2°C below and above the normally used temperature.
6. Check whether the PCR still gives a result at the lowest programmed annealing temperature (A_{AL}) and at the highest programmed annealing temperature (A_{AH}), based on the calibration data, not based on the thermocycler program.
7. Elongation is less critical, but can also be performed.
8. Thermal boundaries: D_{AL}–D_{AH}, A_{AL}–A_{AH}, and E_{AL}–E_{AH}.
9. Calibrate the cycler of unknown performance, check whether it is within D_{AL}–D_{AH}, A_{AL}–A_{AH}, and E_{AL}–E_{AH}.
10. If within defined boundaries, the cycler can be used without any adjustment.
11. If completely or partially outside boundaries, either adjust the protocol to bring the thermocycler within boundaries (see Section 9.7) or qualify as not suitable.
12. Optional: Use the recommended + and –control positions (based on calibration) to check for drifting in time.

9.9.5 CE-IVD Kit Validation, Verification, and Revalidation

Most CE-IVD kits are validated on a large number of samples to check for matrices effects. However, the number of thermocyclers on which the kits are validated is, in general, very modest, often not more than 10 thermocyclers. Given the substantial variability present in the population of a particular model, low numbers of thermocyclers are not statistically representative for the total population. Therefore, the "coldest," "hottest," and "least uniform" thermocyclers, although within specifications, are often missing in these validation studies by the manufacturer. As a result, the kit will produce results in most laboratories, but a small percentage of the laboratories will not be capable of obtaining positive results in all the wells of an instrument or will fail to generate any results with a particular kit. Specifying thermal boundaries in a kit manual would allow an end user to use any model of the thermocycler, as long as the thermocycler with its full nonuniformity lies, within these thermal boundaries. However, the laboratory that uses the kit can also perform an initial verification to check whether the kit produces the result as claimed. If not, the laboratory can also determine the thermal boundaries themselves and can revalidate the kit to meet the requirements of the ISO 17025 or ISO 15189 standard to guarantee the human diagnostic results they produce.

9.10 CONCLUSIONS

Thermocyclers show a substantial variation, not only between brands, but also between models of the same brand, between individual serial numbers of the same model and brand, and even within one thermocycler.

The effects of this variation differ relative to the phase of the PCR. Increased plateau temperatures and high and long overshoots typically cause problems during the denaturation phase due to premature inactivation of the *Taq* polymerase. In the annealing phase, deviating temperatures, either too high or too low, typically cause problems due to nonspecific or inefficient priming. The effects of the final PCR results move on a sliding scale from slightly less-efficient PCRs, which still give a result, to complete failure. The first category is often not noticed but when used for quantification, can lead to incorrect counts.

Owing to the thermocycler variation, it is necessary to calibrate thermocyclers and validate thermocycler–assay combinations.

To perform a calibration that is representative of the process and measures all parameters of thermal performance, including uniformity, accuracy, overshoot, ramp rate, and hold time, a thermocycler calibration should be performed in a dynamic and multichannel manner. To be ISO 17025-compliant, the temperature calibration should be traceable by comparison to the international reference standard ITS-90, performed by trained and qualified persons, under controlled environmental conditions and with a calculated measurement uncertainty. The calibration results can not only be compared to manufacturer specifications, but should also be analyzed in an objective manner. The most critical parameters to analyze are nonuniformity during all phases of PCR and height and length of overshoots. Most thermocyclers can only be calibrated and cannot be adjusted by either the end user or the manufacturer. However, by adapting the programmed protocol, the thermocycler's performance can be adjusted or aligned. To fulfill the requirement of validation under ISO 17025 and ISO 15189 accreditation, either the hot-spot method, thermal boundary method, or approximated thermal boundary method can be used as a validation method. It is most important to realize that thermocyclers do vary and that solutions must be sought to manage this variation in daily use to ensure that correct and reliable data are produced.

In addition to the thermal variation, optical variation in real-time thermocyclers will also contribute to a certain level of variation in real-time PCR data. Indicative studies did show that a certain degree of thermal plus optical variation does exist in real-time thermocyclers[25] but the optical component has not been determined in a traceable manner. Future studies will need to show the individual contribution of the thermal and the optical variation to the final real-time PCR results.

REFERENCES

1. Saiki R.K., Scharf S., Faloona F., Mullis K.B., Horn G.T., Erlich H.A., Arnheim N. 1985. Enzymatic amplification of β-globin genomic sequences and restriction site analysis for diagnosis of sickle cell anemia. *Science* 230: 1350–1354.
2. Rådström P., Knutsson R., Wolffs P., Lövenklev M., Löfström C. 2004. Pre-PCR processing: Strategies to generate PCR-compatible samples. *Molecular Biotechnology* 26(2): 133–146.
3. Kessler H.H., Mühlbauer G., Stelzl E., Daghofer E., Santner B.I., Marth E. 2001. Fully automated nucleic acid extraction: MagNA pure LC. *Clinical Chemistry* 47: 1124–1126.
4. Loens K., Bergs K., Ursi D., Goossens H., Ieven M. 2007. Evaluation of NucliSens easyMAG for automated nucleic acid extraction from various clinical specimens. *Journal of Clinical Microbiology* 45: 421–425.
5. Fleige S., Pfaff M.W. 2006. RNA integrity and the effect on the real-time qRT–PCR performance. *Molecular Aspects of Medicine* 27(2–3): 126–139.
6. Bustin S.A., Nolan T. 2004 Pitfalls of quantitative reverse transcription polymerase chain reaction. *Journal of Biomolecular Techniques* 15: 155–166.
7. Yeung A.T., Holloway B.P., Adams P.S., Shipley G.L. 2004. Evaluation of dual-labeled fluorescent DNA probe purity versus performance in real-time PCR. *Biotechniques* 36(2): 266–275.
8. Livak K.J., Flood S.J., Marmaro J., Giusti W., Deetz K. 1995. Oligonucleotides with fluorescent dyes at opposite ends provide a quenched probe system useful for detecting PCR product and nucleic acid hybridization. *Polymerase Chain Reaction Methods and Applications* 4: 357–362.
9. Tyagi S., Kramer F.R. 1996. Molecular beacons: Probes that fluoresce upon hybridization. *Nature Biotechnology* 14: 303–308.
10. Wittwer C.T., Herrmann M.G., Moss A.A., Rasmussen R.P. 1997. Continuous fluorescence monitoring of rapid cycle DNA amplification. *Biotechniques* 22(1): 130–131, 134–138.
11. Bustin S.A. 2000. Absolute quantification of mRNA using realtime reverse transcription polymerase chain reaction assays. *Journal of Molecular Endocrinology* 25:169–193.
12. Wolffs P., Grage H., Hagberg O., Radstrom P. 2004. Impact of DNA polymerases and their buffer systems on quantitative real-time PCR. *Journal of Clinical Microbiology* 42: 408–411.
13. Abu Al-Soud W., Radstrom P. 1998. Capacity of nine thermostable DNA polymerases to mediate DNA amplification in the presence of PCR-inhibiting samples. *Applied Environment Microbiology* 64: 3748–3753.
14. Kim Y.H., Yang I., Bae Y.S., Park S.R. 2008. Performance evaluation of thermal cyclers for PCR in a rapid cycling condition. *Biotechniques* 44(4): 495–505.
15. Saunders G.C., Dukes J., Parkes H.C., Cornett J.H. 2001. Interlaboratory study on thermal cycler performance in controlled PCR and random amplified polymorphic DNA analysis. *Clinical Chemistry* 47(1): 47–55.
16. Vermeulen J., Pattyn F., De Preter K., Vercruysse L., Derveaux S., Mestdagh S., Lefever S., Hellemans J., Speleman F., Vandesompele J. 2009. Measurable impact of RNA quality on gene expression results from quantitative PCR. *Nucleic Acids Research* 39(9):e63. Epub: February 11, 2011.
17. CEN. 2005. ISO 17205:2005—general requirements for the competence of testing and calibration laboratories.
18. CEN. 2007. ISO 15189:2007—Medical laboratories—particular requirements for quality and competence.
19. Bustin S.A., Benes V., Garson J.A., Hellemans J., Huggett J., Kubista M., Mueller R. et al. 2009. The MIQE guidelines: Minimum information for publication of quantitative real-time PCR experiments. *Clinical Chemistry* 55(4): 611–622.
20. JCGM. 200:2008. *International Vocabulary of Metrology—Basic and General Concepts and Associated Terms* (VIM), 3rd edition. http://www.bipm.org/utils/common/documents/jcgm/JCGM_200_2008.pdf
21. Adams S., Chun H., Uribe M., Mitchell S., Marincola F., Stroncek D. 2004. Use of multichannel dynamic temperature measurement system (MTAS) to determine the efficiency of GeneAmp PCR system 9700. Poster at ASHI 2004 meeting. http://www.ashi-hla.org/docs/pubs/abstracts/abs04/146.html#top
22. Adams S., Russ C., Uribe M., Marincola F., Stroncek D. 2005. PCR validation testing utilizing a novel dynamic measurement system. Poster at ASHI 2005 meeting. http://www.ashi-hla.org/docs/pubs/abstracts/abs05/109.html#top
23. Uribe M.R., Adams S., Marincola F., Stronceck D. 2004. Correlation of data obtained utilizing MTAS thermal cycler validation systems and unexpected primer annealing. Poster at ASHI 2004 meeting. http://www.ashi-hla.org/docs/pubs/abstracts/abs04/76.html#top
24. Carl T. Wittwer C.T., Reed G.H., Gundry C.N., Vandersteen J.G., Pryor R.J. 2003. High-resolution genotyping by amplicon melting analysis using LCGreen. *Clinical Chemistry* 49: 853–860.
25. Herrmann M.G., Durtschi J.D., Bromley L.K., Wittwer C.T, Voelkerding K.V. 2006. Amplicon DNA melting analysis for mutation scanning and genotyping: Cross-platform comparison of instruments and dyes. *Clinical Chemistry* 52(3): 494–503.

10 Ultra-High-Speed PCR Instrument Development

Nick Burroughs and Emmanouil Karteris

CONTENTS

10.1 INTRODUCTION: PCR TEST SPEEDS

Polymerase chain reaction (PCR) testing has come a very long way since its inception by Mullis et al.[1–3] way back in 1983. PCR is now regarded, by most bioscientists, as the "gold standard" method for sample analysis based on genetic markers. The ability to replicate a few strands, or even a single strand, of specific genetic material many times, thus enabling the identification of the marker, has become the core technique for a generation of bioscientists.[4]

Initially, PCR tests required postamplification analysis by methods such as gel electrophoresis with ethidium bromide detection. Subsequently, "real-time" PCR (or qPCR) was developed, which combines the PCR process with a fluorescent probe or a DNA-binding dye detection in the same reaction vessel. This dramatically enhanced the original PCR process giving faster results with more accurate quantification and reduced risks of contamination. As the process has evolved, the applications for it have been increasing exponentially with PCR now forming the basis of thousands of tests, with applications ranging from infectious diseases to paternity identification, and from forensic analysis to food processing. Figure 10.1 shows the numbers of published manuscripts over the last few years giving an indicative measure of the research effort in the PCR field.

However, even at our time of writing, a typical 40-cycle PCR can take around 2 h to complete and that time has struggled to keep pace with other advances in the area. So, some of the potential benefits of this extraordinary process remain limited by the speed of the test. The speed of a PCR test is not just important for its own sake but also because there are applications for which the time to deliver a test result could mean the difference between profit and loss, or even life and death. In a research laboratory where the researcher will have other work to do while they wait for their PCR

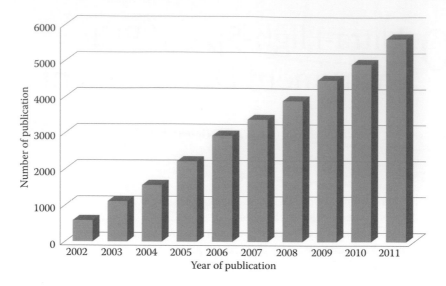

FIGURE 10.1   Manuscripts published in PubMed using real-time PCR technology over the last 10 years.

test to complete, 2 h could be regarded as an inconvenience that slows their progress; however, in a hospital, the difference between that and, for example, 10 min can save lives.[5]

A simple daily example of the importance of speed in diagnostic applications is the issue of MRSA (methicillin-resistant *Staphylococcus aureus*) identification. Patients entering hospital for a planned procedure are routinely tested for MRSA a week prior to their admission; blood samples are sent to a central clinic and typically the results take a few days to process; the actual time required for sample preparation and test is about 5 h. However, for emergency admissions, this process is too long; so, the patient is admitted with the risk of carrying the infection and therefore with the potential to infect others. Having a test that could be administered either at admission to the hospital, or even in an emergency vehicle prior to arrival, and would only take a few minutes would provide a real benefit. It is similar within an industrial setting; if a PCR test is being used to monitor a food process line for contamination, a 10-minute test could be used as an in-line process, whereas the current tests are typically applied retrospectively to daily batches meaning that the potential for costly wastage is much greater.

It is not just applications for which time is critical, that benefit from greater speed. It is far more efficient if a patient can be tested and treated on a single visit to their doctor rather than having to make a journey to the clinic to provide a sample for the test, and then return a week later to receive the results and potential treatment. A single visit would result in the time required away from work being halved, less miles being traveled, more efficient use of health care resources, faster treatment, and reduced patient stress. To make a single visit viable requires these clinic-based tests to provide their results within 30 min. Such locally administered, "while you wait" tests, referred to as "point of care" (POC) tests, can include not only tests for specific diseases but also for the effectiveness of a particular treatment on the patient. Referred to as "companion diagnostic tests" (CDx), these tests can help identify the most appropriate treatment and dosage.

10.2 WHAT LIMITS CURRENT PCR TEST SPEEDS?

The PCR process requires that the test samples are cycled through a temperature profile; a typical profile will see samples cycled between 95°C, 55°C, and 72°C, multiple times. The time taken to change the temperature of the samples between these levels is a key determinant of the speed of the process and thus of the length of a test. Tests often require a number of samples to be cycled simultaneously; for example, for comparison to each other or as process controls, each sample needs to experience the same conditions if these comparative results are to be valid.

The design of many PCR instruments relies on conductive blocks to connect the heating or cooling source(s) to the test samples; heat naturally flows within the blocks to remove any temperature gradients and so, it should, over time, deliver the same conditions across all the test samples. However, this will always be less than perfect due to the variability of cooling across such a block; uniformity of the block temperature is vulnerable to greater heat losses on the edges and surfaces that tend to distort the thermal distribution.[9] The conductivity of these blocks also affects the rate of heat flow and thus affects the thermal uniformity of the samples. That conductivity is directly related to the raw material and its thickness; thicker blocks have better conductivity; however, they also have higher thermal mass. The larger the thermal mass of the block, the greater the amount of heat that needs to be transferred and the longer this will take.

To heat and cool the system, heat needs to be driven in and out of the block. The faster the heat is driven in or out of the system, the less time the conductive block has to even out the temperature distribution and maintain the thermal uniformity. Ultimately, such a system can only maintain its thermal uniformity if the rate of change of the temperature is slower than the time taken by the conductive block to even out the temperature. So, in these type of systems, the need for uniformity of temperature is in direct conflict with the desire for speed; they can deliver one feature or the other but not both. The faster these thermal block systems are driven, the greater the variation of temperature across the test samples. Even the most highly conductive silver blocks are limited to rates of change of temperature of <3°C/s if they are to maintain acceptable conditions across all the test samples.

Systems of this type usually use one or more Peltier cells to provide both the heating and the cooling to the heat-transfer block. Peltier cells can be used to both heat and cool units depending on the direction of current flow. While the control of the heat output of a Peltier cell can be regulated quite precisely, the thermodynamic design of the rest of the system limits its performance. Control of these systems is problematic since a temperature gradient is required to force the temperature flow from the heating source to the test sample; heat flows from high temperatures to lower temperatures; the greater the difference, or gradient, the faster the flow. To achieve quick cycle times, big temperature gradients are applied to the block, which can lead to samples overshooting or undershooting their target temperatures.

These physical limitations of using a "passive heating" method (heat-transfer block) for the PCR process were recognized by BJS Technologies Ltd. As a result, they developed an innovative concept for measuring the temperature of the sample and then providing additional heat where it was needed. It was this technology, called xxpress™, which was used to create the xxpress real-time thermal cycler.

xxpress is an "active heating" methodology that precisely delivers heat to control the temperature of the test plate. It does this by accurately measuring the temperature of the sample and then precisely controlling the amount and location of the additional heating. Low thermal mass is an advantage in a system such as this as it reduces the reaction time from the input of energy to deliver a change in temperature, thus enabling better control. The combination of active heating and low thermal mass works together to enable much faster rates of temperature change while still maintaining uniform temperatures across the sample area. With heating rates of >10°C/s, the system can achieve a thermal uniformity of ±0.3°C. To achieve this performance, the software control algorithm is measuring the temperature of the samples 100 times/s and is calculating where and how much heat to apply.

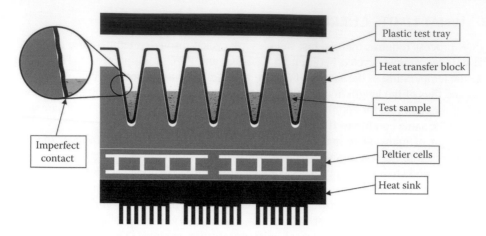

FIGURE 10.2 The layout of a traditional PCR thermal cycler with heat-transfer block showing the major components.

Another key limiting factor to the fast and accurate control of the thermal profiles of the test samples is how those test samples are thermally connected to the heating source. In traditional thermal cycler designs, a thin plastic sample tray, or microtiter plate, sits over the heat-transfer block (see Figure 10.2). This is a disposable test plate and is replaced for every experiment. It makes physical contact with the heat-transfer block and heat flows from the block through the plastic material to heat the sample. The contact from the block to the plastic is not perfect; plastic has a different expansion coefficient to metal and plastic is an insulator so that the heat flow to the samples is restricted and is uneven.[6–8] This leads to a time lag for the sample to achieve the same temperature as the heat-transfer block and that time lag varies from cell to cell depending on the amount of contact between the plastic and the block well. Time lags for conventional PCR block systems can typically be 10 s and while suppliers adopt different strategies to minimize the effects, this factor adds time into each thermal step because the system needs to ensure that each test sample has reached the desired target temperature before the required incubation time can be started. If the test sample could be brought into more intimate contact with the heating source, then the time lag would be shortened, the uncertainty would be reduced, and cycles could be faster.

10.3 HEATING AND COOLING VERY QUICKLY BUT UNDER CONTROL

In the previous sections, the need for faster thermal cycling was identified. In the following sections, the technical challenges that have to be addressed when designing such a system will be described.

The issues around thermal mass and temperature control can be examined by consideration of the physics of heat management within a heating control system; this is essentially what a PCR thermal cycler is. Heat is provided from a source, it is used to raise the temperature of a group of samples to a defined point, and hold them there. There will be heat lost continually from the system by cooling because the temperature of the system is above ambient. There are two states to consider for the system; steady state where the goal is to maintain the temperature of the system and ramping where the temperature is being actively increased or decreased.

At the incubation temperatures, a steady thermal state is required for all samples. In a steady state, the aim is to keep the samples at a chosen temperature and so the points A and B shown in Figure 10.3 should ideally be the same temperature. Assuming that the heat-transfer properties from the block into the sample wells at points X and Y are the same, then to maintain a steady state, the temperatures at X and Y should be the same. But since X is located at the edge of the block, the cooling effect is greater than Y.[9] This results in a temperature difference between X and Y. One way to minimize that difference

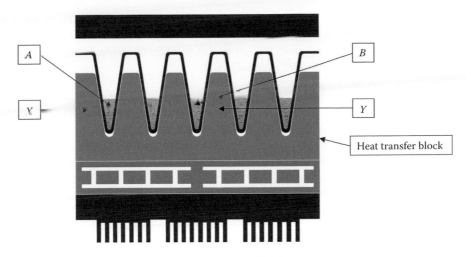

FIGURE 10.3 Schematic figure showing a section through a traditional thermal cycler.

is to ensure that the thermal resistance between X and Y is very low so that the heat flows easily between these points and the temperature is equalized. The heat-transfer block in a PCR thermal cycler is designed to provide low-resistance path; the precise design will depend on the required accuracy of the instrument that will determine the maximum value for the temperature variance across the plate.

In these systems, the greater the thermal resistance of the construct material, the thicker the heat-transfer block needs to be to achieve the thermal uniformity. The thicker the block, the greater the thermal mass, requiring more heat to be moved in or out for the required temperature change. This thermal inertia also slows down the response of the system, making it more difficult to control and leading to greater under- and overshooting of the temperature. So, the thermodynamics of such a system is in conflict; a higher thermal mass is required to ensure that the temperature across the samples is uniform and a lower thermal mass is required to provide accurate control of the plate temperature. To reduce the temperature variation across the plate, the ramp rates need to be low, but there is a desire to ramp the temperature faster to deliver the test results more quickly. Any system that uses a heat-transfer block or heating plate in this way will rely on compromise. A number of systems have been developed that incorporate ways to enhance the heat distribution; the ECO qPCR system from Illumina uses a stirred fluid inside the thermal block and to ensure sample uniformity, they also limit the size of the block. However, these modifications have only made small differences and they do not change the fundamental thermodynamic challenges.[10,11]

Other models, such as the Rotorgene from Qiagen, do not use heat-transfer blocks at all; they rely on hot air to heat the samples. Spinning the samples through this heated air gives good thermal uniformity; however, while faster than heat-transfer block systems, these instruments thermal ramp rates are still limited. The walls of the sample containers are made of insulating material and limit the rate of heat transfer.

An alternative approach was adopted by BJS, who aimed to remove these limitations by using an "active" control process. The concept of an "active" heating system is to divide the area to be thermally controlled into smaller areas or zones. Each of these zones has its own temperature measurement and its own heating system. For each zone, the control system determines how much heating is required to achieve a specific temperature change. The heat is then provided and the temperature is re-measured. The control system uses the ratio between the resulting temperature change and the initial delivery of energy to determine the requirements for the next heating cycle. While this is a simple concept to understand, the complexity of converting this concept into precise instrument performance took the company over 10 years of development. It was also deemed desirable to measure the actual temperature of the sample, or at the very least, something that was intimately linked to it, because traditional systems rely on measurements of the temperature of the thermal transfer medium, the block, or heated air, and these can be quite different from the sample temperature. The following sections contain descriptions of the issues that were encountered during the development of this system and the solutions that were developed.

10.4 TEMPERATURE MEASUREMENT

To deliver an active temperature control process as described earlier, the initial temperature of the plate needs to be established. The measurement of any temperature is a challenge since it is difficult to measure reliably and accurately without affecting the temperature of the material being measured. The xxpress thermal cycler system was developed with the aim of achieving a uniformity between test samples of better than 0.3°C. To control and deliver this, the system must be able to measure test sample temperatures to an accuracy of better than 0.1°C and to do this very quickly to allow precise control. This is a high level of accuracy for such a temperature measurement; contact and noncontact measuring systems were considered.

Contact measurement systems were quickly discounted; placing sensors directly onto a test consumable would add too much to the price of each test and connecting the individual sensors would also be challenging. (At the desired level of accuracy, the difference in the resistance of a connection between operations could be enough to invalidate the results.) A number of methods of placing thermal sensors in contact with the test plates were investigated; however, each method proved to suffer from variations at the contact interface. In addition, there is a potential for the contact to materially change the test plate temperature.

Noncontact sensors that measure the infra-red radiation emitted from a sample seemed to be the most realistic option. Using this approach would not affect the consumable plate or tube, there would be no connections to make or break for each sample, and the same sensors would be used to measure every sample so that this should improve consistency. However, a method of measuring to a higher accuracy than is typical for these sensors had to designed. Working with the U.K.'s National Physical Laboratory (NPL), a specialized calibration method was developed that determines a range of coefficients that can be applied to each of the temperature measurements to achieve the required repeatability and accuracy. Critical to the selection of these sensors were their response times. The aim was for each sensor to make 100 measurements every second as the faster this feedback loop could work, the better would be the control of the plate temperature.

The area of the test plate was subdivided into sections that were measured by a single sensor and could be controlled by the current paths that were available. A number of sizes of array and of thermal sensor were then tested to ensure that the system had sufficient control and that any variation in the temperature over the area covered by the array was much smaller that the accuracy required in the measurement.

From the number of options tested, a design consisting of nine sensors on a 3×3 grid with nine defined heating zones on the test plate proved to give the best uniformity of temperature across the chosen test plate size. This decision also influenced the design of the heating system that is described in the following section.

One of the benefits of this type of temperature measurement system is that in addition to providing the information required by the system to control the temperature of the samples, it also provides a record of the actual profile of temperatures of the samples proving that they have been controlled in the way that was expected. This data provides, a verification log of the actual thermal profile experienced by each of the test samples and is attached to the experiment's data file.

10.5 ACTIVE HEATING AND COOLING

Active heating control requires the ability to direct heating to specific areas or zones of the plate in preference to others and to precisely control the amount of heat delivered. After reviewing different heating approaches, the only practical solution that could give the precision of control and the speed of ramping required for the new system was electrical resistance heating. Electrical resistance heating is the conversion of electrical energy into heat energy when an electrical current flows through a conductor such as a metal. It has two key properties that make it ideal for this technique. First, heating occurs at the atomic level within the material and is directly proportional to the current flowing through the material. This way, it effectively heats from within and the heating is in precise and direct correlation with the electrical current that was passed through it. The second property is that, in a metallic conductor, the shortest path through the conductor has the least resistance, and hence, the greatest current flow. In a situation where heat is proportional to current flow, where the current flow is greater, so is the heating, thus, mapping the current flow through the conductor can be used to control the heating. The technical challenge was then to balance the number of current paths and the resistance of the heating element to deliver the control and accuracy that a PCR system requires.

After a significant investigation requiring thermal modeling and the testing of enumerable samples, a range of relationships were defined that were then used to create a single heating plate (see Figure 10.4) and to map a range of current paths through that plate. One of the key factors in the design was the relationship between the electrical conductivity and thermal conductivity of the plate. The better the thermal conductivity of the plate, the faster the heat can flow to correct any minor variations in the plate temperature and so, fewer heating zones will be required. In a metal, the better the thermal conductivity, the lower the electrical resistance and so, the greater the current that must flow through the plate to provide the required heating. Thus, the choice of material for the consumable plate affects the design of the power supply, the rest of the heating circuit, and the connections within it.

The optimum balance of all the factors described above led to the creation of a simple, thin aluminum plate with six electrical contact fingers, as shown in Figure 10.4. It has a very low thermal mass when compared to the traditional PCR systems and so, it can be heated and cooled much more efficiently, with ramp rates of >10°C/s being possible. Mapping the current through different combinations of these six contact fingers provides a sufficient range of heating paths to enable both the accurate control of the heating of the plate and also the ability to maintain a thermal uniformity of better than 0.2°C. Finally, the resistance of the plate, while quite low was still high enough that an electrical circuit could be practically constructed around it to drive it.

Heating is only one part of the problem when designing a thermal control system for a PCR system; the system also needs to be able to cool to complete a thermal cycle. This was achieved in xxpress by directing jets of cool air onto the base of the thin aluminum plate. These jets are driven by individual, continuously variable pumps so that control of the cooling can also be adjusted to match the cooling required in each area of the test plate. This cooling design can produce rates of cooling of greater than 15°C/s. However, the response time of this type of cooling system is slower than the resistive heating system due to the inertia of the air movement and so it cannot provide the same precision as the heating system. To achieve the same thermal control and uniformity for cooling as for heating, the resistive heating system is used to provide fine control during the cooling process. It provides small amounts of heat into the cooler parts of the test plate to slow their cooling to match the desired rate.

10.6 THERMAL CONTROL ALGORITHM

In the earlier discussions, the ability to provide "active control" of the thermal system is assumed once the physical parameters have been optimized; however, this proved to be one of the key challenges of the development. The system has multiple degrees of freedom and required considerable analysis to develop a control theory and subsequent algorithm.

There are a number of current paths that can be chosen to heat the plate; each path will heat a number of zones, not just a single zone. So, the control algorithm is required to make choices on

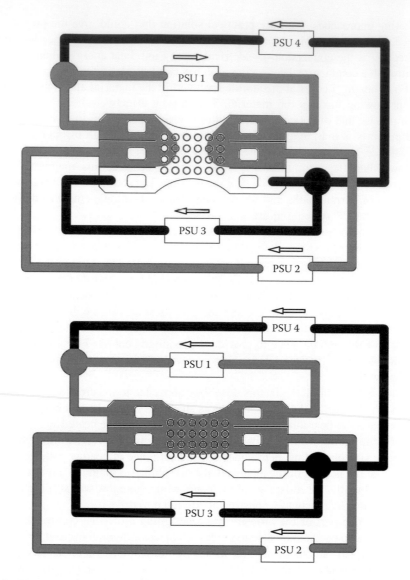

FIGURE 10.4   Two examples of an xxpress™ sample plate showing different current flow patterns.

how best to raise the temperature of a particular zone without detrimentally affecting the temperatures of other zones. Within the same time slice, all the zones will also be changing temperature as they will be above the ambient temperature and subject to cooling; that cooling rate will vary across the plate; the problem is a complex one.

The solution that was developed combined a set of rules for decision making and some high-level mathematical control theory. The decision-making rules guided the algorithm as to the best strategy to use to affect the temperature of the zone based on the desired change. Then, the control theory would calculate the correct electrical energy to deploy to deliver the required change. Finally, the sampling speed of the system was balanced with the thermal inertia of the consumable to give the accuracy required. In the format of the xxpress thermal cycler, that sample speed sees the control loop measuring, calculating a reaction, and delivering the next heating action 100 times every second.

10.7 SAMPLE CONTACT WITH THE HEATING SOURCE

Until this point, the discussion has been focused on controlling the temperature of a test plate or a thermal block but the real requirement of a PCR system is to control the temperature of a small test sample, typically 0.5–40 µL of reaction mix. In traditional PCR instruments, these test samples are contained within multiwell "micro-titer" trays (disposables). These are molded from polypropylene and have a typical wall thickness of 0.2 mm. The individual test wells fit snugly into a mating form on the heat-transfer plate so that heat can travel from the thermal block through the polypropylene to the test sample. However, this process is far from efficient because the walls of the microtiter tray touch the heat-transfer plate intermittently and the thickness of polypropylene used has significant insulating properties. This means that the temperature of the samples generally lag the temperature of the transfer block, often by over 10 s and the degree of temperature difference can vary from sample to sample.

In an effort to minimize this effect, BJS identified a method for coating the aluminum consumable directly with polypropylene so that the plastic would always be in excellent contact with the heat surface. The coating of polypropylene is only 0.01 mm thick that is 20 times thinner than a microtiter plate, allowing more efficient conduction of heat. In addition, they designed the test wells to be flat bottomed to maximize the contact area with the aluminum plate. These critical design steps enable the test samples to follow the temperature of the test plate very closely, resulting in typically less than 0.5 s time lag, and this lag is consistent from well to well. The time lags for all the wells on the tray are over 20 times better than traditional designs. This improvement not only speeds up the reaction time as the system does not need to wait for the sample temperatures to stabilize, but it also provides a much more precise thermal profile for the test samples. As has been shown, PCR experiments give more dependable results when temperatures are brought more quickly to the target temperature.[12]

10.8 UNDERSHOOT AND OVERSHOOT

As previously discussed, the thermal response of a typical PCR system limits its speed and controllability. Thus, as the target temperature of the thermal cycler is approached, the control system has difficulty in delivering the desired temperature both quickly and precisely. As a result, such systems either overshoot or undershoot the desired temperature. Overshooting leads to the samples experiencing temperatures in excess of those planned, often significantly higher, which can damage the sample. Undershooting results in the samples arriving at the desired temperature more slowly than required; the samples do not reach the desired temperature for the requisite time, which may lead to an incomplete reaction step. Both of these issues can be avoided by slowing the ramping rate as the desired temperature is approached, but this extends the length of time required to complete the experiment and, in some situations, going more slowly can promote the formation of unwanted primer–dimer products.[12] Individual manufacturers have differing approaches for combating these factors, but ultimately, the physics of the block temperature regulation, as described earlier, limits the speed at which a stable and uniform temperature can be achieved.

The active temperature control system employed by the xxpress thermal cycler means that even while delivering a 10°C/s temperature ramp rate, it can control the under/over shoot to <0.5°C for <0.5 s. This compares very favorably with a typical performance of Peltier system that can cause an under/over shoot of typically ±4°C for >10 s when delivering only a 4°C/s ramp rate. This performance is a major factor in improving the consistency of the PCR experiment and thus in improving the validity of the results.

10.9 CAN CHEMISTRIES MATCH THE THERMAL CYCLER SPEED?

The xxpress thermal cycler can deliver very rapid and extremely accurate thermal cycling. Thermal ramp rates of 10°C/s save significant time over the current PCR profiles; however, that is not the only component of reaction times. The dwell times at each of the temperatures in the cycle are also a major factor in the length of the entire reaction. Samples are required to dwell at a temperature for a particular reaction to occur; the faster these reactions occur, the shorter the required dwell time and the quicker the cycle.

The dwell time consists of two elements: First, a defined period of time is required to ensure that all the samples are at the required temperature. This time period, the stabilization time, is dependent on the PCR instrument and the chosen ramp profile and not on the reagents. As discussed previously, instruments overshoot and undershoot the desired temperature targets and it can be a number of seconds before the whole plate is within the required temperature window. The second element, the reaction time, is the actual time taken to complete that reaction step. The reaction time is determined by the reagents and the specific experiment design.

Reagent suppliers have made vast improvements in recent years, creating new products that complete the reaction steps quickly, with dwell times of 1 or 2 s being possible, once the reactions are at the correct temperatures.

Figure 10.5 shows a comparison between a traditional thermal cycler and the active control adopted by the xxpress system. The time for each phase of the cycle is reduced when using the xxpress system, leading to dramatic reductions in the length of the experiment. It is clear that it is the combination of the fast ramp rates, good thermal uniformity, and the short stabilization times that enable the latest reagents to perform to their best and deliver a 40-cycle PCR test in less than 10 min. However, even with slower reagents, significant gains can still be achieved by utilizing faster ramping and reduced stabilization times.

10.10   A USER INTERFACE THAT DOES NOT REQUIRE A MANUAL

As has been stated, PCR is the accepted gold standard test for researching DNA and the thermal cycler is the tool that enables that process; however, it should not be a barrier to it. In the past, many thermal cyclers required specialist training to be completed or long user manuals to be read before they can be used. So, in conceiving the xxpress thermal cycler, a key design goal for BJS was that the user, provided they understood the PCR process, should be able to operate their thermal cycler without having to refer to a manual or go on a training course. It was to meet the requirements of this goal that the xxpress touch screen user interface (UI) was created, a UI that would guide them through the process.

The UI on the xxpress thermal cycler is in the form of a "film strip" that moves from right to left across the screen. In the frames of this film strip, the user is asked to make key choices as to what they wish to do or input the specific experimental information. Much of the other information is prepopulated from data files on the reagents chosen. (Many processes for a user of a thermal cycler are repetitive; remembering the user's preferences and past setups make these tasks quicker and less irritating as well as leading to fewer mistakes.)

The process of experimental setup usually consists of just three steps: In the first step, the user identifies the type of qPCR they wish to perform and the reagents that they plan to use; this information is used to create the thermal profile and fluorescence measurement parameters automatically. These values can all be changed if required but they are based upon what the manufacturer of the reagent recommends. The second decision is the choice of the sample size and the number of tests; there are currently three plate formats for the unit: 24, 54, and 96 wells. Interchangeable, each of these fit into the unit without further modification. The unit automatically recognizes the format of the test plate that is put in the machine, makes any necessary adjustments, and checks to see if it matches what the user has requested. Once the test plate format has been chosen, the arrangement of the experiment can be planned. Genes of interest, reference genes, and no-template controls can all be added from the menu. Arrangements are always remembered so that future experiments can use the same layout. Finally, the thermal profile and measurement information is displayed so that any adjustments can be made before starting the program.

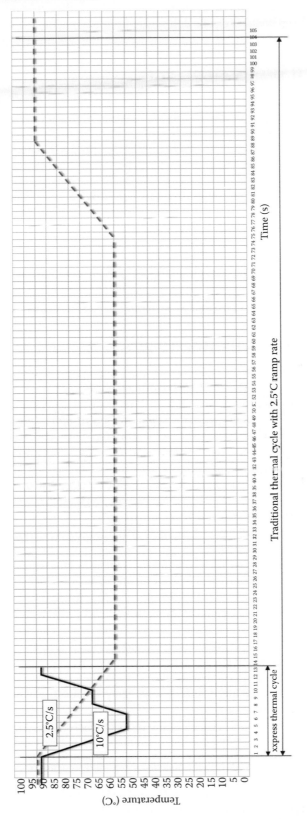

FIGURE 10.5 Comparative timing curves for traditional and xxpress PCR systems.

A similar approach is used to provide analysis of the data once the experiment is complete. This analysis as well as program creation can be done away from the machine to maximize the utilization of the instrument.

10.11 ADDITIONAL DESIGN CHALLENGES

When designing a new instrument, it is important not just to focus on a single strong-performance criterion but also to try to match that higher standard across all the features of the instrument, by delivering improvements in all the areas of performance. This is particularly true when a significant improvement in one feature can highlight a disproportionate limitation in another. The speed of the test process is an important metric, and so too is a total reaction cost. A crucial component of the cost of a PCR test is the reagent. Reagent costs are regularly 20 times the cost of the consumable test plate; minimizing that cost can provide significant savings. Typically reducing reagent costs requires adopting smaller test sample volumes while achieving the same results as in larger volume reactions. To address this challenge, BJS created a range of test plates that could handle different experimental volumes, ranging from 0.5 up to 40 μL using test plates that have a common format and are interchangeable (Figure 10.6). This change in design philosophy over the traditional systems enables an instrument to be used seamlessly by a number of users in a laboratory without creating delays or the potentials for error if the block format requires reconfiguration for each particular experiment.

10.12 BENEFITS OF ULTRA-HIGH-SPEED PCR

PCR has been improving continually ever since its inception approximately 30 years ago and as with all developments, some changes are small and often incremental whereas others occur through larger "breakthrough" improvements that can lead to more radical changes in the way a solution can be applied. Such is the case with ultra-high-speed PCR and the 10-min test; this technology will be disruptive to the accepted convention in that it will cause users to fundamentally review how they use PCR. Until relatively recently, the view of diagnostic testing was that it would be done centrally, in specialist laboratories using larger and evermore expensive equipment. The issues around timescales, the transporting of samples, managing of data, and the need to have patients visit their health care center multiple times, were managed. It was not regarded as inefficient as the alternative would have been not having access to these services at all. However, if these services can be provided closer to the patient on a "while you wait" basis, there will be many benefits to both the patient and the community at large. Access to the latest diagnostic testing could also be made available in less developed areas of the world, in countries where transport infrastructures and sophisticated centralized laboratories, a prerequisite to current solutions, do not exist.

A successful diagnostic test needs to fulfill two key criteria: It needs to be sensitive enough to detect the change that identifies the parameter being tested and it needs to be specific enough to determine that it is that change, and not another, which is the cause. This testing principle applies across all science but is most pertinent in the development of diagnostic tests. Every diagnostic test is assessed for its specificity and sensitivity. Our greater understanding of genetics has been the catalyst for recent breakthroughs in the design of new diagnostic tests; the identification of a genetic marker, DNA, RNA, or otherwise, provides a very specific and hence reliable test. The PCR process lies at the core of these developments because it is highly specific and sensitive, being able to detect as little as a single DNA strand. So, for a number of years, the focus of much of the work on the development of diagnostic tests has been in identifying genetic markers for key diseases and conditions.

An example of the impact of this work is in the area of human immunodeficiency virus (HIV) testing. Until recently, HIV was identified by testing for antibodies that are produced by the body to fight the infection. However, these antibodies can take over 12 weeks to develop, and sometimes as long as 12 months; so, providing a period of uncertainty where the patient may go untreated potentially becomes

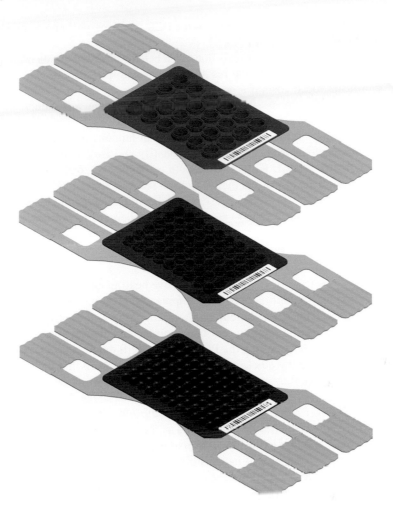

FIGURE 10.6   Three types of xxpress consumable test plates.

a source of infection to others. Recently, a reverse transcription (RT)–PCR-based test was developed that can detect one or more of the several target sequences located in specific HIV genes, such as HIV-I GAG, HIV-II GAG, HIV-env, or HIV-pol. This test enables testing within 3 weeks of possible infection. Such a major improvement enables earlier treatment and reduces significantly the stress from outcome uncertainty; this test is also now routinely used to screen blood supplies.[13]

While PCR-based tests have proved their effectiveness, these types of tests have the potential to provide significantly greater benefits if they could be provided locally where the patient interacts with the health care system so that the patient can be diagnosed and treated at the first visit.[14] One of the barriers to delivering these types of tests closer to their "POC" has been the time taken to get the results. If the results are not going to be available for a number of hours, then is it practical or useful to wait for them? In many cases, it is not. While if the test was to give results in around 20 min, then POC diagnosis becomes a possibility. There are many other elements to delivering fast tests at the point of need, such as the sample preparation that can take many hours and requires specialist laboratory techniques, but the focus on improving these other elements will become much sharper once the current time-limiting step, the PCR, has been redefined. It is not just in the field of medical diagnostics that a time saving in PCR test time could deliver both benefits and drive further development. There are a number of other applications where this level of improvement in the PCR test time will create major revisions in how it can be used.

In food testing, the current PCR testing is done on a batch basis, with items often being shipped out of the production facility prior to the results being available. This can lead to public embarrassment and significant recall costs if a process fails. Clearly, on site, real-time testing providing continuous monitoring of the process, catching issues before shipment, and reducing potential waste would be a huge improvement over the current testing protocols.

Control and monitoring of epidemic outbreaks, both in humans and in animals, relies on the rapid turnaround of sample analysis and results to identify and isolate an outbreak. Examples of such critical situations were the U.K. foot-and-mouth epidemics of 2001 and 2007 where tens of thousands of animals were destroyed and their travel was severely restricted.[15] Quicker identification of the disease in flocks would have isolated the cases earlier reducing the impact to the livestock and the country. PCR test speed improvements can provide clear benefits in time-critical applications.

10.13   SUMMARY AND FUTURE STEPS

The change in philosophy from passive heating to active heating enables a system to be designed with very low thermal mass enabling both fast ramping and accurate temperature control across the whole sample area. The development of the technology for measuring and controlling an active heating system is challenging, but once completed, the benefits are significant. The resulting step change in cycle times for the PCR not only will enable general time savings to be made but could also facilitate the quicker delivery of many tests such that they could be provided in a different manner to how they are currently, on a "while you wait" or "in process" basis.

Looking ahead, the active heating technology is scalable and so can be applied to smaller individual test systems or larger automated batch-based platforms. This process is also significantly more thermally efficient than the traditional systems as the amount of energy being moved in and out of the sample is much less due to the lower thermal masses. What remains to be seen is the level of impact that this increased speed will have within the arena of PCR and against the other technologies available in the wider world of diagnostic testing.

REFERENCES

1. Saiki, R.K., Scharf, S., Faloona, F., Mullis, K.B., Horn, G.T., Erlich, H.A., and Arnheim, N. Enzymatic amplification of beta-globin genomic sequences and restriction site analysis for diagnosis of sickle-cell anemia. *Science*. 1985; 230:1350–1354.
2. Mullis, K.B. The unusual origin of the polymerase chain reaction. *Sci Am*. 1990; 262:56–61
3. Mullis, K.B. and Faloona, F.A. Specific synthesis of DNA *in vitro* via a polymerase-catalyzed chain reaction. *Methods Enzymol*. 1987; 155:335–350.
4. Valasek, M.A. and Repa, J.J. The power of real-time PCR. *Adv Physiol Educ*. 2005; 29:151–159.
5. Kuhlmeier, D. Fraunhofer institute. Fast sepsis test can save lives. *e! Science News*. 2010: Dec 20.
6. Plastic basic material properties, *Plastic Engineering, Manufacturing and Data Handbook*, Springer, 2001, ISBN 978-0792373162.
7. Thermal expansion of solids (*Cindas Data Series on Material Properties*, Vols. 1–4), Ho CY and Taylor RE. (eds), *ASM International*, 1998, ISBN 978-0871706232.
8. Basic heat transfer and some applications in polymer processing (a version of this was published as a book chapter in *Plastics Technician's Toolbox*, Vol. 2, pp. 21–33, SPE 2002) Vlachopoulos J and Strutt D. (eds), ISBN 978-1-904455-72-1.
9. PCR troubleshooting and optimisation: The essential guide. Caister Academic Press, Kenndy S. and Oswald N. (eds), 2011, ISBN 978-1-904455-72-1.
10. Herrmann, M.G, Durtschi, J.D, Bromley, L.K, Wittwer, C.T., and Voelkerding, K.V. Amplicon DNA melting analysis for mutation scanning and genotyping: Cross-platform comparison of instruments and dyes. *Clin Chem*. 2006; 52:494–503.

11. Herrmann, M.G, Durtschi, J.D, Wittwer, C.T., and Voelkerding, K.V. Expanded instrument comparison of amplicon DNA melting analysis for mutation scanning and genotyping. *Clin Chem.* 2007; 53:1544–1548.

12. Kim, Y H, Yang, I., Bae, Y.S., and Park, S.R. Performance evaluation of thermal cyclers for PCR in a rapid cycling condition. *BioTechniques.* 2008; 44:495–505.

13. Defoort, J.P, Martin, M., Casano, B., Prato, S., Camilla, C., and Fert, V. Simultaneous detection of multiplex-amplified human immunodeficiency virus type 1 RNA, hepatitis C virus RNA, and hepatitis B virus DNA using a flow cytometer microsphere-based hybridization assay. *J Clin Microbiol.* 2000; 38(3):1066–1071.

14. Murilo, R. Melo, Samantha C., and Daniel B. Miniaturization and globalization of clinical laboratory activities. *Clin Chem Lab Med.* 2011; 49(4):581–586.

15. Barlic-Maganja, D., Grom, J., Toplak, I., and Hostnik, P. Detection of foot and mouth disease virus by RT-PCR. *Vet Res Commun.* 2004; 28(2):149–585.

11 Dropletization of Bioreactions
How Single-Cell Quantitative Analyses May Improve Gene Measurement and Cancer Diagnosis

Ehsan Karimiani, Amelia Markey, and Philip Day

CONTENTS

11.1 INTRODUCTION

The application of molecular techniques specifically developed for clinical and research settings long predate the Human Genome Project. Molecular pathological methods were used to explicate the genetic basis of many diseases, and these innovations eventually contributed to the discipline of molecular diagnostics. Insights gained from diagnostics have led to improvements in genetic and molecular approaches to the classification of human cancers. This has developed, through multi-parallelized approaches, into the validation of predictive biomarkers for medical response, disease progression, and the susceptibility of individual patients to develop malignancies. Today, molecular diagnostics continues to develop rapidly as diagnosis becomes inextricably linked to therapy.[1,2]

The overall resolution of molecular analysis is improving, particularly as the field reaches a crossroad whereby discrete genetic or other biomarker analysis subsides in the wake of gene-network analyses, enabled through holistic, systems-based, biological modeling of disease-causing pathways. This improvement also comes with a deepening and firmer comprehension of disease mechanisms, the benign term "biomarkers" conceding to more functionally related molecular descriptors.[3]

11.2 IMPLICATIONS OF MATHEMATICAL MODELING

While the implementation of mathematical simulations and network analyses might be thought of as a challenger to the molecular sciences, in fact, the converse is true. Indeed, network approaches might finally lay to rest the biggest Achilles' heel of the molecular sciences' consistent difficulty to reproduce many experimental studies. Modeling networks are complex to produce and derive best meaning when produced from high-grade, wet-laboratory, quantitative data. This takes the experimentalist back to the poor reproducibility scenario that the molecular scientist has endured for so long; however, the systems biologist respects that tissue heterogeneity is a functional feature of life, and tries to understand its role. A, perhaps, outwardly naive perspective held by the largely mathematical community of systems modelers has led the discipline to consider the cell (and its microenvironment) as the unit of life that has consequence for modeling. Furthermore, the units necessary to produce models are discrete molecules, and the cell is the common denominator. This approach can be readily adopted as moles, grams, or relative units can be related in absolute molecules, and since all measured molecular moieties are related to their cells of origin as molecules per cell, then a scenario is derived whereby DNA, miRNA, mRNA, proteins, and all other molecules from a single cell can be integrated through modeling.

The molecular sciences have, for some considerable time, had access to molecular numbers, but have often elected to use rank or relative copy numbers, and have undergone a substantial period, decades long, where unusual terms, including "semi-quantitative" were an acceptable means to describe biological phenomena. The impact and relevance of expressing molecular data as molecules per cell is rapidly gaining popularity, and this is for no other reason than that the unit descriptor incorporates the cell that is readily identifiable and enables reproducible analyses. Particularly for *in vitro* gene-amplification techniques, the timing is fortuitous as single-molecule detection sensitivity is routinely achievable. The limitation that arises is the current inability to isolate a cell that is fully representative of the cell population, or more specifically, a typical disease cell.[4]

11.3 DEFINED SAMPLE HANDLING AND PREPARATION: A PREREQUISITE FOR MEANINGFUL QUANTIFICATION

Network analysis has directly contributed to driving molecular biology analytics toward a point that it has been encircling for some time. This relates to challenges around sample preparation where the objective is to glean a firm comprehension of tissue function through structure and analysis of cellular heterogeneity. Exquisite analytical sensitivity and measurement accuracy of molecules has often been grossly impaired by ill-defined bulk-tissue preparations. Currently, there is a requirement to combine high-sensitivity analyte measurement with a sufficiently high-throughput single-cell capacity, which collectively enables the detailed profiling of cell populations. This, in part, presents a slight departure from the usual gene-analysis approaches. In this circumstance, all cells are not assumed to be equivalent but are known to be a single cell and no more. Only once a cell has been analyzed, is it characterized as being normal, stromal, or disease-related.[5]

This challenge is very demanding but is well placed to steer molecular biology toward the dawning generation of meaningful nucleic acid quantification. Dissecting the heterogeneous nature of tissues becomes plausible and, with it, the prospect of determining how individual cells derive function, both singularly and in terms of organ activity, and their spatial proximity within organs.

11.4 SAMPLE PROCESSING SOLUTIONS PROVIDED BY MICROFLUIDIC TECHNOLOGIES

Molecular research is used to identify the most fundamental genetic and proteomic signatures of the disease. These developments are permitting clinical testing to move toward performing the present day point-of-care (POC) analyses and are laying the basis for personalized medicine for the future. Mutations that are responsible for cancers are being discovered, functionally characterized, and used routinely in molecular diagnostic tests.

A wide variety of microfluidic design concepts have been implemented for the treatment and analysis of biological samples. These designs fall into three broad categories: well-based,[6-8] single-phase continuous flow,[9-13] and droplet-based devices.[14-19] Droplet-based designs possess several, distinct advantages over other designs for the handling of biological samples, most notably in the case of single-cell analysis (Figure 11.1a). First, the throughput of droplet-based devices is superior to that of well-based devices. Multiple wells would need to be manufactured in parallel to produce a high-throughput, well-based design. This has implications for the size of the device and has limited use over the currently available 384- and 1536-well plates in terms of the number of samples processed in a given time. Some commercial manufacturers (such as Fluidigm) have increased the well density still further. However, as both single-phase

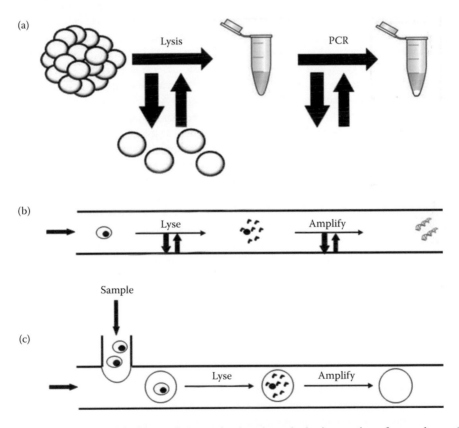

FIGURE 11.1 Reducing sample loss and contamination through the integration of procedures. (a) Conventional sample preparation and PCR with procedures isolated in space and time; the arrows indicate loss and contamination of both cells and nucleic acids. (b) Sample preparation and PCR in a single-phase microfluidic device; the vertical arrows indicate loss and contamination of both cells and nucleic acids between the sample and the walls of the device. (c) Sample preparation and PCR in a dual-phase, droplet-based microfluidic device; the sample is contained and processed with a droplet greatly reducing the likelihood of sample loss and contamination.

and droplet-based systems are a continuous flow, as opposed to static wells, a much higher throughput can be achieved by allowing the sample to travel through the device, making the assay throughput a function of time (see www.raindancetechnologies.com) as opposed to a finite number of available reactors.

A limitation of both well-based and single-phase continuous flow designs is that the sample is in constant contact with the walls of the device.[20] Adsorption of the sample onto the walls of the device may not only hinder the analysis of the sample but may also present a potential source of contamination if the device is to be reused (Figure 11.1b). Problems such as sample loss and contamination can be prevented to some extent by treating the internal walls of the device with a blocking agent (such as bovine serum albumin) or a hydrophobic coating.[12,21–23] However, a more reliable preventative measure against sample loss and contamination is to encapsulate the sample within an aqueous droplet suspended in an immiscible oil-carrier fluid (Figure 11.1c). Optimized droplet generation, along with the use of surfactants, can be used to avoid any interaction between the neighboring droplets while the immiscible oil-carrier phase isolates the droplet contents from interaction with the walls of the device. In this way, each droplet provides an isolated environment for a submicron scale bioreaction.

Another feature of droplet-based, microfluidic systems, which is particularly advantageous to molecular diagnostics applications, is that the contents of each individual droplet can be controlled with precise handling of the fluidics. This can be optimized to allow the isolation of a single cell within each droplet produced on the device.[24,25] In this way, multiple single cells can be processed and analyzed in a high-throughput, contamination-free environment. The analysis of multiple single cells in isolation will not only allow the clinician to dissect the bulk cell population into more meaningful subpopulations of cells, but will also heighten the detection of rare cells within the larger bulk population.[10,26,27] This means of analysis is currently available only if cells are isolated from the bulk population by fluorescence-activated cell sorting (FACS) prior to testing. However, FACS cell sorting is too inaccurate, prone to contamination, and usually does not generate the throughput of analysis required to be feasibly implemented in a clinical setting.

A final requirement of microfluidic systems for bioreactions is that the complete preparation and analysis of the crude sample can be integrated and carried out on a single microfluidic device. This micrototal analytical system (μ-TAS) should be able to prepare the sample from its crude, raw format to a state that is suitable for input into the required biological assay, perform the assay without loss or contamination, and finally detect the output of this assay.[28] Such a goal requires the careful manipulation of both the sample and reagents on the device.

To achieve operations such as purification of the sample from its crude form, cell lysis, PCR, and detection, the sample must be transported through multiple environments involving different buffer conditions, reagent additions, and temperatures. For this to be achieved in a well-based or single-phase continuous flow format, the sample would have to be transported between different regions of the device, which would need to be separated with complex partitioning, gating, or microvalve systems.[6,7,29] A droplet-based system provides the best format for true μ-TAS as droplets can be manipulated in a number of ways, with ease. Through the use of electrodes, the path of droplet travel can be directed, allowing for the splitting,[30–32] merging,[33–37] sorting,[31,37–40] and storage[41–45] of droplets on the device. For example, known quantities of reagents can be added to each sample through the precise merging of reagent-containing and sample-containing droplets.[36]

The integration of all steps of sample preparation and analysis, as well as having a known sample size, that is, a single cell, are important features of a device that is designed for the diagnosis of molecular pathologies in clinical samples. However, to analyze each cell on an individual basis, one must ensure that the cell is shielded from any source of contamination, that none of the limited sample is lost throughout the steps of preparation and analysis, and that cells can be analyzed in a high-throughput manner. As mentioned previously, clinical samples have a tertiary architecture and are often very heterogeneous in nature and as such require the assessment of single cells to comprehend the level of tissue activities, and thus achieve an informative diagnosis. This is particularly important if the goal is to stratify patients into discrete risk groups as the basis of personalized medication. This type of targeted medication for specific risk groups of patients will help to curb the ever increasing drug attrition rate that has been caused by pursuing the "one drug fits all" approach to drug development.[46]

So far, the lab-on-a-chip advances have shed light on the automated, nucleic acid preparation, real-time PCR, and DNA-sequencing processes. However modern, innovative molecular techniques are not without potential disadvantages.[47] Microfluidic methodologies offer many advantages but are often more expensive than the conventional systems. Despite this major concern, a procedure such as miniaturized PCR can be cost-effective if labor, amount of data, ease of data interpretation, and the levels of consumables used are all taken into consideration. Conventional qPCR also necessitates adequate testing facilities with standard operating procedures employing meticulous contamination controls.[48] Accordingly, molecular diagnostic methods based on automated microfluidic devices offer potential advantages as alternative methodologies. Miniaturized procedures often aim to improve sensitivity and specificity compared to the often-times more subjective diagnostic examinations. They may be able to detect nucleic acids in acute disease at POC prior to the formation of a detectable serologic response. The value of rapid POC turn-around time and measurement of *molecules per cell* as the units for quantitative analysis is not necessarily usable or linked to therapy. However, the high-resolution analysis on offer will certainly be pivotal in the implementation of personalized medication that will most likely be based on the identification of molecular biomarker-defined risk groups.

11.5 PROBLEMS ASSOCIATED WITH MEASURING HETEROGENEOUS POPULATIONS

Currently, routine quantitative methodologies are used to analyze the average expression level for a population of cells and are unable to determine the variation between individual cells. Some of the published studies report that the level of mRNA expression among individual cells is lognormally distributed, others report a biomodal two-component lognormal models distribution.[49,50] For that reason, the average expression, of a given transcript, analyzed in a cell population does not represent the transcript expression of any one cell within that population. The averaged data will be strongly biased by a tiny population of individual, active cells with a particularly high or low level of mRNA of a particular gene. Hence, there would be limited value in estimating the dynamics of gene expression in a population of cells or odd single cells at the other end of the spectrum. A high-throughput single-cell analysis system could have a substantial role in the future of molecular diagnostics and, following the current pioneering single-cell studies, would have the added benefit of identifying the minimum number of cells required to be able to draw a conclusion that is accurately predictive of the full spectrum of the gene copy number and hence the associated clinical risk.

11.6 FUTURE PERSPECTIVE

For the present, while miniaturization is making in-roads into molecular biology and the diagnostic sciences, there still exist major obstacles that prevent its full implementation. Until full μ-TAS can be implemented, where the main caveat with μ-TAS relates to sample preparation, the impact of miniaturization will remain limited. Given the current medical research interest to improve molecular diagnostics, and a growing requirement to measure the heterogeneity of single cells in cancers, miniaturization has been applied to address heterogeneity even without μ-TAS. This study fuses FACS, as a front end to supply nucleic acids for two-phase droplet production, to analyze heterogeneity in populations of chronic myeloid leU.K.emia (CML) cells. PCR performed in aqueous droplets was used as an *in vitro*, enzymatic-based chemical amplifier of nucleic acid sequences specific to CML cells.[51]

11.7 METHODS

11.7.1 DROPLET-BASED PCR

Methods for producing water-in-oil microdroplets in miniaturized fluidic devices and PCR reactions within these droplets were investigated. The instruments for droplet fabrication and PCR were based on a previously applied technique reported by this group.[16,52] This platform consists of a "T"-shaped channel, into which water and oil are introduced, as illustrated in Figure 11.2. Under optimum flow conditions, due to shear forces produced by the oil flow, the aqueous stream is sheared into droplets as it enters the oil stream at the "T" junction. The droplets then flow repeatedly through a two-temperature thermocycle to achieve PCR amplification.

11.7.2 FABRICATION AND MACHINING

The PCR fluidic device used in this work is composed of a planar chip for sample injection and a cylindrical, two-temperature aluminum block for the thermo cycling. The sample, after being dropletized in the planar chip, flows through polytetrafluoroethylene (PTFE) tubing that is coiled around the two-temperature cylinder (Figure 11.2). The choice of using PTFE tubing was driven by its superb hydrophobicity, chemical inertness, and ability to withstand high pressures.

The primary planar chip was made from 6-mm-thick polymethyl methacrylate (PMMA) cut to 75×75 mm (RS, Corby, U.K.) prior to machining. First, a rectangular bedding groove for the PTFE tubing (OD = 1.07 mm, ID = 0.55 mm, Microbore PTFE tubing, Cole-Parmer Instrument Co. Ltd., U.K.) was machined and the tubing was inserted. Ultraviolet (UV)-curable epoxy (Norland 68, Norland Products Inc., New Brunswick, New Jersey) was used to fill the gap between the bedding groove and the tubing, and was smoothed after curing. The aqueous sample inlet channel (150 µm wide and deep and 5 mm long) was then machined into the PMMA and the PTFE tubing. Finally, a hole was drilled into the center of the PTFE tubing with a small drill (100 µm diameter) to join the planar aqueous channel with the tubing. The tubing was then coiled around an aluminum cylinder (diameter = 40 mm, length = 65 mm; thread pitch = 1.5 mm, depth = 0.58 mm) that is cut into halves that accommodates heater elements for the thermo cycling. The devices were designed by Dr. Stephan Mohr (University of Manchester) using AUTODESK INVENTOR (Autodesk inc), translated into CNC machine code by EDGECAM software (EdgeCam, Reading, U.K.) and micromachined using a CAT3D M6 CNC milling machine (Datron Technologies Ltd., Milton Keynes, U.K.).

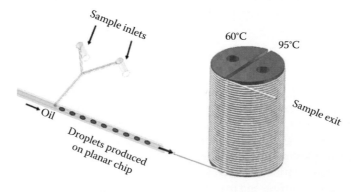

FIGURE 11.2 Device setup. Denaturation and extension steps for the PCR in the aqueous droplets occurred in 95°C and 60°C zones, respectively. The best results were achieved if the droplets dwelled in each temperature zone for 30 s.

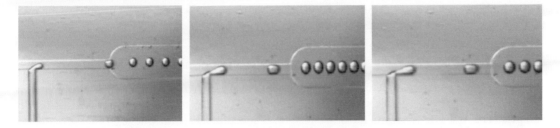

FIGURE 11.3 Droplet formation. An example of aqueous droplets containing the sample are formed in this platform through the shear force generated at the T-junction where the inlet carrying the aqueous phase meets the oil flow. The camera pictures (from left to right) are representative of water droplet production in the oil stream at T-junction. 30 frames/s, water flow rate = 83 nL^{-1} s, and oil flow rate = 1040 nL s^{-1}.

The oil and the aqueous sample were infused by a dual Harvard syringe at different flow rates, causing the sample to be cut off into consistent-sized droplets. An example of aqueous droplets in oil is shown in Figure 11.3 in a planar chip setup.

11.7.3 OPTIMIZATION OF FLOW RATES AND DROPLET FORMATION

Initial experiments were performed to assess the association between fluid flow rates and the size of droplets. In the first set of tests, a steady oil flow rate was introduced to the chip at 0.83 μL s^{-1} while the water stream rate was varied from 83 to 417 nL s^{-1}. Flow rates were optimized to find a balance between the two-temperature exposure times of the droplets and the throughput of the reaction. Before using expensive PCR reagents, water was used to study the effects of the flow rates on the exposure times.

11.7.4 HEATING SYSTEM

The two temperature zones in the coiled aluminum cylinder were generated by using two 50 W cartridge heaters (RS Components Ltd., Corby, U.K.) as shown above in Figure 11.2. To achieve stable temperatures, a K-type thermocouple was used to measure the temperatures and these data were then fed into a proportional–integral–derivative (PID) digital controller (CAL 9900, RS Components Ltd.)

11.7.5 ANALYSIS OF PCR PRODUCTS

A microfluidic-based capillary electrophoresis platform (Agilent 2100 Bioanalyzer, Version A.02.12S1292, Agilent Technology, Illinois) was used to analyze the amplified products.

11.8 RESULTS

11.8.1 DROPLET-BASED PCR

In recent years, the miniaturization of PCR has extensively been reported. This study has attempted to employ a miniaturized PCR platform that would allow single-cell PCR from one individual cell inside one aqueous droplet. Such a system, together with detection optics may allow single-cell PCR to be performed. These high-throughput platforms would be of great value in medical diagnostics where detection of a small population of the cells of interest within a mixed population of cells offers benefit. One example of where such a technique may be carried out is to determine rare cells in CML known as minimal residual disease (MRD). In this study, the application of water-in-oil droplets was analyzed and applied to a miniaturized platform for the amplification of the fusion transcript breakpoint cluster region (BCR)–Abelson murine leU.K.emia (ABL) by splitting the PCR mixture into nanoliter-sized droplets.

11.8.2 Impact of Oil Flow Rates on the Droplet Production and Exposure Times

A feature of this platform is the suspension of aqueous droplets in an immiscible oil-carrier stream. This method allows the droplets to be isolated from each other to avoid cross-contamination of samples. The first step is to form the droplets at the T-junction where the inlet carrying the aqueous phase and oil flow meet at an angle of 90°. The oil carrier fluid flow induces a shear force acting on the aqueous phase. The aqueous solution is sheared into droplets and, due to the surface tension, become spherical. Droplet production can be controlled by adjusting the flow rates of the aqueous phase and the oil flow with the syringe driver.

The flow rates were optimized based on various factors such as the rate of droplet production, size of droplets, the amount of time a droplet spends in each of the PCR temperature zones, and the distance between droplets to avoid cross-contamination. First, the Harvard "33" syringe pump system was used to begin the flow of both the aqueous and oil phases into the chip from two 10 mL open syringes attached to the inlets. Fluorocarbon oil (FC-40, 3 M) was used as an immiscible carrier because of its low viscosity (3.4 centipoise) that can help the generation of faster flow rates with reduced resistance. In addition, FC-40 has a high-quality heat transfer with specific heat of 1050 J kg^{-1}°C^{-1}. [53] The aqueous phase was introduced at right angles to the oil flow, as shown in Figure 11.2, resulting in droplet formation due to sheer stress as noted above. Following formation, the droplets pass forward in the oil stream to a PCR cylindrical device that was divided into two temperature zones (60°C and 95°C) with narrow-bore PTFE tubing coiled around it in a threaded groove (Figure 11.2). PTFE tubing was coiled 40 times around the aluminum core and could be readily replaced with new tubing if needed.

Initially, the oil and aqueous flow rates were set to 50 and 20 nL s^{-1}, respectively. The rates were arbitrarily chosen to ease droplet formation and droplet counting. Denaturation and extension in 95°C and 60°C zones gave successful PCR performance when the droplets spent around 30 s in each temperature zone. The average time that droplets spent in the temperature zones was calculated and plotted against different oil flow rates. Figure 11.4 shows that the exposure time to each temperature zone was decreased with higher oil flow rates. When the flow of the oil phase was set at 50 nL s,$^{-1}$ this resulted in an exposure time close to the desired exposure time of 30 s.

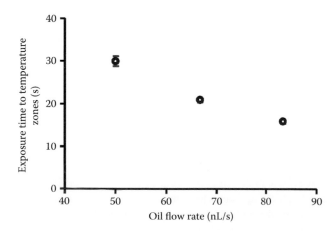

FIGURE 11.4 Relationship between the oil flow rate and incubation time. The aim was to have a balance between the oil and H$_2$O rates to perform a successful PCR. The oil flow was plotted against the exposure time of the droplet to each temperature zone.

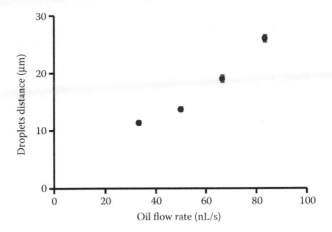

FIGURE 11.5 Relationship between the oil flow rate and the separation of individual droplets. The droplet distances are seen to increase as the oil flow rate increases.

The oil flow rate also has an effect on the distance between adjacent droplets. A set of experiments was performed to assess the relationship between oil flow rate and the distance between droplets. A constant aqueous flow rate was established at 20 nL s⁻¹ while the oil flow rate was varied from 30 to 70 nL s⁻¹. The distance, measured end to end, of 15 droplets under each set of conditions was analyzed in pixels using Scion Image Software (Scion Corporation, Maryland) and then converted into micrometers. The changes in droplet distance with varying oil flow rates and a constant water flow rate are shown in Figure 11.5.

In Figure 11.5, the droplet distance is seen to increase as the oil flow rate increases; the higher the oil flow rate, the further the droplets are apart. Furthermore, droplets were seen to be more spherical with the higher oil flow rates. To balance between the exposure time to each temperature incubation and droplet distances, the oil syringe pump was set to 46.6 nL s⁻¹.

11.8.3 Impact of Aqueous Flow Rates on the Droplet Production

To count the droplets that could be produced in 10 s, droplets were produced and presented into the chip at a range of different aqueous flow rates. The aqueous phase was tested with a range of flow rates starting from 15 to 65 nL s⁻¹, while the oil flow rate was fixed at 46.6 nL s⁻¹. The number of droplets produced per 10 s increased with higher aqueous flow rates, as shown in Figure 11.6. The aim was to have a balance between the oil and water syringe flow rates such that droplets remained spherical, separated to avoid cross-contamination, and with high throughput. It was found that the droplets were optimally produced in the PCR platform when the oil and aqueous flow rates were 46.6 and 11.6 nL s⁻¹ respectively.

11.8.4 Analysis of PCR Products

The BCR–ABL synthetic template oligonucleotide (TO) was then included in the aqueous droplets generated in an oil flow as described. Initially, 10 nM of BCR–ABL TO, which is equal to 10^6 molecules/μL, was supplemented with BCR–ABL probe and primers (as shown in Table 11.1), and PCR master mix composed of FastStart Taq DNA polymerase, PCR buffer, and 3.2 mM $MgCl_2$ (Roche, Switzerland). The PCR mixture was injected via the aqueous phase syringe into the mixing platform using the optimized flow rates. One microliter of PCR products was collected and analyzed using the Agilent Bioanalyzer 2100. This showed a single peak representing the PCR products of BCR–ABL target gene with the expected size of 149 base pairs (Figure 11.7a). Following this, the

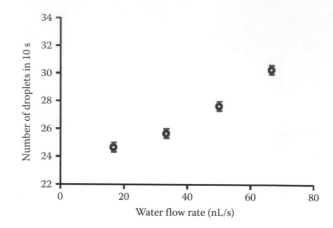

FIGURE 11.6   Relationship between the aqueous phase flow rate and the device throughput. The number of droplets produced per 10 s increased with higher aqueous flow rates.

TO was diluted 10- and 100-fold to contain 1 nM (10^5 molecules/μL) and 100 pM (10^4 molecules/μL) of BCR–ABL and was introduced into the PCR instrument. The analysis demonstrated successful amplification of BCR–ABL (Figure 11.7b) with 1 nM of starting template and failure of amplification with 100 pM of starting template. The results for the no-template control and size determination ladder are shown in Figures 11.7c and d, respectively.

TABLE 11.1
Probe and Primers Employed for Amplifying BCR–ABL from *Homo sapiens*

NCBI Accession Number	AJ131466
Gene used for consensus regions and exon mapping	Chromosomal Translocation *t* (9;22) (q34;q11) (BCR–ABL Translocation)
Probe set used: Roche Universal Probe Library	Universal Probe Library Human # 102

Sequence	Nucleotide Sequence of Primers (5′ → 3′)	Tm	Chromosome	Length	Location
Forward	tccgctgaccatcaataagga	60	9	21	Exon 14, 114998–115018 in BCR
Reverse	ttgagcctcagggtctgagtg	60	22	21	Exon 2, 14451– 144768 in ABL
Probe	ctcagcca (LNA modified)		9	8	
Amplicon				149	
Sequence	gcgaacaagggcagcaaagctacggagaggctgaagaagaagctgtcggagcaggagtcactgctgctgcttatgtctcccagcatg-gccttcagggtgcacagccgcaacggcaagagttacacgttcctgatctcctctgactatgagcgtgcagagtggagggagaacatccgggagcag-cagaagaagtgtttcagaagcttctccctgacatccgtggagctgcagatgctgaccaactcgtgtgtgaaactccagactgtccacagcat**tccgct-gaccatcaataagga**agatgatgagtctccggggctctatgggtttctgaatgtcatcgtcca*ctcagcca*ctggatttaagcagagttcaaa[ ]agcccttcagcggccagtagcatctgact__ttgagcctcagggtctgagtg__aagccgctcgttggaactccaaggaaaaccttctcgctg-gacccagtgaaaatgaccccaacctttcgttgcactgtatgattttgtggccagtggagataacactctaagcataactaaaggtgaaaagctcc-gggtcttaggctataatcacaatggggaatggtgtgaagcccaaaccaaaaatggccaaggctgggtcccaagcaactacatcacgccagtcaa-cagtctggagaaacactcctggtaccatgggcctgtgtcccgcaatgccgctgagtatctgctgagcagcgggatcaatggcagcttcttggtgcgt-gagagtgagagcagtcctggccagaggtccatctcgctgagatacgaagggagggtgtaccattacaggatcaacactgcttctgatggcaagctc-tacgtctcctccgagagccgcttcaacaccctggccgagttggttcatcatcattcaacggtggccgacgggctcatcaccacgctccattatc-cagccccaaagcgcaacaagcccactgtctatggtgtgtcccctaactacgacaagt

Note: ____, Forward primer; ____, reverse primer; *Italic*, probe; [], major break point.

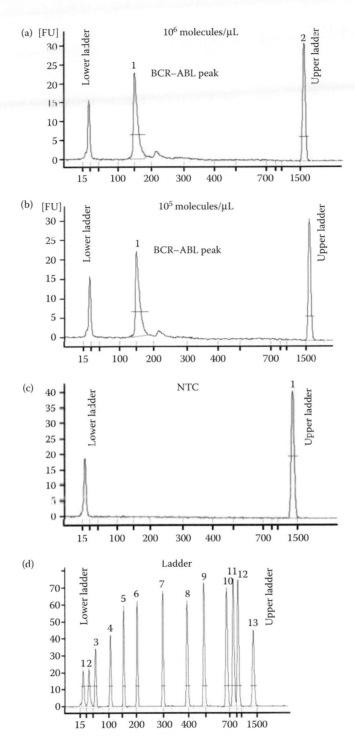

FIGURE 11.7 BCR–ABL PCR product analysis using the Agilent Bioanalyzer 2100. In (a) and (b) the middle peak represents BCR–ABL amplicon with the size of 149 base pairs. 10 nmol and 1 nmol BCR–ABL TO (10⁶ and 10⁵ molecules/μL, respectively) were introduced into the PCR instrument and the analysis showed the successful amplification of BCR–ABL. (c) and (d) show the results for the no-template control (NTC), and the DNA ladder. The Y-axis represents the arbitrary units of fluorescence and the X-axis represents the size of the products in base pairs.

11.9 DISCUSSION

Fully integrated μ-TAS devices for gene analysis are far from being routinely available to date. A number of different devices have been produced that excel in aspects of sample collection, purification,[54–57] treatment, detection, or analysis[58–63] but usually, they do not string these together sufficiently to facilitate a working μ-TAS. For gene analysis, the μ-TAS area where most achievements have been gained resides within PCR and gene-fragment analysis.[8–11,13–15,18,19,64] In many respects, the PCR reactor described in this chapter is another example of the same, except that this application of the two-phase flow has been designed to be compatible with a front-end sample preparation to make possible the analysis of cellular heterogeneity by receiving individual cells and enables the generation of measurements describing the molecules per cell unit definition of gene activity.[65] In addition, the dropletization of PCR is amenable to the application of digital PCR, where the input sample is diluted sufficiently to place one cell or aliquot of the sample into each droplet to permit quantitative determination of the analyte in each droplet. These BCR–ABL PCRs worked at a detection limit of 10^5 molecules per 1 μL PCR product coalesced from PCR-amplified droplets. This resembles near-single-molecule detection sensitivity if scaled-down to the volume of the cell.

The preliminary data we demonstrate are highly supportive of the hypothesis that droplet PCR packaged in two-phase flow will provide a suitable means to interface with the sample preparation to enable analysis of single cells. Of course, a major benefit of the two-phase flow is that it is innately suited to high throughput, with our system being optimized at approximately 1000 droplets/h. Higher flow rates tended to cause droplet merger as the intervening oil barrier between droplets became too narrow to be sustained. The demonstrated droplet PCR system has a reasonably high throughput of PCRs per hour, may offer single molecule per cell (molecules/volume) sensitivity, and offers quantitative data if supplemented with a single-cell capacity front end, such as digital PCR. In the future, this approach is envisaged to be useable for detailed and reproducible analysis of biological phenomena such as meaningful threshold values for the disease, including MRD, and will provide sufficiently high-resolution quantitative data to permit the generation of accurate systems models of disease.

REFERENCES

1. Erickson, D., X. Liu, U. Krull, and D. Li. 2004. Electrokinetically controlled DNA hybridization micro-fluidic chip enabling rapid target analysis. *Anal Chem 76*: 7269–7277.
2. Molinaro, R.J. 2011. Diabetes cases on the rise: Current diagnosis guidelines and research efforts for a cure. *MLO Med Lab Obs 43*: 8, 10, 12 passim; quiz 16–17.
3. Coupland, P. 2010. Microfluidics for the upstream pipeline of DNA sequencing—A worthy application? *Lab Chip 10*: 544–547.
4. Stahlberg, A. and M. Bengtsson. 2010. Single-cell gene expression profiling using reverse transcription quantitative real-time PCR. *Methods 50*: 282–288.
5. Bengtsson, M., A. Stahlberg, P. Rorsman, and M. Kubista. 2005. Gene expression profiling in single cells from the pancreatic islets of Langerhans reveals lognormal distribution of mRNA levels. *Genome Res 15*: 1388–1392.
6. Marcy, Y., T. Ishoey, R.S. Lasken, T.B. Stockwell, B.P. Walenz, A.L. Halpern, K.Y. Beeson, S.M. Goldberg, and S.R. Quake. 2007. Nanoliter reactors improve multiple displacement amplification of genomes from single cells. *PLoS Genet 3*: 1702–1708.
7. Marcy, Y., C. Ouverney, E.M. Bik, T. Losekann, N. Ivanova, H.G. Martin, E. Szeto et al. 2007. Dissecting biological "dark matter" with single-cell genetic analysis of rare and uncultivated TM7 microbes from the human mouth. *Proc Natl Acad Sci USA 104*: 11889–11894.
8. Wilding, P., M.A. Shoffner, and L.J. Kricka. 1994. PCR in a silicon microstructure. *Clin Chem 40*: 1815–1818.
9. Chien, L.J., J.H. Wang, T.M. Hsieh, P.H. Chen, P.J. Chen, D.S. Lee, C.H. Luo, and G.B. Lee. 2009. A microcirculating PCR chip using a suction-type membrane for fluidic transport. *Biomed Microdevices 11*: 359–367.

10. Crews, N., C. Wittwer, and B. Gale. 2008. Continuous-flow thermal gradient PCR. *Biomed Microdevices* 10: 187–195.

11. Frey, O., S. Bonneick, A. Hierlemann, and J. Lichtenberg. 2007. Autonomous microfluidic multi-channel chip for real-time PCR with integrated liquid handling. *Biomed Microdevices* 9: 711–718.

12. Sugumar, D., A. Ismail, M. Ravichandran, I. Aziah, and L.X. Kong. 2010. Amplification of SPPS150 and *Salmonella typhi* DNA with a high throughput oscillating flow polymerase chain reaction device. *Biomicrofluidics* 4(2): 024103.

13. Sun, Y., Y.C. Kwok, and N.T. Nguyen. 2007. A circular ferrofluid driven microchip for rapid polymerase chain reaction. *Lab Chip* 7: 1012–1017.

14. Gonzalez, A., D. Ciobanu, M. Sayers, N. Sirr, T. Dalton, and M. Davies. 2007. Gene transcript amplification from cell lysates in continuous-flow microfluidic devices. *Biomed Microdevices* 9: 729–736.

15. Kiss, M.M., L. Ortoleva-Donnelly, N.R. Beer, J. Warner, C.G. Bailey, B.W. Colston, J.M. Rothberg, D.R. Link, and J.H. Leamon. 2008. High-throughput quantitative polymerase chain reaction in picoliter droplets. *Anal Chem* 80: 8975–8981.

16. Markey, A.L., S. Mohr, and P.J. Day. 2010. High-throughput droplet PCR. *Methods* 50: 277–281.

17. Mohr, S., Y.H. Zhang, A. Macaskill, P.J.R. Day, R.W. Barber, N.J. Goddard, D.R. Emerson, and P.R. Fielden. 2007. Numerical and experimental study of a droplet-based PCR chip. *Microfluidics Nanofluidics* 3: 611–621.

18. Ohashi, T., H. Kuyama, N. Hanafusa, and Y. Togawa. 2007. A simple device using magnetic transportation for droplet-based PCR. *Biomed Microdevices* 9: 695–702.

19. Schaerli, Y., R.C. Wootton, T. Robinson, V. Stein, C. Dunsby, M.A. Neil, P.M. French, A.J. Demello, C. Abell, and F. Hollfelder. 2009. Continuous-flow polymerase chain reaction of single-copy DNA in microfluidic microdroplets. *Anal Chem* 81: 302–306.

20. Kuncova-Kallio, J. and P.J. Kallio. 2006. PDMS and its suitability for analytical microfluidic devices. *Conf Proc IEEE Eng Med Biol Soc* 1: 2486–2489.

21. Kricka, L.J. and P. Wilding. 2003. Microchip PCR. *Anal Bioanal Chem* 377: 820–825.

22. Shoffner, M.A., J. Cheng, G.E. Hvichia, L.J. Kricka, and P. Wilding. 1996. Chip PCR. I. Surface passivation of microfabricated silicon–glass chips for PCR. *Nucleic Acids Res* 24: 375–379.

23. Wang, W., H.B. Wang, Z.X. Li, and Z.Y. Guo. 2006. Silicon inhibition effects on the polymerase chain reaction: A real-time detection approach. *J Biomed Mater Res A* 77: 28–34.

24. Brouzes, E., M. Medkova, N. Savenelli, D. Marran, M. Twardowski, J.B. Hutchison, J.M. Rothberg, D.R. Link, N. Perrimon, and M.L. Samuels. 2009. Droplet microfluidic technology for single-cell high-throughput screening. *Proc Natl Acad Sci USA* 106: 14195–14200.

25. Um, E., S.G. Lee, and J.K. Park. 2010. Random breakup of microdroplets for single-cell encapsulation. *Appl Phys Lett* 97: 153703.

26. Chen, C.L., K.C. Chen, Y.C. Pan, T.P. Lee, L.C. Hsiung, C.M. Lin, C.Y. Chen, C.H. Lin, B.L. Chiang, and A.M. Wo. 2011. Separation and detection of rare cells in a microfluidic disk via negative selection. *Lab Chip* 11: 474–483.

27. Zimmerlin, L., V.S. Donnenberg, and A.D. Donnenberg. 2011. Rare event detection and analysis in flow cytometry: Bone marrow mesenchymal stem cells, breast cancer stem/progenitor cells in malignant effusions, and pericytes in disaggregated adipose tissue. *Methods Mol Biol* 699: 251–273.

28. Manz, A., N. Graber, and H.M. Widmer. 1990. Miniaturized total chemical-analysis systems—A novel concept for chemical sensing. *Sens Actuators B-Chem* 1: 244–248.

29. Brivio, M., W. Verboom, and D.N. Reinhoudt. 2006. Miniaturized continuous flow reaction vessels: Influence on chemical reactions. *Lab Chip* 6: 329–344.

30. Link, D.R., S.L. Anna, D.A. Weitz, and H.A. Stone. 2004. Geometrically mediated breakup of drops in microfluidic devices. *Phys Rev Lett* 92: 054503.

31. Link, D.R., E. Grasland-Mongrain, A. Duri, F. Sarrazin, Z. Cheng, G. Cristobal, M. Marquez, and D.A. Weitz. 2006. Electric control of droplets in microfluidic devices. *Angew Chem Int Ed Engl* 45: 2556–2560.

32. Song, H., J.D. Tice, and R.F. Ismagilov. 2003. A microfluidic system for controlling reaction networks in time. *Angew Chem Int Ed Engl* 42: 768–772.

33. Chabert, M., K.D. Dorfman, and J.L. Viovy. 2005. Droplet fusion by alternating current (AC) field electrocoalescence in microchannels. *Electrophoresis* 26: 3706–3715.

34. Christopher, G.F., J. Bergstein, N.B. End, M. Poon, C. Nguyen, and S.L. Anna. 2009. Coalescence and splitting of confined droplets at microfluidic junctions. *Lab Chip* 9: 1102–1109.

35. Mazutis, L., J.C. Baret, and A.D. Griffiths. 2009. A fast and efficient microfluidic system for highly selective one-to-one droplet fusion. *Lab Chip* 9: 2665–2672.

36. Tan, W.H. and S. Takeuchi. 2006. Timing controllable electrofusion device for aqueous droplet-based microreactors. *Lab Chip 6*: 757–763.

37. Tan, Y.C., J.S. Fisher, A.I. Lee, V. Cristini, and A.P. Lee. 2004. Design of microfluidic channel geometries for the control of droplet volume, chemical concentration, and sorting. *Lab Chip 4*: 292–298.

38. Ahn, K., C. Kerbage, T.P. Hunt, R.M. Westervelt, D.R. Link, and D.A. Weitz. 2006. Dielectrophoretic manipulation of drops for high-speed microfluidic sorting devices. *Appl Phys Lett 88*: 024104.

39. Baret, J.C., O.J. Miller, V. Taly, M. Ryckelynck, A. El-Harrak, L. Frenz, C. Rick et al. 2009. Fluorescence-activated droplet sorting (FADS): Efficient microfluidic cell sorting based on enzymatic activity. *Lab Chip 9*: 1850–1858.

40. Kawano, T., K.N., S. Ando, M. Yamamoto, J. Fujiwara, T. Torii, and T. Higuchi. In *Proceedings of the MicroTAS* 2004 (eds., N.J. Laurell T., Jensen K., Harrison D.J., and Kutter J.P.) 2004, pp. 144–146 (the Royal Society of Chemistry) Malmo, Sweden.

41. Beer, N.R., K.A. Rose, and I.M. Kennedy. 2009. Monodisperse droplet generation and rapid trapping for single molecule detection and reaction kinetics measurement. *Lab Chip 9*: 841–844.

42. Sgro, A.E., P.B. Allen, and D.T. Chiu. 2007. Thermoelectric manipulation of aqueous droplets in microfluidic devices. *Anal Chem 79*: 4845–4851.

43. Shi, W., J. Qin, N. Ye, and B. Lin. 2008. Droplet-based microfluidic system for individual *Caenorhabditis elegans* assay. *Lab Chip 8*: 1432–1435.

44. Stan, C.A., G.F. Schneider, S.S. Shevkoplyas, M. Hashimoto, M. Ibanescu, B.J. Wiley, and G.M. Whitesides. 2009. A microfluidic apparatus for the study of ice nucleation in supercooled water drops. *Lab Chip 9*: 2293–2305.

45. Wang, W., C. Yang, and C.M. Li. 2009. On-demand microfluidic droplet trapping and fusion for on-chip static droplet assays. *Lab Chip 9*: 1504–1506.

46. Myers, S. and A. Baker. 2001. Drug discovery—An operating model for a new era. *Nat Biotechnol 19*: 727–730.

47. Napoli, M., J.C. Eijkel, and S. Pennathur. 2010. Nanofluidic technology for biomolecule applications: A critical review. *Lab Chip 10*: 957–985.

48. Bustin, S.A., V. Benes, J.A. Garson, J. Hellemans, J. Huggett, M. Kubista, R. Mueller et al. 2009. The MIQE guidelines: Minimum information for publication of quantitative real-time PCR experiments. *Clin Chem 55*: 611–622.

49. Hebenstreit, D. and S.A. Teichmann. 2011. Analysis and simulation of gene expression profiles in pure and mixed cell populations. *Phys Biol 8*: 035013.

50. Stahlberg, A., D. Andersson, J. Aurelius, M. Faiz, M. Pekna, M. Kubista, and M. Pekny. 2011. Defining cell populations with single-cell gene expression profiling: Correlations and identification of astrocyte subpopulations. *Nucleic Acids Res 39*: e24.

51. Okochi, M., H. Tsuchiya, F. Kumazawa, M. Shikida, and H. Honda. 2010. Droplet-based gene expression analysis using a device with magnetic force-based-droplet-handling system. *J Biosci Bioeng 109*: 193–197.

52. Taniguchi, T., T. Torii, and T. Higuchi. 2002. Chemical reactions in microdroplets by electrostatic manipulation of droplets in liquid media. *Lab Chip 2*: 19–23.

53. Acota. Fluorinert™ Electronic Liquids. [cited; available from: http://www.acota.co.U.K./products/fluorinert-electronic-liquids].

54. Chen, X., D.-F. Cui, C.-C. Liu, and H. Li. 2007. Microfabrication and characterization of porous channels for DNA purification. *J Micromech Microeng 17*: 68–75.

55. Poeckh, T., S. Lopez, A.O. Fuller, M.J. Solomon, and R.G. Larson. 2008. Adsorption and elution characteristics of nucleic acids on silica surfaces and their use in designing a miniaturized purification unit. *Anal Biochem 373*: 253–262.

56. Toriello, N.M., C.N. Liu, R.G. Blazej, N. Thaitrong, and R.A. Mathies. 2007. Integrated affinity capture, purification, and capillary electrophoresis microdevice for quantitative double-stranded DNA analysis. *Anal Chem 79*: 8549–8556.

57. Ueberfeld, J., S.A. El-Difrawy, K. Ramdhanie, and D.J. Ehrlich. 2006. Solid-support sample loading for DNA sequencing. *Anal Chem 78*: 3632–3637.

58. Lee, J.G., K.H. Cheong, N. Huh, S. Kim, J.W. Choi, and C. Ko. 2006. Microchip-based one step DNA extraction and real-time PCR in one chamber for rapid pathogen identification. *Lab Chip 6*: 886–895.

59. Lee, S.H., S.W. Kim, J.Y. Kang, and C.H. Ahn. 2008. A polymer lab-on-a-chip for reverse transcription (RT)–PCR based point-of-care clinical diagnostics. *Lab Chip 8*: 2121–2127.

60. Legendre, L.A., J.M. Bienvenue, M.G. Roper, J.P. Ferrance, and J.P. Landers. 2006. A simple, valveless microfluidic sample preparation device for extraction and amplification of DNA from nanoliter-volume samples. *Anal Chem 78*: 1444–1451.

61. Wen, J., C. Guillo, J.P. Ferrance, and J.P. Landers. 2007. Microfluidic chip-based protein capture from human whole blood using octadecyl (C18) silica beads for nucleic acid analysis from large volume samples. *J Chromatogr A 1171*: 29–36.

62. Wen, J., C. Guillo, J.P. Ferrance, and J.P. Landers. 2007. Microfluidic-based DNA purification in a two-stage, dual-phase microchip containing a reversed-phase and a photopolymerized monolith. *Anal Chem 79*: 6135–6142.

63. Witek, M.A., S.D. Llopis, A. Wheatley, R.L. McCarley, and S.A. Soper. 2006. Purification and preconcentration of genomic DNA from whole cell lysates using photoactivated polycarbonate (PPC) microfluidic chips. *Nucleic Acids Res 34*: e74.

64. Li, S., D.Y. Fozdar, M.F. Ali, H. Li, D. Shao, D.M. VykoU.K.al et al. 2006. A continuous-flow polymerase chain reaction microchip with regional velocity control. *J Microelectromech Syst 15*: 223–236.

65. Kamme, F., R. Salunga, J. Yu, D.T. Tran, J. Zhu, L. Luo, A. Bittner et al. 2003. Single-cell microarray analysis in hippocampus CA1: Demonstration and validation of cellular heterogeneity. *J Neurosci 23*: 3607–3615.

Part IV

Design, Optimization, QC, and Standardization

12 Assay Design for Real-Time qPCR

Gregory L. Shipley

CONTENTS

12.1 INTRODUCTION

The majority of the material that is described in this chapter will be devoted to demonstrating how to design and validate a new quantitative PCR (qPCR) assay. However, before reaching the point of designing a new assay, it is important to consider whether to take the plunge and design a custom assay or to order a premade assay from a commercial source. The first step will be to go through that thought process, keeping the various potential goals of the research project in mind. Whether the decision is to purchase a predesigned assay or to make a new one, the validation steps described below will still need to be carried out. There are varying amounts of information about assay quality from the major sources of premade assays and even when it is available, you cannot assume that the assay will perform as expected in your hands, using your qPCR instrument with the reverse transcription (RT) and polymerase chain reaction (PCR) reagents that you have assembled.

At first glance, designing a new, real-time qPCR assay would seem to be a simple matter of obtaining the desired sequence and, using assay design software, choosing a pair of primers and possibly a probe, purchasing these oligos from a reputable supplier, making the stock solutions, and running the experiment. It is almost that easy, but there are a few more steps that should be inserted into this workflow. This chapter contains an outline of the complete assay design process with examples and tips along the way. By following this tried-and-tested work flow, it should be possible to design qPCR assays that perform well every time. Following this procedure will also result in acquiring all the data that describe the assay and that is required for publication to satisfy the new minimum information for publication of quantitative real-time PCR experiments (MIQE) guidelines for real-time qPCR.[1]

There are three important hallmarks of a good qPCR assay: First, the assay must have the expected template specificity. Nonspecific detection of similar target sequences will produce signals that confound your data analysis, almost certainly leading to incorrect conclusions. Second, the assay must have good template sensitivity. An assay that is not sensitive will not be useful for all possible applications. Awareness of the assay sensitivity is critical for any kind of template quantification. The term semiquantitative is truly an oxymoron when applied to real-time qPCR! Finally, the assay should have high PCR efficiency. There are usually additional problems resulting in an assay with low PCR efficiency and it is important to avoid these, particularly since an inefficient assay will also be insensitive.

There are several assay formats available and an important first step is to decide the most appropriate one for the project. The most popular assay detection choices require fluorescent probes or DNA-binding dyes such as SYBR Green I. Examples of data using both these assay formats will be shown throughout this chapter. There are many fluorescent probe detection choices available, as well as alternative dyes to SYBR Green I, which will be covered elsewhere, but a thorough understanding of how to design and validate a probe- or dye-based assay can then be applied to the process of assay design and validation when using any of the other detection modalities. Although the assay should fulfill the criteria outlined above, there is always a cost–benefit trade-off for any laboratory expenditure. Should you design your own assay or purchase one? Importantly, what kind of qPCR assay best fits the project and will give you the best value for your money? Predesigned hydrolysis probe assays (may also be referred to as TaqMan®) are commercially available from several vendors, including Roche Applied Science or Applied Biosystems/Life Technologies, hybridization probe-based assays through Thermo-Fisher/Dharmacon, Sigma, and others while assays based on SYBR Green I dye detection can be purchased from Qiagen and other vendors. If you need to run many different qPCR assays on a small sample set, do not plan to run all or most of those assays again in the future and the data quickly; the commercial assays are probably the best option. The main benefit of making your own assay is that you will be able to run thousands of assays per set of primers/probe from a single purchase; the current cost per reaction on a 384-well plate is about $0.25 for both primers and the probe, when running 20 μL reactions. In addition, you have full control over the design location and a complete understanding of any compromises that may affect the quality of your data. There are situations where a custom design is the only option, such as detection of a specific splice variant or for investigators working on the biology of organisms other than human, mouse, and rat.

12.2   INTRODUCTION TO QPCR ASSAY DESIGN

When designing a project that requires a new qPCR assay, it is essential to determine the requirements of the study as a whole. It is instructive to outline that process here so that you can follow this path to determine your specific needs for each study.

1. *What role does qPCR play within the project?*
   There are multiple possibilities here but they can be rolled into two broad categories:
   A validation role for a microarray study, *in situ* hybridization observations, siRNA knock down result, or a biochemical observation at the protein level.
   An initial study to investigate a hypothesis or extend a previous study where the **qPCR data will be the heart of the data set.**
2. *How many samples will be involved, over the entire project, for each assay?*
   For validation studies, the number of samples tends to be **relatively small**, under 50. When qPCR plays a more major role in the project, the sample number **will be much larger**, 100s to 1000s.
3. *How many target sequences are there; therefore, how many qPCR assays will be required?*
   For assay validation experiments, the number of qPCR assays/sample tends to be large. In the second case above, where qPCR will provide the core of the experimental data, the **number of qPCR assays/sample tends to be low** with high sample numbers.
4. *Will the new assay(s) be used over a long period of time, perhaps with future projects, or only for this specific project?*
   In the case of validation studies, the answer is almost always **no**; once this project is completed, the assay will not be used again. When qPCR is at the heart of the project, the investigator will usually want to use the assay over a **long period of time**, if the initial experiment has been successful.

· 5. *Is a quantification value desirable for each unknown or would a relative fold difference suffice?*

In validation experiments, the goal is to determine quickly whether targets under investigation in experimental samples are up- or downregulated when compared to a control sample. Here, measurement of the fold difference is appropriate. When qPCR is at the heart of the study, although a fold difference may be the final goal, it is more useful to have a quantification value for each target in each sample so that the full complement of statistical analyses can be performed.

Once the goal of the project is clear, the assay type that will work best can be selected. For projects that fall into the validation category, that is, those with large target numbers and small sample sizes, it is often best to purchase assays from a company. Another option is to let your local Core Lab or other commercial laboratories which offer qPCR services, design, and validate the assay for you. Some Core Labs will work with investigators from outside their parent institutions and many companies offer support for some or all the design and validation processes. If the budget would not support commercial options, developing the skills required for qPCR assay design and validation is a favorable option. For validation experiments, we generally use either the Roche Universal Probe Library (UPL) assays (as we have the entire probe library in our Core Lab) or the primer-based SYBR Green I assays. In practice, any assay type will work but these assay types are quickest for the investigator to do because they remove the uncertainty around the probe by merely focusing on the primer function and arguably these are the least expensive options.

For projects that fall into the second category, where they will be used for a long time in the laboratory and will be applied to a larger project containing many samples, it is preferable to use a probe-based assay choosing standard hydrolysis probes as the first choice and then considering a probe modification such as locked nucleic acid (LNA), minor groove binding (MGB), or an alternative probe structure if required due to the sequence context. Assays are best run with defined standard curves alongside the samples. The probe is to ensure template specificity and the standard curve is for precise sample quantification. However, there are many investigators using SYBR Green I dye-based assays for long-term projects with great success. It is important to note, however, that there are specific requirements for the validation of these assays to ensure template specificity.

Having selected the most appropriate assay for your project, the next step is to either acquire the desired qPCR assay(s) from a commercial vendor (skip to Section 12.2.3) or to design your own qPCR assay(s).

There are four basic steps to qPCR assay design:

1. *In silico* assay sequence selection and analysis using BLAST
2. Assay design (i.e., primer and probe selection) utilizing software and BLAST confirmation
3. Order and preparation of oligonucleotide stocks
4. Empirical testing of all assay reagents

Each of these steps is critical for the design of a successful qPCR assay. Each step is described below, with examples as required, in the hope of making the process as painless as possible.

12.2.1 *IN SILICO* SEQUENCE SELECTION AND ANALYSIS

Sequence selection and analysis may be the most important step in the design of a new qPCR assay and yet it is the one that is either skipped altogether or performed inadequately. The examples presented are specific to sequence analysis when applied to transcript detection and quantification. However, the same principles hold for the detection of specific regions within a genome. For those working with microbial or viral genomes, whether environmental or pathogenic, the extent to which you can query the main sequence databases may not be as complete as for those investigating human or mouse transcriptomes, for example. However, this line of investigation should be pursued as far as possible, prior to the assay design. There is still a lot of useful sequence information available in public databases and these resources are growing rapidly.

Prior to the completion of the human genome project, the sequence selection and evaluation step began by downloading every available sequence for a transcript from the NCBI database and then performing an alignment. Regions of sequence where there were anomalies were marked and avoided during the assay design. To find information about splicing required a trip to the library to find a publication containing alternative splicing data, making a photocopy of the sequence around the splice sites, translating those back to your alignment, and marking those regions with a highlighter pen. Ah, the good old days! In this age of enlightenment, the rate-limiting step to sequence analysis is the identification of the appropriate Refseq accession number. This will have the form NM_xxxxxx, or alternatively NR_xxxxxx if it is describing an untranslated RNA such as 18S rRNA. Using this, you simply download the entire Refseq sequence file for the required transcript, ensuring that the currently curated version is selected (each file has a date). All the splicing information will be shown in the sequence header, along with a lot more useful data.

Having identified the correct target sequence, the next step is to perform a BLAST analysis at the NCBI, UCSC, DDBJ, or EMBL websites (see Table 12.1 for a list of URLs) to identify regions of homology between your target sequence and any other genes in the database. Select the database you want to search, for example, human, mouse, and so on, and note that there is also a link for microbes within this initial list. If your desired organism is not listed, select the link for "list all genomic BLAST databases" to find the one you are looking for. On the second screen, the default database is for the genome of the organism. If you want to look at transcripts, click on the drop down menu and choose "RefSeq RNA." Next, select "Somewhat similar sequences (blastn)" instead of the default "highly similar sequences (megablast)." Selecting BLASTN will yield a much broader hit list. Enter the accession number or sequence into the box at the top of the page and click on the "BLAST" button near the bottom of the web page.

Ideally, this search will result in just the desired target transcript being listed. However, you will often find multiple hits, usually splice variants of the parent sequence and because you chose BLASTN, other sequences with low homology will also be shown. Any sequence with a high homology, even if only in a single region of the transcript, should be downloaded. To check how much sequence homology there is between the target sequence and the one found by BLAST, click on the link for that alignment and the region of homology will be shown with the alignment. If the region of homology has stretches of near, or exact, identity, you will want to download those transcript sequences. Mostly you will identify transcripts representing genes in a related family with the target sequence. However, you may also find regions of homology where two disparate genes share a common function; the adenosine triphosphate (ATP)-binding region of kinases, for example. It is important to download these as well for reference during the assay design.

Once you have your cadre of sequences, perform a multiple or clustal *W* alignment (http://www.clustal.org/). If you do not have access to suitable alignment software, a less convenient method to view the aligned sequences is to copy and paste each pair-wise alignment from the BLAST results.

TABLE 12.1
Websites for Genomic Databases

Website	URL
NCBI	http://www.ncbi.nlm.nih.gov/
UCSC	http://genome.ucsc.edu/
EMBL	http://www.ebi.ac.uk/embl/
DDBJ	http://www.ddbj.nig.ac.jp/
Clustal *W*	http://www.clustal.org/
Mfold	http://mfold.rna.albany.edu/

The multiple alignment plot is used to (1) avoid regions of homology with nonrelated sequences. These are usually, but not always, transcripts within the same gene family or that share the same function and (2) to find the potential locations for the assay. If you want to detect a single splice variant, ensure that you cross an exon–exon junction that is unique to the desired variant. If you want an assay that will detect all splice variants (i.e., a generic assay), select regions within the sequence that are common to all sequence variants and target these regions for assay development.

Having identified the regions of sequence that may be suitable for assay development, the next step of the design process is to check for secondary structure within the target sequence. I can tell you from previous, unhappy, personal experience that there are few things that will make an apparently good assay perform poorly, or even not perform at all; at the top of this list is locating a primer, or a probe, on top of a large stem structure within the target sequence. Most assay design software packages do not consider secondary structure as part of the assay design process. The exceptions are the Premier Biosoft assay design software packages AlleleID and Beacon Designer. Using either software package, a secondary structure search via the m-fold server is performed from within the software and the subsequent assay design process avoids the folded regions. When using other design packages, it is necessary to run an m-fold analysis directly over the web (http://mfold.rna.albany.edu/). When entering sequence information, it is important to select folding for either RNA or DNA target sequences. When running the unix OS or an Mac-running OSX (under the Terminal app), the m-fold application can be downloaded and can be run locally. There are advantages to using the older version of m-fold (version 2.3) because this allows the temperature for the folding analysis to be set. For RNA, select 55°C to reflect the temperature of the RT reaction conditions, particularly if using gene-specific priming. If you select to run the immediate return part of the program where the folding result is returned to your e-mail address immediately after finishing, there is a restriction of 1200 bases per folding run. The trick is to make sure the format of the fold is the one that makes sense to you. There are many choices. What you want is one where the folding pattern is linear, going left to right. Since most transcripts and genes are longer than 1200 bases, it is best to run all regions in succession, with a 100–200 base overlap. There are usually regions within the transcript or gene that are filled with secondary structure, some that are almost devoid of any folding and, more commonly, those that have a mixture of both. If you cannot specify the length of sequence for the assay design software to use, perform an m-fold analysis over the entire sequence and select the region where the folding is the least problematic. Of course, this has to be done in conjunction with the sequence alignment information derived from the initial sequence analysis and any targeting requirements you might have set at the outset. An example of an m-fold readout for a 1200 base region of the human Frizzled 7 transcript can be seen in Figure 12.1.

By using a multisequence alignment in conjunction with the m-fold analysis of the target sequence, problem areas can be excluded and the qPCR assay is positioned within a template-specific region with little-to-no issues with the secondary structure or homology to other templates in the same sample. These precautions may not guarantee a perfect assay, but they will ensure that there are no seminal template issues with the design location.

12.2.2 ASSAY DESIGN

Armed with the alignment and m-fold information obtained after following the steps described above, the next step is to design the qPCR assay. A common design request is "can you design a probe-based assay that will work with human and mouse, human and rat or all 3 species?" While this may appear to be a reasonable request and an effective way to minimize effort and resources, in reality, it is not a good idea to try to force an assay into multiple species use unless there is high-sequence homology among the individual transcripts. The risk is that a single assay would be designed that works poorly for all three species. A safer design philosophy is to design the best assay possible for each target and species. If multiple species can be accommodated, all the better, however, it should not be a dominant feature of the initial assay design goal. As an example, in Figure 12.2, there is a portion of the clustal W alignment for the human, mouse, and rat matrix metalloproteinase 2 (MMP2) transcripts. It is

Accession number: +000081
Folding bases: 1001–2200
Total bases = 1200
dG = −41.08

Linear DNA folding at
        Temperature = 55.0
        [Na$^+$]   = 50.0 nH
        [Mg^{++}]   = 5.0 nH

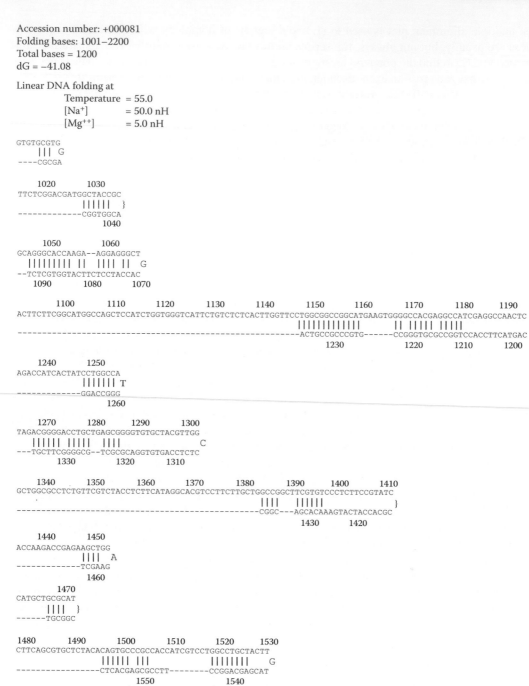

FIGURE 12.1 m-Fold analysis of a region of the human Frizzled 7 transcript. Long stem stretches should be avoided within the PCR product for the assay and should definitely not be a region where a primer/probe binds. Many transcripts will have long stretches with no folding. Choose these regions of the sequence if possible. Also keep in mind that a forward primer with a 5′-base starting at position 1290 above will not amplify the corresponding region of stem formation seen in the 1320–1335 span and thus would be a suitable region for assay placement.

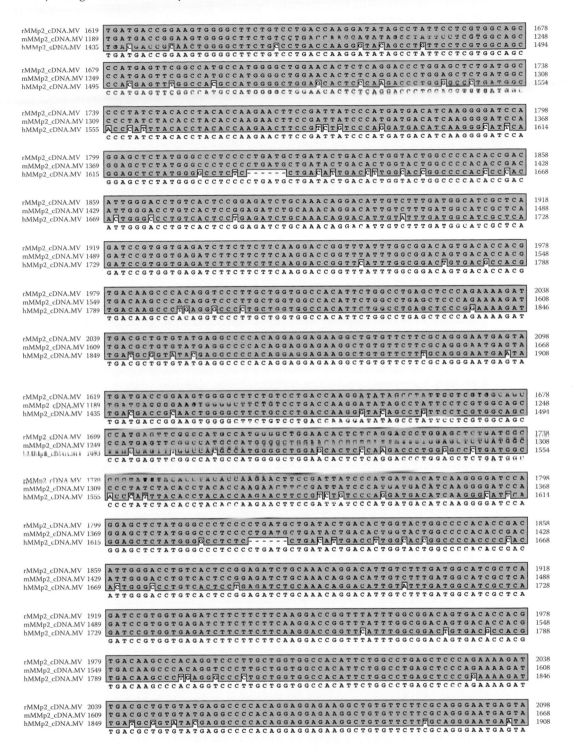

FIGURE 12.2 Alignment of the human, mouse, and rat MMP2 transcripts showing a region of high homology between the mouse and rat, less homology with the human sequence. In this case, it may be possible to make a mouse/rat MMP2-combined assay but it would be difficult to make a three-species assay including the human transcript. Minor base changes can affect the T_m of the primers/probe and thus, simply changing a few bases from a successful mouse/rat assay to match the human sequence may or may not be a successful strategy. Making a dedicated human assay would be the best approach.

clear that designing an assay for both mouse and rat would not pose too much of a problem within this region, due to high-sequence homology. However, there are enough differences between the rodent and the human sequences to make the design of a three-species assay difficult. After further consideration of the m-fold results for the mouse–rat and human sequences within this region, along with the sequence region selected by the design software for the best assay design, a three-species assay is likely to be impossible. This is an example of reasonably close alignment among the species. However, there are situations when gene sequences from the human and rat or human and mouse, or all three, share long stretches of complete homology. From this example, it is clear that sequence homology and folding must be investigated prior to the final assay design.

There are also times when related sequences for a transcript are found among the gene family members. For example, the human Frizzled 2 and 7 family members share a section of high homology (Figure 12.3). These common sequence regions usually represent areas encoding common functional motifs within the transcript or gene. When the desired target sequence is a single transcript of a gene family (e.g., Frizzled 2), it is necessary to locate the design in the last exon within the coding region or to move into the 3′-UTR where there is more sequence divergence among family members. However, this will negate crossing an exon–exon junction that is the usual

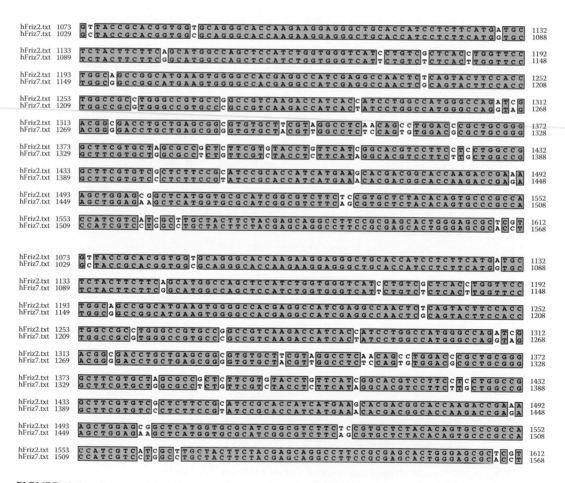

FIGURE 12.3 Portion of the alignment of human Frizzled family members 2 and 7 showing high-sequence homology. Alignments and BLAST searches are invaluable tools in qPCR assay design. In making a hFrizzled 7 assay, all Frizzled family member sequences detected following a BLAST search were downloaded and aligned. Frizzled 2 was found to have high homology compared to Frizzled 7. From this information, a human Frizzled 7-specific assay was made.

practice to avoid detection of genomic DNA contamination. When quantifying highly expressed, single-copy genes, the risk of gDNA contamination in a purified (and DNAse-treated) RNA sample effecting the quantification is significantly lower than for a gene with low expression. In all cases, it is prudent to use high-quality RNA that is as free of gDNA contamination as possible. Keep in mind that even if gDNA contamination is not detected by qPCR due to long intronic sequences separating the primers in gDNA, assay resources will still be used in each cycle to prime and extend from DNA templates. RNA samples that have been properly treated with DNase should not have significant DNA detected by any assay unless the transcript level is near the limit of quantification (LOQ) for the assay. When working with animal species other than mammalian, plant, fungal, microbial, or viral sequences, whether at the DNA or RNA level, you are bound to be dependent upon the analysis of published sequences in the various databases in an effort to avoid cross-species, sequence-related family member, or strain homologies. The only way to be entirely sure that the correct sequences are being detected is to run a selection of samples in a qPCR experiment using the new assay and sequence the final amplified product(s) to make sure that the sequence is solely that of the intended target. For some mammalian and many nonmammalian targets, transcripts may not undergo splicing, again stressing that RNA samples should be as free from gDNA contamination as possible.

For the actual design and selection of the oligos for qPCR assays, it is essential that you use the software for selecting compatible primer pairs (and probes). These programs will ensure that there are minimal interactions of the primers with themselves (folding) or with each other (interprimer or primer–probe interactions) that will have an adverse effect on PCR efficiency. This is particularly important when multiplex assays are being defined. There are several commercial assay design software packages in the market and others are freely available. The more popular software packages are listed in Table 12.2. This is not an exhaustive list nor is it intended to promote one software package over another. As with all software, the critical factor is to find one that you feel comfortable using and that gives desirable results, and then stick to it. The more you use a design software package, the more expert you will become with its application to qPCR assay design.

After thorough analysis of the alignment and m-fold data as outlined above, you will have identified one or more suitable regions within the target sequence for placing the new assay. With that decision being made, the next consideration concerns the reaction conditions for the assay, primarily the annealing temperature for the primers. The most commonly used annealing temperature in qPCR programs is 60°C. Most software programs use a default of 59°C for assay design that probably goes back to the first assay design software available from Applied Biosystems: Primer Express. The first commercial version of this program that I used was the labeled version 0.5. Needless to say, it has come a long way since then, as have several other commercial programs now available. The most popular freeware design package is Primer3 from MIT (Table 12.2). However, there are other newer and really excellent programs now and some of them are available on the web as well (Table 12.2).

TABLE 12.2

Real-Time qPCR Assay Design Software

Commercial Software	Company	URL
Beacon Designer	Premier Biosoft	www.premierbiosoft.com/
AlleleID	Premier Biosoft	www.premierbiosoft.com/
Primer Express	Applied Biosystems/Life Tech	http://www.appliedbiosystems.com/
Visual OMP	DNAsoftware	http://dnasoftware.com/

Free Software	Source	URL
Oligo	Architect	http://www.sigmaaldrich.com/
Real Time Design	Biosearch Technologies	http://www.biosearchtech.com/
PCR Design Tool	Integrated DNA Technologies	http://www.idtdna.com/
Primer3	MIT	http://frodo.wi.mit.edu/primer3/

It is important to bear in mind that qPCR assay design software is not perfect. Although they all do the job in terms of identifying good primer pairs for PCR in the correct location, the calculations for T_m are generally inaccurate. The newer versions of the Premier Biosoft programs, however, appear to have a T_m estimation that more closely reflects the function of the oligo in PCR. Of course, designing primers with a T_m of 60°C and finding out the real T_m 65°C is not a tragedy from a functional point of view, but if the functional T_m is lower than the predicted T_m, the primers would function with lower efficiency, or not at all, at the desired annealing temperature. In general, aiming for a T_m between 55°C and 60°C (in the software setting) will give primers that work well enough for further optimization using a 60°C annealing temperature in the qPCR instrument software.

The next consideration is the desirable PCR amplicon length. In general, amplicon length in qPCR is significantly shorter than that used for conventional PCR where a gel will be used for visualization of the PCR product(s). For probe-based assays, the recommended range is 60–90 bases; the shorter the PCR amplicon length, the easier it is to get the optimal PCR efficiency. It is advisable to have amplicons a little longer, from 90 to 150 bases when using primer-based qPCR assays, such as those using SYBR Green I dye for detection. The longer amplicon is more easily differentiated from any primer–dimers in the postrun melt analysis. It is possible to use primers that were designed to run in a probe-based assay, where the amplicon is as short as 65 bases with SYBR Green I dye detection but determination of specificity using a postamplification melt curve can be more difficult if the primer–dimers and target amplicons melt at very similar temperatures.

Most qPCR design packages offer a bewildering array of potential settings for assay design. A list of the most common settings for both probe-based and primer-based assays with reasonable parameter settings for both are presented in Table 12.3. Some settings relate to binding energy (ΔG) values for multiple calculations and should be left as the default setting until you are confident about changing them. Regardless of the settings in the software, at the end of the day, what matters is how the assay performs empirically. This cannot always be predicted; thus, a certain amount of empirical testing is required (see Section 12.2.4).

When a suitable assay cannot be designed within your sequence boundaries, it is possible to try changing some of the search criteria in the software settings. Relatively safe modifications to the search parameters are: annealing T_m, number of bases (including zero) for the G/C clamp setting, and occurrence of T on the 3′-end of the primer. Table 12.4 contains a list of some of the factors that can be used as guidance for good assay design of primers or probes. Assay success is most likely when the primers and probes selected fall within these boundaries. It is feasible that the first assay design on the list is not necessarily the best one when run in the laboratory. It is advisable to choose at least two forward and two reverse primers for the same target and order all of them for testing. Make sure that the 2 forward and 2 reverse primers are selected by the software such that they can be empirically tested in all 4 possible combinations. I should also point out that there are times when the software will find only a single primer with multiple choices for the primer partner. After exhausting all the tricks in the book and this is still the case, ordering a combination of three primers for one position and the single primer for the partner position is still better than a single primer pair. In this case, there will only be three unique primer combinations but that will still give you the comparison you need to pick the best combination during empirical assay testing (described below).

Finally, it is critical to confirm that the final PCR amplicon is specific for the intended target template. This is a verification of the original homology search performed to identify a suitable target region using the BLAST function (http://blast.ncbi.nlm.nih.gov/). Select the appropriate species database to use for the BLAST search. Copy the entire PCR amplicon sequence and paste into the sequence box. Under "Database" leave at genome if working with DNA or change to "RefSeq RNA" for transcripts. Under "Program" select BLASTN then hit "Begin Search" and the window will switch to the next view. These settings are left at their default values. The results will appear in the same window of your browser. There should only be a single hit consistent with the original template. If there is more than one, hold the mouse pointer over each aligned sequence and the

TABLE 12.3
Basic Settings in Assay Design Software

Probe-Based Assays	Setting	Comment
Percentage of G/C content	30–80%	50% is ideal
Max primer pair mismatch	1–2°C	Want both primers to be in the same range
Primer length range	18–25 bases	Varies with the percentage of G/C content
Primer T_m	50–59°C	Varies with the percentage of G/C content of the target
Probe T_m	5–10°C > primer	The probe has to anneal before primers
Monovalent salt	50–100 mM	Can vary with PCR buffer composition
Mg^{+2}	3–5 mM	Can vary with the manufacturer, test empirically
Primer concentration	300–400 nM	Setting may be intermediate—250 nM
Probe concentration	100 nM	Do not need any more probe than this
PCR amplicon length	60–90 bases	Short amplicons are the best, high PCR efficiencies
Primer-Based Assays	**Setting**	**Comment**
Percentage of G/C content	30–80%	50% is ideal
G/C clamp on 3′-end	0–2 bases	Can be helpful but not always necessary
Max primer pair mismatch	1–2°C	Want both primers to be in the same range
Primer length range	18–25 bases	Varies with the percentage of G/C content
Primer T_m	50–59°C	Varies with the percentage of G/C content of the target
Monovalent salt	50–100 mM	Can vary with PCR buffer composition
Mg^{+2}	1.5–3 mM	Can vary with the manufacturer, test empirically
Primer concentration	50–100 nM	Setting may be intermediate—250 nM
PCR amplicon length	60–90 bases	Short amplicons are the best, high PCR efficiencies

identity of each line will show in the box above the alignment graphic. The links to the aligned sequences are in the box below. The number under "Max Score" represents how close the hit is to the original sequence. If the homology is very high between your target and one of the sequences found by BLAST, the design is not in a good location. This should have been evident from the initial BLAST search. If you want to retrieve the homologous sequence located by BLAST, click on the "Accession" number and you will be directed to the sequence report for that gene or transcript. If there are multiple splice variants detected by the assay or if the assay targets all members of a gene family, multiple hits would be expected. Once you have cleared this final hurdle, you are ready to move onto the empirical phase of the assay design.

TABLE 12.4
Rules for Primer and Probe Design

Primers	Comment
Max primer mismatch—2°C	Adjust the annealing temperature of the assay—same for both primers
Max primer–probe distance	Can be none up to 20–30 bases
Three G/C in last five bases max	Over three G/Cs can lead to inappropriate priming, false products
Three Gs in a row max	G residues will cause a bend in a short oligo, affects binding
Probes	**Comment**
First base not a G	Guanine can quench fluorescein-based reporter dyes
Max length = 30 bases	Longer distances lead to poorly quenched probes
More C than G residues	Most important for probes with large imbalance, G > C bases
Three Gs in a row max	G residues will cause a bend in a short oligo, affects binding
T_m 5–10°C above primers	The probe has to bind before primers, optimal hydrolysis, or signal

12.2.3 Acquisition and Preparation of Oligonucleotide Stocks

All experiments require that your adjustable pipettes are clean (RNase-free) and are in good working order. They should be routinely cleaned and sent out for calibration at least once a year, depending on how much you use them. The P-10 or equivalent should be checked more often since this is the most used pipette when setting up qPCR assays and is also the most critical because it is usually used for adding the template to the reaction. If your pipette is not working well for template addition, you do not have a prayer of getting good results, no matter how well the qPCR assay is designed. Also, be wary of fly-by-night calibration offers that seem to be very low priced. These folks cannot be found if something goes wrong and often use cheap, off-brand parts that will fail quickly. Practice your pipetting skills using exercises such as: dispense 1 mL water into several tubes and then weigh them. How close to 1 g are your aliquots? Once you are comfortable with a large volume, move on to the following: Make up a 7-log serial dilution series of any PCR template, ranging from 10 to 10^7 molecules or copies, to prepare a standard curve for any qPCR assay (i.e., already validated and known to work well). Selecting PCR product as the template is ideal, but any PCR template that you can dilute over a large concentration range will suffice (see contamination discussion below before you begin!). If you have a working SYBR Green I dye assay, use the amplified template for making the standard curve. Start by diluting PCR product 1/10,000 in $2 \times 1/100$ dilution steps. Make each of the serial dilutions for the standard curve by adding 3 µL of template to 27 µL of carrier (*Escherichia coli* or yeast molecular biology grade tRNA at 30–100 ng/µL in nuclease-free H_2O). The volume used for PCR will depend on what plate format you are using (i.e., 96-, 384-well). Pipette a volume, say 18–20 µL of PCR master mix into each well followed by 2–4 µL of each dilution of the standard curve into triplicate wells for the qPCR assay. When all the points for each dilution of the standard curve fall on a straight line, you are officially an excellent pipette operator. Do not despair when this does not happen the first time. It takes practice, concentration, and a steady hand. Do not pipette from the bottom of the tube (ever); take solutions from just below the meniscus and follow the liquid level down as you aspirate. Bring the liquid into the tip slowly, with a steady release of the plunger; never "pop" the plunger with a rapid release. Dispense the liquid the same way. The tip should not have any liquid on the sides when you are finished. Mix well (vortexing or taping the bottom of the tube is better than pipetting up and down) and change the tips between each new concentration for the dilution curve. Aerosols of concentrated solutions containing the template will spread into open primer (probe) stocks as well as sample tubes, reagents, and so on and will contaminate them. Always handle concentrated template stocks as far from primers, probes, and samples as possible (preferably in another room or a distant laboratory bench) and only when these are in closed tubes. Prevent aerosols during pipetting by avoiding the use of the "blowout" feature of the pipette. Aspirate the liquid by pushing the plunger all the way to the bottom (past the first stop) and then dispense it into the original tube to the first stop (no blowout). Then aspirate the same way again that will bring extra liquid into the tip. Then dispense into the target vessel, again only to the first stop, and on the side of the tube or well, not into any liquid. Repeat for all destination tubes or wells that are to contain the same concentration and then dispose of the tip with the extra volume. By avoiding the blowout feature, you avoid any aerosols and air-borne contamination. In general, this is a more accurate way to pipette solutions and is referred to as "reverse pipetting" by some.

Ordering of primers, probes, and DNA/RNA standards is primarily done over the web for most companies. The main rule here is: NEVER type in a sequence, ever. Typing sequences will lead to mistakes. When a typo does happen, the time and money wasted and subsequent anguish trying to figure out why things are not working is not worth it. This point cannot be stressed enough. Copy and paste the sequences from a text file into the website order form. For qPCR primers, purification using standard desalting is adequate; however, for probes, high-performance liquid chromatography (HPLC) purification is preferable, though it is not always necessary. Some oligo houses offer less expensive probes that are not completely free of the reporter dye that is not conjugated to the oligo probe, and these can work very well nonetheless. However, unbound label will result in a higher background signal and may compromise detection if there is insufficient data within the linear range of the detection capability of the instrument. When using a DNA oligo as the standard, it is not necessary to order extra bases beyond the primer-binding sites; it is sufficient to order the complete amplicon sequence. However, following synthesis, it is critical that the oligo must be PAGE-purified to remove as many failure products as possible. A further precaution to adopt, when using synthetic DNA standards, is to ensure that these are synthesized and purified in a completely different production facility, even using a different company, if necessary. However carefully the oligo house synthesizes and purifies their oligos, there will be contamination issues if they are synthesized in the same facility. If the assay is based on a standard assay, for example, SPUD[2] or a diagnostic sequence, it is also worth verifying that the standard has not been previously manufactured before allowing synthesis of the primers at the same facility. When the oligos arrive in the laboratory, keep the concentrated DNA standards separate and unopened until the primers/probe stock solutions have been prepared. Again, this will reduce the chance for assay contamination of the stock solutions. Good laboratory practice for PCR in general dictates that amplified DNA products should never be pipetted or even unsealed/lidded in the same laboratory where assay reagents are made up and reactions are prepared. Further, the "dirty room" should have its own set of pipetters, tips, water, and any other resource required and these supplies and equipment should never come back into the "clean room." If this separation is not possible, the use of a special hood can help but it is best to keep the hood, where reactions are set up and primer/probe stocks are made, as far away from possible PCR contamination as possible. Real-time qPCR is arguably the most sensitive quantitative assay around. If you are not careful at every step, it will show up in the final data set. Hence, there is no substitution or a shortcut for good technique and excellent laboratory practice.

When making up dried oligo stocks, use the mass values on the label as a guide, not as the gospel truth because the variation in the amount of oligo can vary by as much as 10-fold (less) than that stated on the label. This is a known issue with the oligo houses and they are working on it but for now, better to be safe than sorry. Add sufficient nuclease-free H_2O or buffer at pH 7.5 (the pH is critical since this value can severely impact on the A_{260} reading so that water should be pure and from a commercial supplier for greatest security) to the primers/probe that will result in a calculated 200 µM stock (twice the desired final concentration). Vortex the tube very well, let it stand for 30 min at room temperature (probably overkill but it works), vortex well again, and tap spin to collect the liquid at the bottom of the tube. Then, read the actual A_{260} values for each oligo using a spectrophotometer or a nanodrop instrument. For consistency, the same instrument should be used for this purpose to ensure consistency from assay to assay. Add nuclease-free H_2O to give a final 100 µM stock concentration to each oligo using the A_{260} reading, values for the molecular weight, and microgram/A_{260} and nanomoles from the specifications sheet that comes with the oligos to do the calculations. The reporter dyes on probes will not cause a problem for an accurate reading at A_{260} because optimal absorbance is at longer wavelengths.

The 100 µM oligo stocks are the master stocks and can be stored in the tubes that the oligos were provided in from the company. To make a working stock solution, make up tubes with up to 100 µL of 20 µM final concentration in nuclease-free H_2O. Store the master stocks at −80°C and the working stocks at −20°C. Master stocks can be aliquotted into replicate tubes, if required.

The presence of a standard template material, PCR product, a plasmid clone, or a synthesized DNA oligo, at high concentration in the vicinity of the reaction tubes poses a huge contamination risk. These DNA molecules can contaminate the universe. It is important to follow some simple precautions to prevent problems. Identify a neutral site where the standards can be handled at high concentrations (e.g., more than 20 pg/µL). This is best done in a "dirty room" if possible and must be a laboratory space that is separated from the set up site for RT–qPCR or qPCRs. If there is only one laboratory room to work in, investigate the possibility of using part of a bench in a neighboring laboratory on a temporary basis where no one is performing qPCR. Next, set aside a separate set of pipetters, tip boxes, and nuclease-free H_2O aliquot, for setting up the standards. Nothing used in making up DNA standard stocks should find its way back to your laboratory and workbench. Unlike primers/probes, DNA standards are made up at nanograms/microliters rather than a molar concentration. This is because the copy number calculations are based on the mass of standard to be added to the reaction. Make up long oligo standard stocks as described for the primers and probes above. After the stock has been made, calculate the concentration in nanograms/microliter from an A_{260} reading and put that value on the stock tube label. Use the information on the label as a guide and, as with the primers/probes, add a volume of H_2O that will give a more concentrated solution than desired. The amount for a synthetic oligo will differ over a large range depending on how well the synthesis and PAGE purification worked. In general, a stock concentration of 100–400 ng/µL is a convenient concentration range but if the solution cannot fit into the original tube, a higher standard concentration may be necessary. For plasmids or genomic DNA, use an A_{260} reading to find the concentration and label the stock tube accordingly. When using an RNA standard (synthetic oligo, *in vitro* transcribed, or total RNA), there is little chance of contaminating the qPCR. However, the total RNA can be challenging to use as the standard as it does not dilute as expected possibly due to intermolecular interactions of the single-stranded RNA in solution. A synthetic RNA may work better as it would be a single species and has the advantage, as with a DNA oligo or a plasmid, of having a known, high concentration of the target of interest. Further, the amount of target transcript within the total RNA preparation can vary depending upon the source and the conditions under which the cells or animals were grown. Store DNA (or RNA) standard stocks at −80°C. DNA standards constructed from oligos are generally in the 400–1000 µg/µL concentration range whereas plasmids tend to have very high concentrations but are larger in size as well. In the dirty room, make dilutions down to a working stock of 20 pg/µL for oligo standards in the 60–90 base length range. For plasmids, you will have to calculate the mass that correlates to around 2×10^7 copies (remember that the former is ssDNA and the latter is dsDNA). Formulae for these calculations, as well as others commonly used in real-time qPCR are presented within (Table 12.5). When diluting the material to make a standard curve, it is recommended that a carrier is used. Various options are available such as nuclease-free molecular biology grade, *E. coli,* tRNA, or yeast tRNA can also be used for a carrier depending on the target species as well as linearized acrylamide or even PEG. The aim is to pipette more than 2 µL of any solution, for example, a suitable standard 10-fold dilution series is 3 µL of oligo into 27 µL of carrier.

I would like to make a comment about using standards in qPCR and the term "absolute quantification." As you can see from the description above, standards are quantified using an A_{260} determination. This rapid and inexpensive method is an estimate of the sample DNA or RNA quantity but is certainly not an absolute value. The best method would be to use a quantification technique based on fluorescent dyes specific for the target (e.g., picogreen, ribogreen or oligreen). Incorporating these methods into the work flow would provide a much more accurate estimate of the standard concentration but that would not only be more time-consuming but would also be more expensive. Further,

---

TABLE 12.5
Useful Formulas for Real-Time qPCR

Purpose	Formula
Assay amplification	$10^{-(1/slope)}$
Assay PCR efficiency	$10^{-(1/slope)} - 1$
Sample quantity	$(amplification)^{-dCq}$ (single sample)
Sample quantity	$(amplification)^{-ddCq}$ (relative to another sample)
ssDNA copy number[a]	$(2 \times 10^{-12}/(DNA\ length\ (bases)) \times 330\ (ave\ MW\ dNTP)) \times 6.023 \times 10^{23}$
dsDNA copy number[a]	$(2 \times 10^{-12}/(DNA\ length\ (bases)) \times 660\ (ave\ MW\ 2\ dNTP)) \times 6.023 \times 10^{23}$
ssRNA copy number[a]	$(2 \times 10^{-12}/(RNA\ length\ (bases)) \times 340\ (ave\ MW\ NTP)) \times 6.023 \times 10^{23}$

[a] In 12 pg of the respective nucleic acid.

it would not guarantee that the final result was "absolute" due to the inherent errors in performing any technique (pipetting, etc.). Thus, the values obtained from using a standard curve are valuable in that you have a value for each sample that allows much more flexibility during data analysis and, as with what is commonly referred to as "relative quantification" provides quantitative data within the context of the experiment. But, the copy number values obtained using the standard curve are only a good estimate of the gene or transcript copy number in the sample. The only way to approach an absolute gene or transcript copy number value for a sample would be for everyone to use the same assay, instrument, and reagents with a standard from an agency such as NIST, in the United States, which was quantified using an independent method to have x number of copies per unit volume.

12.2.4 EMPIRICAL TESTING

Empirical testing of the new assay is the first step in determining whether all the previous preparation during the assay design will pay off. In my experience, the vast majority of assays I design will pass validation (quality control or QC) the first time but there are times when it is necessary to order alternative primers for an assay. Reassuringly, this is the exception to the norm. Testing a new assay requires a reliable template so that QC experiments can be performed and the QC data obtained can be directly correlated to the assay (without questions regarding the template quality). Purchased synthetic oligo standards, a plasmid construct bearing the specific template (both described above), *in vitro* transcribed RNA, PCR product from a previous RT–PCR, or genomic DNA are potentially suitable QC templates. Using any one of these templates, determine the amount of template required for approximately 2×10^3 copies per reaction from an estimate of the total nucleic acids based on A_{260} (see Table 12.5). If the target is a viral, microbial or lower eukaryotic DNA, an aliquot of one sample can be used. To use an RNA sample as a starting point for making a standard, perform an RT–qPCR of one sample using the primer pair from the set of four newly designed primers that results in the longest PCR amplicon. If there is not an available sample RNA to use as a template, the best source for most transcripts is testis RNA. This is not a panacea of course, but it does contain the largest range of transcripts in one tissue that I have found so far. Testis RNA can be purchased for the common mammalian species. If the target is a cytokine/chemokine assay, it is important to select samples that have had the target transcript induced as the mRNA levels for many of these transcripts are not detectable in normal samples. There are also companies that sell RNA from a large range of tissues or cell lines that work well, although it is important to check with the manufacturer, prior to purchase, that your template is present within the total RNA mix.

Test your new qPCR assay using one of the templates listed above using a probe or SYBR Green I PCR master mix remembering to use the primer pair that yields the longest PCR product. This will be your first look at the assays performance under the qPCR conditions you have set or the manufacturer has recommended.

Once you have completed this initial qPCR run, make a note of the C_q values. These should be nearly identical for replicates. The next test is to see how the four primers perform in all four possible combinations. For this, use a template concentration that will yield a C_q value from 24 to 27 cycles. This will strain the assay, but not to the point where it would not give a decent amplification signal. To make the test dilution series, make a 10,000-fold dilution of the original template in a carrier solution. Unless the C_q value was very low (too much template for this experiment), this dilution factor will result in an initial C_q value around 13 cycles. To get an approximate C_q value of 25 would require four 10-fold serial dilutions, for example. The recommended cycle value does not have to be exact but you want to make sure the assay has to go over 24 cycles before a detectable signal is seen on your qPCR instrument. Otherwise, if there is an excess of template and the signal comes up with a C_q in the teens, all the primers will look comparable. In this test, you are looking for the primer pair with the best amplification curve (most acute) and usually, but not always, for the lowest C_q value. A primer pair with a strong amplification curve can be preferable to the one that has the lowest C_q value if the corresponding amplification curve with the lowest C_q value is not robust. Figure

12.4 shows an example of four primer pair combinations that gave about the same result. When running this test with several primers, I have seen examples where one primer pair combination gave no signal at all. Sometimes, one primer pair clearly sticks out as the best as can be seen in Figure 12.5. Interestingly, neither of these primer pairs were the ones selected by the assay design software as being the best. Both these assays are probe-based. An example of an SYBR Green I assay for *Macaca* Collagen 1A1 using dilutions of RT–qPCR product as the template can be seen in Figure 12.6. Again, there is a difference in how well each primer pair amplified the target sequence. As my rule #1 for research says: "You don't know until you know." That is certainly true in this case. By having more than one amplification curve to compare, the best primer pair becomes obvious. If you had ordered just one primer pair and it was not the best pair, the assay would still work but you may not have an acceptable LOQ or PCR efficiency and you would end up ordering another primer set in any case. The time saved in ordering and accessing multiple primers per assay at one time more than offsets the low cost of the extra primers. If money is an issue and you do not have many assays to keep tabs on (we have well over 1000), you may be able to improve on the performance of the best primer pair combination found by the software by trying them in a matrix with differing primer concentrations so that the two primers no longer have the same concentration. Hold one steady and try higher and lower concentrations of one using the lowest C_q as the measure and then hold that primer steady at the optimal concentration you have found and test the second primer in the same way. If this improves performance (lowers the C_q value with a better amplification curve geometry), test using the standard curve again with the new asymmetric primer concentrations. A quick way to see if the primer matrix might work is to try lowering the annealing temperature from 60°C in 2°C decrements. A gradient thermocycler makes this kind of experiment particularly simple. The downside of just using a lower annealing temperature is that you can lose template specificity if the annealing temperature gets too low. I would not recommend an annealing temperature below 55°C. This experiment is one reason why the recommended T_m difference between the primers was kept tight at 2°C. Another way to increase the binding efficiency of primers is by increasing the $MgCl_2$

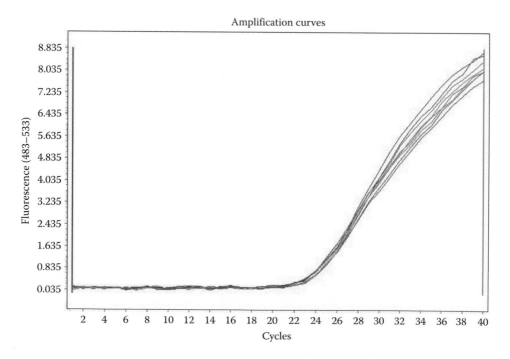

FIGURE 12.4 **(See color insert.)** Human PTENP1 pseudogene assay for hPTEN—results for four primer combinations around a single probe. The red primer pair was selected in this case.

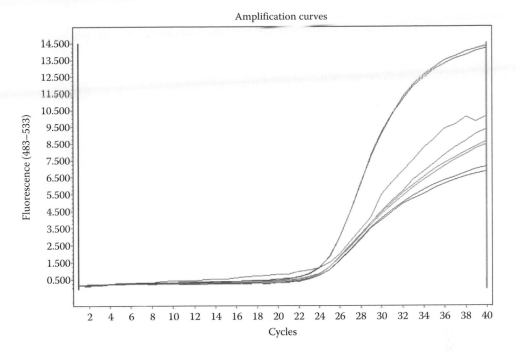

FIGURE 12.5 **(See color insert.)** Human PTEN transcript assay—results for four primer combinations around a single probe. The purple primer pair is clearly the best for this assay.

concentration in the PCR master mix. This may not be an available option if using a commercial master mix. Check the manual that comes with the master mix. Usually, an $MgCl_2$ concentration of 5–6 mM is about as high as I would go. Using too much Mg^{2+} can lead to loss of template specificity. We routinely use 5 mM $MgCl_2$ for probe-based assays and 3 mM for SYBR Green I assays but this is not a hard-and-fast rule.

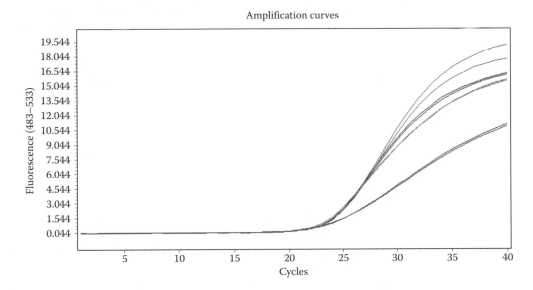

FIGURE 12.6 **(See color insert.)** *Macaca* assay for Collagen 1A1—results for four primer combinations for an SYBR Green I assay using diluted PCR product as the template. The blue primer pair was selected in this case as giving the best amplification curve although the C_q values were not the lowest.

Hopefully, you now know which primer combination gives the best results for the assay. The final QC experiment is a determination of assay efficiency and LOQ using a standard curve. Let me digress just a moment here to say that even if you do not plan to run standard curves for the analysis of your test samples, you still need to run this QC standard curve to qualify the assay. You may also decide that using some of the diluted PCR template as a sample on each plate would be a good surrogate for a standard curve. Using this DNA template on each plate can help with post-run data analysis; in that the C_q value for this sample should remain nearly identical on all plates for that assay and thus the unknowns will then be comparable from plate to plate. Having a common sample, whether replicates (3) of a standard or an abundant sample on each plate, in multiple-plate experiments, is critical to marrying the data into a cohesive data set. Such samples can serve as critical interplate calibrators during post-run data analysis.

From the initial dilution of the test template, you should now be able to make a dilution that will give a C_q signal around 13–14 cycles when your working sample volume is added to the qPCR master mix. Using this dilution as the starting concentration, construct seven more standard templates in 10-fold decrements of the starting template dilution. You want to make sure that you have a dilution series that will surely lie beyond the detection limit of the assay. Make a PCR master mix for all samples and controls and add to 18 wells. Use the reaction volume you usually use or that is recommended by the assay manufacturer for the plate and instrument you are using. Add the template to two wells/standard dilution starting with the second well in a row that will fill 16 wells. Add nuclease-free H_2O to the first and 18th well in each row as no-template controls (NTC). By now, you have had ample opportunity to contaminate your working stocks and this is an excellent opportunity to see if that has happened. In the instrument software, label the standard curve, putting in copy number values for each dilution to match the dilution series, and save the file with the days date. If the exact copy number is unknown (e.g., using the total RNA), use relative dilution factor values in the copy number. As a guide, a C_q value of 30 should be ≈200 copies if the slope is close to ideal (−3.32). Use the rule of thumb for estimating differences, 1 cycle ≈2-fold difference, 2 cycles ≈4–5-fold difference, and 3 cycles ≈10-fold difference. With this information, you should be able to label the copy numbers on the x-axis to get close to reality. This can be done following the run using the instrument software.

Run the plate under the same master mix and cycling conditions that were used in the primer combination test or those determined to be optimal using the primer matrix. This should result in a series of amplification curves that are regularly spaced and all of them are parallel to the first curve. Analyze the data using the instrument software and look at the standard curve and the numerical data associated with the curve. First, notice how many dilutions are represented as points on the line drawn by the software. The last dilution is expected to either fall off the curve (high) or not show any amplification in 40 cycles of PCR, depending on where the dilution series started. For a good assay, the first seven dilutions should be part of the linear standard curve. Depending on your pipetting prowess, all the replicate points should be on the line with very little deviation. If this is not the case, repeat the experiment and validation steps (if required) until all the points are on the line. Otherwise, the slope, lowest LOQ, and lowest limit of detection (LOD) for the assay will not be reliable and should not be used to perform a real-time qPCR experiment with unknowns. It should go without saying that the two NTC reactions should have no signal. If they do, assay reagents have been contaminated. There are three basic reagents that can get contaminated: The primers/probe, the PCR master mix, and the H_2O. You can determine which of these is at fault and obtain fresh or you can toss out everything and move on. However, remember that the contamination is most likely only for one assay. Thus, you may be able to use the reagents for all other assays but the one in question. This too will require testing, of course, if funds are a concern. Otherwise, just start over with all new reagents and working stock aliquots of the primers/probe.

There are three important values to be gained from the standard curve QC experiment. The first is the slope of the standard curve. The acceptable range commonly adopted is −3.2 to −3.5. That translates into 105–93% PCR efficiency (Table 12.5). If the assay does not fit into that range, look at the primer optimization experiment again. If there was another primer pair that you almost selected, try that primer pair and repeat the standard curve QC experiment. If that does not pass either, one approach is to return to the assay design software and find another pair of primers that can be combined with the first four and expand QC experiment #1 to include nine unique primer combinations (three forward, three reverse primers). This approach has proved to be sufficient to bring an assay into the acceptable range of PCR efficiency with the desired LOQ and keeps the qPCR-cycling conditions consistent from assay to assay. An alternative approach is to stay with the same primer pair and try to optimize those using various methods. One reason the current primer pair is not working optimally may be due to an inability for both the primers to work efficiently under the same primer concentration and/or annealing temperature. Thus, experiments designed to try a matrix of different primer concentrations for each primer (asymmetric primer concentrations) as well as lower annealing temperatures (2°C increments) may lead to optimal assay performance as outlined above.

The second important diagnostic value is the limit of quantification (LOQ).[3] This is defined by the lowest dilution that is still coincident with the rest of the standard curve. If you are using a standard curve for quantification of unknown samples on your assay plates, define the LOQ in terms of the copy number. When using the ddCq method, or a derivative thereof for quantification of the target in unknown samples, record the LOQ as a cycle number (C_q value). Either way, the LOQ defines the lowest quantifiable limit for this assay. An acceptable LOQ is in the 5–10 copies for qPCR for any assay to pass QC when there are no known circumstances that would affect the assay sensitivity. That translates into a C_q value between 34 and 35 cycles, depending on the PCR efficiency of the assay. Under these circumstances, the target quantity in any sample with a C_q value that is higher than the LOQ cannot be quantified using the assay. If the LOQ value is not low enough, for example, 100s instead of 10s of copies, or 30 versus 33–35 cycles, for example, then go again and start at QC experiment #1 as described above. The LOQ should not be confused with the limit of detection (LOD). It is quite possible that the assay has the capacity to record a measurable C_q value that is lower than 40 cycles, but that C_q value is higher than the LOQ. That higher, but nonquantitative C_q value is the LOD. It is this capacity to give a value that is not quantitative (LOD) that makes determining the LOQ for each assay so important. The literature is undoubtedly littered with bad data points based on LOD rather than LOQ values for many types of assays.[3]

The third value that is an indicator of assay quality is the y-intercept on a graph of C_q versus template concentration on a log scale (log [template]). This value is only meaningful if the number of copies of template going into the assay is known. The intercept on the y-axis represents the theoretical cycle number corresponding to the detection of one copy of template (using a log scale on the x-axis where concentration has not been log transformed for a linear x-axis scale). Or, in terms of $y = mx + b$, the value of b has not been log transformed. For assays that have optimal PCR efficiencies (at or near 100%), the value of b will be around 37–38 cycles, if the copy numbers have been calculated correctly. The value of the y-intercept results from a combination of the C_q value of the sample and the PCR efficiency of the assay. Stable y-intercept values between different plates are indicative that there has been no drift in either the amount of sample (standard) or PCR efficiency of the assay on successive plates bearing the same qPCR assay. If you are running the standard curves on your plates, the y-intercept is the most informative value for monitoring consistency from plate to plate. The 7-log standard curves for the human PTENP1 and PTEN assays can be seen in Figures 12.7 and 12.8. In this case, a probe was used to detect the amplicon. The standard curve for the *Macaca mulata* SYBR Green I assay using diluted PCR product can be seen in Figure 12.9. As can be seen in all these figures, the results are about the same, regardless of the detection method used.

The two values from this QC exercise that are required for publication are the PCR efficiency and the LOQ of the assay. Of course, the LOQ is critical for deciding whether the determined concentration in an unknown sample can be quantified using your assay. One other important factor that should be set for the new assay is the threshold setting that will be used for data analysis. For each assay, it is desirable to use the same threshold setting for all plates within the same project. The threshold value may be different for different qPCR assays within an experiment but it should

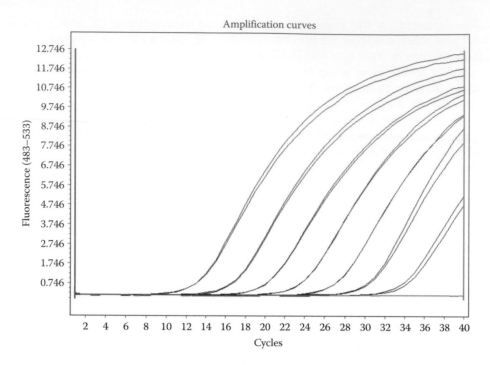

FIGURE 12.7 **(See color insert.)** 7-Log standard curve for the hPTENP1 assay. The PCR efficiency is 94%, the slope of −3.47 with a *y*-intercept of 38.42. The green line is the NTC.

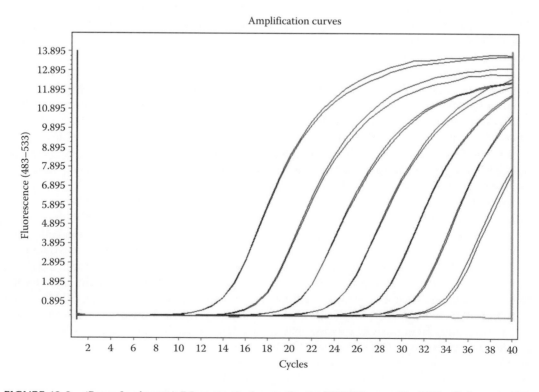

FIGURE 12.8 **(See color insert.)** 7-Log standard curve for the hPTEN assay. The PCR efficiency is 98%, the slope of −3.37 with a *y*-intercept of 37.53. The green line is the NTC.

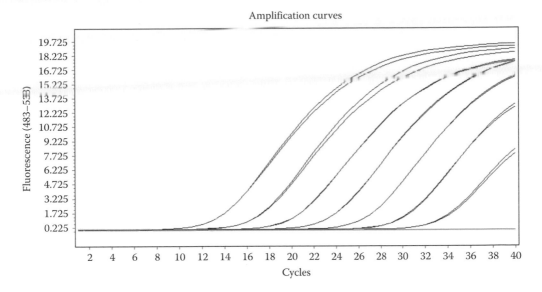

FIGURE 12.9   (**See color insert.**) 7-Log standard curve for the *Macaca* Collagen 1A1 assay. The PCR efficiency is 99%, the slope of −3.31 with a *y*-intercept of 37.69. The detection used SYBR Green I. The green
line is the NTC.

be the same throughout for the same qPCR assay. There are alternative postrun data analysis algorithms available from some of the instrument manufacturers that do not use the threshold method
for determination of C_q. However, they all retain a threshold software algorithm and there are times
when it is still the preferred method of postrun C_q determination.

12.3   FINAL COMMENTS

Outlined above is a method for real-time qPCR assay design that has worked successfully for the
author for over 15 years. This approach is the result of much empirical, trial and error, success,
and failure over that time span. There is a technical side to any laboratory technique that cannot
be taught well in a textbook. Therefore, it is not possible to guarantee initial success. However, this
chapter contains a presentation of how to perform some of the most technically critical elements
in new qPCR assay design and validation. Hopefully the job has been well done and you will have
great success.

It is down to each scientist to decide what the allowable boundaries are for PCR efficiency and
LOQ for new assays in your laboratory. It is best to set these values prior to performing QC experiments and to stick to them no matter what the empirical outcome. Set high standards for assay quality and your real-time qPCR data will never disappoint you, from a technical point of view. A good
qPCR assay is a joy to use forever, a bad qPCR assay can be a reoccurring nightmare.

REFERENCES

1. Bustin, S. A., Benes, V., Garson, J. A., Hellemans, J., Huggett, J., Kubista, M., Mueller, R. et al. 2009.
   The MIQE guidelines: Minimum information for publication of quantitative real-time PCR experiments.
   Clin Chem 55, 611–622.
2. Nolan, T., Hands, R.E., Ogunkolade, W., and Bustin, S. A. 2006. SPUD: A quantitative PCR assay for the
   detection of inhibitors in nucleic acid preparations. *Anal Biochem*, 351, 308–310.
3. Linnet, K. and Kondratovich, M. 2004. Partly nonparametric approach for determining the limit of detection. *Clin Chem, 50*(4), 732–740.

13 Primer Design for Large-Scale Multiplex PCR and Arrayed Primer Extension

Maido Remm, Kaarel Krjutškov, and Andres Metspalu

CONTENTS

13.1 INTRODUCTION

Variations in the sequences of the human genome play a central role in determining disease susceptibility and the results of drug therapy. Understanding this process will be a cornerstone of future "personalized medicine."[1] However, the detailed structure of the human genome is still not fully understood despite getting clearer.[2,3] During the last few years, genome-wide association studies (GWAS) have produced hundreds of new susceptibility loci for many common diseases.[4] This recent progress could lead to a situation where the genetic risk assessment for certain diseases and traits becomes feasible and small diagnostic genotyping platforms might be needed instead of the current high-throughput discovery genotyping and sequencing technologies.[5] It would not be unrealistic to expect that only about 100–1500 genetic markers will be analyzed for a certain diagnostic test and this information, together with "traditional" information about the age, sex, environment, and lifestyle, could be used to predict the disease, its progression, or the potential dose of a suitable drug. This is already within reach using the currently available medium-scale genotyping technology. Here, we describe polymerase chain reaction (PCR) and genotyping protocols that allow us to solve problems at this genomic scale.

13.2 BIOINFORMATICS SUPPORT FOR LARGE-SCALE GENOMIC PCR

Custom genotyping platforms allow for testing of thousands of single-nucleotide polymorphisms (SNPs) from each individual. However, most of these genotyping platforms require PCR amplification to achieve sufficiently strong signals. Amplification of thousands of fragments from the human genome is not a trivial task. Automated methods for primer design are necessary to achieve this within a reasonable time frame. Achieving high-quality primer design with automated methods is typically more difficult than with manual design. However, automated methods can surpass the speed and quality of the manual design if the primer parameters are properly selected. We have collected PCR data from a genotyping experiment in which 1014 different markers were amplified and genotyped from chromosome 22. The primer properties that could be used to predict PCR success or failure were assessed from this dataset using statistical methods (type I generalized linear models). The results were verified on an independent dataset of 300 primer pairs, designed to random template regions of the human genome.

13.2.1 OPTIMAL PARAMETERS FOR PCR PRIMER DESIGN

Most of the PCR primers used in this analysis were designed automatically with the program PRIMER3.[6,7] All PCR primer pairs were tested at least 10 times on human genome samples. PCR results were recorded as negative when the predicted fragment was absent, there were multiple or smeared fragments, or bands that were of the wrong size. A list of 236 different parameters was then generated that describe PCR primers or primer pairs, which could possibly be used to predict the PCR success rate. Using statistical models, we searched for the parameters that are the best indicators for the prediction of PCR primer success rate (Table 13.1). In general, longer and AT-rich primers work slightly better for PCR amplification from genomic DNA targets. A "GC-clamp" is often designed into the 3'-end of primers to achieve higher specificity. We did not observe any benefits from the addition of guanine cytosine (GC) clamps when primers were used for amplifying the human genomic DNA.

13.2.2 GENOME TEST

In addition to traditional parameters, the relationship between PCR quality, the number of primer binding sites, and the number of products derived from the human genome sequence was tested. A string consisting of the last 16 nucleotides from the 3' end of the primer was used to search for matches in the human genome. Only sites with 100% identity to the primer sequence were counted as a primer binding site. A potential PCR product was counted if the sense and antisense primer binding sites were found in the genome within 1000 bp from each other. The results in Table 13.2 show that, for successful PCR, both primers should bind to less than 100 unique locations and this

TABLE 13.1
Primer Design Parameters and Their Effect on the PCR Quality

Primer Property	Effect of PCR Failure Rate	Suggested Range
Length	Weak	Both primers >23 nt
GC% of the whole primer	Medium	Average GC 20–60%
GC% of the 3-half	Weak	Average GC 20–60%
GC% within last 3-nucleotides of both primers	None	
PCR product length	Medium	<400 nt

Note: Results of an analysis of 1014 primer pairs.

TABLE 13.2
Relationship between Primer Specificity and Assay Quality

Primer Property	Effect of PCR Failure Rate	Suggested Range
Number of binding sites in the genome (higher value of both primers)	Very strong	<10 binding sites
Number of PCR products generated from the genome	Strong	<5 products

demonstrates convincingly that testing the primers against the genome is essential for high-quality PCR from the genomic DNA.

The process of determining the genomic locations of PCR primer binding sites is referred to as the *genome test*. If all possible binding sites for both primers are known, counting the number of alternative PCR products is a trivial task. Although the results in Table 13.2 show that primers that could generate up to five products from the genome might work in a PCR reaction, it is wise to avoid primer pairs that generate more than one product. For example, in genotyping experiments, any additional PCR product could interfere with the genotyping signal from the real SNP marker, so that the observed signal is, in fact, a mixture from two different PCR products. Although this situation might be detectable as heterozygote excess in the Hardy–Weinberg test, the results from such markers cannot be used in further analyses.

It has long been the practice to perform similarity searches between the primer sequence and the sequence to be amplified, to avoid primer binding on repeated motifs. However, such similarity searches typically return only a local alignment with the best matching area. From this local alignment, it is hard to make sure that the chosen primers are really unique and would generate a unique product from the genome. Furthermore, running BLAST on NCBI server against the typical "nr" database[8] is slow, can be difficult to interpret, and does not allow for a proper estimation of the number of products. Our solution to the problem is to directly localize and enumerate all possible binding sites in the genome. We have tested different string and similarity search programs to find the most efficient solution. A search against the 3×10^9 nucleotides of the human genome cannot be performed on a simple desktop computer. Even if only matches with 100% identity (no gaps or mismatches) are allowed, a very large memory size is required. The search for primer binding sites can be accelerated by running BLAST over the Internet[8] or by a specific software such as MEGABLAST,[9] SSAHA,[10] or BLAT.[11]

One important parameter is the length of the string ("word length") from the 3′ end of the primer chosen for searching binding sites. Ideally, this should be the length of the primer that is sufficient for specific binding to genomic DNA at a given T_m under specified salt, primer, and template concentrations. The word length that is sufficient might be different for different primers; longer for AT-rich primers and shorter for GC-rich primers. In practice, it is easier to use equal word length for all primers. We have tested different word lengths on the test set mentioned above to see which gives the best separation between "good" and "bad" primers. Our results show that word lengths between 15 and 18 nucleotides, taken from the 3′ end of the primer give the best predictive power in the genome test. The results shown in Table 13.2 are calculated with word length 16.

We have implemented these findings in the computer software GenomeTester.[12] This software is capable of determining genomic locations of up to 100,000 primer pairs within a few minutes. Single primer pairs can be analyzed in a few seconds. An online version of the program is available for testing over the Internet.[12]

13.2.3 MULTIPLEXING PCR PRIMERS

Simultaneous amplification of many marker regions in a multiplex reaction offers significant savings of template DNA, time, and money. Setting up a multiplex PCR with consistent quality for all targets is not trivial. Often, significant efforts are made to optimize reagent sources and concentrations for multiplex PCR.[13,14] However, testing for primer–primer and primer–product interactions in each group is equally important. Primer–dimer formation can be avoided by designing identical nonpalindromic dinucleotides to the end of each primer,[14] but this is often difficult to achieve when working with a large number of primers. We have created a computer program, MULTIPLX,[15] which tests all primer pairs for interactions, including dimer formation, and automatically generates compatible primer groups for multiplex PCR. In principle, the primers in one multiplex group should conform to the same criteria as a primer pair in a single PCR reaction. They should not form primer–dimers with each other and they should not bind to each other's products. Therefore, the same criteria that are used to avoid primer–primer interactions in PRIMER3 could be appropriate for checking primer–primer interactions in MULTIPLX. The PRIMER3 program uses two parameters to check for possible primer–primer interactions: PRIMER_SELF_ANY and PRIMER_SELF_END. These are calculated as local and global alignment scores between two primers or between the primer and its complement. We added two more parameters: PRIMER_PRODUCT_ANY and PRIMER_SELF_END1. PRIMER_PRODUCT_ANY is similar to PRIMER_SELF_ANY except that it calculates the highest score of the primer–product local alignment, instead of primer–primer local alignment. When calculating these parameters, thermodynamic modeling of DNA–DNA interactions is used. With a rather stringent set of parameters, we have been able to achieve 8-plex to 16-plex PCR without any decrease in the genotyping call rate.

13.3 SNP GENOTYPING USING APEX AND APEX-2

13.3.1 TEMPLATE DNA AMPLIFICATION

During recent decades, single-base extension (SBE) has been widely accepted as the standard nucleic acid detection method and has been applied to different assay formats and detection platforms.[16–23] We have developed this further, resulting in the arrayed primer extension (APEX) method and have adjusted it for small-to-medium scale genotyping projects. The APEX protocol contains six steps: The nucleic acid is (i) amplified from the genomic DNA using PCR, (ii) fragmented and purified, and (iii) hybridized to oligonucleotides, which have been arrayed on a solid surface. Hybridization and (iv) SBE are simultaneous reactions on the array surface. After that, (v) fluorescent signals are detected and (vi) converted into genotypes using specific data analysis software.

The APEX reaction takes place on a solid surface after template amplification and purification. This is a large, multiplex reaction enabling hundreds to thousands of simultaneous hybridizations and SBEs.[24] However, the template preparation by single- or low-level multiplex PCR does not match the efficiency of the current microarray-based methods, making PCR the limiting step in the assays.[25] The first APEX applications were based on a single-plex PCR template-preparation protocol.[26,27] This principle was a labor-intensive and uneconomical procedure. Low-level multiplex PCR seemed more cost-effective, but knowledge about effective parameters of reliable multiplex PCR has only been gained in recent years.[28–30] Prior to these reports, multiplexing relied upon *in silico* testing to determine where primer specificity was ensured but there was a lack of knowledge about the ideal amplicon length, reaction, and cycling conditions. However, 7-plex has been the maximum level achieved by the APEX assay, where 50 loci were analyzed but only 41 gave 99.85% genotype concordance with the HapMap data, at a call rate of 94.9%.[30] Using a higher multiplex level, 50-plex, yielded a 100% call rate and >99.9% accuracy was achieved after reducing amplicon sizes to <200 bp.[25] In addition, 18 bp common linker sequences were designed and added to the 5′ end of all PCR primers. These linkers have two properties: a balanced and reasonably high GC content to increase the melting temperature of the primer, and a unique sequence not found in the human DNA template.[30,31]

13.3.2 APEX-2 MULTIPLEX PCR

APEX-2 is a further development of APEX wherein the limiting step of multiplex PCR for template preparation is minimized. Improved primer construction and two-step amplification enabled the robust amplification between 124[32] and 640[33] genetic markers simultaneously. This is a single-tube

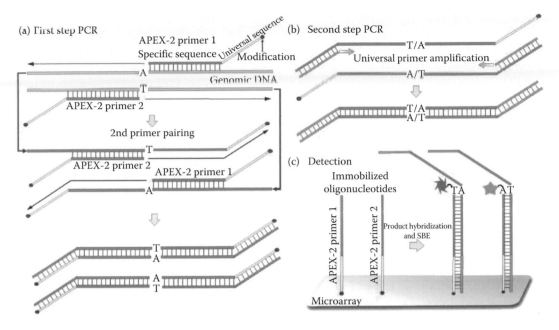

FIGURE 13.1 Principle of the APEX-2 assay. (a) Primers bind to template DNA immediately upstream of the position of interest. After primer extension, the synthesized sequence contains the complementary strand of the respective APEX-2 primer and the position of interest. (b) The universal primer hybridizes to the 3′ end of the previously generated products. (c) The purified universal primer-generated PCR products hybridize to the immobilized APEX-2 primers that have a 5′ amino modification. The 5′ amino group facilitates the attachment of oligos on the solid microarray surface. Genotyping is performed as a four-color single-base extension reaction.

PCR amplification protocol where only two oligonucleotides are needed per locus of interest, for amplification and detection using APEX. Each primer contains a genome-specific region and a common 5′ tail for amplification (Figure 13.1). Primers are designed so that the 3′ end of the oligonucleotides resides one nucleotide before the studied position. As a result of the first amplification reaction, only the studied polymorphism is novel in the amplicon. The second amplification is performed using the common tails and primers complementary to them. This step, called universal primer amplification, is essential for multiplication of all previously formed products and to raise the amplicon concentration to a level at which later APEX detection is possible. The basic requirements for this innovation include: (i) generation of short products (~100 bp) due to the close proximity of hybridizing APEX-2 primers and (ii) incorporation of the universal primer in the APEX-2 oligonucleotide sequence, enabling phase-2 PCR amplification under optimal conditions, which is not feasible when using PCR primers containing only genome-unique sequences. Although significant changes were made in the template-amplification phase, the primer extension protocol for APEX-2 contains only two minor modifications: (i) printed oligonucleotides are the same as those used in the multiplex PCR and (ii) primers are longer (median 45 bp).

13.3.3 APEX REACTION ON MICROARRAY

On a microarray, APEX, just as APEX-2, utilizes two oligonucleotides per analyzed base pair; one for the sense and the other for the antisense direction. In this format, the high information content of oligonucleotide arrays and the specificity of molecular recognition by DNA polymerase are combined. All four dye terminators are used in the same reaction, allowing for simultaneous evaluation of all possible nucleotide changes. APEX can be used for identification of different types of mutations and polymorphisms[26,27,34] and can also be used for gene resequencing.[24] We used 24 × 60 mm aminosilanized glass microscope slides coated with phenylenediisothiocyanate as the support material for microarrays (Asper Biotech Ltd.). The oligonucleotides were equipped with C6 or C12 amino linkers at their 5′ ends. Primers were diluted to a final concentration of 50 μM in 100 mM sodium carbonate/bicarbonate buffer (pH 9.0) and spotted onto an activated glass surface. After spotting, the slides were blocked with 1% ammonia solution and were stored at 4°C until needed. Washing steps were performed before APEX reactions to reduce the background fluorescence and to remove unbound oligonucleotides: 95°C water for 2 min, 10 min in 100 mM NaOH (not optional in APEX-2), and 3 times for 2 min in 95°C ddH$_2$O.

Double-stranded template DNA to be used for APEX is amplified by PCR, which allows the sequence from both DNA strands to be read simultaneously. A critical parameter in the APEX assay is the length of the target. Amplicon lengths of ~100 bp were optimal, and PCR products of up to 200 bp could be used directly in the assay. Longer PCR products needed fragmentation before use in the APEX reaction. To achieve this, a fraction of the dTTP is replaced by dUTP in the amplification mix, allowing for later treatment with uracil N-glycosylase (UNG) and heat treatment.

Single-nucleotide extension reactions can work only if no deoxyribonucleotide triphosphates are carried over from the amplification mix. A reliable method to inactivate dNTPs left over from PCR is enzymatic digestion with shrimp alkaline phosphatase (SAP). This can be performed simultaneously with the UNG treatment and must be followed by thermal inactivation of both enzymes prior to the APEX reaction.

Approximately 1.5–2 μg of fragmented amplification product mix is used for the APEX assay. Engineered DNA polymerases are able to incorporate the dye terminators quite efficiently. The reaction mix contained 5 U of thermo sequenase DNA polymerase (Amersham Biosciences), thermo sequenase reaction buffer, and a final concentration of 1.2 μM of each fluorescently labeled ddNTP: Texas Red-ddATP, Cy3-ddCTP, fluorescein-ddGTP, and Cy5-ddUTP. The DNA in the buffer was denatured at 95°C for 10 min. The enzyme and dye were immediately added to other components and the whole mix was applied to prewarmed slides at 58–62°C. The reactions were allowed to proceed for 20 min under coverslips and were stopped by washing twice at 95°C in ultrapure water. Incorporation of labeled terminators is a very quick reaction, but hybridization is an equilibrium process and thus it needs time to achieve strong enough signals. After reaction and washing, a droplet of SlowFade® Light Antifade Reagent (Molecular Probes, Eugene, OR) was applied to the chips to limit bleaching of the fluorescein, and then the chips were covered with coverslips and imaged with the Genorama™ imaging system, at 20 μm resolution. All genotypes were identified by Genorama 3.0 genotyping software (Asper Biotech Ltd.). The signal intensities of grayscale pictures from different fluorochromes were first equalized. The strongest signal represents the base that was called. If the next strongest signal from both strands had an intensity level higher than 30–50% of the strongest signal, the position was called heterozygous. The sequence was compared with a reference and diverging bases were indicated. All divergences and heterozygous positions were verified manually by comparing signals from different images by eye and using histogram values.

13.4   DISCUSSION OF PROBLEMS AND PRACTICAL HINTS

13.4.1   PRACTICAL HINTS FOR THE PRIMER DESIGN

We routinely use the modified mPRIMER3 program[7] with the following advanced settings:

- TABLE_OF_THERMODYNAMIC_PARAMETERS = SantaLucia 1998
- SALT_CORRECTION_FORMULA = Owczarzy 2004
- USE_LOWERCASE_MASKING = YES
- PRIMER_MISPRIMING_LIBRARY = HUMAN

Masking repetitive template regions with the GenomeMasker program[35] before primer design might be a good idea. Using RepeatMasker[36] is less practical, because it tends to mask long contiguous areas, even if the region contains nonrepeated segments, suitable for the primer design. None of the currently available masking methods guarantee uniqueness of the designed PCR product; so, the genome test with GenomeTester[37] or electronic PCR is necessary in any case.

13.4.2   PRACTICAL HINTS FOR FINDING THE PRIMER BINDING SITES

We suggest using GenomeTester[12] for predicting the number of primer binding sites and for predicting the number of PCR products. GenomeTester uses preindexed hash tables to look up the locations of strings in the genome. It is significantly faster than the other available programs, if the number of primer pairs in the test exceeds 100.

The suggested command-line parameters for the GenomeTester are

- Creating hash tables of the genomic sequence:
  - Gindexer chr1.fas chr1
- Predicting the number of primer binding sites and PCR products:
  - Gtester my_primers.txt index_files.txt
- For other software, the suggested command-line parameters are
  - BLAST: blastall -p blastn -i input.fas -d chr1.fas -F F
  - MEGABLAST: megablast -i input.fas -d contigs.fas -F F -W 12 -D 2
  - SSAHA INDEXES: ssaha chr1.fas -ph -sf fasta -sn chr1 -wl 10 -sl 1
  - SSAHA SEARCH: ssaha input.fas chr1 -qf fasta -sf hash -pf -mp 15

GenomeTester is also available online.[37] BLAST and MEGABLAST should be run with the DUST filter turned off by defining '-F F'; otherwise, some of the primer binding sites might overlap the masked areas and might remain undetected.

Note: The '-F F', parameters can be specified in the BLAST command line. Switching the DUST filter off means that we will count the binding sites in the entire genome, not only the nonrepeated regions of the genome. On the web version of BLASTN, it is shown within "Algorithm parameters" section, "Filter: Low complexity regions" switched off.

13.4.3   PRACTICAL HINTS FOR GENOTYPING

To perform PCR amplification simultaneously at hundreds of loci,[32,33] Mg^{2+} concentration needs to be higher (4.5–6 mM) than in traditional PCR. It is believed that high total primer concentration (40 µmol and average oligonucleotide length is 75 bp) may decrease free Mg^{2+} concentration in the buffer solution and may cause suboptimal conditions for DNA polymerase. First-step PCR stringency depends on the lowest temperature in the annealing step and some optimization is recommended if different primer T_m calculation software has been used. The last enzymatic step, primer extension, is also Mg^{2+} sensitive. Thus, if the molar amount of the template is increased (large arrays with over 1000 extension reactions), the Mg^{2+} concentration has to be adjusted to a higher value.

13.5   CONCLUSIONS AND FUTURE DIRECTIONS

For optimal PCR results, primer binding sites should be rare in the genome and the generated product should be unique. Our data show that this is best achieved by using optimal primers for the genomic PCR that are longer than 20 nucleotides and are AT-rich. The uniqueness of the product can be achieved by counting all possible binding sites and products in the genome for any given primer pair. The programs MEGABLAST,[9] SSAHA,[10] BLAT,[11] or GenomeTester[37] are recommended for the prediction of primer-binding sites in large genomes. On a smaller scale,

BLAST[8] is a slower but simpler alternative. BLAST can also be used to report binding sites with mismatches, which slightly increases its sensitivity in the genome test. Unfortunately, there is, at the moment, no method for the genome test, which would permit appropriate modeling of primer binding sites. Thermodynamic algorithms[38] for predicting primer binding sites would yield a higher sensitivity in the genome test, but are computationally extremely intensive and, so far, impractical. Most of the current primer design methods use identity-based alignments to measure the strength of primer–primer interactions. Therefore, the ultimate test is the real "wet lab" experiment. If in this test, a particular SNP test will not work, it should be discarded and a new one picked from the same linkage disequilibrium (LD) region. If an SNP assay does not work perfectly, it will cause enormous problems later when DNA quality is often variable and then the base calling will suffer. In our experience, about 10% of the selected SNP assays need to be changed after this first "wet lab" experiment.

ACKNOWLEDGMENTS

This study was supported by Targeted Financing from the Estonian Ministry of Education and Research (SF0180142s08 to A.M. and SF0180026s09 to M.R.), Estonian Science Foundation grant ETF7859, EU FP7 OPENGENE (#245536 A.M.), and by EU via the European Regional Development Fund grant to the Centre of Excellence in Genomics, Estonian Biocentre, and University of Tartu.

REFERENCES

1. Ginsburg, G. S. and McCarthy, J. J. 2001. Personalized medicine: Revolutionizing drug discovery and patient care. *Trends Biotech*, **19**, 491.
2. The International HapMap 3 Consortium. 2010. Integrating common and rare genetic variation in diverse human populations. *Nature*, **467**, 52–8.
3. http://www.1000genomes.org.
4. Manolio, T. A. 2010. Genome-wide association studies and assessment of the risk of disease. *N Engl J Med*, **363**, 166–76.
5. www.illumina.com.
6. Rozen, S. and Skaletsky, H. 2000. Primer3 on the WWW for general users and for biologist programmers. *Methods Mol Biol*, **132**, 365–86.
7. Kõressaar, T. and Remm, M. 2007. Enhancements and modifications of primer design program Primer3. Bioinformatics, the advanced parameters can be used online at http://www.bioinformatics.nl/cgi-bin/primer3plus/primer3plus.*cgi.*, **23**(10), 1289–91.
8. Altschul, S. F. et al. 1997. Gapped BLAST and PSI-BLAST: A new generation of protein database search programs. *Nucleic Acids Res*, **25**(17), 3389–402.
9. Zhang, Z. et al. 2000. A greedy algorithm for aligning DNA sequences. *J Comput Biol*, **7**(1–2), 203–14.
10. Ning, Z., Cox, A. J., and Mullikin, J. C. 2001. SSAHA: A fast search method for large DNA databases. *Genome Res*, **11**(10), 1725–29.
11. Kent, W. J. 2002. BLAT—The BLAST-like alignment tool. The BLAT server is available at ENSEMBL (http://www.ensembl.org/Multi/blastview). *Genome Res*, **12**(4), 656–64.
12. Andreson, R. et al. 2006. GENOMEMASKER package for designing unique genomic PCR primers. Online version available at http://bioinfo.ut.ee/genometester/. *BMC Bioinf*, **7**, 172.
13. Henegariu, O. et al. 1997. Multiplex PCR: Critical parameters and step-by-step protocol. *Biotechniques*, **23**(3), 504–11.
14. Zangenberg, G., Saiki, R. K., and Reynolds, R. 1999. Multiplex PCR: Optimization guidelines. In *PCR Application: Protocols for Functional Genomics*, Innis, M. A., Gelfand, G. H., and Sninsky, J. J., eds., Academic Press, San Diego.
15. Kaplinski, L. et al. 2005. MultiPLX: Automatic grouping and evaluation of PCR primers. Online version available at http://bioinfo.ut.ee/multiplx/. *Bioinformatics*, **21**(8), 1701–2.
16. Syvänen, A. C. et al. 1990. A primer-guided nucleotide incorporation assay in the genotyping of apolipoprotein E. *Genomics*, **8**(4), 684–92.
17. Pastinen, T. et al. 2000. A system for specific, high-throughput genotyping by allele-specific primer extension on microarrays. *Genome Res*, **10**(7), 1031–42.

18. Oliphant, A. et al. 2002. BeadArray technology: Enabling an accurate, cost-effective approach to high-throughput genotyping. *Biotechniques*, Supplement: S, 56–8, 60–1.

19. Chen, X., Levine L., and Kwok, P. Y. 1999. Fluorescence polarization in homogeneous nucleic acid analysis. *Genome Res*, **9**(5), 492–8.

20. Gunderson, K. L. et al. 2005. A genome-wide scalable SNP genotyping assay using microarray technology. *Nat Genet*, **37**(5), 549–54.

21. Hardenbol, P. et al. 2003. Multiplexed genotyping with sequence-tagged molecular inversion probes. *Nat Biotechnol*, **21**(6), 673–8.

22. Pastinen, T. et al. 1997. Minisequencing: A specific tool for DNA analysis and diagnostics on oligonucle-otide arrays. *Genome Res*, **7**(6), 606–14.

23. Ross, P. et al. 1998. High level multiplex genotyping by MALDI-TOF mass spectrometry. *Nat Biotechnol*, **16**(13), 1347–51.

24. Tõnisson, N. et al. 2002. Evaluating the arrayed primer extension resequencing assay of TP53 tumor sup-pressor gene. *Proc Natl Acad Sci USA*, **99**(8), 5503–8.

25. Syvänen, A. C. 2005. Toward genome-wide SNP genotyping. *Nat Genet*, **37** , S5–10.

26. Dawson, E. et al. 2002. A first-generation linkage disequilibrium map of human chromosome 22. *Nature*, **418**(6897), 544–8.

27. Jaakson, K. et al. 2003. Genotyping microarray (gene chip) for the ABCR (ABCA4) gene. *Hum Mutat*, **22**(5), 395–403.

28. Sanchez, J. J. et al. 2006. A multiplex assay with 52 single nucleotide polymorphisms for human identi-fication. *Electrophoresis*, **27**(9), 1713–24.

29. Tebbutt, S. J. and Ruan, J. 2008. Combining multiple PCR primer pairs for each amplicon can improve SNP genotyping accuracy by reducing allelic dropout. *Biotechniques*, **45**(6), 637–8, 640, 642 passim.

30. Podder, M. et al. 2008. Robust SNP genotyping by multiplex PCR and arrayed primer extension. *BMC Med Genom*, **1**, 5.

31. Wang, D. G. et al. 1998. Large-scale identification, mapping, and genotyping of single-nucleotide poly-morphisms in the human genome. *Science*, **280**(5366), 1077–82.

32. Krjutskov, K. et al. 2009. Evaluation of the 124-plex SNP typing microarray for forensic testing. *Forensic Sci Int Genet*, **4**(1), 43–8.

33. Krjutskov, K. et al. 2009. Development of a single tube 640-plex genotyping method for detection of nucleic acid variations on microarrays. *Nucleic Acids Res*, **36**(12), e75.

34. Kurg, A. et al. 2000. Arrayed primer extension: Solid-phase four-color DNA resequencing and mutation detection technology. *Genet Test*, **4**(1), 1–7.

35. Andreson, R., Puurand, T., and Remm, M. 2006. SNPmasker: Automatic masking of SNPs and repeats across eukaryotic genomes. Online version available at http://bioinfo.ut.ee/snpmasker/. *Nucleic Acids Res*, **34** (Web Server issue), W651–5.

36. Smit, A. F. and Green, P. RepeatMasker. (http://www.repeatmasker.org/).

37. Andreson, R. et al. 2006. GENOMEMASKER package for designing unique genomic PCR primers. *BMC Bioinf*, **7**, 172.

38. SantaLucia, J., Jr. 1998. A unified view of polymer, dumbbell, and oligonucleotide DNA nearest-neigh-bor thermodynamics. *Proc Natl Acad Sci USA*, **95**(4), 1460–5.

14 Development and Use of qPCR Assays for Detection and Study of Neglected Tropical and Emerging Infectious Diseases

Ashley R. Heath, Norha Deluge, Maria Galli de Amorim, and Emmanuel Dias-Neto

CONTENTS

14.1 INTRODUCTION

Neglected tropical and emerging infectious diseases present a continuing, global challenge for detection/diagnosis, treatment, and, ultimately, eradication.[1,2] The diseases run the entire gamut of possibilities from viral (e.g., the recent H1N1 epidemics), to parasitic, bacterial, and even fungal. In this chapter, we will focus the discussion mainly on two parasitic diseases, schistosomiasis and neurocysticercosis (NCC). In addition, a description of some work concerning hepatitis B virus (HBV) is also included.

14.1.1 SCHISTOSOMIASIS

Schistosomiasis is a neglected tropical disease that is caused by parasitic trematodes of the genus *Schistosoma*, including *Schistosoma mansoni*. It is estimated that this disease affects about 200 million people, living in 74 countries of Africa, Middle East, Asia, and South America.[3] The parasite's life cycle includes a nonsexual multiplication and differentiation inside the intermediate mollusk host, which is followed by the infection of the definitive mammalian host, where immature worms migrate from the skin to the lungs, and then to the mesenteric veins that drain the intestines. There the young schistosomula develop into adult worms with sexual dimorphism. Male and female worms develop, mate, and lay eggs that are released into the environment, where the life cycle is continued. A portion of the eggs remain trapped inside the human body and may cause serious conditions, including severe liver damage (due to granuloma formation), as well as some presentations of neuroschistosomiasis, which include transverse myelitis, paralysis, and brain microinfarcts.

The detection of eggs in the feces is a simple, effective, and low-cost method to detect the infestation with *S. mansoni*. However, in some situations, this approach is not sensitive enough. This is particularly true after treatment, or in low-prevalence areas where the low incidence of infection leads to patients with low egg counts and an underdiagnosed disease. Another situation of particular interest is the "prepatent period," an intermediate period immediately after the initial infestation, in which the parasites are present but the worms have not yet matured, and thus have not started to produce and release eggs into the feces. These are examples of situations where more sensitive diagnostic methods are an absolute requirement. As the current control strategy recommended by the World Health Organization for schistosomiasis is based on the treatment of infected patients,[4] the prompt identification of infected individuals is remarkably important.

The development of specific and more sensitive diagnostic methods has been described and these are based on the detection of repetitive DNA elements by polymerase chain reaction (PCR),[5,6] followed by the visualization of the amplicons in silver-stained agarose gels.[7] In the above-mentioned publications by Pontes et al., a high analytical sensitivity was achieved, as the target in the parasite's genome was a tandemly repeated sequence (GenBank M61098) that comprises at least 12% of the parasite's genome.[8,9] Using this repetitive element as a target, a detection level of 1 fg of *Schistosoma* DNA was achieved. In artificial mixtures of parasite DNA and feces, the level of detection was 2.16 eggs/g, which makes this approach 10 times more sensitive than stool examination by the Kato-Katz examination (see review in Ref. 10). The detection of *S. mansoni* DNA by real-time quantitative PCR (qPCR) has been reported more recently. Using this approach, the amplified rDNA repeat is visualized by SYBR Green I binding dye, which allowed the detection of 10 fg of the parasite's DNA.[11,12] The 18S rDNA amplification together with a TaqMan® probe detection was used by Zhou et al.,[13] who reported the same detection limit of 10 fg. More recently, other targets for molecular detection have been unraveled by transcriptome and genome sequencing projects of *S. mansoni*.[14–16]

14.1.2 Neurocysticercosis

Neurocysticercosis (NCC) is the most important parasitic disease of the human central nervous system and the main cause of acquired epilepsy worldwide. The infection of the human host starts with the ingestion of raw or undercooked food; usually salads, fruits, or vegetables, contaminated with human feces carrying eggs of the cestode pork tapeworm *Taenia solium*. Inside the human host, these embryonated eggs circulate, hatch in different areas of the body, and the larvae (cysticerci) mature. When these larvae are localized in the brains, NCC develops. The disease may cause a wide range of symptoms, usually triggered by inflammatory responses elicited by antigens that are released by the degenerating vesicles, and depend on the number and on the location of the larvae in the brain. Whereas seizures are the main manifestation of NCC, its clinical presentation is largely variable. Patients are asymptomatic in about 50% of the cases,[17] but can also present mild or severe headache episodes, increased intracranial pressure, involuntary movements, psychosis, and sudden death.[18,19]

The disease is endemic in many poor areas of the world, including South America, Asia, and Africa. NCC is expanding globally due to higher multilateral migration, tourism, and overall globalization.[20] The gold standard for NCC diagnosis is the demonstration of the cysticerci in the brain, with a visible scolex, by magnetic resonance imaging or computerized tomography scans. However, while imaging approaches allow precise diagnosis, the elevated cost of the equipment and of the procedures involved precludes its large-scale adoption in the low-income high-prevalence endemic areas. Alternative diagnostic approaches include the detection of antibodies raised against the larvae antigens or the identification of antigens released by the parasite. Unfortunately, antibody/antigen detection protocols do not result in the desired sensitivity and specificity, and alternative diagnostic approaches are urgently required.

Almeida et al.[21] were the first to demonstrate that *T. solium* DNA could be found in the cerebrospinal fluid (CSF) of NCC patients, and that this could be the basis of the first PCR-based method to detect NCC. To achieve the high sensitivity required for a diagnostic method, these authors used a highly repetitive DNA element found in the *T. solium* genome, named pTsol9 (GenBank accession code: U45987), and showed that it is possible to detect DNA amounts that are equivalent to 3% of the genome content of a single parasite cell.

The protocol described by Almeida et al.[21] was further evaluated by Michelet et al.[22] in a study that involved patients and controls from Mexico and France, and compared the sensitivity and the specificity of PCR-based assays with immunodiagnostic assays. The authors observed, "PCR exhibited the highest sensitivity (95.9%) and variable specificity (80% or 100%) depending on the controls used." Interestingly, the PCR-based method showed a lower specificity (80%) in the Mexican non-NCC controls cohort. As NCC is endemic in Mexico, the authors conclude that the PCR diagnosis data indicate the possibility of underdiagnosis of NCC, which could not be detected by available immunological or imaging tests. These authors found that this PCR-based diagnostic approach showed the highest sensitivity when compared to any other tests (EITB and ELISA) for all groups of patients.

Also working with CSF, Yera et al.[23] recently described the use of qPCR to diagnose NCC in nine patients living in France. In this small cohort, this method showed 100% specificity and 83.3% sensitivity. These are examples that demonstrate that DNA-based approaches are useful, specific, and sensitive enough for the diagnosis of this disease in endemic areas. The costs involved are small and the methods can be adopted by a number of diagnostic centers in the different continents affected by NCC.

14.1.3 HEPATITIS B VIRUS

Since the conclusive demonstration of its existence in the late 1960s, it has been accepted that HBV is a cause of acute as well as chronic hepatitis, the latter leading to the possibility of the development of cirrhosis and liver cancer. The WHO estimates that in excess of 350 million people worldwide suffer from this infection, for which there is preventative treatment in the form of vaccination but no truly successful therapeutic approach once infection has taken hold. Molecular diagnosis of the HBV is complicated by the fact that at least eight genotypes of the virus have been shown to exist.[24] In this study, we attempted to develop a generic qPCR assay that would give positive results for all eight genotypes.

Common to all of these diseases is the need for rapid, sensitive, and specific methods of detection. It is here that the application of qPCR techniques plays a major role. In this chapter, we shall describe the application of qPCR for the detection of the three entities, *S. mansoni*, *T. solium*, and HBV, using a combination of computer-aided design, assay testing using synthetic amplicons, and, ultimately, testing of natural clinical samples.

14.2 DESIGN OF qPCR ASSAYS

The general principles behind assay design and validation for qPCR are well described elsewhere, including Chapter 12 of this book. In the case of this work, we opted to utilize assays based on dual labelled hydrolysis probes[25] essentially because experience has shown that these tend to be straightforward to develop and highly robust in their use. Additionally, costs for synthesis of dual labelled hydrolysis probes, in general, lower than for other specialized probe types, which is an important factor to be considered for assays that are being developed for use in situations where resources may be limited. With the exception of the HBV assay described below, all primer and probe design was carried out using Beacon Designer from Premier Biosoft International.[26] All primer and probe designs are compliant with the recently proposed minimum information for publication of quantitative real-time PCR experiments (MIQE) guidelines for qPCR standardization.[27]

In recent years, improvements in the procedures and processes used for the chemical synthesis of oligonucleotides have resulted in significant increases in the total length that may be achieved for any particular sequence. At SIGMA Custom Products, for example, oligonucleotides of up to 150 bases in length are now routinely synthesized. As the optimal amplicon size for a 5′ nuclease qPCR assay is typically 150 base pairs or less, this means that for most, if not all, assays it is now possible to synthesize an artificial amplicon. Such amplicons may then be used to test and optimize new assays, prior to their deployment in the field, thus affording the potential for a considerable savings of precious samples and time and effort while providing robust data with which to analyze the quality of the qPCR assay. This was the general approach taken in the present study.

14.2.1 SCHISTOSOMA MANSONI

Over the past 20 years or so, a number of endpoint PCR-based methods requiring post-PCR gel analysis for the detection of *S. mansoni* in patient samples have been described. As described above, Pontes et al.[5] and Gomes et al.[10,11] have developed an assay for the detection of the parasite in human serum and fecal samples, using, as the basis for the target, a highly repetitive element that was originally discovered by Hamburger et al.[8] The sequence of this element used as a target for the PCR assay is shown in Figure 14.1a. We sought to develop a probe-based qPCR assay as an extension of these studies. Two approaches were used. First, an assay was developed using the Hamburger target sequence (*S. mansoni* 1) and the existing PCR primers. Second, an assay was developed using the Hamburger target sequence and *de novo* designed primers and probes (*S. mansoni* 2). The primer and probe sequences for the two assays are shown in Table 14.1.

14.2.2 TAENIA SOLIUM

Building upon the previously mentioned study by Almeida et al.,[21] primers and probes were designed against part of the repetitive pTsol9 element. Once again, using Beacon Designer, one primer and probe set was chosen using the existing PCR primers (*T. solium* 2) and another assay was developed using a *de novo* design (*T. solium* 1). In the case of *T. solium*, both forward primers have the same sequences, but the reverse primers and probes differ (Table 14.1).

(a) 5′
  GATCTGAATCCGACCAACCGTTCTATGAAAATCGTTGTATCTCCGAAACCACTGG
  ACGGATTTTTATGATGTTTGTTTTAGATTATTTGCGAGAGCGTGGGCGTTAATAT 3′

(b) 5′
  CAGGGTGTGACGTCATGGCAGGCAGCCTGGCCAATGCGCCTGTCGCCGATCT
  GGGTCATGTGCCGCAGTCCACACGGCAAAGGACAGCCTCCGAG 3′

(c) 5′
  TATCGCTGGATGTGTCTGCGGCGTTTTATCATATTCCTCTTCATCCTGCTGCTATGCC
  TCATCTTCTTGTTGGTTCTTCTGGACTACCAAGGTATGTTGCCCGT TTGTCCTCTA 3′

FIGURE 14.1 (a) *Schistosoma mansoni* target sequence used for the development of both standard and quantitative PCR assays. (b) *Taenia solium* target sequence used for the development of both standard and qPCR assays. (c) Hepatitis B virus target sequence used for the development of qPCR assay.

TABLE 14.1

Primer and Probe Sequences Used for the Detection of Each Target Organism

Organism	Forward Primer 5′ to 3′	Reverse Primer 5′ to 3′	Dual Labeled Probe 5′ to 3′
S. mansoni 1	GATCTGAATCCGACCAACCG	ATATTAACGCCC ACGCTCTC	6FAM-CGTTGTATCTCCGAAACC ACTGGACGG-BHQ1
S. mansoni 2	CCGACCAACCGTTCTATGA	CACGCTCTCGCAAA TAATCTAAA	6FAM-TCGTTGTATCTCCG AAACCACTGGACG-BHQ1
T. solium 1	CAGGGTGTGACGTCATGG	CTCGGAGGCTG TCCTTTG	6FAM-CGCCTGTCGCC GATCTGGGTC-BHQ1
T. solium 2	CAGGGTGTGACGTCATGG	AGGAGGCCAGT TGCCTAGC	6FAM-CGATCTGGGTCATG TGCCGCAGTCCAC-BHQ1
Hepatitis B virus	GATGTTCTGCGGCGTTTATC	GAGGACATACGGG CAACATAC	6FAM-CA [+T] CC [+T] G [+C] TG[+C]T [+A]TG [+C] CTC-BHQ1

Note: 6FAM, 6-carboxyfluorescein; BHQ1, black hole quencher 1; I, inosine; [+N], locked nucleic acid base.

14.2.3 Hepatitis B Virus

A total of 1025 HBV target sequences were downloaded from Genbank. After removal of short, poor-quality, and redundant sequences a total of 888 sequences were aligned in groups using ClustalW from EMBL-EBI (http://www.ebi.ac.uk/Tools/msa/clustalw2/). From the alignment results of the ClustalW analysis, consensus regions with suitable lengths for primer design were identified using Perl scripts. Initial primers and probe design was carried out using Beacon Designer[26]; however, the presence of mixed base positions in the regions corresponding to both forward and reverse primers required the incorporation of inosine bases that have the capability to base pair with any of the naturally occurring DNA bases and so allow effective base pairing of the primers to their targets irrespective of the specific base at that site (see Table 14.1). Also, since there were only short regions of sequence that were available for a probe and to ensure maximum specificity and strength of hybridization for this specific assay, locked nucleic acid (LNA) bases were incorporated into the probe sequence.

14.3 MATERIALS AND METHODS

The following reagents were all purchased from SIGMA Aldrich (St. Louis, MO): JumpStart™ Taq ReadyMix (P2893), DNAse/RNAse-free water (W4502), magnesium chloride (MgCl$_2$, M8787), and albumin solution from bovine serum, BSA (B8667). All oligonucleotides were manufactured by SIGMA Custom Products (The Woodlands, TX). Synthetic amplicons were purified by polyacrylamide gel electrophoresis (PAGE), dual-labeled probes were high-performance liquid chromatography (HPLC)-purified, and primers were standard desalted (DST) preparations.

14.3.1 Amplification of Synthetic Targets

The qPCR reactions for the synthetic (oligonucleotide) targets were carried out using a LightCycler® II qPCR instrument (Roche). All pipetting was done using dedicated pipettes and pipette tips with barrier filters. The primers, probes, and amplicons were resuspended in DNAse-/RNase-free water to 100 µM concentration as determined by absorbance measurements at 260 nm using a SpectraMax spectrophotometer. Based on the calculated concentrations, working stock solutions were prepared at 10 µM. The reaction mixtures were prepared by mixing the different reagents down to these final concentrations in the LightCycler® capillaries: MgCl$_2$ (3 mM), BSA

(0.1 mg/mL), probe 200 nM, and equimolar concentrations of the forward and reverse primers, based on primer optimization.

For each assay, standard curves were generated with eight data points, each data point being run in triplicate, by mixing 10 μL of the JumpStart™ Taq Ready Mix, 5 μL of the reaction mixture, and 5 μL of the corresponding template solution (from 10^8 to 10 copies based on serial dilutions of the templates).

For each assay primer, the concentration was first optimized by testing different concentrations of forward and reverse primer (100 nM/100 nM; 150 nm/150 nm; 200 nm/200 nm)[28] while keeping the other components constant (200 nM probe, 10^6 copies of template, to generate a Cq of around 20). Cycling conditions were 95°C for 2 min, followed by 45 cycles of 95°C for 10 s, 60°C for 20 s, 72°C for 1 s (detection). The C_q values and the fluorescence differential ($F_{max} - F_{min}$) were used to determine the optimal primer concentration, defined as the concentration yielding the lowest C_q value with highest fluorescence differential.

Subsequently, using the optimal primer concentrations determined for each assay (200 nM for *S. mansoni* 1, *S. mansoni* 2, and *T. solium* 1; 250 nM for *T. solium* 2) standard curves were generated for each assay in triplicate, using the same thermal profile as used for primer optimization.

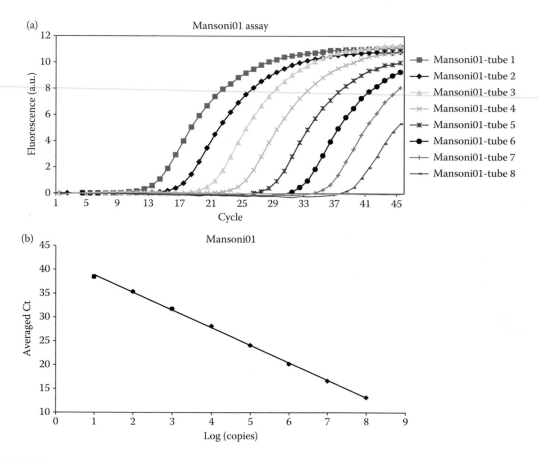

FIGURE 14.2   (a) Fluorescence curves from standard curve of *S. mansoni* assay 1 showing the range of responses from the most concentrated (10^8 copies) to the most dilute (10 copies). (b) Standard curve result for *S. mansoni* assay 1. For this standard curve, the slope = −3.681, the *y* intercept = 42.574, $r^2 = 0.999$, and the efficiency = 1.869.

14.3.2 AMPLIFICATION OF *S. MANSONI* DNA

Using DNA extracted from adult worms, qPCR was carried out on an ABI 7500 Fast system. The final probe concentration was 250 nM, and the concentration of both forward and reverse primers was 1 µM. The mastermix used was JumpStart™ Taq Ready Mix and the total reaction volume was 20 µL.

14.4 RESULTS

The initial results showed good assay sensitivity (detection down to 10 copies), the curves were linear, and reproducibility was good. However, reaction efficiency required optimization as the initial efficiency levels were around 85–87%. Optimization may be done by modifying the cycling conditions, as demonstrated here for the HBV assay. As described previously, primer concentration was first optimized and the best conditions found to be from using 200 nM. Standard curves were then generated in triplicate (from 10^8 copies down to 10 copies) this time using the following cycling parameters: 95°C for 2 min, followed by 45 cycles of 95°C for 15 s, 60°C for 30 s, and

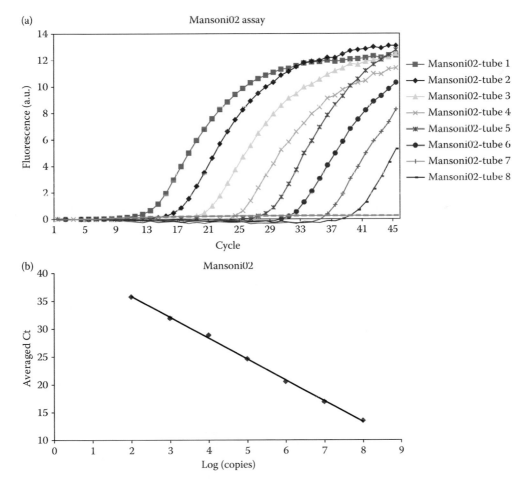

FIGURE 14.3 (a) Fluorescence curves from standard curve of *S. mansoni* assay 2 showing the range of responses from the most concentrated (10^8 copies) to the most dilute (10 copies). (b) Standard curve result for *S. mansoni* assay 02. The slope = −3.752, the *y* intercept = 43.290, r^2 = 0.999, and the efficiency = 1.847.

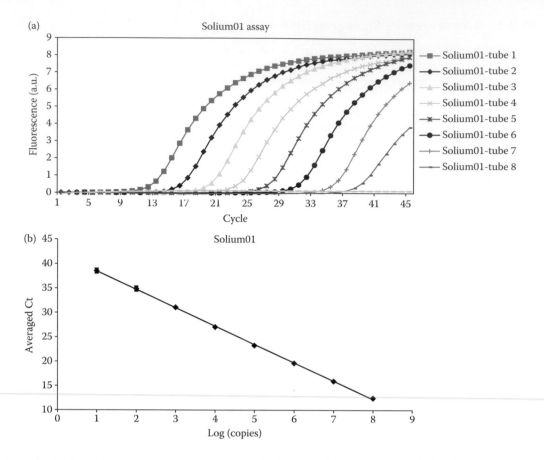

FIGURE 14.4 (a) Fluorescence curves from standard curve of *T. solium* assay 1 showing the range of responses from the most concentrated (10^8 copies) to the most dilute (10 copies). (b) Standard curve results for *T. solium* assay 1. The slope = −3.749, the *y* intercept = 42.241, r^2 = 1.000, and the efficiency = 1.848.

72°C for 1 s. The results obtained were good in terms of sensitivity, linearity, reproducibility, and efficiency (97%).

Figures 14.2a through 14.6 present the results of amplification of the synthetic target amplification for all the assays developed. In Figure 14.7, representative fluorescence curves for *S. mansoni* assay 02 are shown, using DNA extracted from adult worms (Figures 14.3 through 14.7).

14.5 DISCUSSION

The work presented here demonstrates the utility of designing qPCR assays based upon previous results using classical PCR, and in conjunction with preliminary assay development using synthetic targets. If appropriate classical PCR studies are available for any specific disease of interest, the process of creating a qPCR assay for that disease will be relatively straightforward and rapid; synthesis of the synthetic targets may be accomplished within 2–3 days, including any necessary purification steps, and initial testing of the primers and probes may be carried out within a day. Thus, it is possible in principle to have a high-quality functional qPCR assay available for neglected and emerging diseases in rather less than 1 week. In the case of HBV, it was necessary to first apply bioinformatics methods to determine the relevant target sequences, and also to incorporate modified bases into the primers and probes. Using

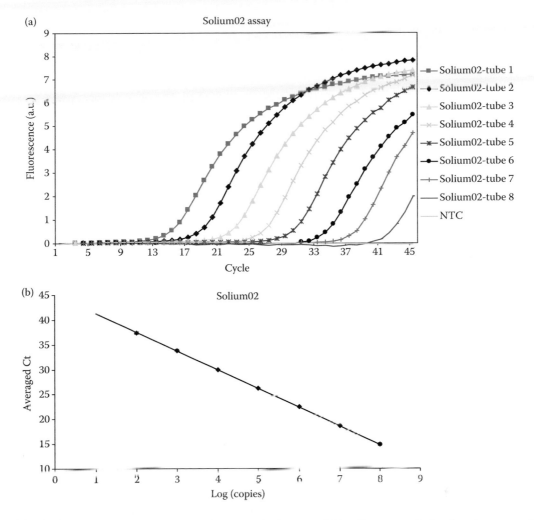

FIGURE 14.5 (a) Fluorescence curves from standard curve of *T. solium* assay 2 showing the range of responses from the most concentrated (10^8 copies) to the most dilute (10 copies). (b) Standard curve results for *T. solium* assay 2. The slope $= -3.759$, the y intercept $= 44.997$, $r^2 = 1.000$, and the efficiency $= 1.845$.

bioinformatics methods added to the time required to complete the overall process, though not significantly. Modified bases such as LNAs[29] do add to the cost of producing the required oligonucleotides but that cost increase is more than compensated for by the increase in efficiency of the resulting assay. The HBV assay was in fact successfully used in a study comparing it against several other HBV primer and probe sets, including those available commercially.[30] The *S. mansoni* and *T. solium* assays are currently being tested against patient samples and we await those results with interest.

ACKNOWLEDGMENTS

Thanks are due to Tao Zhao and Fei Zhong of the SIGMA Aldrich Bioinformatics group for their analysis of the HBV target sequences and to Carlos Martinez of SIGMA Custom Products for access to the LightCycler® II system. E.D.N. thanks Associação Beneficente Alzira Denise Hertzog Silva (ABADHS) for their continuous support LIM-27.

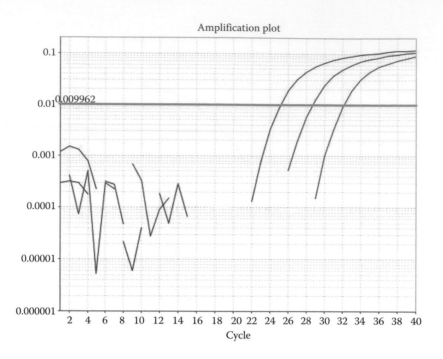

FIGURE 14.6   Fluorescence curves obtained by real-time amplification of *S. mansoni* DNA using *S. mansoni* assay 2. Target dilutions were 1:10,000, 1:100,000, and 1:1,000,000 with Cqs 25.03, 28.63, and 32.02, respectively.

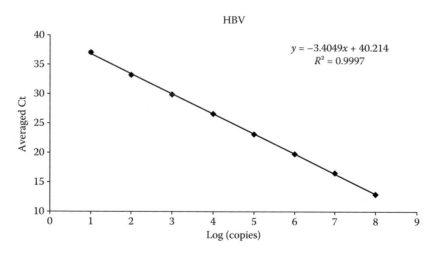

FIGURE 14.7   Standard curve results for optimized HBV assay from the most dilute (10 copies) to the most concentrated (10^8 copies).

REFERENCES

1. Feldmann, H. Truly emerging—A new disease caused by a novel virus. *N. Engl. J. Med.*, 364(16), 1561, 2011.
2. Savioli, L. et al. *Working to Overcome the Global Impact of Neglected Tropical Diseases. First WHO Report on Neglected Tropical Diseases.* World Health Organization, Geneva, Switzerland, 2010.
3. Bergquist R. et al. Blueprint for schistosomiasis vaccine development. *Acta Trop.*, 82:183–192, 2002.

4. Harder A. Chemotherapeutic approaches to schistosomes: Current knowledge and outlook. *Parasitol. Res.*, 88:395–397, 2002.
5. Pontes, L.A. et al. Detection by polymerase chain reaction of *Schistosoma mansoni* DNA in human serum and feces. *Am. J. Trop. Med. Hyg.*, 66:157–162, 2002.
6. Pontes L.A. et al. Comparison of a PCR assay and the Kato-Katz technique for diagnosing *Schistosoma mansoni* infection. *Am. J. Trop. Med. Hyg.*, 68:652–656, 2003.
7. Sanguinetti C.J. et al. Recovery and reamplification of PCR products from silver stained polyacrylamide gels. *Biotechniques*, 17(5):914–921, 1994.
8. Hamburger J. et al. Highly repeated short DNA sequences in the genome of *Schistosoma mansoni* recognized by a species-specific probe. *Mol. Biochem. Parasitol.*, 44:73–80, 1991.
9. Hamburger, J. et al. Development and laboratory evaluation of a polymerase chain reaction for monitoring *Schistosoma mansoni* infestation of water. *Am. J. Trop. Med. Hyg.*, 59:468–473, 1998.
10. Rabello A. et al. Recent advances in the diagnosis of Schistosoma infection: The detection of parasite DNA. *Mem. Inst. Oswaldo Cruz*, 97:171–172, 2002.
11. Gomes A.L. et al. Development of a real time polymerase chain reaction for quantitation of *Schistosoma mansoni* DNA. *Mem. Inst. Oswaldo Cruz*, 101 Suppl 1:133–136, 2006.
12. Gomes, L.I. et al. Further evaluation of an updated PCR assay for the detection of *Schistosoma mansoni* DNA in human stool samples. *Mem. Inst. Oswaldo Cruz*, 104:1194–1196, 2009.
13. Zhou L. et al. A highly sensitive TaqMan real-time PCR assay for early detection of *Schistosoma* species. *Acta Trop.*, 120:88–94, 2011.
14. Verjovski-Almeida S. et al. Transcriptome analysis of the acoelomate human parasite *Schistosoma mansoni*. *Nat. Genet.*, 35:148–157, 2003.
15. Verjovski-Almeida S. et al. Insights into Schistosome metazoan and parasitic functions: Perspectives for functional genomics. *Trends Parasitol.*, 20:304–308, 2004.
16. Berriman et al. The genome sequence of the blood fluke *Schistosoma mansoni*. *Nature*, 460:352–358, 2009.
17. Almeida S.M. and Torres L.F. Neurocysticercosis—Retrospective study of autopsy reports, a 17-year experience. *J. Community Health*, 36:698–702, 2011.
18. Moskowitz J. and Mendelsohn G. Neurocysticercosis. *Arch. Pathol. Lab. Med.*, 134:1560–1563, 2010.
19. Carabin H. et al. Clinical manifestations associated with neurocysticercosis: A systematic review. *PLoS Negl. Trop. Dis.*, 5:e1152, 2011.
20. Wallin M.T. and Kurtzke J.F. Neurocysticercosis in the United States: Review of an important emerging infection. *Neurology*, 63:1559–1564, 2004.
21. Almeida, C.R. et al. *Taenia solium* DNA is present in the cerebrospinal fluid of neurocysticercosis patients and can be used for diagnosis. *Eur. Arch. Psychiatry Clin. Neurosci.*, 265:307–310, 2006.
22. Michelet L. et al. Human neurocysticercosis: Comparison of different diagnostic tests using cerebrospinal fluid. *J. Clin. Microbiol.*, 49:195–200, 2011.
23. Yera H. et al. Confirmation and follow-up of neurocysticercosis by real-time PCR in cerebrospinal fluid of patients living in France. *J. Clin. Microbiol.*, 49:4338–4340, 2011.
24. Vivekanandan, P. and Singh, O.V. Molecular methods in the diagnosis and management of chronic hepatitis B. *Expert Rev. Mol. Diagn.*, 10:921–935, 2010.
25. Livak, K.J. et al. Oligonucleotides with fluorescent dyes at opposite ends provide a quenched probe system useful for detecting PCR product and nucleic acid hybridization, *PCR Methods Appl.*, 4(6), 357, 1995.
26. Beacon Designer v7.9, Premier Biosoft International, http://www.premierbiosoft.com/index.html.
27. Bustin, S.A. et al. The MIQE guidelines: Minimum information for publication of quantitative real-time PCR experiments. *Clin. Chem.*, 55:611–622, 2009.
28. Nolan, T. et al. Quantification of mRNA using real-time RT-PCR. *Nat. Protoc.*, 3:1559–1582, 2006.
29. Petersen, M. and Wengel, J. LNA: A versatile tool for therapeutics and genomics. *Trends Biotechnol.*, 21:74–81, 2003.
30. Wendt, S.F. et al. Evaluation of three new HBV DNA quantitation assays for detection of genotypes A through H. *25th Clinical Virology Symposium*, Daytona Beach, Florida. 2009.

15 MIQE
Guidelines for Reliable Design and Transparent Reporting of Real-Time PCR Assays

Tania Nolan and Stephen A. Bustin

CONTENTS

15.1 INTRODUCTION

Quantitative real-time PCR (qPCR) is often referred to as the gold standard for nucleic acid detection and quantification.[1] However, qPCR is an umbrella term for techniques encompassing a range of quite disparate chemistries, specialized enzymes, reagents, and instruments, as well as numerous protocols and data-analysis methods. Its latent power and versatility have encouraged a growing enthusiasm toward more complex applications, which have displaced the practical simplicity of the initial qPCR, making the concept of a single "gold standard" difficult to sustain. In its simplest form, qPCR continues to be used with DNA-binding dyes as a low-throughput method for the detection or quantification of limited numbers of individual nucleic acids in relatively few samples. Slightly more complex applications use high-resolution melt curves or probe-based assays to detect single-nucleotide polymorphisms (SNPs) or pathogens. At its most complex application, multiple targets are detected in a single tube (multiplex) or numerous parallel reactions are carried out in high-density PCR arrays. Most recently, DNA purification, concentration, and qPCR have been integrated into a real-time micro-PCR chip.[2] Coupled with the development of innovative, portable PCR instruments,[3] this has resulted in the development of hand-held devices that allow local and immediate assays for point-of-care testing applications.[4] Hence, the technology is moving from a research setting where, in general, qPCR assays are designed, performed, and analyzed by experienced researchers, into a wider arena, where high-volume custom assays are performed as rapidly as possible by less expert staff relying on bespoke criteria for interpretation of results. Modern qPCR has become a very high-throughput technology using minimal cycling times to generate vast numbers of data points in a very short time. The problem is that it is very easy to generate quantitative results, but these results may be artifacts caused by inappropriate data analysis and so are often not biologically or clinically relevant.

15.2   FLAWS IN CURRENT qPCR ASSAY DESIGN AND REPORTING

The original limited options of detection chemistry (SYBR Green I-binding dye or 5′-nuclease assay), enzyme (*Taq* polymerase), and instrument (Lightcycler® or ABI 7700) have been replaced with extensive alternatives of methodology, methods, instruments, and reagents, giving rise to numerous protocol choices with the potential to generate technique-dependent conflicting data.[5–8] Any underlying tendency toward inconsistency is aggravated by the significant probability of variability associated with the numerous steps that make up a qPCR assay[9] (Figure 15.1). Preanalysis variability derives from issues such as poorly defined sample selection or handling, patchy nucleic acid quality, inconsistent use of controls, poor assay design, as well as nonexistent optimization and validation. Inconsistencies with post-qPCR data processing are caused by a lack of quality control of the qPCR data, especially unsuitable methods of normalization, misguided data-analysis procedures, and challenges associated with applying the correct statistical methods.[10]

For RT-qPCR, in particular, these problems are critical and have been discussed for a long time.[5,11–15] Poorly designed, validated, and executed assays may, and indeed do, generate a wealth of measurements, but the results can be highly variable and thus inaccurate, ultimately reducing the precision of the measurement or, worse, introduce bias that generates statistically significant, yet biologically incorrect results. Hence, the persistent plea is aimed at developing a set of standard and objective quality-control measures.[6–9,16–20]

The most reliable means of judging the plausibility of a body of work is to study the "Materials and Methods" section for the relevant information, particularly that relating to the latter aspect of the experimental protocol. A recent survey confirms that this section is avidly read by researchers,[9] but a second survey shows that the detail of the information provided is wholly inadequate.[21] Unfortunately, not only are many assays not designed well, but many publications utilizing qPCR technology also provide insufficient information to allow the reader to assess the assay and evaluate the validity of conclusions derived from the qPCR data.[9]

The common omissions that could easily be remedied include simple things such as an accession number for mRNA or genomic DNA sequences and specifics of the experimental procedure followed for the reverse transcription (RT) step.[21] The latter is especially exasperating, since cDNA priming method and choice of RT have long been known to have a significant impact on the results.[22, 23] Even if the methods state that "the experimental protocol recommended by the manufacturer was followed," individual researchers frequently introduce subtle, yet consequential variations that, if unpublished, can lead to results not being reproducible. The lack of information with regard to other parameters is even more serious:

1. *Information about the sample*: Many publications utilizing qPCR technology omit basic information with respect to the samples under investigation. This is of particular importance when considering gene-expression analyses from tissue biopsies, where sample selection, acquisition, handling, and storage can significantly affect quantification results. It is also essential to provide details of sample-processing procedures, since samples pass through numerous preparative steps prior to the qPCR assay, every one of which can introduce additional variability.[24,25]

2. *Template quality*: Assessment of RNA purity and integrity is a basic requirement for reliable quantification of RNA[26–28]; yet, very few publications even mention the term RNA quality.[21] Inhibition of the RT or PCR steps should be checked by dilution of the sample or by the use of a universal inhibition assay such as SPUD.[29] Microfluidics-based devices allow automated, rapid, and standardized quality assessment of very small amount of total RNA with quality metrics such as the RIN (Agilent) or RQI (BioRad) to represent the level of degradation in a sample. However, the integrity of rRNA may not extrapolate to mRNA integrity[30] and methods such as the 3′:5′ mRNA-specific integrity assessment may be more suitable.[15]

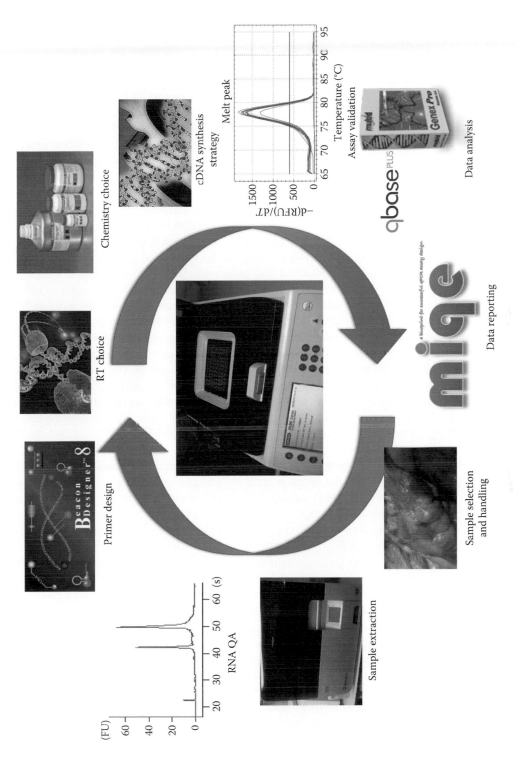

FIGURE 15.1 qPCR workflow. A successful RT-qPCR experiment requires completion of a series of steps, each one of which requires careful consideration.

3. *PCR efficiency*: Relative expression levels of mRNA of genes of interest (GOI) are frequently reported relative to those of reference genes using the comparative Cq method.[31] For this approach to be reliable, it requires a reaction with 100% efficiency and hence requires a constant doubling of the GOI as well as the reference gene(s) amplicons with each cycle, or equal amplification efficiency at the very least. However, the actual reaction efficiency can range between 60% (which cannot really be termed exponential) and 100%, depending on assay quality, sample, and target characteristics, reagents as well as instrument variability. Hence, small differences between measurements of GOI and reference genes can lead to huge differences in relative expression ratios and can generate distorted results. Again, it is remarkable that (i) this has been a well-known fact for more than 15 years and (ii) there are several efficiency-corrected relative quantification models that provide an efficient and reliable means of quantifying nucleic acids[32,33]; yet, a majority of publications blithely report ΔΔCqs without any knowledge of PCR efficiencies. Incidentally, popular software packages such as Genex,[34] qBasePlus,[10] and Relative Expression Software Tool (REST)[35] all make use of efficiency-corrected models and algorithms.

4. *Normalization*: Normalization increases confidence in the data by reducing technical variability and controlling for experimental error introduced during the multistage process required to extract and process the RNA. Accurate normalization must be matched to sample size, RNA quantity, and expected fold differences and, with abundant evidence that a single reference gene cannot be used reliably when detecting small changes in the target copy numbers[36,37]; the general recommendation is to use around three validated internal controls for the final normalization.[38] Even so, despite there being several published methods clearly detailing the methods for selection of appropriate sets of reference genes,[39–41] most publications continue to normalize the target gene copy numbers against single, unvalidated reference genes. Unfortunately, this frequently supports the apparent reporting of small changes (e.g., differences of less than threefold), making it impossible to conclude whether the findings are really the differences in the expression of the GOI, the references gene, or a combination of both. This introduces statistically significant measurement bias and results in the publication of incorrect findings.[42]

In recent years, there has been a significant increase in the number of retracted publications, with the most common reason being the operator error or nonreplicable findings.[43,44] Furthermore, the frequency of retraction varies among journals and shows a strong correlation with the journal impact factor.[45] Worryingly, subsequent rebuttals have no effect on the number of times that a publication is cited, and even if rebuttals are cited, on average, the citing papers have neutral views of the original article, or they believe that the rebuttal agreed with the original article.[46] This recent evidence confirms longstanding doubts about the opinions published in research papers[47] and the lack of scientific soundness of peer-reviewed papers.[48] As a result, the peer-reviewed literature is plagued by contradictory, irrelevant, wrong, or even fraudulent publications that combine to pose a challenge to the integrity of the scientific literature, with serious consequences not just for basic research, but also potentially calamitous implications for drug development and disease monitoring.[9] Indeed, this problem is abundantly clear to any reader of the literature concerning molecular biomarkers, as has been referred to repeatedly in the course of this chapter. One of the most consequential examples of the enormous implications for the health and lives of individuals that result from inappropriate use of this technology is provided by the use of RT-qPCR data that purported to demonstrate the presence of measles virus (MeV) RNA in the intestines of children with autism.[49] It provided sustenance to the controversy surrounding the triple measles, mumps, and rubella (MMR) virus vaccine, as the data were interpreted as providing evidence for a link between MMR, gut pathology, and autism. However, a detailed analysis of the raw data underlying that report carried out by one of the authors (SAB) acting as an expert witness to the UK High Court and the US Vaccine Court revealed that these data were obtained among a catalog of mistakes, inaccuracies, and inappropriate analysis methods as well as contamination and poor assay performance.[50] A reanalysis of the data concluded that the assay had been detecting DNA contamination and since MeV is an RNA-only virus, the RT-qPCR data had been erroneously interpreted, a conclusion confirmed elsewhere.[51–54] Peer-reviewed publication of these findings was only possible because the paper did not provide any information that would have allowed a competent reviewer to judge the quality of the data being reported.

15.3 MINIMUM INFORMATION FOR PUBLICATION OF QUANTITATIVE REAL-TIME PCR EXPERIMENTS

The "Minimum Information for Publication of Quantitative Real-Time PCR Experiments" (MIQE) guidelines[55] address these concerns. Indeed, the publication has provided a focus for a growing consensus that acknowledges the need to improve published information with relevant experimental detail, covering every aspect important to the qPCR assay, as well as issues relating to pre- and postassay parameters. The paper has become the most read publication in *Clinical Chemistry* (Figure 15.2a) and is in the top 10 of cited papers, with a rapidly increasing number of citations in the peer-reviewed literature (Figure 15.2b). There is even an iPAD/iPOD/iPhone app available from the iTunes store.

The MIQE guidelines have two aims:

- Transparency: Specify the minimum information required for the reader to be able to assess the technical validity
- Assay design: Provide a blueprint for rational design, optimization, and validation of new qPCR assays

MIQE consists of nine sections (experimental design, sample, nucleic acids, reverse transcription, target, primers and probes, assay details, PCR cycling, and data analysis), with 85 parameters that constitute the minimum information required to allow for potential reproduction, as well as unambiguous quality assessment, of a qPCR-based experiment. All these parameters relate to information that should be obtained as a matter of course during the experimental design, optimization, and validation stages of assay development. The use of commercial assays or those obtained from the published literature avoids having to design primers, but optimization and validation should still be carried out. Importantly, there is a clear hierarchy with some parameters, labeled "E" (essential), indispensable for an adequate description of the qPCR assay, whereas other components, labeled "D" (desirable), are more peripheral, yet constitute an effective foundation for the realization of best-practice protocols. Of course, these parameters are not set in stone and are open for discussion; indeed, a core set of sections is being implemented by the BMC group of open-access journals.[56]

Following their publication, the most contentious aspects of the original MIQE guidelines related to the requirement that publications must divulge the sequences of primers used and should ideally also report probe sequences. The rationale behind releasing the primer sequences was straightforward; an experiment is more difficult to reproduce if one of the principal reagents is unavailable and there is evidence that high-quality assay designs to the same target can result in apparently different quantities. Lack of access to a probe sequence, on the other hand, does not preclude the analysis of an assay's specificity, efficiency, and sensitivity. Some commercial qPCR assays are not supplied with primer/probe sequences, since most vendors consider this as commercially sensitive information; usually, there are also no details provided on empirical validation of each individual assay. Publications utilizing such assays could not satisfy the original MIQE requirements, placing limits on a universal acceptance of MIQE. Consequently, an amendment to the original guidelines was published that no longer insists on primer-sequence disclosure, but accepts the provision of a clearly defined amplicon-context sequence.[57] This guidance was issued based on the assessment that in the absence of full primer-sequence disclosure, it is possible to achieve an adequate level

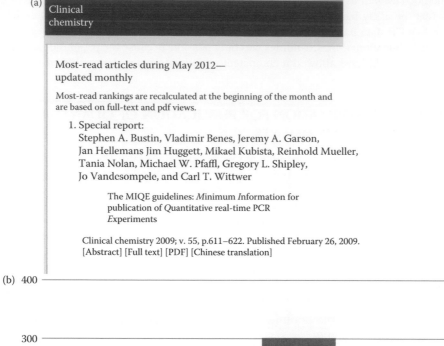

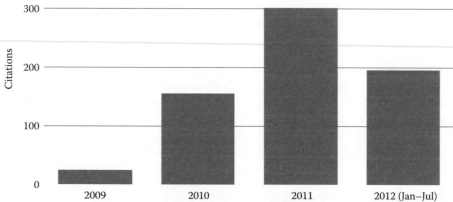

FIGURE 15.2 (a) MIQE paper has become the most read and is among the top 10 cited papers in *Clinical Chemistry*. (b) Citations of the MIQE paper in the peer-reviewed literature. The number of citations stood at 702 by the end of July 2012. (Data courtesy of Web of Knowledge, www.webofknowledge.com.)

of transparency, but only if there is an appropriate level of background information and disclosure of validation results in the qPCR assay. Consequently, if primer sequences are not disclosed, an MIQE-compliant publication should institute the same validation criteria used for assays reporting primer/probe sequences. Specifically, when reporting a precise fold change for a transcript, it remains an essential requirement that the PCR efficiency, analytical sensitivity, and specificity of each individual assay must be determined. This information should be verified by the investigator for the actual assay that is being reported using the conditions and personnel in their laboratory and not extrapolated from validation data for commercial assays provided by the vendors.

Data analysis is a critical final step in the qPCR assay workflow and there are a large number of statistical tools that can be used to address and minimize the variability discussed above. Several studies discuss the identification and handling of outliers and precision associated with calibration curves,[58] the relative merits of obtaining Cqs from the threshold method or sigmoidal functions,[59–61] and limits of detection modeled from sample replication and C_q values.[62] Importantly, tools have become available that allow management and analysis of qPCR data.[10,63–66] However, a systematic evaluation of the various qPCR data-analysis methods has shown that they differ substantially in their performance;[67] hence, the absence of guidelines or standards for data handling and interpretation means that the use of multiple statistical tools can add to variability and discordance. Hence, MIQE guidelines include a discussion regarding the importance of providing detailed information on the methods of data analysis and confidence estimation, especially identifying the statistical methods used to evaluate variances. The problem is clearly illustrated by using the commonly aired question of how many technical replicates should be run. Since increasing the sample size can increase the power of a statistical test, technical repeats are a quick and relatively inexpensive means of increasing statistical significance. Technical replicates help reduce measurement error and provide an assessment of the researchers' ability to pipette accurately. Hence, they have a place when optimizing a PCR reaction and when generating dilution curves. However, the high precision of an optimal qPCR assay (typically less than 10% coefficient of variation) means that an extended set of technical replicates can distort the statistics of determining confidence in experimental data rather than increasing the reliability of results. For RT-qPCR, technical replicates must be at the level of the more variable RT reaction rather than performing a single RT and replicating the PCR step on the cDNA sample. This action has surely led to numerous publications of very precise yet rather biased results.

Instead, biological replication is essential if findings are meant to be valid in the context of a conceptually large population from which the subjects were sampled, rather than only for the particular individuals considered in the experiment.[68] Since biological variability is larger than technical variation, increasing biological replication usually translates into more effective gains in power. However, increasing the sample size generally leads to added cost and increased time for performing the experiments. In addition, some biological replication cannot be increased, for example, when comparing large numbers of healthy individuals with a limited number of patients with a particular disease. Hence, as always, it is essential to combine the best practice with common sense.

15.4 CONCLUSIONS

qPCR and RT-qPCR are dominant technologies that lie at the core of many of the advances made in our understanding of the basic biological and disease processes; both are also increasingly used for clinical diagnostic purposes. However, the combination of ease of use and lack of rigorous standards of practice has resulted in widespread misinterpretation of data and consequent publication of erroneous conclusions. Any solution to the challenge of how to make PCR-based assays more reliable requires both an appreciation and an understanding of numerous attributes that include biological concepts, statistics, mathematical modeling, technical know-how, and a willingness to share this knowledge. This range of fundamental variables must be addressed by guidelines that permit a shift of focus from questions regarding the technological relevance underlying a publication's conclusion to the actual biological or diagnostic issues being addressed. MIQE constitutes a reference framework for communication within the research community, instrument and reagent manufacturers, and publishers that promise to deliver guidelines that promote transparency of experiments and confidence in results and conclusions that advance, rather than impede our knowledge.

REFERENCES

1. Bustin S.A. 2010. Developments in real-time PCR research and molecular diagnostics. *Expert Rev Mol Diagn* 10:713–715.
2. Min J. et al. 2011. Functional integration of DNA purification and concentration into a real time micro-PCR chip. *Lab Chip* 11:259–265.
3. Qiu X. et al. 2011. A portable, integrated analyzer for microfluidic-based molecular analysis. *Biomed Microdevices* 13:809–817.

4. Won B.Y. et al. 2011. Investigation of the signaling mechanism and verification of the performance of an electrochemical real-time PCR system based on the interaction of methylene blue with DNA. *Analyst* 136:1573–1579.

5. Bustin S.A. and Nolan T. 2004. Pitfalls of quantitative real-time reverse-transcription polymerase chain reaction. *J Biomol Tech* 15:155–166.

6. Murphy J. and Bustin S.A. 2009. Reliability of real-time reverse-transcription PCR in clinical diagnostics: Gold standard or substandard? *Expert Rev Mol Diagn* 9:187–197.

7. Bustin S. 2008. Molecular medicine, gene-expression profiling and molecular diagnostics: Putting the cart before the horse. *Biomarkers Med* 2:201–207.

8. Bustin S.A. 2008. Real-time quantitative PCR—Opportunities and pitfalls. *Eur Pharm Rev* 4:18–23.

9. Bustin S.A. 2010. Why the need for qPCR publication guidelines?—The case for MIQE. *Methods* 50:217–226.

10. Hellemans J. et al. 2007. qBase relative quantification framework and software for management and automated analysis of real-time quantitative PCR data. *Genome Biol* 8:R19.

11. Bustin S.A. 2000. Absolute quantification of mRNA using real-time reverse transcription polymerase chain reaction assays. *J Mol Endocrinol* 25:169–193.

12. Bustin S.A. 2002. Quantification of mRNA using real-time reverse transcription PCR (RT-PCR): Trends and problems. *J Mol Endocrinol* 29:23–39.

13. Bustin S.A. 2004. Meaningful quantification of mRNA using real-time PCR, in T.W. Griffin, H.G. Griffin (Eds.) *PCR Technology Current Innovations*, pp. 225–233. Boca Raton, FL: CRC Press.

14. Bustin S.A., Benes V., Nolan T., and Pfaffl M.W. 2005. Quantitative real-time RT-PCR—A perspective. *J Mol Endocrinol* 34:597–601.

15. Nolan T., Hands R.E., and Bustin S.A. 2006. Quantification of mRNA using real-time RT-PCR. *Nat Protoc* 1:1559–1582.

16. Bustin S.A. and Mueller R. 2005. Real-time reverse transcription PCR (qRT-PCR) and its potential use in clinical diagnosis. *Clin Sci (Lond)* 109:365–379.

17. Bustin S.A. 2006. Nucleic acid quantification and disease outcome prediction in colorectal cancer. *Personalized Med* 3:207–216.

18. Bustin S.A. and Mueller R. 2006. Real-time reverse transcription PCR and the detection of occult disease in colorectal cancer. *Mol Asp Med* 27:192–223.

19. Murphy J., Dorudi S., and Bustin S.A. 2007. Molecular staging of colorectal cancer: New paradigm or waste of time? *Expert Opin Med Diagn* 1:31–45.

20. Bustin S.A. 2008. Real-time polymerase chain reaction—Towards a more reliable, accurate and relevant assay. *Eur Pharm Rev* 6:19–27.

21. Huggett J. and Bustin S.A. 2011. Standardisation and reporting for nucleic acid quantification. *Accredit Qual Assur* 16:399–405.

22. Stahlberg A., Kubista M., and Pfaffl M. 2004. Comparison of reverse transcriptases in gene expression analysis. *Clin Chem* 50:1678–1680.

23. Stahlberg A. et al. 2004. Properties of the reverse transcription reaction in mRNA quantification. *Clin Chem* 50:509–515.

24. Hammerle-Fickinger A. et al. 2009. Validation of extraction methods for total RNA and miRNA from bovine blood prior to quantitative gene expression analyses. *Biotechnol Lett* 32:35–44.

25. Tichopad A. et al. 2009. Design and optimization of reverse-transcription quantitative PCR experiments. *Clin Chem* 55:1816–1823.

26. Perez-Novo C.A. et al. 2005. Impact of RNA quality on reference gene expression stability. *Biotechniques* 39:52, 54, 56.

27. Fleige S. and Pfaffl M.W. 2006. RNA integrity and the effect on the real-time qRT-PCR performance. *Mol Asp Med* 27:126–139.

28. Fleige S. et al. 2006. Comparison of relative mRNA quantification models and the impact of RNA integrity in quantitative real-time RT-PCR. *Biotechnol Lett* 28:1601–1613.

29. Nolan T., Hands R.E., Ogunkolade B.W., and Bustin S.A. 2006. SPUD: A qPCR assay for the detection of inhibitors in nucleic acid preparations. *Anal Biochem* 351:308–310.

30. Vermeulen J. et al. 2011. Measurable impact of RNA quality on gene expression results from quantitative PCR. *Nucleic Acids Res* 39:e63.

31. Livak K.J. and Schmittgen T.D. 2001. Analysis of relative gene expression data using real-time quantitative PCR and the 2(-Delta Delta $C(T)$) method. *Methods* 25:402–408.

32. Pfaffl M.W. 2001. A new mathematical model for relative quantification in real-time RT-PCR. *Nucleic Acids Res* 29:E45.

33. Peirson S.N., Butler J.N., and Foster R.G. 2003. Experimental validation of novel and conventional approaches to quantitative real-time PCR data analysis. *Nucleic Acids Res* 31:e73.
34. Bergkvist A. et al. 2010. Gene expression profiling—Clusters of possibilities. *Methods* 50:323–335.
35. Pfaffl M.W., Horgan G.W., and Dempfle L. 2002. Relative Expression Software Tool (REST) for group-wise comparison and statistical analysis of relative expression results in real-time PCR. *Nucleic Acids Res* 30:e36.
36. Tricarico C. et al. 2002. Quantitative real-time reverse transcription polymerase chain reaction: Normalization to rRNA or single housekeeping genes is inappropriate for human tissue biopsies. *Anal Biochem* 309:293–300.
37. Dheda K. et al. 2004. Validation of housekeeping genes for normalizing RNA expression in real-time PCR. *Biotechniques* 37:112–119.
38. Huggett J., Dheda K., Bustin S., and Zumla A. 2005. Real-time RT-PCR normalisation, strategies and considerations. *Genes Immun* 6:279–284
39. Vandesompele J. et al. 2002. Accurate normalization of real-time quantitative RT-PCR data by geometric averaging of multiple internal control genes. *Genome Biol* 3:0034.1–0034.11.
40. Pfaffl M.W., Tichopad A., Prgomet C., and Neuvians T.P. 2004. Determination of stable housekeeping genes, differentially regulated target genes and sample integrity: BestKeeper—Excel-based tool using pair-wise correlations. *Biotechnol Lett* 26:509–515.
41. Andersen C.L., Jensen J.L., and Orntoft T.F. 2004. Normalization of real-time quantitative reverse transcription–PCR data: A model-based variance estimation approach to identify genes suited for normalization, applied to bladder and colon cancer data sets. *Cancer Res* 64:5245–5250.
42. Dheda K. et al. 2005. The implications of using an inappropriate reference gene for real-time reverse transcription PCR data normalization. *Anal Biochem* 344:141–143.
43. Wager E. and Williams P. 2011. Why and how do journals retract articles? An analysis of Medline retractions 1988–2008. *J Med Ethics* 37:567–570.
44. Steen R.G. 2011. Retractions in the scientific literature: Is the incidence of research fraud increasing? *J Med Ethics* 37:249–253.
45. Fang F.C. and Casadevall A. 2011. Retracted science and the retraction index. *Infect Immun* 79:3855–3859
46. Banobi J.A., Branch T.A. and Hilborn R. 2011. Do rebuttals affect future science? *Ecosphere* 2:1–11.
47. Horton R. 2002. The hidden research paper. *JAMA* 287:2775–2778.
48. Baltic S. 2001. Conference addresses potential flaws in peer review process. *J Natl Cancer Inst* 93:1679–1680.
49. Uhlmann V. et al. 2002. Potential viral pathogenic mechanism for new variant inflammatory bowel disease. *Mol Pathol* 55:84–90.
50. Bustin S.A. 2008. RT-qPCR and molecular diagnostics: No evidence for measles virus in the GI tract of autistic children. *Eur Pharm Rev Dig* 1:11–16.
51. Afzal M.A. et al. 2006. Absence of detectable measles virus genome sequence in blood of autistic children who have had their MMR vaccination during the routine childhood immunization schedule of UK. *J Med Virol* 78:623–630.
52. D'Souza Y., Fombonne E., and Ward B.J. 2006. No evidence of persisting measles virus in peripheral blood mononuclear cells from children with autism spectrum disorder. *Pediatrics* 118:1664–1675.
53. D'Souza Y. et al. 2007. No evidence of persisting measles virus in the intestinal tissues of patients with inflammatory bowel disease. *Gut* 56:886–888.
54. Hornig M. et al. 2008. Lack of association between measles virus vaccine and autism with enteropathy: A case–control study. *PLoS ONE* 3: e3140.
55. Bustin S.A. et al. 2009. The MIQE guidelines: Minimum information for publication of quantitative real-time PCR experiments. *Clin Chem* 55:611–622.
56. Bustin S.A. et al. 2010. MIQE precis: Practical implementation of minimum standard guidelines for fluorescence-based quantitative real-time PCR experiments. *BMC Mol Biol* 11:74.
57. Bustin S.A. et al. 2011. Primer sequence disclosure: A clarification of the MIQE guidelines. *Clin Chem* 57:919–921.
58. Burns M.J., Nixon G.J., Foy C.A., and Harris N. 2005. Standardisation of data from real-time quantitative PCR methods—Evaluation of outliers and comparison of calibration curves. *BMC Biotechnol* 5:31.
59. Rutledge R.G. 2004. Sigmoidal curve-fitting redefines quantitative real-time PCR with the prospective of developing automated high-throughput applications. *Nucleic Acids Res* 32: e178.
60. Rutledge R.G. and Cote C. 2003. Mathematics of quantitative kinetic PCR and the application of standard curves. *Nucleic Acids Res* 31:e93.
61. Rutledge R.G. and Stewart D. 2008. A kinetic-based sigmoidal model for the polymerase chain reaction and its application to high-capacity absolute quantitative real-time PCR. *BMC Biotechnol* 8:47.

62. Burns M.J. and Valdivia H. 2008. Modelling the limit of detection in real-time quantitative PCR. *Eur Food Res Technol* 226:1513–1524.
63. Gallup J.M. and Ackermann M.R. 2008. The "PREXCEL-Q Method" for qPCR. *Int J Biomed Sci* 4:273–293.
64. Muller P.Y., Janovjak H., Miserez A.R., and Dobbie Z. 2002. Processing of gene expression data generated by quantitative real-time RT-PCR. *Biotechniques* 32:1372–1379.
65. Jin N., He K. and Liu L. 2006. qPCR-DAMS: A database tool to analyze, manage, and store both relative and absolute quantitative real-time PCR data. *Physiol Genomics* 25:525–527.
66. Ritz C. and Spiess A.N. 2008. qPCR: An *R* package for sigmoidal model selection in quantitative real-time polymerase chain reaction analysis. *Bioinformatics* 24:1549–1551.
67. Karlen Y. et al. 2007. Statistical significance of quantitative PCR. *BMC Bioinform* 8:131.
68. Mehta T., Tanik M., and Allison D.B. 2004. Towards sound epistemological foundations of statistical methods for high-dimensional biology. *Nat Genet* 36:943–947.

Part V

Data Analysis

16 Hypothesis-Driven Approaches to Multivariate Analysis of qPCR Data

Anders Bergkvist, Azam Sheikh Muhammad, and Peter Damaschke

CONTENTS

16.1 INTRODUCTION

Quantitative real-time PCR (qPCR) has emerged as the touchstone for quantification of minute amounts of nucleic acid.[1] The increasing trend toward smaller reaction volumes and instruments with increased capacity to handle simultaneously large number of qPCR reactions is resulting in large amounts of data being generated. This opens up interesting avenues for analyzing complex biological questions with improved precision, but requires particular care to avoid any over- or misinterpretation of conclusions. There is a need for multidisciplinary understanding of issues, including the underlying biology, technical handling of qPCR assays, and statistical analysis. In this chapter, we aim to highlight some emerging analysis issues while also suggesting further expansion of existing techniques.

16.2 STATISTICAL DATA ANALYSIS

It may be argued that, in general, the scientific method centers on the formulation and testing of hypotheses.[2] In this context, we distinguish between exploratory and confirmatory studies (Figure 16.1). The purpose of the exploratory study is to analyze data with one or several different techniques in order to substantiate a plausible hypothesis. The data set may be redefined and/or different analysis techniques may be employed repeatedly in order to support one or several hypotheses. The exploratory study is thus very much adaptable to the specifics of any scientific question. However, the repeated probing for hypotheses on one data set may lead to multiple testing issues that often undermine statistical conclusions. The exploratory study is therefore often combined with a confirmatory study.

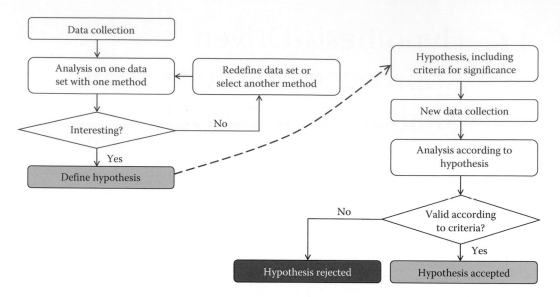

FIGURE 16.1 Flowchart illustrating operations involved in exploratory and confirmatory statistical analyses. The left-hand side of the figure, before the dashed arrow, shows operations in an exploratory statistical study. The right-hand side of the figure, after the dashed arrow, shows operations in a confirmatory statistical study.

A confirmatory study needs to be conducted under strict statistical criteria. First of all, the hypothesis of study, including criteria for significance, needs to be defined before the collection of data and before the analysis. Furthermore, a new data set needs to be collected. It is statistically incorrect to reuse the data set from the exploratory study in the confirmatory study since that data set would inherently favor the proposed hypothesis. The end result is a rejected or accepted hypothesis according to the prestated criteria.

Scaling or transformation of the raw data can often facilitate data analysis. This is perfectly acceptable, under the condition that these manipulations are included in the statistical criteria for the confirmatory study. Autoscaling is a common transformation for qPCR data.

16.3   AUTOSCALING

For the analysis of qPCR data, we are often concerned with the discovery of regulated gene-expression levels. However, the baseline expression level of a particular gene may differ very much from the baseline expression level of another gene. In addition, the natural variation around the baseline expression level may also differ to a large extent between different genes in a particular assay. These gene-to-gene differences may obscure comparison of the regulatory components of gene expression.

To highlight the regulatory aspects of the gene expression and to facilitate comparison, we may subtract the mean expression level (such as the relative quantities or the C_q values) from individual measurements and divide by the standard deviation.[3]

$$\mathrm{FD}_{AS} = \frac{\mathrm{FD} - \overline{\mathrm{FD}}}{\mathrm{SD}} = \frac{\mathrm{FD}_{MC}}{\mathrm{SD}}$$

These new data are called autoscaled. Although this may be an important step in order to arrive at sensible comparisons of gene-expression levels and identification of regulatory patterns, statistical tests are required to quantify significances of any associated hypotheses.

16.4   STATISTICAL TESTS

We contend that it is not possible to conclusively prove a scientific theory without prior assumptions. An alternative approach, which is used in statistical testing, is to analyze the likelihood that an observed phenomenon occurred by random chance. This is called the Null hypothesis.[4] If the observed phenomenon is rare according to the Null hypothesis, we conclude that it is unlikely that the Null hypothesis is valid; we then reject the Null hypothesis and accept the alternative hypothesis as significant.

The estimated likelihood that the observed phenomenon occurred by random chance is called the p-value. The p-value is measured in a range from 0 to 1, or equivalently in percentage units. The statistical criteria for a confirmatory study include an alpha cutoff under which calculated p-values would indicate significance for the observed phenomenon. An alpha cutoff of 5% is commonly used, although this must be adjusted to fit the desired and necessary criteria specific to the subject of study.

Many algorithms have been developed for calculating p-values under various assumptions and for various purposes. A common algorithm is the Student's t-test. The Student's t-test calculates a p-value based on the difference in the mean values between two groups of data. The main assumption of Student's t-test is that the two groups of data are independent and conform to normal distributions. An advantage of the Student's t-test is that it is powerful, compared to nonparametric statistical tests.[5] A nonparametric test that is equivalent to the Student's t-test is also one of the most well-known nonparametric statistical tests; the Wilcoxon rank-sum test (sometimes called Mann–Whitney U test; not to be confused with Wilcoxon signed-rank test used to compare two *paired* groups). Nonparametric statistical tests, such as the Wilcoxon rank-sum test, have an advantage over parametric statistical tests, such as the Student's t-test, in that they do not depend on prior assumptions of the data set distributions.

In addition to choice of algorithm for p-value calculation, data sets that are fed to the p-value calculation algorithm may be manipulated to facilitate observation of desired properties in the data set. The combination of raw data manipulation steps and choice of p-value calculation algorithm is part of building a hypothesis model.

16.5   HYPOTHESIS MODELS

A fundamental task in hypothesis finding is to identify relevant attributes in a data set.[6–9] Since, in principle, all information can be reduced to yes/no decisions, it suffices in many instances to consider the case of Boolean attributes. (This also relates to the case study later in this chapter.) The setting can be formally described as follows: A Boolean function f of n variables is given. That is, f assigns outcome $f(x) = 0$ or 1 to each vector x of n bits, which are the Boolean values of attributes. Function f is unknown, but we have data in the form of several pairs $(x, f(x))$. Moreover, we have reason to conjecture that $f(x)$ is determined by the values of a "junta" of only r specific attributes, where r is some small number: $r = 1, 2, 3$, and so on. The problem is to find these r attributes based on the data, or more moderately, to identify the candidate sets of r attributes that solely determine the outcome. Naïve, exhaustive search on all subsets R of r attributes solves the problem: For each R, we only have to check that $f(x) = f(y)$ for any two data vectors x and y that agree on R. Algorithms that are faster than naïve search can be developed under some circumstances, but for the applications we have in mind here, numbers n and r are barely that large. In the case study below, the n attributes are genes, data x are patient data, where each bit means that the gene is low or high expressed, $f(x)$ is "healthy" or "diseased," and we are to find r marker genes that have sufficient difference in expression level to determine the disease status. It is important to notice that f could be any Boolean function of the r relevant attributes, and their number is doubly exponential in r (despite some trivial cases and symmetries). Already for $r = 2$, the function can be AND, OR, XOR, EQUALITY, IF–THEN, and so on, meaning that both marker genes are high expressed, at least one/exactly one is high expressed, both genes are coexpressed, high expression of the first marker causes high expression of the second, and so on. For $r = 3$, many further combinations are conceivable, including the majority function where at least two of the three marker genes are highly expressed. Furthermore, if we relax the condition $f(x) = f(y)$ for any two data vectors x and y that agree on R, and allow outliers, we get some noise resilience which might be more appropriate for biological data. For example, instead of EQUALITY we would look for similarly expressed genes, which in turn is a process that is related to clustering. We will not go into more details since aspects of inferring Boolean functions from data are well studied in algorithmic learning theory, we only want to emphasize that similarity (that gives rise to clustering) is, by far, not the only interesting type of relationship. Finding relevant variables may also be interpreted as a Boolean version of principal components analysis (PCA).

Owing to the vast range of options for statistical hypothesis models and the potentially complex task of designing suitable models, specific methods have been developed as standardized tools to characterize data properties and these have gained popularity both for their powerful features as well as for their ease of use. Popular methods include hierarchical clustering and PCA. In this chapter, we would also like to demonstrate how alternative hypothesis models may be constructed and what benefits may be gained from such an effort.

16.6   HIERARCHICAL CLUSTERING

Arguably one of the easiest and useful methods to characterize data is plotting the data in a scatterplot (e.g., plotting measured C_q values of one gene against the corresponding C_q values of another gene for a set of biological samples in a 2D plot). Plots in one or two dimensions are conveniently visualized by the human eyes. Plots in three dimensions may also be possible with appropriate tools, but higher-dimensional plots are significantly harder to visualize and this becomes progressively harder with higher dimensionality. However, for exploratory studies, the data set is inherently multidimensional and scatterplots of whole data sets may thus become impractical. From a qPCR data set, we may, for example, have several genes and/or several types of biological samples represented.

A popular, alternative way of characterizing and visualizing data from exploratory studies is to analyze measures of distances between data points in the scatterplot. Different distance measures exist, including Euclidean, Manhattan, and Pearson distance measures; and with computational power, it is straightforward to calculate distances, even for multidimensional data of much higher dimensionality than three dimensions. For agglomerative hierarchical clustering, the following iterative process is performed: (1) find the two closest objects and merge them into a cluster, (2) define the new cluster as a new object through a clustering method, (3) repeat from (1) until all objects have been combined into clusters.[10] Alternatives for clustering methods include Ward's method, single linkage, and average linkage.[11] A dendrogram is often used to visualize results from hierarchical clustering.

Interpretation of qPCR hierarchical clustering dendrograms often results in conclusions about gene-expression profile similarities. In an exploratory study, these similarities may then be used to formulate hypotheses about gene-expression coregulation, which may be accepted or rejected in subsequent confirmatory studies. The advantages of hierarchical clustering dendrograms include the clarity by which similarity relationships are visualized. On the other hand, the strong emphasis on similarity measures may be perceived as limiting with respect to formulating hypotheses, since similar expression profiles may be redundant attributes in hypotheses. It may be of higher value to identify sets of expression profiles that complement each other in a specific combination to answer the desired hypothesis.

16.7   PRINCIPAL COMPONENT ANALYSIS

Another popular alternative way of characterizing and visualizing data from exploratory studies is to take advantage of the information contained in the whole, multidimensional data set, select desired properties, and present it in a lower-dimensional scatterplot, such as a 2D or 3D plot. This can be done by principal component analysis (PCA).[12–15] Here, the original coordinate system of the data set (i.e., the expression profiles measured by qPCR) is transformed onto a new multidimensional space where new variables (principal components: PC or factors) are constructed. Each PC is a linear combination of the subjects in the original data set.

By mathematical definition, the PCs are extracted in successive order of importance. This means that the first PC explains most of the information (variance) present in the data, the second less, and so forth. Therefore, we can use the first two or three PC coordinates (termed scores) to obtain a projection of the whole data set onto a conveniently small dimension, suitable for visualization in a 2D or 3D plot. By using the first two or three PCs for representation, we obtain the projection that accounts for the most variability in the data set. Variance from experimental design conditions is expected to be systematic, while confounding variance is expected to be random, so this representation may be desired under appropriate conditions.

As previously noted for hierarchical clustering, the interpretation of qPCR PCA often results in conclusions about gene-expression profile similarities. Although PCA and hierarchical clustering may yield complementary insights into gene-expression coregulation patterns, both techniques focus on gene-expression profile similarities. This places limitations on the types of hypotheses that can be found, in exploratory studies, using these techniques alone.

In contrast to the capabilities of PCA and hierarchical clustering, we have found that it might be useful to identify hypotheses where gene-expression patterns combine in a nonredundant way to correlate with phenotypic indicators. The following case study demonstrates how this can be accomplished.

16.8   CASE STUDY

A simulated data set was prepared to illustrate a biologically relevant case. The basic premise was laid out under the assumptions that (1) three genes jointly contribute to some binary measurable phenotypic indicator, such as healthy/diseased, (2) the gene products act in a binary fashion, that is, genes are either activated or deactivated, (3) in one state of the phenotypic indicator, two or more of the genes are activated and in the other state of the phenotypic indicator at most one of the genes are activated.

Random baseline C_q values were assigned to deactivated states of each gene in the range 17–31 cycles. Random activation ΔC_q values were assigned to activated states of each gene in the range 2–3 cycles in addition to the baseline C_q values of the deactivated states. Each gene was furthermore overlaid with random, normal distributed noise at a standard deviation of 0.3–1.7 cycles.

For the case study, a group of 10 samples (samples 1–10) were generated to phenotypic indicator 0 and another 10 samples (samples 11–20) were generated to phenotypic indicator 1. Three genes (GOI) were setup so that they were sensitive to the phenotypic indicator according to the basic premise above, randomly determining which single gene would be active for phenotypic indicator 0 and randomly determining which two genes that would be active for phenotypic indicator 1. Another seven genes (Other) were setup to be insensitive to the phenotypic indicator by all being setup as deactivated throughout the samples.

A result of this random generation process is shown in Table 16.1 and illustrated in the 3D line plot in Figure 16.2a. Since three genes contribute to the phenotypic indicator, it is not apparent from looking at each gene individually that they hold information regarding the phenotypic indicator. Figure 16.3 shows that although genes 1_GOI and 2_GOI (together with gene 10_Other) have expression tendencies that may be used to differentiate between the two phenotypic indicators, due to the large variations in the data, the p-values are all very large (>0.17) indicating nonsignificant relations (Table 16.2).

Hierarchical clustering is often used for characterizing multivariate data relationships for exploratory study purposes. The fundamental hypothesis of hierarchical clustering is

TABLE 16.1
Randomly Generated qPCR C$_q$ Values for the Case Study

	1_GOI	2_GOI	3_GOI	4_Other	5_Other	6_Other	7_Other	8_Other	9_Other	10_Other	#Class
S01	24.57	19.65	24.79	22.76	26.28	15.87	30.25	24.36	20.31	21.57	0
S02	22.47	21.53	27.71	22.60	21.54	18.25	30.90	24.41	20.10	21.89	0
S03	22.01	26.88	25.56	21.96	21.42	19.02	31.57	24.74	21.28	22.24	0
S04	22.60	21.44	27.02	21.55	22.95	17.89	30.45	24.07	20.08	21.34	0
S05	24.01	18.35	24.93	20.94	23.72	18.30	29.97	23.90	21.06	20.82	0
S06	25.24	19.11	26.64	22.23	22.36	17.83	30.65	24.84	21.88	21.18	0
S07	22.41	20.75	26.85	23.02	22.77	18.44	29.80	24.08	20.97	21.04	0
S08	25.79	20.60	24.08	21.67	22.46	17.08	27.27	24.10	20.09	21.25	0
S09	21.95	18.78	29.58	22.84	22.15	19.08	30.87	24.17	21.70	20.72	0
S10	22.08	20.87	27.84	24.14	20.57	18.16	28.30	23.88	20.98	20.46	0
S11	24.69	21.18	25.12	20.83	20.77	16.80	29.96	24.73	20.73	19.85	1
S12	24.81	18.69	28.92	23.36	20.75	18.13	29.88	23.91	20.13	21.23	1
S13	22.89	21.30	27.69	24.79	23.90	18.17	31.35	23.96	21.01	20.12	1
S14	24.13	24.21	24.24	24.17	21.81	18.99	30.66	24.41	21.73	22.20	1
S15	25.08	22.22	25.42	21.87	22.51	18.44	28.69	23.89	20.74	21.30	1
S16	24.64	20.39	27.21	21.61	23.66	18.67	30.16	24.40	20.00	19.53	1
S17	25.10	23.14	26.61	22.03	22.22	17.71	30.28	23.93	20.35	20.75	1
S18	22.12	23.90	27.18	22.23	22.73	17.13	30.54	24.15	20.47	20.19	1
S19	22.79	24.44	28.71	25.12	23.97	19.77	29.22	24.69	21.24	20.76	1
S20	22.92	22.01	28.06	20.10	22.31	18.70	29.75	24.05	20.51	22.17	1

Note: Values were simulated to illustrate effects of cooperatively regulated expression levels and their correlation to a phenotypic indicator (class). Details are described in the text.

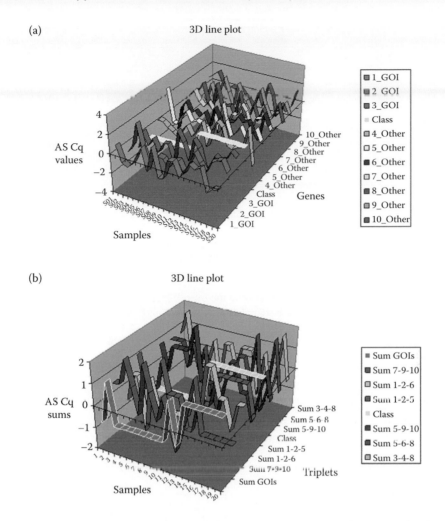

FIGURE 16.2 **(See color insert.)** (a) A 3D line plot of the randomly generated data set of qPCR C_q values used in the case study. The phenotypic indicator represented by the Class 3D line. It is worth noting the lack of one-on-one correlation between any of the gene-expression profiles, including those of the genes of interests, and the phenotypic indicator. (b) A 3D line plot of a selection of the triplet sum combinations, including the ones with lowest p-value. The phenotypic indicator represented by the Class 3D line. We observe the apparent good correlation between the genes of interest's triplet sum combination and the phenotypic indicator.

degree of similarity. The location of branch points in the hierarchical clustering tree structure is related to the distance between pairs of gene-expression profiles.[11] Using autoscaled data allows us to compare expression profiles with the binary phenotypic indicator itself. However, in cases, like the current model case study, where several genes contribute to the phenotypic indicator, hierarchical clustering may not be the most appropriate approach. Figure 16.4a shows that although gene 2_GOI shows the most similarity to the phenotypic indicator, it is a very weak similarity and no other particular relationship is apparent from the relationship tree structure.

PCA is another technique that is used for characterizing multivariate data relationships for exploratory study purposes. Again the fundamental hypothesis of PCA is degree of similarity. A 2D scatter plot representation of the first and second principal components represents a projection of the original multidimensional case study data set, taking the largest variations into account.[3]

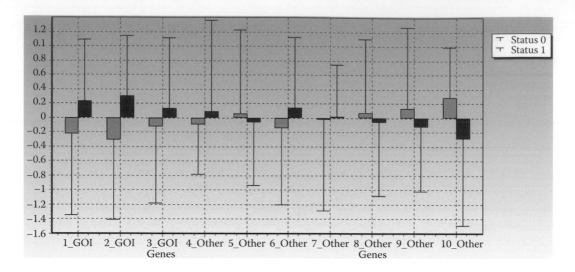

FIGURE 16.3 Bar diagram showing deviation from the overall mean (autoscaled) among samples in each gene from the simulated data set in the case study between the two classes of phenotypic indicators: status 0 and status 1. It is worth noting that there are tendencies for genes 1_GOI, 2_GOI, and 10_Other to distinguish between the two classes of the phenotypic indicator, but in all cases data variability is very large, confounding any conclusions of statistical significance.

Figure 16.4b shows that neither of the GOI genes have a projection that fall particularly close to that of the phenotypic indicator and all of the genes are spread out without any particular pattern similarities. Both methods, hierarchical clustering and PCA, thus fail to detect the relationship between the triplicate 1_GOI, 2_GOI, and 3_GOI, on the one hand, and the phenotypic indicator, on the other hand.

TABLE 16.2

Student's *t*-Test *p*-Values for an Unpaired, Two-Tailed *t*-Test Comparing Mean Gene Expression to the Phenotypic Indicator from the Simulated Data Set in the Case Study

Gene	*p*-Value
1_GOI	0.3116
2_GOI	0.1750
3_GOI	0.5732
4_Other	0.6962
5_Other	0.7988
6_Other	0.5330
7_Other	0.9236
8_Other	0.7715
9_Other	0.5750
10_Other	0.2080

Note: Bonferroni correction at α 0.05 for 10 genes is 0.00512. We observe that individually gene of interest 2_GOI comes closest to qualify for a hypothesis of gene expression correlated with the phenotypic indicator. However, the observed *p*-value is very high (0.1750), indicating that it would be unlikely to find this a significant correlation compared to the random variation in 2_GOI's expression levels.

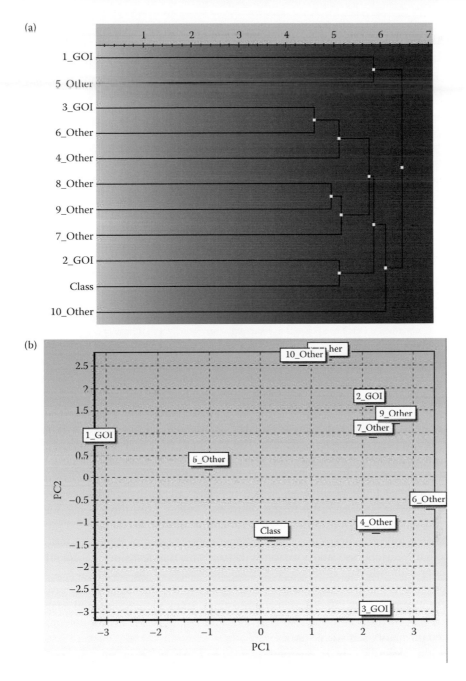

FIGURE 16.4 (a) Dendrogram of a hierarchical clustering of the expression profiles of the 10 genes from the simulated data set from the case study together with the phenotypic indicator (Class). Data were autoscaled; Euclidean distance measure and average linkage clustering methods were used. Calculation was performed and presentation was produced in the analysis software GenEx from MultiD Analyses AB. (b) Scatterplot of the two first principal components of the expression profiles of the 10 genes from the simulated data set from the case study together with the phenotypic indicator (Class). Data were autoscaled. PCA was performed and presentation was produced in the analysis software GenEx from MultiD Analyses AB.

TABLE 16.3

List of a Selection of Calculated Wilcoxon Rank-Sum Test *p*-Values Comparing Mean Triplet Sum Gene Expressions to the Phenotypic Indicator

Gene Triplet Combination			*p*-Value
1_GOI	2_GOI	3_GOI	0.0025045
7_Other	9_Other	10_Other	0.044134
1_GOI	2_GOI	6_Other	0.04854
1_GOI	2_GOI	5_Other	0.048684
5_Other	9_Other	10_Other	0.10021
⋮	⋮	⋮	⋮
3_GOI	4_Other	8_Other	1
5_Other	6_Other	8_Other	1

Note: The analysis was performed on the simulated data set of the example case study. The list was ordered according to the lowest *p*-values among all combinations. Bonferroni correction at α 0.05 for 120 genes is 0.00043. We observe that the combination of 1_GOI, 2_GOI, and 3_GOI comes out as the most significant combination. The observed *p*-value for this combination (0.0025) is not lower than the Bonferroni-corrected α cutoff, but in the context of an exploratory study, it nevertheless identifies a concrete hypothesis that may warrant a confirmatory study before rejection or validation.

16.9   HYPOTHESIS-DRIVEN SOLUTION

Here, we present an alternative to the many analysis-driven approaches for multivariate data analysis, such as hierarchical clustering and PCA. The idea is to start with a hypothesis model with a single relevant variable that we would like to explore for our data set. In the current case study, we may define the hypothesis based on the variable as follows: Autoscale the measured C_q values. Define the data set such that positive autoscaled C_q values are assigned 1 and negative autoscaled values are assigned 0. Take the sum of the autoscaled values from three genes for each sample measurement. Autoscale the sum of the three genes across all samples. Divide the samples into two groups according to the classes given with the input data. Perform a statistical test of data in the two groups; in the current case study, we use Wilcoxon rank-sum test since it is an efficient statistical test on nonparametric data of this type. Obtain the *p*-value of the statistical test. Repeat this in an exhaustive search for all combinations of three genes in the data set; in the current case study of 10 genes that computes to 120 different combinations. Rank the *p*-values and analyze the triplet combinations that correspond to the lowest *p*-values for hypothesis leads for potential selection to subsequent confirmatory statistical studies.

The algorithm described above was implemented in MATLAB® and run on the simulated data set of the example case study. The list of triplet sums ranked according to Wilcoxon rank-sum test *p*-values are shown in Table 16.3. The top triplets with *p*-values below 0.05 together with three other illustrative example triplets were selected for display in the 3D line plot of Figure 16.2b. As can be seen in Table 16.3 and Figure 16.2b, the hypothesis-driven algorithm was able to correctly identify the three genes 1_GOI, 2_GOI, and 3_GOI as interesting candidates for a triplet sum hypothesis as described in the algorithm above.

16.10   CONCLUSIONS

In this chapter, we have shown that a hypothesis-driven, custom-designed algorithm may identify biologically relevant hypotheses that may otherwise be missed by commonly used techniques for multivariate data analysis. This demonstration illustrates the importance of distinguishing between exploratory and confirmatory statistical studies and keeping in mind the hypothesis that is the subject of study.

We would also like to call for researchers to bridge the interdisciplinary gap on this subject matter. Biologically oriented researchers may identify and communicate hypotheses that hold potential biological relevance. Mathematically and/or computationally oriented researchers may formulate and implement algorithms that address these specific hypotheses.

It is our conviction that many relations in the complex network of molecular dependencies in living organisms may share common themes that may be described in standardized sets of relationship hypotheses. On the other hand, the complexity of living systems may require a rich palette of such standardized hypotheses. We look forward to a lot of interesting developments in this multidisciplinary field of research.

REFERENCES

1. Bustin, S., Benes, V., Garson, J.A., Hellemans, J., Huggett, J., Kubista, M., Mueller, R. et al. 2009, The MIQE guidelines: Minimum information for publication of quantitative real-time PCR experiments, *Clin. Chem. 55*, 611–622.
2. *Merriam-Webster Dictionary*, 2011, http://www.merriam-webster.com (accessed January 2011).
3. Bergkvist, A., Rusnakova, V., Sindelka, R., Andrade Garda, J. M., Sjögreen, B., Lindh, D., Forootan A., Kubista, M., 2010, *Gene Expression Profiling—Clusters of Possibilities, Methods 50*, 323–335.
4. Fisher, R.A., 1966, *The Design of Experiments*. 8th ed. Hafner: Edinburgh.
5. Motulsky, H., 1995, *Intuitive Biostatistics*. New York, NY: Oxford University Press, ISBN: 0-19-50-8607-4.
6. Arvind, V., Köbler, J., Lindner, W., 2009, Parameterized learnability of juntas. *Theor. Comput. Sci. 410*, 4928–4936.
7. Feldman, V., 2006, On attribute efficient and non-adaptive learning of parities and DNF expressions. Electronic colloquium on computational complexity 13-066. Available from: http://eccc.hpi-web.de/eccc/.
8. Hofmeister, T., 1999, An application of codes to attribute-efficient learning. Computational Learning Theory, 4th European Conference EuroCOLT 1999, Nordkirchen, Germany, Lecture Notes in Computer Science, volume 1572, pp. 101–110.
9. Servedio, R.A., 1999, Computational sample complexity and attribute-efficient learning. 31st ACM Symposium on Theory of Computing STOC 1999, Atlanta, Georgia, pp. 701–710.
10. Ward, J. H., 1963, Hierarchical grouping to optimize an objective function, *J. Am. Statist. Assoc. 58*, 236–244.
11. Lance, G.H. and Williams, W.T., 1966, A general theory of classificatory sorting strategies, I. Hierarchical systems. *Comput. J., 9*(4), 373–380.
12. Jolliffe, I.T., 2002, *Principal Component Analysis, Series: Springer Series in Statistics*. 2nd ed., New York: Springer. (ISBN: 978-0-387-95442-4).
13. Rao, C.R., 1964, The use and interpretation of principal components analysis in applied research. *Sankhya, Ser. A, 26*, 329–357.
14. Hotelling, H., 1933, Analysis of a complex statistical variable into principal component. *J. Edu. Psy., 24*, 417–441 and 498–520.
15. Pearson, K., 1901, On lines and planes of closest fit to systems of points in space. *Philos. Mag., 2*(11), 559–572.

17 Experiment Design, Data Management, and Univariate Statistical Analysis of Gene-Expression Data Obtained by Real-Time Quantitative PCR

Ales Tichopad, Tzachi Bar, Malin Edling, Boel Svanberg,
Tania Nolan, and Anders Bergkvist

CONTENTS

17.1 INTRODUCTION

The most notable, recent developments in quantitative polymerase chain reaction (qPCR) technology have been achieved by improving technological elements of PCR. Modern cyclers have a more uniform temperature over the plate, new DNA polymerases are able to synthesize across more complex secondary structures of DNA, and novel detection chemistries are capable of more specific amplification and monitoring of the amplified DNA. Jointly, these advances have provided users with solid instrument performance, readily achieving precision as low as 0.25 C_q among PCR technical replicates.[1] In general, the issue of precision that is achievable for technical replicates has been emphasized largely by the instrument manufacturers, often ignoring the importance of biological heterogeneity among subjects of research and the effect of the sample preprocessing steps (see Chapter 1). This may have contributed to a perception that quantification of gene expression by real-time PCR (qPCR) is challenged by technical imperfection at the PCR level. A recent study in which the C_q precision of replicates is tested[2,3] revealed that the PCR component of a gene-expression experiment yields, by far, the highest precision when compared to all the preceding steps of sample acquisition and processing. Hence, there is a greater potential to reduce data variability and improve accuracy by putting more focus on improving the experimental design. The parameters within the experimental design that may benefit from additional attention include the required number and distribution of biological subjects within experimental groups and the number of replicates employed for individual sample-processing operations and the points of the workflow at which these are taken. In our experience, strictly adhering to the use of appropriate statistics that was defined prior to taking the pipette in hand further increases the overall validity of results. The power of a statistical test depends on several factors, including (1) the statistical significance criteria, (2) the sample size, (3) the size of experimental effects, and (4) the magnitude of variability in experimental measurements. The aim of this contribution is to provide the reader with concise guidelines for design and analysis of gene-expression experiments and to advice on how to undertake appropriate analyses. We will present ways to visualize data to highlight all the factors that demonstrate the power of the statistical test. Throughout all stages, the objective is to achieve the strongest possible biological inference.

17.2   NEUROSEARCH EXAMPLE DATA

To illustrate our approach, we use a data set that was kindly provided to us by NeuroSearch, Sweden, AB.

Twenty rats, divided into four groups, were used for the data collection. One group received saline control treatment and three groups received treatments of three different doses of the compound (confidential information).

After treatment, the rat brains were dissected into a left and a right part. The left part was analyzed for gene expression with qPCR. The left part of the brains was dissected into four different areas: limbic system (containing nucleus accumbens, most parts of the olfactory tubercle, ventral pallidum, and amygdala), striatum, frontal cortex, and hippocampus.

Total RNA was prepared by the guanidinium isothiocyanate method.[4] RNA pellets were dissolved in RNase-free water by repeated pipetting up and down a few times and then were put into a heating block at 55°C for 10 min followed by immediate cooling on ice. The dissolved RNA was subsequently stored at −80°C. The sample concentration was determined spectrophotometrically by a NanoDrop ND-1000 (Thermo Scientific). A quality indicator number and an integrity number of rRNA were measured with an Experion instrument (Bio-Rad).

Reversed transcription (RT) was performed by using a SuperScript III Kit (Invitrogen). One microgram of total RNA was reverse-transcribed with 5 μL of 2× RT reaction mix, 1 μL of RT enzyme mix, and the volume was adjusted to 10 μL with diethylpyrocarbonate (DEPC)-treated water. The following temperature protocol was used: 25°C for 10 min, followed by 50°C for 30 min, followed by 85°C for 5 min, and subsequent cooling on ice. After the addition of 0.5 μL (1 U) of *Escherichia coli* RNase H, the reaction was incubated at 37°C for 20 min and then at 85°C for 5 min. cDNA was diluted 40 times and stored at −20°C.

For qPCR measurements, 0.7 μL of the cDNA reaction was amplified in a 25-μL reaction mixture containing 1× PCR buffer, 0.2 mM deoxyribonucleotide triphosphate (dNTP), 3.7 mM $MgCl_2$, 0.15 mM SYBR Green, 0.4 μM of each primer, and 1 U JumpStart *Taq* DNA polymerase. Progress of the qPCR assay was measured on a CFX96 (Biorad) using the following settings for all genes: 60-s preincubation at 95°C followed by 40 cycles of denaturation at 95°C for 10 s, annealing at 56°C for 10 s, and elongation at 72°C for 10 s.

Correct PCR products were confirmed by agarose gel electrophoresis (2%), and PCR products were purified with a PCR purification kit from Qiagen (Valencia, CA). All amplicons were sequenced at MWG, Germany, and were routinely evaluated by the melting-curve analysis. The amounts of amplification product of interests were normalized using the geometric mean of the amounts of the two reference genes *Hprt* and *Cyc*.

The data set consists of qPCR measurements of six different genes of interest (*Arc*, *Bdnf*, *Fos*, *Gad*, *Glud1*, and *Penk*) together with two reference genes (*Hprt* and *Cyc*) in four different rat brain sections (limbic system, striatum, frontal cortex, and hippocampus). Before sample collection, rats were treated with a compound (confidential information) at one of three different doses (12, 37, and 110; arbitrary units) or were subjected to the procedure with a saline injection as a control. Five biological replicates were used for each dose/brain section for a total of 20 samples for each gene and brain section.

Even though the overall study included six genes of interest, only the *Fos* gene was studied in detail for this report. Thus, the designed primer sequences for the gene targets used in this chapter were as follows:

Rattus norvegicus Hypoxantine phosphoribosyl transferase (Hprt) (accession number AF001282)
    Sense 5'-GGC CAG ACT TTG TTG GAT TTG-3' (start 38)
    Antisense 5'-CCG CTG TCT TTT AGG CTT TG-3' (start 181)
    Product size: 144
Rattus norvegicus Cyclophilin A (Cyc) (accession number M19533)
    Sense 5'-GTC TCT TTT CGC CGC TTG CT-3' (start 17)
    Antisense 5'-TCT GCT GTC TTT GGA ACT TTG TCT G-3' (start 143)
    Product size: 127
Rattus norvegicus proto-oncogen (Fos) (accession number DQ089699)
    Sense 5'- CAG AGC ATC GGC AGA AGG-3'
    Antisense 5'- AGT TGA TCT GTC TCC GCT TGG-3'
    Product size: 155

Before statistical data analysis, the raw C_q values were preprocessed in the following way: The C_q values from the two reference genes were averaged using the arithmetic mean and were subtracted from the corresponding C_q values of the genes of interest. Relative quantities were calculated by setting sample with the lowest expression for each gene and brain section to one (defining this sample as the calibrator) and relating the calculated quantities of the other samples to this one. There are other approaches for relating quantities (that were not used here), which include taking a mean value of one group, the maximum sample value, or any other sample value. The different relational approaches do not change the relationship between the compared samples; they merely lead to different ways of visualizing the data. Relative quantities were subsequently transformed into fold change ($\log_2$ scale). The data preprocessing was performed using the GenEx data-analysis software (MultiD Analyses AB).

Throughout this chapter, we will use these data to illustrate aspects of visualization techniques, analysis techniques, and some examples of interpretation of results.

17.3   EXPLORATORY AND CONFIRMATORY APPROACH IN STATISTICS

In the context of statistical data analysis, we distinguish between exploratory and confirmatory studies (see Chapter 16). The purpose of the exploratory study is to analyze data with one or several different techniques to generate one or several plausible hypotheses. Since in an exploratory study, we make no attempt to prove a hypothesis, the rigor of the statistics that we use is of secondary importance. The primary focus is to disclose structures in the data that may form plausible hypotheses. In contrast, the purpose of the confirmatory study is to determine whether a specifically formed, prior hypothesis should be rejected or accepted based on evidence accompanying the experimental data, employing predetermined criteria of inferential statistics. In the case of a confirmatory study, it is therefore critical to adhere to rigorous statistics. The definition of rigorous statistics implies evaluation of data so that false-positive discoveries (referred to as the type I error in statistical terminology) are avoided with a known probability. In certain applications, it is also possible to employ a statistical technique to design experiments in a way that false-negative discoveries (type II errors) are avoided. Employing this technique, called power analysis, however, requires prior knowledge of the effect being studied along with the variance among the measurements.

The first step in research design is to define the clear prediction that the study is testing, the experimental hypothesis. The hypothesis is used to define a predicted relationship between two events or variables. It is critical that the predicted outcome is precisely stated before the research commences and that the methods of measurement are also determined. As an example, it may be expected that administration of drug X to cells Y grown in culture would result in a change in the expression of gene Z. For the experimental hypothesis, it is important to define the level of change that constitutes a change in expression, for example, "2 fold change" (up or down) or "an increase by 10,000 copies" may be suitable measures of significant change for the given system but all hypotheses must be determined according to the experiment. Within the definition, the direction of change may also be stated.

The concept described above is based on ideas by Ronald A. Fisher[5] and Karl Popper[6] who established the research tradition of statistical inference and falsifiable research. Fisher coined the term *null hypothesis* to describe a technique that offers a deterministic, mechanical, and objective scheme that leads to clear-cut *yes–no* decisions. The null hypothesis is accepted when the observed differences between variables could be expected by chance fluctuations and there is no evidence to support the experimental hypothesis.

17.4 STATISTICAL TESTS AND CONFIDENCE INTERVALS

A statistical test consists of a mathematical calculation that generates a *test statistics* value. This, seemingly arbitrarily, constructed value has an important property; that it can be interpreted with reference to the probability of an event, giving rise to the *p*-value. The *p*-value is a measure of the risk that the experimental result obtained from a data set may not support the experimental hypothesis. Expressed in a strict terminology, the *p*-value quantifies the chance that the observed effect, for example, difference in gene expression between two drug doses, may have been obtained by chance from a population where no such difference can be justified and should not have been observed as a genuine difference, for example, the expression of a gene in the cells after one drug dose does not actually differ from the gene expression in the cells after another drug dose. Such an event is referred to as false-positive error and is a result of the significance level chosen when the hypothesis was defined; for example, significance of 95% results in a 5% chance of a false-positive error. False negative is another additional type of error that may occur after data analysis. A false negative occurs when a truly existing difference in a studied biological response was not observed in the experiment. Typically, poor experimental design and low precision and accuracy yield such outcomes. Failure in resolving a genuine effect will result from an experiment with a sample size that is too small for the studied effect, improper sample handling, or inadequate use of replicates. In addition, improper or fully missing data management may jeopardize statistically significant resolution of existing effects.

17.5 POWER ANALYSIS

In addition to sample selection, decisions about the sample size will benefit from qualified statistical computation performed prior to the experiment. The required technique is referred to as a power analysis and is used to provide an estimate of the necessary sample size required to maintain the false-negative error risk under a defined probability, frequently under 20%. The power analysis requires that the researcher provides several parameters for the calculation of the statistical power, the ability to resolve the existing experimental effect, and derive the sample size. One of the most important, and often an unknown parameter, is the size of experimental effect, that is, the difference in the response variable that is manifested between studied groups of subjects. Where previous results (e.g., publications or pilot studies) exist, prior information about the size of the experimental effect may be available. This also applies to another parameter that is used for the power analysis; the magnitude of variability in the experimental measurements. Unfortunately, in a large number of gene-expression studies, the researcher has only limited or no knowledge of the size of the experimental effect expected, or its scatter among subjects. In such situations, it is recommended that a physiologically similar gene that is reported in an existing source is considered. The power analysis can then be performed using the effect size and scatter [in terms of standard deviation (SD) or variance] reported for this gene. Alternatively, it may be necessary to assume a biologically relevant difference in expression and to take this as the effect size (e.g., twofold). Likewise, it may be necessary to assume a variance and take this as the common variance due to a sample procession, often SD $(C_q) = 1$ (see, e.g., Oleksiak et al.[7] and Storey et al.[8]). In reality, the actual biologically relevant difference in expression and the real variability in the experimental measurements will depend on the specifics of the experimental assay. Realistic assumptions may nevertheless help provide useful illustrations of the structure of the measured data in exploratory studies.

There are a myriad of methods that are used for power analysis, based on the type of statistical test used. Nevertheless, arguably the most common and powerful technique, Student's *t*-test, is derived from a case of two independent samples. A simple experimental design, limited to two normal distributed groups, allows for the application of this technique and leads to straightforward yes–no determinations of significance.

17.6　TRADITIONAL VISUALIZATION TECHNIQUES

As there are many analysis methods available, there are also many data-visualization techniques available. With multivariate data, such as, for example, the previously referred to NeuroSearch data, multivariate data-analysis techniques, such as hierarchical clustering and principal component analysis (described in Chapter 16) are the options.

Arguably the most common visualization technique is a simple bar diagram with associated error bars (Figure 17.1). Even though this is a common and simple visualization technique, we find that there are issues with this technique that are worth emphasizing. First of all, error bars may mainly illustrate two different types of variability: either the inherent variability of the data (SD) or the precision by which the mean value has been determined. Second, the precision by which the mean value has been determined can be illustrated in different ways, but it ultimately depends on a combination of the inherent variability of the data together with the number of samples, and in its raw form, it is called the standard error of the mean (SEM):

$$SEM = \frac{SD}{\sqrt{N}}$$

However, the SEM is not a very intuitive measure and it is usually not straightforward to compare SEMs from different experiments in a meaningful way. A more popular way of illustrating the precision of the estimated mean and, mainly, indicating the statistical significance in a graphical way is the confidence interval:

$$CI = \left(\bar{C}_q - t^* \cdot \frac{SD}{\sqrt{N}} \right) to \left(\bar{C}_q + t^* \cdot \frac{SD}{\sqrt{N}} \right)$$

The confidence interval is based on the SEM. The lower limit of the confidence interval is constructed by subtracting the SEM multiplied by a percentile of a *t*-distribution from the mean. The upper limit of the confidence interval is constructed by adding the SEM multiplied by a percentile of a *t*-distribution from the mean. The confidence level of the confidence interval is set by the confidence level associated with the critical value t^*, typically a 95% confidence level.

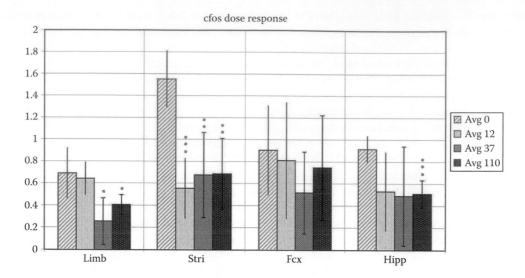

FIGURE 17.1 Fold change (log$_2$) expression of gene of interest *Fos* relative to reference genes *Hprt* and *Cyc*, relative to the least expressed sample within each brain section. The bar heights indicate the mean expression of five samples in groups of nontreated (avg0) samples or samples treated at one of the three different drug doses (avg12, avg37, and avg110). The error bars indicate 95% confidence interval estimates of the mean expressions. One asterisk indicates statistically significant difference between the means of a treated sample set compared to the mean of the nontreated sample set to 5%, two asterisks indicate statistically significant difference to 1%, and three asterisks indicate statistically significant difference to 0.1%.

Figure 17.1 shows the 95% confidence interval within each experimental group, highlighting the uncertainty associated with the mean estimate for the *Fos* gene expression after each drug dose in different brain sections. In addition, the *t*-test statistical significance *p*-values are shown for the difference in expression for the control samples and each of the three different dose responses indicated by means of an asterisk notation. It is customary to have one asterisk corresponding to a *p*-value below 0.05, two asterisks corresponding to a *p*-value below 0.01, and three asterisks corresponding to a *p*-value below 0.001.

Given that the asterisk notation hides the absolute value of *p*, it is often encouraged to include a table with the absolute values of *p*, as shown in the example in Table 17.1. One reason behind this is that a *p*-value of, for example, 0.032 is only slightly more "significant" than a *p*-value of 0.055. Borderline cases like this can lead to some confusion when deciding precisely what cut-off to use

TABLE 17.1

Significance Estimates of Difference of Means between the Means of a Treated Sample Set Compared to the Mean of the Nontreated Sample of Gene of Interest *Fos* Relative to Reference Genes *Hprt* and *Cyc*, Relative to the Least Expressed Sample within Each Brain Section

	Fos p-Values			
	Limb	**Striatum**	**Fcx**	**Hippocampus**
12 versus 0	0.70274	0.00034***	0.78194	0.05551
37 versus 0	0.01379*	0.00295**	0.20956	0.07582
110 versus 0	0.03180*	0.00157**	0.61582	0.00075***

Note: Student's *t*-test was used to produce the *p*-values.

when classifying data as significant. In realistic cases, a p-value of 0.051 could be just as significant as a p-value of 0.049; yet, a strict (although fundamentally arbitrary) cut-off of 0.05 would classify one as significant and the other as not significant.

Recall that statistical tests such as the t-test were designed as simple *yes–no* determinations of significance.[9] They were not designed as a way to *quantify* significance. It could therefore be argued that data visualizations and presentations as in Figure 17.1 and Table 17.1 overemphasize significance while neglecting the other factors that are associated with the power of a statistical test: Sample size, size of experimental effects, and magnitude of variability in experimental measurements. In the rest of this chapter, we demonstrate an alternative technique that may do a better job of visualizing all the factors associated with the power of the statistical test.

17.7   ALTERNATIVE VISUALIZATION TECHNIQUE AND THE CONFIDENCE INTERVAL OF THE DIFFERENCE BETWEEN MEANS

The heights of the bars in Figure 17.1 represent the fold change of gene-expression quantities relative to the sample with the lowest expression for each gene and rat brain section. Regrettably this parameter is of minor importance for analysis of the results. It is rather the *effect size* that is the desired parameter for the analysis. The effect sizes are visualized as the difference between the bar heights of the control samples and bar heights of each of the different dose responses. However, the effect size can be further emphasized by subtracting the control samples, bar heights from the dose–response bar heights and then plotting the difference (Figure 17.2). In this case, the variation of the difference, between the bar height of the control sample and the bar height of each of the different dose responses, needs to compound the variability within both the control sample and within the respective dose–response sample. To take this into account, we need to calculate the confidence interval of the difference between means.[10]

$$CI = (\Delta \overline{C}_q - t^* \cdot D) \, to \, (\Delta \overline{C}_q + t^* \cdot D)$$

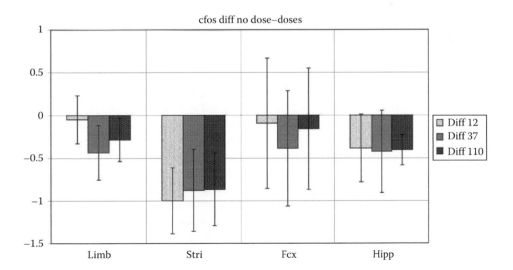

FIGURE 17.2   Bar diagram showing the difference between means of the nontreated sample set and one of the treated sample sets. The error bars show the confidence interval of the difference between means. The error bars that do not cross the x-axis indicate that the corresponding means comparison is statistically significant to 5% in a t-test.

where

$$\Delta \bar{C}_q = \bar{C}_q(\text{dose}) - \bar{C}_q(\text{control})$$

and

$$D = \text{SD}_{\text{joined}} \cdot \sqrt{\frac{1}{n_{\text{control}}} + \frac{1}{n_{\text{dose}}}}$$

and

$$\text{SD}_{\text{joined}} = \sqrt{\frac{\sum\limits_i \left(C_q(\text{control})_i - \bar{C}_q(\text{control})\right)^2 + \sum\limits_i \left(C_q(\text{dose})_i - \bar{C}_q(\text{dose})\right)^2}{(n_{\text{control}} - 1) + (n_{\text{dose}} - 1)}}$$

An interesting feature of the confidence interval of the difference between means is that a confidence interval that encompasses the zero difference between means corresponds exactly to a nonsignificant result in a Student's t-test. Conversely, a confidence interval that does not encompass the zero difference between means corresponds exactly to a significant result at the confidence level corresponding to the p-value cut-off (5% in Figure 17.2). The confidence interval of the difference between means thus nicely illustrates both the data variability and significance in one elegantly featured parameter. Therefore, analysis of confidence intervals avoids some of the problems associated with classical significance tests.[11]

17.8 INTERPRETATION OF THE VISUALIZATION

The confidence intervals in Figure 17.2 cross the zero difference between means and this agrees with the p-values of Table 17.1. By clearly visualizing effect size, data variability, and significance, Figure 17.2 thus provides a good tool for us to interpret the NeuroSearch data for *Fos*. First of all, all brain sections seem to experience some decreased expression of *Fos* in the drugged samples when compared to the control samples. It appears that the drug has a clearly significant (at 5% cut-off) effect on *Fos* expression in stria. However, the large variability and the small effect size in the frontal cortex make this trend insignificant in the frontal cortex. The effects of the drug on *Fos* expression in the limbic system and hippocampus are less apparent. There is no clear dose–response trend in any of the brain sections. With a great deal of good will, we may interpret the response in the limbic system as being activated at a dose between the low dose, 12 and the intermediate dose, 37. According to the p-values in Table 17.1, only the highest dose (dose 110) in the hippocampus has a significant effect on the *Fos* expression. However, it is apparent that a large, if not the major contributing, factor to that significance is the low variability at the highest dose in the hippocampus (Figure 17.2). In fact, it is rather curious that the variability of one drug dose would be different than another drug dose for a given brain section and gene of interest. Under the assumption that the drug affects the mean gene expression, but not gene-expression variability, we may conclude that the evidence for reduced expression of *Fos* for the highest dose (dose 110) in the hippocampus is no more significant than the evidence for reduced expression of *Fos* for the other two drug doses. On that note, the lower two doses are close to significant at a 5% cut-off level. If we were performing a confirmatory study and our hypothesis called for a 5% cut-off, we would need to conclude that those two drug doses failed to satisfy our hypothesis. However, as we have not defined a prior hypothesis, it is more appropriate to consider this an exploratory study, and in such case, we are looking to identify trends and hypotheses that are plausible rather than trends and hypotheses that satisfy specific criteria. In this context, it seems reasonable to conclude that there is a potential hypothesis calling for a drug effect on *Fos* expression in the hippocampus, independent of any of the doses tested in this exploratory study.

17.9   MULTIPLE TESTING

As has been emphasized in this chapter, when an exploratory study is performed, it is not necessary to ensure that hypotheses satisfy specific criteria, since the purpose of an exploratory study is only to generate new hypotheses for subsequent confirmation. On the other hand, when performing a confirmatory study, it is of critical importance to correctly set up and adjust the significance criteria. One important consideration relates to compensation of the increased risk of false positives when testing multiple hypotheses.

Multiple testing refers to the fact that a statistical test with several independent hypotheses is more likely to yield a positive significance and that the chance of this increases with the more hypotheses that are tested, even if the underlying probability distributions are identical. For example, consider that the expression of none of the six genes tested in the NeuroSearch data is affected by the drug. A statistical test comparing drugged and control samples for a single gene at a 5% significance level will still have a 5% risk of inappropriately showing significance (false positive) that only appears due to random chance. If the same test (at a 5% significance level) is repeated for all six of the genes, then the overall risk that at least one of the tests yields a false-positive result is 26% (Table 17.2). To maintain an overall risk of a false-positive result at 5%, we thus need to compensate for multiple testing. One of the most common correction techniques for multiple testing is the Bonferroni correction.

The Bonferroni correction maintains an overall risk of false-positive results by reducing the significance cut-off for each individual hypothesis. For example, to maintain an overall risk that at least one of the tests yields false-positive results at 5%, the Bonferroni correction calls for a 0.85% significance level for each test comparing drugged and control samples for a single gene. The risks with multiple testing and the appropriate Bonferroni corrections for a range of different number of hypotheses are listed in Table 17.2.

Even though an exploratory study is performed, a multiple testing correction may serve a useful purpose as a benchmark for what may subsequently be considered plausible hypotheses. In the case

TABLE 17.2
Bonferroni Correction

Number of Independent Null Hypotheses (N)	Probability (P^*) of Obtaining One or More p-Values Less Than 0.05 by Chance (%)	Alpha (α^*) to Keep Overall Risk of False-Positive Error Equal to 0.05
1	5	0.0500
2	10	0.0253
3	14	0.0170
4	19	0.0127
5	23	0.0102
6	26	0.0085
7	30	0.0073
8	34	0.0064
10	40	0.0051
20	64	0.0026
50	92	0.0010
100	99	0.0005

of the NeuroSearch data, there are, in fact, 72 independent hypotheses (six genes multiplied by four brain sections multiplied by three different drug doses) and the corrected significance level (assuming an overall significance level of 5%) is therefore 0.07% for each individual hypothesis. It is therefore not surprising that we have several individual "significant" results at the 5% significance level for the NeuroSearch data. Among the *Fos* data, only the lowest drug-dose observation (12 vs. 0) for the stria brain section is indicated as significant. For a plausible hypothesis, we would expect higher drug doses to have an equal if not larger effect on *Fos* expression. Nevertheless, the fact that there are significances down to Bonferroni-corrected significance cut-offs adds weight to a conclusion that there are real biological effects in the experimental design that can be validated in subsequent confirmatory studies. The key to validating these effects will be to determine the magnitude of the biological effects with greater precision, that is, to reduce the variability of the data.

While the Bonferroni correction is the most commonly applied correction, it does, however, assume independence among the tested hypotheses (here genes), an assumption that rarely holds true within a complex biological environment. Furthermore, the Bonferroni correction is known to be the least-sensitive method, and thus, it is very restrictive in confirming hypotheses. Therefore, new methods have been suggested, including the most acknowledged alternative, the false-discovery rate (FDR). The FDR of a test is defined as the expected proportion of false-positive results among the declared significant results.[12–15] Owing to this straightforward interpretation, FDR is a more convenient scale to work with instead of the *p*-value scale. For example, if expression is measured in 100 genes with a maximum FDR of 0.10 as the cut-off for the genes to be considered to be differentially expressed (DE), then it would be expected that a maximum of 10 genes would be false positives. No such interpretation is available when calculating the *p*-value. Methods have been proposed either to transform *p*-value into an FDR or to compute FDR directly.[16–19]

17.10 MINIMIZING SPURIOUS EFFECTS AND CONFOUNDING FACTORS

The qPCR method is prone to error due to its extreme sensitivity (down to single-copy detection) and nonlinear principle of quantification. Inhibition of the PCR or RT steps due to contaminants in the samples (see Chapters 1, 2, and 4), primer–dimer generation or suboptimal assay design and optimization (see Chapters 12 and 15) are just some of the factors that are frequently responsible for invalid data. In addition, extreme, yet valid, measurements may be obtained from randomly selected subjects for an experiment. Altogether, experimental group data may show variation that extends the scale of the studied effect and hence confounds its resolution. It is therefore important that the researcher pays increased attention to the data quality before any statistical technique is applied. Several methods of data validation have been suggested. In general, the methods may be categorized into two groups: Plausibility-based validation and performance-based validation.

When using plausibility-based validation, the aim is to exclude C_q values that are outside of an arbitrary limit, which is often set by observing other related values. Vandesompele and coworkers[20] suggest that C_q values of technical replicates should be excluded when they are above a given factor of the mean of the group of technical replicates. Similarly, adoption of a Grubbs test to identify outliers has also been recommended.[21] Certainly these are pragmatic approaches that can be adopted to prevent some of the erroneous reactions from confounding the group mean estimate. However, this approach can only be used to control for later-stage pipetting errors or amplification failures. Most errors infiltrate the experimental process before the PCR step. Hence, failures in sampling biological material, storage, purification, or RT will not be regarded if only PCR replicates are considered. Another disadvantage of this method is the fact that, in most cases, three technical PCR replicates or even fewer are employed in the qPCR. Three replicates do not provide a sufficiently reliable base for a probabilistic decision about exclusion of the C_q for one reaction that may be deviant from the mean of the remaining reactions.

The amplification compatibility test,[22,23] also referred to as multivariate kinetics outlier detection (MKOD),[24] is a method to detect kinetics outliers regardless of their C_q value. It tests the kinetics

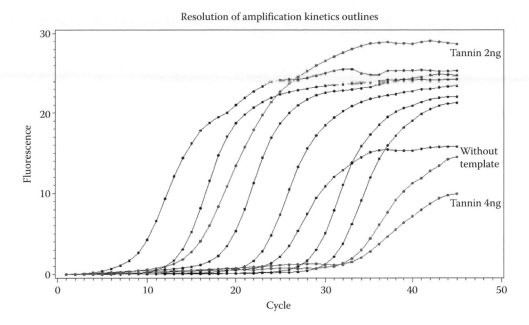

Resolution of amplification kinetics outlines

FIGURE 17.3 Example of a set of amplification curves that were evaluated by calculating multivariate distances using the MKOD method. The amplification curves marked "Tannin 2ng," "Tannin 4ng," and "Without template" were indicated as outliers.

similarity among compared reactions, a prerequisite for qPCR. MKOD is based on comparison of a few kinetic parameters describing a test reaction to those describing a set of reference reactions. It is a direct way to test the compatibility of the amplification kinetics captured in the shape of the amplification curve. These methods allow comparison of reactions throughout the entire set of qPCR samples that are amplified within the experiment, which usually assures a sufficient size of the quality reference. Solely on the basis of data analysis, these tools complement measures to improve and control preanalytics. The amplification compatibility test is used for routine procedures such as diagnostics where a high level of standardization is required (Figure 17.3). The MKOD method is available in the Kineret software (Labonnet Ltd.).

Methods such as these are useful to identify candidates for outlier correction or removal. However, when considering outliers, it is desirable, if possible, to also include a trace back to the collection of the sample in question and the processing procedures it has been exposed to. Ideally a plausible cause of the outlier deviation can be identified in the sample-handling process. If no plausible cause can be identified, great caution should be exercised before correcting or removing any outlier since excessive outlier correction or removal will artificially increase data precision and then any conclusions drawn from the results will not be an accurate representation of reality.

17.11 MORE COMPLEX EXPERIMENTAL DESIGN

The drive by scientific ambitions to discover promising trends in data results in experiments that are designed with more than one factor to be studied. Either one or more factors such as a genotype, gender, or tissue type, or one or more metric variables (also called covariates) such as age, body weight, dose, or exposure time accompany the data. In these cases, a more advanced test method, based on a suitable data model, should be involved to perform a statistical test on all such terms simultaneously. This requires extensive skills in statistics. Further, an interaction between factors and covariates is often of scientific interest as it may disclose an interesting biological phenomenon, raising new hypotheses. Unfortunately, recognizing the need for an approach such as a multiway analysis of variance, analysis of covariance, or, generally, methods jointly referred to as generalized linear models (or *mixed model* in its more recent advanced modification) often lies out of the scope of the researchers performing such experiments. This contribution is certainly not the place to discuss these techniques in detail; nevertheless, we will try to provide the user with a comprehensive guide on how to treat more complex experiment setups in a pragmatic way when more than one factor or covariate is involved.

17.11.1 PILOT STUDY

The pilot study concept involves studying a secondary classification by first employing one or more smaller experiments, prior to the main experiment that is aimed at addressing the primary objective. The secondary classification can be studied descriptively or it can be explored using a statistical test, when the outcome is not interpreted as a proof of hypothesis, but is used to guide the direction of further experimentation. An example could be: a study where the effect of a drug in varying doses is tested on a group of subjects where the main objective is to test whether the drug has an effect differing from placebo. The setup can be either understood as a single integrated experiment with all classifications analyzed simultaneously by an appropriate statistical model or it can be decomposed to smaller experiments that are conducted on smaller sets of data, where each experiment is conducted for each dose, with the aim of determining the dose generating the strongest response. Then, knowing the strongest response, the experiment may progress to analyzing a confirmatory study where the difference of the drug from the placebo is examined, using the dose that generated the strongest response. The advantage of such an approach is that the statistical tests employed are simple enough to be well mastered by anybody with a basic knowledge of statistics. Further, this approach may save money because the extension of the pilot is substantially smaller than the main experiment. This approach is somewhat comparable to the three-phase approach used in clinical drug testing where a similar objective is approached in a sequential fashion, but each subsequent phase involves increasing the number of subjects while maintaining the reduced dimensionality of the study and therefore the associated risk of failure in confirming the hypothesis by, for example, inefficient dose chosen. A further, and more general, advantage of the pilot study is that a suitably designed experiment may also allow for validating the reference genes to be used, optimizing the number of replicates at various stages of the sample preanalytics,[2] and generally, focusing the study on the most responsive genes. It is important that experimental effects that are identified in the pilot as not significant are no longer tested in the main experiment. Otherwise, the selected significance level will be inappropriate due to the increased chance of false positivity by repeating the test.

17.12   CONCLUSIONS

The example data set used within this contribution was analyzed with a rather simple technique. This allowed a clear illustration of how a scientific hypothesis can be approached by designing a suitable experiment series and interpreting the results. We stress that a simple experimental design should be considered as a major advantage over cases where too many factors are studied simultaneously within a single experimental setup, generating a need for complex tests that are often associated with multiple testing problems. We have shown how improvements in the performance of qPCR instrumentation drives research interest toward more complex, precise, and potentially rewarding research issues, thus moving the bottlenecks of research projects to the sample collection and data-analysis steps. As the practice of more advanced data analysis gets more common, the production of more and more statistical analyses risks undermining the power of the statistical tests and appropriate corrections need to be applied. This chapter provides an overview of the underlying fundamental features of statistical analysis. It cautions the reader to distinguish between exploratory and confirmatory statistical analyses. In addition, the traditional concepts and parameters of statistical tests are described, such as the null hypothesis, the p-value, and power analysis. The power of a statistical test depends on several factors including (1) the statistical significance criteria, (2) the sample size, (3) the size of experimental effects, and (4) the magnitude of variability in experimental measurements. The p value is a powerful criterion of significance and it is often used in the scientific literature to evaluate the power of a hypothesis. However, the p-value was not designed to *quantify* significance. To evaluate the power of a particular hypothesis, it is therefore of great benefit to also consider the effects of the sample size, the size of the experimental effects, and the variability in the experimental measurements. In this chapter, we propose that the approach of taking advantage of the confidence interval of the difference between means in a bar plot, of the difference between means of control, and treated samples are in order to highlight all these factors. We applied this technique to the data generated by NeuroSearch, Sweden, AB. The resulting visualization of the NeuroSearch data generated a lot of interesting hypotheses as proposals for further analysis. As the field of qPCR research continues to develop, we propose that this may be a useful technique to analyze the data.

REFERENCES

1. Bustin, S., Benes, V., Garson, J. A., Hellemans, J., Huggett, J., Kubista, M., Mueller, R. et al. 2009, The MIQE guidelines: Minimum information for publication of quantitative real-time PCR experiments, *Clinical Chemistry* 55, 611–622.
2. Tichopad, A., Kitchen, R., Riedmaier, I., Becker, C., Stahlberg, A., Kubista, M., 2009, Design and optimization of reverse-transcription quantitative PCR experiments, *Clinical Chemistry* 55, 10.
3. Kitchen, R. R., Kubista M., Tichopad, A., 2010, Statistical aspects of quantitative real/time PCR experiment design, *Methods* 50, 231–236.
4. Chomczynski, P., Sacchi, N., 1987, Single-step method of RNA isolation by acid guanidinium thiocyanate–phenol–chloroform extraction, *Analytical Biochemistry* 162(1), 156–159.
5. Fisher, R. A., 1937, *The Design of Experiments*, London: Oliver and Boyd.
6. Popper, K., 1959, *The Logic of Scientific Discovery*, New York, NY: Basic Books.
7. Oleksiak, M. F., Churchill, G. A., Crawford, D. L., 2002, Variation in gene expression within and among natural populations, *Nature Genetics* 32, 261–266.
8. Storey, J. D., Madeoy, J., Strout, J. L., Wurfel, M., Ronald, J., Akey, J. M., 2007, Gene expression variation within and among human populations, *American Journal of Human Genetics* 80, 502–509.
9. Cohen, J., 1990, Things I have learned (so far), *American Psychologist* 45(12), 1304–1312.
10. Blom, G., 1984, Statistikteori med tillämpningar, Lund, Studentlitteratur, ISBN 91-44-05592-7.
11. Brandstaetter, E., 1999, Confidence intervals as an alternative to significance testing, *Methods of Psychological Research* 4(2), 33–46.
12. Benjamini, Y., Hochberg, Y., 1995, Controlling the false discovery rate: A practical and powerful approach to multiple testing, *Journal of the Royal Statistics Social Series B* 57, 289–300.
13. Keselman, H. J., Cribbie, R., Holland, B., 2002, Controlling the rate of type I error over a large set of statistical tests, *British Journal of Mathematical Statistics Psychology* 55, 27–39.
14. Pawitan, Y., Michiels, S., Koscielny, S., Gusnanto, A., Ploner, A., 2005, *Bioinformatics*, 21(13), 3017–3024.
15. Benjamini, Y., Hochberg, Y., 2000, On the adaptive control of the false discovery rate in multiple testing with independent statistics. *Journal of Educational Behaviour and Statistics* 25, 60–83.
16. Storey, J. D., Tibshirani, R., 2003, Statistical significance for genomewide studies, *Proceedings of the National Academy of Sciences USA* 100, 9440–9445.
17. Storey, J. D., 2002, A direct approach to false discovery rates, *Journal of the Royal Statistics Social Series B* 64, 479–498.
18. Aubert, J., Bar-Hen, A., Daudin, J. J., Robin, S., 2004, Determination of the differentially expressed genes in microarray experiments using local FDR, *BMC Bioinformatics* 5, 125–133.
19. Reiner, A., Yekutieli, D., Benjamini, Y., 2003, Identifying differentially expressed genes using false discovery rate controlling procedures, *Bioinformatics* 19, 368–375.
20. Hellemans, J., Mortier, G., De Paepe, A., Speleman, F., Vandesompele, J., 2007, qBase relative quantification framework and software for management and automated analysis of real-time quantitative PCR data, *Genome Biology* 8(2), R19.

21. Burns, M. J., 2005, Standardisation of data from real/time quantitative PCR methods—Evaluation of outliers and comparison of calibration curves, *BMC Biotechnology* 5, 31.
22. Tichopad, A., Bar, T., Pecen, L., Kitchen, R. R., Kubista, M., Pfaffl, M., 2010, Quality control for quantitative PCR based on amplification compatibility test, *Methods* 50, 308–312.
23. Sisti, D., Guescini, M., Rocchi, M. B. L., Tibollo, P., D'Atri, M., Stocchi, V., 2010, Shape-based kinetic outlier detection in real-time PCR, *BMC Biotechnology* 11, 186.
24. Bar, T., Kubista, M., Tichopad, A., 2012. Validation of kinetics similarity in qPCR, *Nucleic Acids Research* 40(4), 1395–1406.

18 Biomarker Discovery via RT-qPCR and Bioinformatical Validation

Christiane Becker, Irmgard Riedmaier, and Michael W. Pfaffl

CONTENTS

18.1 INTRODUCTION

There is a growing interest in life science research in the use of expressed transcripts that form the basis of biological markers (biomarkers) and in addressing some of the challenging statistical issues that arise when attempting to validate them. Biomarkers have extensively been used across diagnostic and therapeutic areas of many life science disciplines, including clinical, physiological, biochemical, developmental, morphological, and molecular applications.[1] Biomarkers have been defined as "cellular, biochemical or molecular alterations that are measurable in biological media such as human tissues, cells, or fluids."[2] The official definition, developed by the "Biomarkers definitions working group" of the NIH is[3] "A biomarker is a characteristic that is objectively measured and evaluated as an indicator of normal biologic processes, pathogenic processes, or pharmacologic responses to a therapeutic intervention." More recently the definition has been broadened to include more biological characteristics that can be objectively measured and evaluated as a biological indicator.[4] A biomarker can refer to any measurable molecular, biochemical, cellular, or morphological alternations in biological media such as human tissues, cells, or fluids.[5]

18.1.1 BIOMARKERS AT VARIOUS MOLECULAR LEVELS

Advances in genomics, proteomics, transcriptomics, and metabolomics have generated many candidate biomarkers with the potential for diagnostic and clinical value. Current efforts are focused on biomarker discovery, reliable detection, and early diagnosis, for example, in cancer biology through the application of various–omics technologies. The success of biomarker identification depends on many factors, such as the type of molecule (e.g., gene, transcript, protein, metabolite), the intensive validation across a heterogeneous population and its variations (age, sex, species, breed), the quality and integrity of the biological sample, the size of the dataset(s) used, and the statistical methods that were applied for validation. Probably the most significant factor leading to success is the number of variables and conditions being tested, because, what appears to be specific in a given biological dataset may not necessarily be so in a larger set or even in the entire population. Therefore, the more conditions and variables being evaluated; the better will be the outcome of the prediction and the validity of the discovered biomarkers.[6,7]

The integration of different technologies on various–omics levels for data collection and their analyses are pivotal for biomarker identification, characterization, validation, and successful usage. The application of integrative functional informatics represents a novel direction in such biomarker discovery and brings a new dimension to molecular diagnostics.[1] These markers can represent the combination of multiple pieces of information on various biological levels, such as genes, their mutations, SNPs, gene methylation pattern, alternative gene transcripts (mRNA and miRNA), posttranslational modified proteins, metabolites, morphological changes, or altered physiological responses.

The first step in all biological and physiological processes is the transcription of specific genes into mRNAs and noncoding RNAs as prerequisite for the generation of functional proteins. Gene expression is a dynamic process that adapts rapidly to physiological changes or exogenous stimuli and thus the transcriptome with its enormous number of alternative spliced mRNAs, large and small noncoding RNAs reflects the current physiological situation in different tissues, organs, and even in single cells.[8] Therefore, monitoring the transcriptome is, potentially, a very promising approach for detecting biomarkers for specific physiological situations, diseases, or treatments.

Further biomarkers can be discovered at the level of the proteome and the metabolome. To investigate the complex proteome, applied proteomic technologies are used to separate, identify, and characterize a global set of proteins. In addition, information should be provided about the protein concentration, tissue, or cellular location, any modifications or functional attachments, and interactions, for example, protein–protein, protein–DNA, or protein–ligand.[9] The proteome, unlike the "fixed" genome, possesses an intrinsic complexity and is in a constant state of flux. The benefit of protein analysis is the ability to take into account posttranslational modifications, which can markedly alter the function, activity or half-life of a protein. In addition, the final amount of the active protein can differ greatly from the initial amount of mRNA transcribed and present in the cell.

Metabolomics is a relatively new discipline that can facilitate rapid *in vivo* screening of various factors, including drug efficacy and/or toxicity and underlying physiological processes. The metabolomic approach is complementary to the other–omic profiling technologies and can provide a chemical and biochemical profile of a specific body fluid, organ, or tissue during a continuous time-course analysis.[9] Overall, metabolomics can facilitate the determination of metabolic profiles and the mapping of interactions between metabolic pathways across organisms.

As described, there are various ways to discover biomarkers: at the level of the genome, transcriptome, proteome, or metabolome. Herein, we will focus on biomarkers identified in the transcriptome at the RNA level, with the current focus on mRNA and miRNA. Each gene has its set of characteristic expression profiles and alternative splice variants, that is, in which cells or tissues and at what time it is expressed and how it responds to environmental stimuli. For research and biomedical purposes, only a few genes may be sufficiently reliable to be used as indicators of healthy or diseased biological states.[7] Approaches centered on transcriptomics consist of various methods to measure the expression of genes, including microarray analysis, RT-PCR-based methods or holistic assumption-free methods such as next-generation sequencing technologies.[10,11] Quantitative RT-PCR and microarray-based analysis have significantly expanded the throughput of expression studies, and numerous examples of potential microarray-based biomarkers have been published.[12–14]

18.1.2 BIOMARKERS AT THE mRNA LEVEL

There are different methods available to quantify single transcripts. In general, these methods differ in the number of quantifiable genes. Northern blot is the classical approach for the detection of different mRNAs. A more precise method for the quantification of gene expression is quantitative real-time RT-PCR (RT-qPCR). With both methods it is possible to quantify single genes or multiple gene sets in one run. Using qPCR arrays, up to 384 different transcripts can be analyzed in parallel. There are also other methods available for a holistic screen of gene-expression changes. Until recently, microarray analysis has been the screening method of choice for most gene-expression experiments at the mRNA and miRNA level. With this approach, a sample can be screened for the expression of all known transcripts present in the gene database. But how can new, unknown genes, alternative splice variants, or miRNA intermediates be measured? RNA sequencing (RNA-Seq) is a new method which permits the sequencing as well as the quantification of the whole transcriptome of a biological sample. It is a very sensitive approach; a single transcript of a given gene is detectable, and since it is assumption-free it is also possible to discover new transcripts or unknown splice variants.[10,11]

The application of transcriptomics to biomarker research has successfully been used in various fields of life science. In molecular medicine, it has been shown that changes in the expression pattern of specific genes are indicative of different pathological processes. It is also possible to distinguish between different types and stages of diseases, for example, various forms of cancer, heart disease, neuropsychiatric disorders, and the causes of infertility.[8,14] Another application in molecular medicine is pharmacogenomics, the analysis of gene expression to predict the response of a patient to treatment with specific drugs, thus enabling the choice of the most appropriate treatment for each individual patient.[8,15]

The use of gene-expression biomarkers for the detection of specific, external stimuli is a further field of application. Our group is interested in the misuse of drugs for growth-promoting purposes in human sports and animal husbandry. There are numerous reports about gene-expression changes caused by the use of anabolic substances in different tissues and species. In cattle, several promising candidate genes have been proposed for the detection of the misuse of anabolic substances; IGF-1, for example, has highly abundant expression in liver and muscle.[16–19] A further group of promising candidate genes are the receptors for specific substances, for example, the steroid hormone receptors or the β-adrenergic receptors in different tissues.[20–22] A lot of promising organs for biomarker discovery after hormone application have been reported: for example, uterus, ovary, prostate, vaginal epithelial cells, liver, muscle, and blood.[23–29] Most of these tissue samples have to be taken at the slaughterhouse or in the surgery room. In human sports, there is only a limited number of tissues available to trace the misuse of anabolic substances, for example, blood, urine, and hair. However, there are also reports regarding gene expression in human hair follicle cells and primate blood, suggesting that analysis of gene-expression changes caused by anabolic substances is feasible in humans.[27,29,30] All these examples demonstrate the potential of biomarker research at the mRNA level in different veterinarian and human research fields.

18.1.3 BIOMARKERS AT THE miRNA LEVEL

miRNAs are small, regulatory RNA molecules that are involved in the regulation of mRNA expression and hence influence almost all physiological processes and metabolic pathways.[31] Dysregulation of miRNAs could be correlated with several different human pathologies, for example, diabetes, liver disease, or human cancer.[12,32–34] In this context, recent studies have revealed that specific miRNAs could be "the" upcoming biomarkers in clinical diagnostics. miRNAs show good suitability for biomarker research as they appear to be expressed in a developmental, disease, and tissue-specific manner, which is not the case for other established biomarkers. In contrast to mRNAs, miRNAs are more stable[35] and less sensitive to RNAse exposure and, besides the transcriptional processing forms, from primary miRNA (pri-miRNA) to pre-miRNA to mature miRNA, no further modifications have been described. In contrast to the possibility of mRNA splice variants occurring from a single gene, for the mature miRNA no variants of the same molecule are known, thus facilitating accurate detection.[13,36] In cancer diagnostics, expression profiles of the so-called "oncomirs" (miRNAs, which are implicated in the formation of malignancies) have already proved their superiority over mRNA profiles.[37,38] For example, Lu and coworkers[12] were able to discriminate gastrointestinal cancer tissue from nongastrointestinal cancer tissue by characterization of specific miRNA profiles. This was not possible when screening the same biological samples for around 16,000 mRNAs. Besides diagnosis, screening of miRNA expression gives exceptional insights in to disease progression, for example, differentiation stages, developmental lineage of tumors, or response to therapy.[36,38]

miRNAs have also been discovered as circulating cell-free nucleic acids in the body fluids (e.g., blood, breast milk) of healthy and diseased individuals.[39–42] In addition, levels of circulating miRNAs in plasma have been linked to cancer (e.g., prostate, B-cell lymphoma) and other diseases (e.g., inflammatory bowel disease).[43–45] Mitchell et al.[43] showed that it is possible to identify prostate cancer patients by measuring the plasma levels of miR-141. In B-cell lymphoma, miR-21 was proposed as a promising biomarker because its serum abundance appears to be associated with patients' survival.[44] In blood, miRNAs are thought to be secreted from normal or tumor cells in microvesicles.[40] These findings could pose a breakthrough in the field of medical diagnostics as this would offer a possibility for prognostic information and early disease detection that is of minimal invasion.

Most studies that address circulating miRNAs as disease markers are targeting those that are originating from tumor cells as secretory products. However, these could also have physiological and regulative functions. For example, a notable number of miRNAs, which are known to play important functions in the immune system are found in breast milk (e.g., miR-155, miR-181a, miR-181b).[46,47] It is believed that miRNAs together with other immune-related agents contained in breast milk, like IgA and leucocytes, are responsible for the development of the immune system of the newborn baby by influencing the intestine.[41] Even though the underlying processes are not yet clarified, once more, the positive effect of breastfeeding for the health of the offspring can be supported by those observations.

18.2 miRNAS AS BIOMARKERS IN SURVEILLANCE OF ILLEGAL USE OF ANABOLIC STEROIDS

Even though the most common field of interest concerning miRNAs lies in human medical research, the concept of establishing biomarkers has also been introduced into veterinary medicine, for example, in the surveillance of illegal use of steroidal growth promoters.[19,24–27] Steroid hormones are known to alter gene expression and might also influence the expression of miRNAs. Recently, an innovative study investigated the effects of the anabolic combination of trenbolone acetate plus estradiol on miRNA abundance in bovine liver.[48] miRNA RT-qPCR arrays for gene-expression screening followed by statistical validation of results established an expression profile characterized by an upregulation of miR-29c, miR-130a, and miR-103 and a downregulation of miR-34a, miR-181c, miR-20a, and miR-15a. Using principal components analysis (PCA) as the biostatistical method of choice for pattern recognition (see Chapter 16), a separation on the basis of the miRNA expression profile between the untreated control group and treatment group could be shown. The significance of the group separation can be maximized, when integrating additionally significant mRNA expression results together with the miRNA.[48] It can be seen from these results, that the combination of gene expression results from mRNA and miRNA might be an upcoming integrative approach to use for the specific generation of gene-expression patterns as biomarkers for anabolic treatment screening.

18.3   TECHNICAL CHALLENGES IN miRNA BIOMARKER RESEARCH

RT-qPCR is the current gold standard for sensitive and reproducible miRNA gene-expression analysis. It is also established as the method of choice for validating results from holistic approaches, such as high-throughput sequencing (e.g., NGS), microarrays, or PCR array experiments.[36,49] However, the nature of miRNA molecules poses a challenge for reliable analytics.

The combination of short length of mature miRNAs (~22 nt) and a heterogeneous GC content poses a challenge for cDNA synthesis and primer and probe design since these results in significant difference in the melting temperatures of different miRNAs.

The sequence of interest is not only present in the mature miRNA, but also in the precursor sequences, the pri-miRNA and pre-miRNA.

The members of one miRNA family (e.g., let-7 family) usually differ by just one nucleotide, mostly at the 3′ end of the sequence.

There are no specific guidelines for miRNA expression data analysis and normalization.

Strategies to deal with those challenges have been published and are being intensively discussed (summarized in Refs. [35], [36], [49–51]). Not only are the properties of the molecule challenging for established technical procedures, but also sample matrices pose additional problems. Especially in clinical research, patient samples (e.g., tumor samples) are frequently available as formalin-fixed paraffin-embedded (FFPE) tissues. From mRNA expression studies it is known that this type of tissue conservation is challenging because RNA from FFPE tissue is often cross-linked or degraded. Also, qPCR efficiency could be inhibited by formalin fixation.[52,53] Fortunately, various studies show that, in contrast to mRNAs, miRNAs seem to be less affected and more stable. Specialized extraction kits for FFPE samples could be used for miRNA expression analysis and reliable qPCR results could be gained.[53,54] Also, sample preparation from blood to perform RT-qPCR analysis of circulating miRNAs is challenging. Therefore, specified RNA isolation protocols and optional pre-amplification steps are required to deal with the low amounts of miRNAs present in plasma or serum. Additionally, inhibitors of qPCR present in blood (e.g., albumin) must be removed. An established method for analyzing cell-free miRNAs in plasma and serum is presented elsewhere.[42]

18.4   BIOINFORMATICS

As already mentioned, there is no single gene-expression biomarker for any given disorder or clinical situation. In most cases, multiple biomarkers must be present to distinguish between specific diseases, disease states, or treatments, hence a biomarker pattern consisting of various mRNA and miRNA transcripts, must be available. An important question is how to deal with these data to get the desired information. The best way seems to be the construction of clusters using methods for dimension reduction combined with pattern recognition technologies to visualize the gene-expression pattern in two- or three-dimensional graphs.[25,46] There are different multivariate analysis methods available, which are used for biomarker selection and validation, namely hierarchical clustering analysis (HCA) and PCA.

18.4.1   HIERARCHICAL CLUSTER ANALYSIS

The most popular method for the visualization of gene groups or treatment patterns is hierarchical clustering. An advantage of hierarchical clustering compared to the direct visualization methods is that a high dimensionality of the data set, represented by a large number of genes and samples, is reduced to a convenient two-dimensional representation of subject similarities.[55,56] HCA is the classification of similar objects into different groups, or more precisely, the partitioning of a data set into subsets, called clusters. The goal is to create clusters that share some common trait that is a matchable expression pattern. Hierarchical clustering can be performed either for the genes (comparing biological sample expression profiles) or for the biological samples (comparing

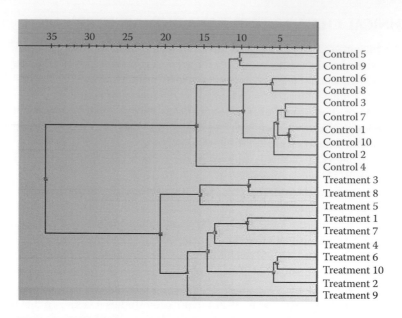

FIGURE 18.1　HCA of a set of 10 significantly regulated genes between 10 untreated control animals and 10 animals treated with steroid hormones.

gene-expression profiles). HCA uses distance measure to identify pairs of animals showing high similarity based on the expression of different genes (Figure 18.1). Within many steps the animals with the highest similarity are merged in a cluster, and then the process is repeated. The result of the analysis is a tree dendrogram displaying the distances between the individuals based on the expression of genes.[56,57]

Using hierarchical clustering a tree dendrogram can either be designed for the measured genes (in all samples) or for the samples (based on all measured expressed genes). Using a heatmap analysis these two classifications can be combined, resulting in a two-dimensional color-coded description of the whole experimental matrix. It displays in a very convenient way all samples versus gene expression where each tile is colored with a different intensity according to all available data. Figure 18.2 shows a heatmap created from a set of 10 regulated genes in 20 animals (10 untreated control calves and 10 calves treated with steroid hormones). In both figures applying clustering methods, a clear separation of the two treatment groups underlying hormone-dependent physiological expression pattern changes upon selected biomarkers are visualized. In the two-dimensional heatmap, additionally the gene clusters with comparable regulation kinetics are obvious.

18.4.2　Principal Components Analysis

A further useful biostatistical and visualization method to group data is principal component analysis (PCA). PCA is a mathematical procedure that converts a multidimensional data set into a lower number of variables called principal components (PC).[57,58] The classification of the genes is based on unscaled Cq values and the overall changes of the gene-expression magnitudes.[59] The first principal component (PC1) represents the most significant PC, while gene-expression changes or variations in expression profiles are contained in the subsequent PCs. Inspecting the PC2, we see that treated and untreated individuals form two clusters that reflect the common biological functions and physiological processes of its members (Figure 18.3). Each analyzed animal will be represented by one spot which results from diminishing all significantly regulated genes of one specific sample to two PC. Variance from experimental study design conditions is

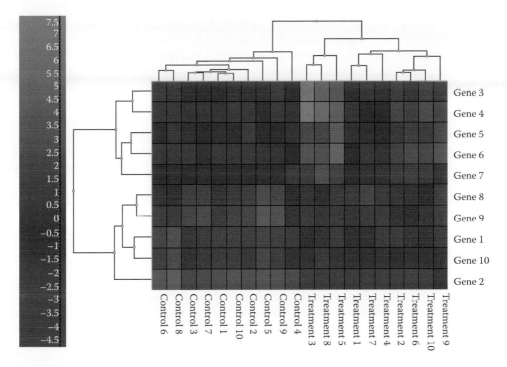

FIGURE 18.2 **(See color insert.)** Heatmap analysis of a set of 10 significantly regulated genes between 10 untreated control animals and 10 animals treated with steroid hormones.

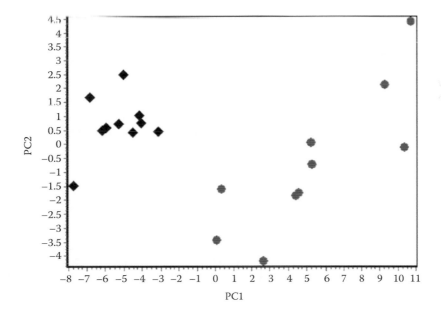

FIGURE 18.3 PCA of a set of 10 significantly regulated genes between 10 untreated control animals (black diamonds) and 10 animals, treated with steroid hormones (grey dots).

expected to be systematic, while confounding variance is expected to be random. Since the last PCs are derived from a very small amount of information, they can be considered to include noise or random information and can, therefore, be ignored. In this way, PCA can be a very efficient tool to separate systematic effects from noise.[56]

PCA has effectively been employed to visualize a treatment pattern in bovine tissues.[24–26,48,60] In bovine liver, PCA results obtained from mRNA and miRNA expression in combination showed a better separation between the groups than by employing the results from each individual transcript type. In Figure 18.3, a PCA of a set of 10 significantly regulated genes between 10 untreated control animals and 10 animals treated with steroid hormones is displayed. The clear separation of the two treatment groups indicate that PCA is a good tool for pattern recognition in gene-expression biomarker research.

The advantage of the PCA in comparison to the HCA methods is obvious. PCA allows a much clearer recognition and more precise differentiation of the treatment groups, because the commonalities in gene-expression pattern are visualized by the symbol interspaces in two dimensions.

18.5   SOFTWARE TOOLS

There are multiple software tools available to perform HCA, heatmaps, or PCA, either as stand-alone software or as packages freely available on the Internet.

The "Genex" software package offers a lot of tools to analyze mRNA and miRNA expression data in a correct and MIQE compliant way (according to Ref. [61]). Genex with its multiple functions helps to find and validate stable biomarkers (MultiD, Gothenburg, Sweden). More about the software, its functionality and the application of multidimensional data analysis is explained on the programmer's webpage (www.multid.com).

The "Genevestigator" software tool (www.genevestigator.com) aims at detecting specific patterns of expression in a multi-dimensional expression space by including a very large number of conditions processed from thousands of microarrays. The intuitive interface allows users easily to obtain lists of potential biomarker genes that can then be further validated using Genevestigator tools or in the laboratory. The classical clustering method (HCA) is for the grouping of genes according to their global pattern. Genevestigator provides several tools for clustering array data or meta-profiles.[6,7] The gene similarity is measured across all arrays or conditions. A dendrogram is applied to the clustered matrix and indicates relationships between clusters. More advanced biclustering is a method that identifies groups of genes that have similar profiles in a subset of conditions, irrespective of their profile similarity in the other conditions. Recent studies have shown that biclustering performs better than methods that require similarity over all conditions.[6,7]

A further method to discover and validate expressed biomarkers is to use R programming language, summarized in the "Bioconductor" project database (www.bioconductor.org). Bioconductor is an open source, open development software project to provide tools for the analysis and comprehension of high-throughput genomic data.[62] There are multiple packages and meta-data packages available, which provide the analysis of various data sources, for example, DNA, mRNA, miRNA, transcriptomics, microarray, real-time RT-PCR, sequence, or SNP. The broad goal is to provide widespread access to a full range of powerful statistical and graphical methods for the analysis of transcriptomics data. For real-time PCR data analysis and normalization a bundle of projects are available, for example, "HTqPCR," "qpcrNorm," "SLqPCR," or "ddCT" (summarized in "The qPCR library—Analysis of real-time PCR data using R"—http://www.dr-spiess.de/qpcR.html).[63] Further specialized packages for multidimensional expression analysis, PCA, HCA, or biomarker discovery are available in the database, for example, "BioMark" or "optBiomarker" project (http://www.bioconductor.org/help/search/index.html?q=biomarker).

Comparing the software packages, "Genex" and "Genevestigator" are the more user friendly, because they are working on a windows-based environment. The "Bioconductor" packages expect an advanced operator who is able to handle and modify the text-based input script lines. Further, the graphical output of the results is limited in style and generally very simple in appearance.

18.6 CONCLUSION

We have described how biomarkers can be discovered from quantitative mRNA and miRNA transcript studies using RT-qPCR data obtained from various hormone treatment experiments in farm animals. The application of new transcriptomics technologies has resulted in the discovery of new, regulated transcripts and yielded potential biomarkers or biomarker patterns. But one critical point in biomarker discovery is the heterogeneity in the population and the variance of the biological samples itself. The application of integrative functional informatics as a novel direction of biomarker identification and validation seems to be very promising. Hence, the quantity of analyzed transcripts, on various levels and in multiple organs, in combination with the applied statistical method will have an impact on the informative value and the validity of the biomarkers.

Despite this enormous potential, so far none of the biomarker candidates described is included in veterinary screening or routine diagnostics. There is still a lack of validation of these discovered candidates in multiple organisms, in various environments, under changing conditions, and for veterinary research in multiple breeds. The existence of potential biomarkers is opening new insights in molecular diagnostics, an auspicious track to individualized treatment, or translated to human studies to future personalized medicine.

REFERENCES

1. Ilyin, SE, Belkowski, SM, Plata-Salaman, CR. 2004. Biomarker discovery and validation: Technologies and integrative approaches. *Trends Biotechnol.* 22:8, 411–6.
2. Hulka BS. Overview of biological markers. 1990. In: *Biological Markers in Epidemiology.* eds Hulka BS, Griffith JD, Wilcosky TC. Oxford University Press, Oxford, UK.
3. Biomarkers Definitions Working Group. 2001. Biomarkers and surrogate endpoints: Preferred definitions and conceptual framework. *Clin. Pharmacol. Ther.* 69:3, 89–95.
4. Mayeux, R. 2004. Biomarkers: Potential uses and limitations. *NeuroRx.* 1:2, 182–8.
5. Naylor, S. 2003. Biomarkers: Current perspectives and future prospects. *Expert. Rev. Mol. Diagn.* 3:5, 525–9
6. Hruz, T, Laule, O, Szabo, G, Wessendorp, F, Bleuler, S, Oertle, L, Widmayer, P, Gruissem, W, Zimmermann, P. 2008. Genevestigator v3: A reference expression database for the meta-analysis of transcriptomes. *Adv. Bioinformatics.* 2008, 420747.
7. Zimmermann, P, Laule, O, Schmitz, J, Hruz, T, Bleuler, S, Gruissem, W. 2008. Genevestigator transcriptome meta-analysis and biomarker search using rice and barley gene expression databases. *Mol. Plant* 1:5, 851–7.
8. Sandvik, AK, Alsberg, BK, Norsett, KG, Yadetie, F, Waldum, HL, Laegreid, A. 2006. Gene expression analysis and clinical diagnosis. *Clin. Chim. Acta* 363:1–2, 157–64.
9. Robosky, LC, Robertson, DG, Baker, JD, Rane, S, Reily, MD. 2002. *In vivo* toxicity screening programs using metabonomics. *Comb. Chem. High Throughput. Screen.* 5:8, 651–62.
10. Wang, Z, Gerstein, M, Snyder, M. 2009. RNA-Seq: A revolutionary tool for transcriptomics. *Nat. Rev. Genet.* 10:1, 57–63.
11. Marioni, JC, Mason, CE, Mane, SM, Stephens, M, Gilad, Y. 2008. RNA-seq: An assessment of technical reproducibility and comparison with gene expression arrays. *Genome Res.* 18:9, 1509–17.
12. Lu, J, Getz, G, Miska, EA, Alvarez-Saavedra, E, Lamb, J, Peck, D, Sweet-Cordero, A et al. 2005. MicroRNA expression profiles classify human cancers. *Nature* 435:7043, 834–8.
13. Wang, K, Zhang, S, Marzolf, B, Troisch, P, Brightman, A, Hu, Z, Hood, LE, Galas, DJ. 2009. Circulating microRNAs, potential biomarkers for drug-induced liver injury. *Proc. Natl. Acad. Sci. USA* 106:11, 4402–7.
14. Quackenbush, J. 2006. Microarray analysis and tumor classification. *N. Engl. J. Med.* 354:23, 2463–72.

15. Seo, D, Ginsburg, GS. 2005. Genomic medicine: Bringing biomarkers to clinical medicine. *Curr. Opin. Chem. Biol.* 9:4, 381–6.
16. Johnson, BJ, White, ME, Hathaway, MR, Christians, CJ, Dayton, WR. 1998. Effect of a combined tren-bolone acetate and estradiol implant on steady-state IGF-I mRNA concentrations in the liver of wethers and the longissimus muscle of steers. *J. Anim. Sci.* 76:2, 491–7.
17. Pfaffl, MW, Daxenberger, A, Hageleit, M, Meyer, HH. 2002. Effects of synthetic progestagens on the mRNA expression of androgen receptor, progesterone receptor, oestrogen receptor alpha and beta, insulin-like growth factor-1 (IGF-1) and IGF-1 receptor in heifer tissues. *J. Vet. Med. A Physiol Pathol. Clin. Med.* 49:2, 57–64.
18. White, ME, Johnson, BJ, Hathaway, MR, Dayton, WR. 2003. Growth factor messenger RNA levels in muscle and liver of steroid-implanted and nonimplanted steers. *J. Anim. Sci.* 81:4, 965–72.
19. Reiter, M, Walf, VM, Christians, A, Pfaffl, MW, Meyer, HH. 2007. Modification of mRNA expression after treatment with anabolic agents and the usefulness for gene expression-biomarkers. *Anal. Chim. Acta* 586:1–2, 73–81.
20. Beermann DH. 2002. Beta-adrenergic receptor agonist modulation of skeletal muscle growth. *J. Anim. Sci.* 80:18–23.
21. Pfaffl, MW, Lange, IG, Meyer, HH. 2003. The gastrointestinal tract as target of steroid hormone action: Quantification of steroid receptor mRNA expression (AR, ERalpha, ERbeta and PR) in 10 bovine gastro-intestinal tract compartments by kinetic RT-PCR. *J. Steroid Biochem. Mol. Biol.* 84:2–3, 159–66.
22. Sato, S, Nomura, S, Kawano, F, Tanihata, J, Tachiyashiki, K, Imaizumi, K. 2008. Effects of the beta2-agonist clenbuterol on beta1- and beta2-adrenoceptor mRNA expressions of rat skeletal and left ventricle muscles. *J. Pharmacol. Sci.* 107:4, 393–400.
23. Toffolatti, L, Rosa, GL, Patarnello, T, Romualdi, C, Merlanti, R, Montesissa, C, Poppi, L, Castagnaro, M, Bargelloni, L. 2006. Expression analysis of androgen-responsive genes in the prostate of veal calves treated with anabolic hormones. *Domest. Anim. Endocrinol.* 30:1, 38–55.
24. Riedmaier, I, Tichopad, A, Reiter, M, Pfaffl, MW, Meyer, HH. 2009. Identification of potential gene expression biomarkers for the surveillance of anabolic agents in bovine blood cells. *Anal. Chim. Acta* 638:1, 106–13.
25. Riedmaier, I, Becker, C, Pfaffl, MW, Meyer, HH. 2009. The use of omic technologies for biomarker development to trace functions of anabolic agents. *J. Chromatogr. A* 1216:46, 8192–9.
26. Riedmaier, I, Tichopad, A, Reiter, M, Pfaffl, MW, Meyer, HH. 2009. Influence of testosterone and a novel SARM on gene expression in whole blood of *Macaca fascicularis*. *J. Steroid Biochem. Mol. Biol.* 114:3–5, 167–73.
27. Riedmaier, I, Reiter, M, Tichopad, A, Pfaffl, MW, Meyer, HH. 2011. The potential of bovine vaginal smear for biomarker development to trace the misuse of anabolic agents. *Exp. Clin. Endocrinol. Diabetes* 119:2, 86–94.
28. Reiter, M, Pfaffl, MW, Schonfelder, M, Meyer, HH. 2008. Gene expression in hair follicle dermal papilla cells after treatment with stanozolol. *Biomark. Insights.* 4:1–8.
29. Reiter, M, Luederwald S, Pfaffl, MW, Meyer, HH. 2008. First steps towards a new screening method for anabolic adrogenic agents in human hair follicle. *Doping Journal* 5:3.
30. Reiter, M, Pfaffl, MW, Schoenfelder M, Meyer, HH. 2009. Gene expression in hair follicle dermal papilla cells after treatment with stanozolol. *Biomark. Insights.* 4:1–8.
31. Lee, RC, Feinbaum, RL, Ambros, V. 1993. The *C. elegans* heterochronic gene lin-4 encodes small RNAs with antisense complementarity to lin-14. *Cell* 75:5, 843–54.
32. Esquela-Kerscher, A, Slack, FJ. 2006. Oncomirs—MicroRNAs with a role in cancer. *Nat. Rev. Cancer* 6:4, 259–69.
33. Pandey, AK, Agarwal, P, Kaur, K, Datta, M. 2009. MicroRNAs in diabetes: Tiny players in big disease. *Cell Physiol. Biochem.* 23:4–6, 221–32.
34. Farazi, TA, Spitzer, JI, Morozov, P, Tuschl, T. 2011. miRNAs in human cancer. *J. Pathol.* 223:2, 102–15.
35. Becker, C, Hammerle-Fickinger, A, Riedmaier, I, Pfaffl, MW. 2010. mRNA and microRNA quality control for RT-qPCR analysis. *Methods* 50:4, 237–43.
36. Benes, V, Castoldi, M. 2010. Expression profiling of microRNA using real-time quantitative PCR, how to use it and what is available. *Methods* 50:4, 244–9.
37. Bartels, CL, Tsongalis, GJ. 2009. MicroRNAs: Novel biomarkers for human cancer. *Clin. Chem.* 55:4, 623–31.
38. Heneghan, HM, Miller, N, Kerin, MJ. 2010. MiRNAs as biomarkers and therapeutic targets in cancer. *Curr. Opin. Pharmacol.* 10:5, 543–50.

39. Laterza, OF, Lim, L, Garrett-Engele, PW, Vlasakova, K, Muniappa, N, Tanaka, WK, Johnson, JM et al. 2009. Plasma MicroRNAs as sensitive and specific biomarkers of tissue injury. *Clin. Chem.* 55:11, 1977–83.
40. Kosaka, N, Iguchi, H, Ochiya, T. 2010. Circulating microRNA in body fluid: A new potential biomarker for cancer diagnosis and prognosis. *Cancer Sci.* 101:10, 2087–92.
41. Kosaka, N, Izumi, H, Sekine, K, Ochiya, T. 2010. MicroRNA as a new immune-regulatory agent in breast milk. *Silence.* 1:1, 7.
42. Kroh, EM, Parkin, RK, Mitchell, PS, Tewari, M. 2010. Analysis of circulating microRNA biomarkers in plasma and serum using quantitative reverse transcription-PCR (qRT-PCR). *Methods* 50:4, 298–301.
43. Mitchell, PS, Parkin, RK, Kroh, EM, Fritz, BR, Wyman, SK, Pogosova-Agadjanyan, EL, Peterson, A et al. 2008. Circulating microRNAs as stable blood-based markers for cancer detection. *Proc. Natl. Acad. Sci. USA* 105:30, 10513–8.
44. Lawrie, CH, Gal, S, Dunlop, HM, Pushkaran, B, Liggins, AP, Pulford, K, Banham, AH et al. 2008. Detection of elevated levels of tumour-associated microRNAs in serum of patients with diffuse large B-cell lymphoma. *Br. J. Haematol.* 141:5, 672–5.
45. Ng, EK, Chong, WW, Jin, H, Lam, EK, Shin, VY, Yu, J, Poon, TC, Ng, SS, Sung, JJ. 2009. Differential expression of microRNAs in plasma of patients with colorectal cancer: A potential marker for colorectal cancer screening. *Gut* 58:10, 1375–81.
46. Thai, TH, Calado, DP, Casola, S, Ansel, KM, Xiao, C, Xue, Y, Murphy, A et al. 2007. Regulation of the germinal center response by microRNA-155. *Science* 316:5824, 604–8.
47. Bala, S, Marcos, M, Szabo, G. 2009. Emerging role of microRNAs in liver diseases. *World J. Gastroenterol.* 15:45, 5633–40.
48. Becker, C, Riedmaier, I, Reiter, M, Tichopad, A, Pfaffl, MW, Meyer, HH. 2011. Changes in the miRNA profile under the influence of anabolic steroids in bovine liver. *Analyst* 136:6, 1204–9.
49. Meyer, SU, Pfaffl, MW, Ulbrich, SE. 2010. Normalization strategies for microRNA profiling experiments: A "normal" way to a hidden layer of complexity? *Biotechnol. Lett.* 32:12, 1777–88.
50. Schmittgen, TD, Lee, EJ, Jiang, J, Sarkar, A, Yang, L, Elton, TS, Chen, C. 2008. Real-time PCR quantification of precursor and mature microRNA. *Methods* 44:1, 31–8.
51. Mestdagh, P, Van, VP, De, WA, Muth, D, Westermann, F, Speleman, F, Vandesompele, J. 2009. A novel and universal method for microRNA RT-qPCR data normalization. *Genome Biol.* 10:6, R64.
52. Masuda, N, Ohnishi, T, Kawamoto, S, Monden, M, Okubo, K. 1999. Analysis of chemical modification of RNA from formalin-fixed samples and optimization of molecular biology applications for such samples. *Nucleic Acids Res.* 27:22, 4436–43.
53. Andreasen, D, Fog, JU, Biggs, W, Salomon, J, Dahslveen, IK, Baker, A, Mouritzen, P. 2010. Improved microRNA quantification in total RNA from clinical samples. *Methods* 50:4, S6–9.
54. Li, J, Smyth, P, Flavin, R, Cahill, S, Denning, K, Aherne, S, Guenther, SM, O'Leary, JJ, Sheils, O. 2007. Comparison of miRNA expression patterns using total RNA extracted from matched samples of formalin-fixed paraffin-embedded (FFPE) cells and snap frozen cells. *BMC. Biotechnol.* 7:36.
55. Lee, G, Rodriguez, C, Madabhushi, A. 2008. Investigating the efficacy of nonlinear dimensionality reduction schemes in classifying gene and protein expression studies. *IEEE/ACM. Trans. Comput. Biol. Bioinform.* 5:3, 368–84.
56. Bergkvist, A, Rusnakova, V, Sindelka, R, Garda, JM, Sjogreen, B, Lindh, D, Forootan, A, Kubista, M. 2010. Gene expression profiling—Clusters of possibilities. *Methods* 50:4, 323–35.
57. Beyene, J, Tritchler, D, Bull, SB, Cartier, KC, Jonasdottir, G, Kraja, AT, Li, N, et al. 2007. Multivariate analysis of complex gene expression and clinical phenotypes with genetic marker data. *Genet. Epidemiol.* 31 Suppl 1:S103–9.
58. Kubista, M, Andrade, JM, Bengtsson, M, Forootan, A, Jonak, J, Lind, K, Sindelka, R, et al. 2006. The real-time polymerase chain reaction. *Mol. Aspects Med.* 27:2–3, 95–125.
59. Vandesompele, J, Kubista, M, Pfaffl, MW. 2009. Reference gene validation software for improved normalization. In: *Real-Time PCR: Current Technology and Applications.* eds Logan J, Edwards K, Saundes N. Caister Academic Press, Norfolk, UK.
60. Becker, C, Riedmaier, I, Reiter, M, Tichopad, A, Groot, MJ, Stolker, AA, Pfaffl, MW, Nielen, MF, Meyer, HH. 2011. Influence of anabolic combinations of an androgen plus an estrogen on biochemical pathways in bovine uterine endometrium and ovary. *J. Steroid Biochem. Mol. Biol* 125(3):192–201.
61. Bustin, SA, Benes, V, Garson, JA, Hellemans, J, Huggett, J, Kubista, M, Mueller, R, et al. 2009. The MIQE guidelines: Minimum information for publication of quantitative real-time PCR experiments. *Clin. Chem.* 55:4, 611–22.

62. Gentleman, RC, Carey, VJ, Bates, DM, Bolstad, B, Dettling, M, Dudoit, S, Ellis, B, et al. 2004. Bioconductor: Open software development for computational biology and bioinformatics. *Genome Biol.* 5:10, R80.

63. Ritz, C, Spiess, AN. 2008. qpcR: An R package for sigmoidal model selection in quantitative real-time polymerase chain reaction analysis. *Bioinformatics* 24:13, 1549–51.

19 Optimization of Quantitative Real-Time PCR for Studies in Cartilage Mechanobiology

Oto Akanji, Donald Salter, and Tina Chowdhury

CONTENTS

19.1 INTRODUCTION

Articular cartilage is a mechanosensitive tissue containing chondrocytes that respond to biomechanical signals by upregulating synthetic activity and/or production of inflammatory mediators.[1–3] This process, termed mechanotransduction, refers to the way in which chondrocytes respond to mechanical loading and convert biomechanical signals into a cellular response. The mechanotransduction process, in turn, influences the composition and structure of the extracellular matrix, enabling chondrocytes to adapt to their physical environment. Over nearly three decades, researchers of cartilage mechanobiology have identified several key biochemical pathways that are involved in the signal transduction process.[4] However, the identity of the mechanosensors and the way in which they contribute to changes in gene expression are complicated by the types of *in vitro* models and bioreactor systems used to study the cell signaling process.[5–6] It is now widely accepted that anabolic and catabolic transcriptional activities in chondrocytes are tightly regulated by distinct mechanical signals.[7–8] The general consensus, derived from previous *in vitro* mechanical loading studies, is that static compression which mimics excessive loading or an injurious response, inhibits some gene expression and biosynthesis of matrix proteins.[9–12] In contrast, dynamic compression, which could be interpreted as a physiological environment of cartilage, causes an increase in expression of extracellular matrix components.[13–15] Several research groups, including our own, have shown a strong interplay between the signal transduction pathways induced by both mechanical loading and interleU.K.in-1β (IL-1β) in chondrocytes seeded in 3D agarose gels and in monolayer culture.[4,16–20] There is recent evidence that demonstrates the involvement of the mitogen-activated protein kinase (MAPK) and nuclear factor kappa B (NFκB) pathways in mediating the transcriptional response of chondrocytes to mechanical loading and/or IL-1β.[21–23] Nevertheless, the molecular mechanisms underlying the specific mechanotransduction pathways are complex and will vary depending on the pathological environment of the tissue.[1,7–8] Furthermore, the nature of the mechanical stimulus will additionally determine how mechanical loading controls the activation or inhibition of anabolic- and catabolic-associated genes in chondrocytes.[24] For instance, frequent bursts of intermittent compression for longer time periods favored expression of matrix synthesis whereas shorter bursts of intermittent compression had the opposite effect. A similar response was described by other research groups who showed that gene expression associated with chondrocyte anabolic and catabolic activities were dependent on the duration and type of compression regime employed.[11,15] Ultimately, elucidation of the intracellular pathways in response to physiologically relevant mechanical loading conditions will therefore facilitate the successful identification of mechanotherapeutic agents to treat degenerative joint disorders like osteoarthritis. These types of therapies are termed chondroprotective agents because they are focused on preventing the loss of extracellular matrix and/or regeneration of damaged tissue. However, identification of the relevant signaling events has proved difficult due to cross talk with other signaling pathways and in some instances led to contradictory results. The differences in gene-expression data are in part due to the wide range of *in vitro* models and bioreactors available as tools to explore mechanotransduction pathways. For instance, it is relatively easy to extract high yields of total RNA from isolated cells cultured in monolayer when compared to 3D models such as agarose. The latter is complicated by the low cell density and presence of large amounts of agarose, leading to inadequate RNA yields and less robust PCR assay performance. However, one of the key challenges encountered in cartilage mechanobiology research is the analysis of real-time quantitative PCR (qPCR) data for monitoring mRNA gene-expression profiles of chondrocyte anabolic- and catabolic-associated genes. The results can be significantly different for studies with multiple experimental test conditions involving mechanical loading, chemical and time-dependent variables since the data are dependent on the appropriate selection of normalization strategy. Thus, inadequate data normalization and errors in statistical analysis makes it difficult for the researcher to obtain meaningful results and compare or even repeat experiments from previously published studies. Additional drawbacks include poor optimization and validation of conditions for qPCR and unreliable choice of reference genes for each experimental test condition. This chapter will therefore contain a discussion of the steps required to analyze qPCR data for mechanotransduction studies utilizing a well-characterized *in vitro* model and bioreactor system; namely the chondrocyte/agarose model in conjunction with a compressive strain bioreactor. The time-dependent effect of IL-1β and mechanical loading on the relative expression of well-known chondrocyte anabolic (aggrecan, collagen type II) and catabolic (iNOS, COX-2) genes will be presented using a comparative quantification cycle (C_q) PCR approach, as described by others.[25] In addition, data normalization and validation will be discussed using relevant statistical methods.

19.2 METHODS

19.2.1 Chondrocyte/Agarose Culture and Experimental Conditions

Chondrocytes were isolated from the bovine metacarpalphalangeal joint, seeded in 3% agarose type VII at a cell concentration of 4×10^6 cells/mL and equilibrated in culture for 24 h, as previously described.[26,27] Chondrocyte/agarose constructs were subsequently cultured under free-swelling conditions in 1 mL DMEM + 1× ITS supplemented with 0 or 10 ng/mL IL-1β in the presence and absence of 10 μM p38 MAPK inhibitor (SB203580) for 0, 0.75, 1.5, 3, 6, 12, 24, and 48 h. In separate experiments, a compressive-strain bioreactor was used to apply dynamic compression (15%, 1 Hz frequency) to constructs cultured with 0 or 10 ng/mL IL-1β and/or 10 μM p38 MAPK inhibitor for 6, 12, and 48 h, as previously described.[17] At the end of each experiment, the constructs were snap-frozen in liquid nitrogen and stored at −70°C before mRNA extraction. Total RNA was isolated from individual constructs using protocols described in the QIAquick® Spin gel extraction and RNeasy® kits (Qiagen, West Sussex, U.K.). Following manufacturer's instructions, Ambion's DNA-*free* DNase treatment and removal reagents were used to eliminate any contaminating DNA from the RNA sample (Ambion Applied Biosystems, Warrington, U.K.). RNA was quantified on the Nanodrop ND-1000 spectrophotometer (LabTech, East Sussex, U.K.) and stored in 40 μL RNasefree water at −80°C until reverse transcription could be performed. RNA quality was determined by agarose gel electrophoresis and samples without degradation selected for study. Identical amounts of total RNA were used for reverse transcription using the manufacturer's protocols from the Stratascript™ First-Strand cDNA synthesis kit (Agilent, Stratagene, Amsterdam, the Netherlands). Briefly, 200 ng of total RNA was reverse transcribed in a 20 μL reaction volume using the manufacturer-supplied oligo(dT) primers. To screen for contamination of reagents or false PCR amplification by genomic DNA, the following controls were included in the PCR experiments: No reverse transcriptase (NoRT) controls were prepared by omitting the Stratascript™ reverse transcriptase; PCR controls for each sample were prepared with identical reaction mixtures except for the addition of the template (no template control, NTC). All cDNA preparations were diluted 1:10 prior to carrying out the real-time qPCR reaction.

19.2.2   OPTIMIZATION OF QUANTITATIVE REAL-TIME PCR ASSAYS

To maximize performance of qPCR assays, all assay conditions were optimized and validated. Primers and Molecular Beacons used in qPCR experiments were designed using the Beacon Designer software (Premier Biosoft Interntaional, California), as previously described[17] and synthesized by Sigma Genosys Limited, (Cambridge, U.K.). For primer optimization, dilutions of forward and reverse primers ranging from 50 to 600 nM were prepared following guidelines in Nolan et al [28] and qPCR was run in Brilliant SYBR Green I Master Mix (Stratagene, Amsterdam, the Netherlands). All reactions were run in duplicate on the qPCR Mx3000P instrument using a three-step protocol which was suitable for SYBR green I chemistry. For probe optimization, dilutions ranging from 50 to 600 nm[33] were prepared in Brilliant RT-qPCR Master Mix containing optimal forward and reverse primer concentrations. Reactions were run in duplicate using a three-step protocol, which was suitable for primers coupled with Molecular Beacon probe detection. The real-time PCR efficiencies of amplification for each primer pair and/or probe were derived from three standard curves of C_q against log target concentration of the template. The slope and R^2 values were determined and the amplification efficiencies (E) were defined according to the following relationship: $E = 10(−1/$ slope). Validation of primer/template systems was achieved by generating a standard curve using a 10-fold serial dilution of cDNA sample that represented the untreated control samples. In addition, the expression stability of the validated common reference gene, GAPDH was determined for all treatment conditions using the same concentration of target DNA.

19.2.3   QUANTITATIVE REAL-TIME PCR ASSAYS

Real-time qPCR assays coupled with Molecular Beacons were performed in 25 μL reaction mixtures containing 1 μL diluted cDNA, 12.5 μL Brilliant® qRT-PCR Master Mix, optimal primer pairs, and probes as listed in Table 19.1 and nuclease-free PCR grade water to 25 μL. Reactions for each sample were run in duplicate and efficiencies for the target sample were derived from a standard curve. Thermocycling conditions comprised an initial polymerase activation step at 95°C for 10 min, followed by 35 cycles at 95°C for 30 s, at 55°C for 1 min, and at 72°C for 1 min. In addition, NTC and NoRT controls were included in each PCR assay.

TABLE 19.1

Description of the Beacon Designer Sequences Used to Quantify Gene Expression and Real-Time Reaction Efficiencies of qPCR Assays

Gene	Accession No.	Sequences	nM	Efficiency
iNOS	U14640	Probe: 5′-FAM-<u>CGCGATC</u>CCTGCTTGGTGGCGAAG ATGAGC<u>GATCGCG</u>-DABCYL-3′	200	1.98 ± 0.06
		Forward: 5′-GTAACAAAGGAGATAGAAACAACAGG-3′	200	
		Reverse: 5′-CAGCTCCGGGCGTCAAAG-3′	200	
COX-2	AF031698	Probe: 5′-FAM-<u>CGCGATC</u>GTCAGAAATTCGGGTGT GGTACAGTT<u>GATCGCG</u>-DABCYL-3′	200	1.99 ± 0.03
		Forward: 5′-CGAGGTGTATGTATGAGTGTAGG-3′	300	
		Reverse: 5′-GTTGGGAGTGGGTTTCAGG-3′	300	
Aggrecan	U76615	Probe: 5′-FAM-<u>CGCGATC</u>CACTCAGCGAGTTGTCA GGTTCTGA<u>GATCGCG</u>-DABCYL-3′	200	1.97 ± 0.05
		Forward: 5′-TGGTGTTTGTGACTCTGAGG-3′	100	
		Reverse: 5′-GATGAAGTAGCAGGGGATGG-3′	200	
Collagen type II	X02420	Probe:5′-FAM-<u>CGCGAT</u>GCGTCAGGTCAGGTCAGC CAT<u>ATCGCG</u>-DABCYL-3′	200	2.00 ± 0.05
		Forward: 5′-AAACCCGAACCCAGAACC-3′	100	
		Reverse: 5′-AAGTCCGAACTGTGAGAGG-3′	100	
GAPDH	U85042	Probe: 5′-HEX-<u>CGCGATC</u>CACCATCTTCCAGGAGC GAGATCC<u>GATCGCG</u>-DABCYL-3′	200	2.03 ± 0.01
		Forward: 5′-TTCAACGGCACAGTCAAGG-3′	200	
		Reverse: 5′-TTCAACGGCACAGTCAAGG-3′	200	

Note: Primer concentrations were determined using a matrix of forward and reverse primers ranging from 50 to 600 nM. Combinations which gave the lowest C_q and the highest ΔRn values were selected and used to determine optimal probe concentration. Amplification efficiencies were determined from three standard curves using a cDNA sample which represented the untreated control. The amplification efficiencies for qPCR assays with optimal primer and probe concentrations are shown above. Standard curves confirmed that the slope was between −3.2 and −3.5 and the R^2 value exceeded 0.9998. In addition, the Beacon Designer software was used to design forward and reverse primer and probe sequences for Molecular Beacon applications and were synthesized by Sigma Genosys Ltd, Cambridge, U.K.. Secondary structures were avoided using the Mfold program and sequences were analyzed using BLAST to verify specificity. Probes contain FAM or HEX as a 5′-reporter dye and DABCYL as 3′-quencher. Note that the arm sequences are underlined.

19.2.4 DATA ANALYSIS AND NORMALIZATION

Fluorescence data for qPCR experiments were collected on the Mx3000P qPCR instrument (Agilent, Stratagene, Amsterdam, the Netherlands) and ratios calculated by normalizing each target to the reference gene and to the calibrator sample using a comparative C_q approach. For the free-swelling experiments, the difference in cycle threshold (ΔC_q) for the target was calculated by subtracting the mean C_q value for the untreated, time zero control (calibrator) from the C_q value of the target sample. For the mechanical loading studies, ΔC_q was calculated by subtracting the mean C_q value for the unstrained, no treatment control (calibrator) from the target sample. The ΔC_q of the target was then normalized to the ΔC_q of the reference gene, namely GAPDH. Thus, for each sample, the ratio of the relative expression level of target ΔC_q and reference ΔC_q was calculated, as shown by Equation 19.1.[29] All data were subsequently analyzed using appropriate statistical methods (Section 19.3.3) to examine differences between multiple treatment groups.

$$\text{Ratio} = \frac{(1 + E_{\text{Target}})_{\text{Target}}^{\Delta C_q}(\text{MEAN Calibrator} - \text{Sample})}{(1 + E_{\text{Reference}})_{\text{Reference}}^{\Delta C_q}(\text{MEAN Calibrator} - \text{Sample})} \tag{19.1}$$

E represents the efficiencies obtained for the target and reference gene. $\Delta C_{\text{qtarget}}$ represents the difference in C_q values for the mean calibrator or sample for the target gene. $\Delta C_{\text{qreference}}$ represents the difference in C_q values for the mean calibrator or sample for the reference gene, GAPDH.

19.3 RESULTS

19.3.1 OPTIMIZATION AND VALIDATION OF qPCR CONDITIONS

Total RNA was quantified on the Nanodrop ND-1000 spectrophotometer and all samples contained amounts up to 2 µg from a single chondrocyte/agarose construct with A_{260}/A_{280} values ranging between 1.6 and 1.9. The yields and absorbance values were broadly similar for samples cultured under free-swelling and mechanical loading conditions and the levels did not vary significantly at the different time points examined. Each construct contained approximately 4×10^5 cells so the total yield found was typical for the 3D agarose model. RNA integrity was further verified by examining the 28S and 18S ribosomal RNA bands using a 1.5% denaturing agarose gel loaded with 2 µg RNA and stained with ethidium bromide after electrophoresis. gDNA contamination was checked by qPCR analysis of the NoRT and NTC control samples and since these were all negative, the RNA samples were regarded as suitable for qPCR assays.

To improve reaction sensitivity and ensure reproducibility of replicates for qPCR experiments, the concentration of primers and probes were optimized for each assay. The results of the primer optimization experiments are presented in Table 19.1. On close examination of the melt curves and amplification plots, samples which showed the lowest C_q values and highest end point fluorescence (ΔRn) were selected. The amplification plots were further inspected and showed no evidence of primer–dimer formation on the post-PCR melt curve and the absence of a fluorescence signal in the NTC samples. Standard curves confirmed the quality of the qPCR assay and revealed high amplification efficiencies close to 100% which covered the C_q range for all samples. The results of the probe optimization experiments fulfilled similar selection criteria with concentrations listed in Table 19.1.

19.3.2 SELECTION OF REFERENCE GENES

The expression of the reference gene used for normalization should remain constant under all experimental test conditions. In the current study, the time-dependent effect of IL-1β and dynamic compression on the expression of the common reference gene, GAPDH was examined in chondrocyte/agarose constructs. Data were exported as tab lineated files and analyzed in Microsoft Excel for 30 replicate samples for each test condition. The Grubbs test was used to examine the distribution of C_q values under all treatment conditions and data which showed more than a twofold difference were removed as outliers. Analysis was performed by displaying the data as Box and Whisker plots and is shown in Figure 19.1. In the absence and presence of IL-1β, GAPDH did not exceed more than 1 C_q value difference between free-swelling constructs or constructs subjected to dynamic compression and culture for up to 48 h. Incubation with the p38 MAPK inhibitor and/or stimulation with dynamic compression broadly did not influence

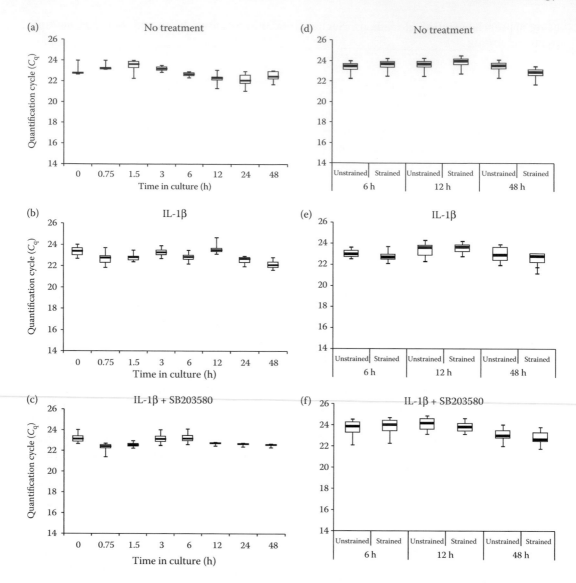

FIGURE 19.1 Distribution of quantification cycle (C_q) values for GAPDH by chondrocyte/agarose constructs cultured under free-swelling conditions (a–c) or in unstrained and strained constructs subjected to dynamic compression (d–f). Constructs were cultured with 0 or 10 ng/mL IL-1β ng/mL and/or 10 μM SB203580 for 48 h. The median values are shown as lines, the 25–75th percentile as boxes and range as whiskers for 30 replicates.

C_q values for GAPDH in IL-1β-treated constructs. In summary, the current data can be used to demonstrate that GAPDH remained stable with no significant changes detected for free-swelling or mechanical loading experiments (Figure 19.1). GAPDH expression was subsequently calculated by comparing the C_q values for each sample within each experiment with the mean C_q value for the untreated controls using the equation $E^{-\Delta C_q}$ (E is the efficiency and ΔC_q is the difference in C_q values). Further analysis using a two-way ANOVA with post hoc Bonferroni corrected t-test revealed that GAPDH expression was not significantly influenced by IL-1β, the p38 MAPH inhibitor and/or dynamic compression. The current data therefore support the use of GAPDH as a suitable reference gene in the chondrocyte/agarose model for the test conditions examined.

19.3.3 Data Normalization and Statistical Analysis

Several mathematical models have been developed to examine the expression ratio based on real-time PCR efficiency and C_q values of an unknown sample and a control sample. Models such as the $\Delta\Delta C_q$ method rely on calculating events between one control and one sample with presumed efficiencies of target and reference gene, using simple linear regression or multiple regression models for statistical analysis.[30] For a more reliable ratio calculation, the relative expression software tool (REST), enables pairwise comparisons of up to 16 values in a given control or sample group using the fixed reallocation randomization test to obtain the significance level.[29] It has the advantage of considering several interactions between gene, treatment, or sample number, making no distributional assumptions about the data organization and is therefore, highly suitable for gene comparisons between several targets and multiple reference genes.[31] Additionally, the purpose of the pairwise variation eliminates differences between two genes and therefore removes any nonspecific variation in normalized factors. However, the calculations are based on the group means and can only compare two treatment groups and not multiple variables of more than six experimental test conditions when using REST-MCS.[28] The current study involves a two-dimensional model (mechanical loading vs. treatment or time), so the only way to analyze more than two factors is via a two-way ANOVA. In this setting, an ANOVA coupled to the post hoc t-test, with Bonferroni correction factor, allows us to establish differences among the multiple experimental groups with no limit on test conditions (e.g., time, treatment, compression) or data points. However, before a parametric test could be performed, outliers were removed and the Gaussian distribution for all treatment groups examined. We found that the data did not follow a normal distribution with equal variance and was therefore unsuitable for further parametric tests. To compress the variance and ensure a normal distribution, ratios were logarithmicaly transformed (log base 2) and analyzed using a post-hoc t-test. Since the current data involved multiple comparisons, the p-value was corrected using a Bonferroni correction factor. This method allows us to establish pairwise comparisons among multiple experimental groups with no limit on test conditions or data points (see Figures 19.2 and 19.3 for examples of results).

19.4 CONCLUDING REMARKS

The data from the current study illustrate the importance of performing optimization experiments of qPCR conditions in studies involving cartilage mechanobiology as described by others in cartilage research.[32] The validation process will be dependent on the normalization strategy used to analyze the data and the expression stability of the reference gene used as a control to normalize the data.[31–33] Although real-time PCR is a powerful tool to compare relative transcriptional abundance, methods for rigorous data normalization and statistical analyses have lagged behind applications.[34] To date, there are no software tools, which consider the effect of more than two factors in multiple treatment groups greater than 6, in a statistically relevant manner. The strategies and problems used to analyze qPCR data are a subject of much debate.[35] Therefore, choosing the appropriate model for normalization and statistical analyses will be dependent on the experimental design and the scientific questions being asked.

ACKNOWLEDGMENTS

This work was supported by a project grant from the Wellcome Trust (073972). Dr. Chowdhury would like to thank Drs. Oto Akanji and Shah Arghandawi for qPCR data.

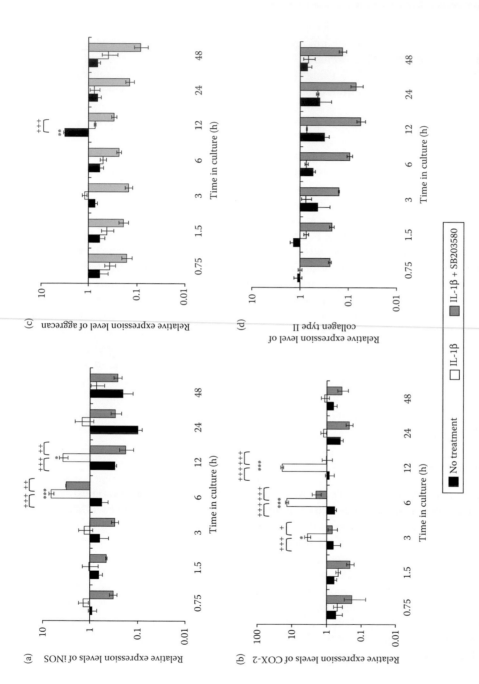

FIGURE 19.2 Time-dependent profile of IL-1β on iNOS (a), COX-2 (b), aggrecan (c), and collagen type II (d) expression by chondrocyte/agarose constructs cultured under free-swelling conditions with 0 or 10 ng/mLIL-1 β and/or 10 μM SB203580. Bars represent the mean and SEM of six replicates from three separate experiments. Ratios for the target were calibrated to the mean value at time = 0 and normalized to the reference gene, GAPDH. Ratios were log transformed and expressed on a logarithmic scale (arbitrary units). Two-way ANOVA with post hoc Bonferroni-corrected t-test was used to compare data under the different treatments, where * indicates $p < 0.05$ for comparisons between time zero with IL-1β; + indicates $p < 0.05$ for comparisons between untreated with IL-1β or IL-1β with IL-1β + SB203580.

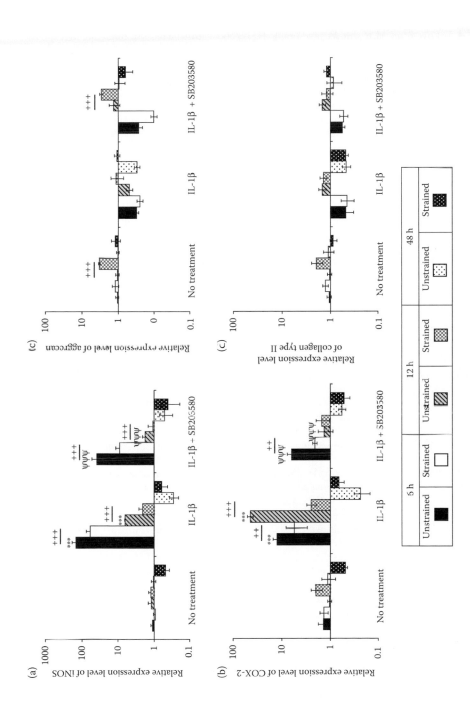

FIGURE 19.3 **(See color insert.)** Effect of dynamic compression (15%, 1 Hz) on iNOS (a), COX-2 (b), aggrecan (c), and collagen type II (d) expression in unstrained and strained constructs cultured under no-treatment conditions or with 10 ng/mL IL-1β and/or 10 μM SB203580 for 6, 12, and 48 h. Ratios for the target were calibrated to the mean unstrained (untreated) control value and normalized to GAPDH. Ratios were log transformed and expressed on a logarithmic scale (arbitrary units). Bars represent the mean and SEM of 16–18 replicates from four separate experiments. Two-way ANOVA with post hoc Bonferroni-corrected *t*-tests was used to compare data. Significant differences are indicated as follows: + indicates comparisons for unstrained versus strained; * indicates comparisons in unstrained samples for no treatment versus IL-1β; ψ indicates comparisons in unstrained samples for IL-1β versus IL-1β + SB203580.

REFERENCES

1. Guilak, F., Fermor, B., Keefe, F.J., Kraus, V.B., Olson, S.A., Pisetsky, D.S., Setton, L.A., and Weinberg, J.B. 2004. The role of biomechanics and inflammation in cartilage repair and injury. *Clin. Orthop. Relat. Res.* 423, 17–26.

2. Grodzinsky, A.J., Levenston, M.E., Jin, M., and Frank, E.H. 2000. Cartilage tissue remodeling in response to mechanical forces. *Annu. Rev. Biomed. Eng.* 2, 691–713.

3. Butler, D.L., Goldstein, S.A., and Guilak, F. 2000. Functional tissue engineering: The role of biomechanics. *J. Biomech. Eng.* 122, 570–575.

4. Millward-Sadler, S.J., and Salter, D.M. 2004. Integrin-dependent signal cascades in chondrocyte mechanotransduction. *Ann. Biomed. Eng.* 32, 435–446.

5. Freed, L.E., Guilak, F., Guo, X.E., Gray, M.L., Tranquillo, R., Holmes, J.W., Radisic, M., Sefton, M.V, Kaplan, D., and Vunjak-Novakovic, G. 2006. Advanced tools for tissue engineering: Scaffolds, bioreactors, and signaling. *Tissue Eng.* 12, 3285 3305.

6. Martin, I., Wendt, D., and Heberer, M. 2004. The role of bioreactors in tissue engineering. *Trends Biotechnol.* 22(2), 80–86.

7. Loeser, R.F. 2006. Molecular mechanisms of cartilage destruction: Mechanics, inflammatory mediators, and aging collide. *Arthritis Rheum.*, 54(5), 1357–1360.

8. Goldring, M.B. and Goldring, S.R. 2010. Articular cartilage and subchondral bone in the pathogenesis of osteoarthritis. *Ann. N Y Acad Sci.* 1192(1), 230–237.

9. Sah, R., Kim, Y.J., Doong, J.Y.H., Grodzinsky, A.J., Plaas, A.H.K., and Sandy, J.D. 1989. Biosynthetic response of cartilage explants to dynamic compression. *J. Orthop. Res.* 7, 619–636.

10. Bachrach, N.M., Valhmu, W.B., Stazzone, E., Ratcliffe, A., Lai, W.M., and Mow, V.C. 1995. Changes in proteoglycan synthesis of chondrocytes in articular cartilage are associated with the time-dependent changes in their mechanical environment. *J. Biomech.* 28, 1561–1569.

11. Valhmu, W.B., Stazzone, E.J., Bachrach, N.M., Saed-Nejad, F., Fischer, S.G., Mow, V.C., and Ratcliffe, A. 1998. Load-controlled compression of articular cartilage induces a transient stimulation of aggrecan gene expression. *Arch. Biochem. Biophys.* 353, 29–36.

12. Ragan, P.M., Badger, A.M., Cook, M., Chin, V.I., Gowen, M., Grodzinsky, A.J., and Lark, M.W. 1999. Down-regulation of chondrocyte aggrecan and type-II collagen gene expression correlates with increases in static compression magnitude and duration. *J. Orthop. Res.* 17, 836–842.

13. Blain, E.J., Mason, D.J., and Duance, V.C. 2001. The effect of cyclical compressive loading on gene expression in articular cartilage. *Biorheology* 40, 111–111.

14. Lee, J.H. and Fitzgerald, J.B., Dimicco, M.A., and Grodzinsky, A.J. 2005. Mechanical injury of cartilage explants causes specific time-dependent changes in chondrocyte gene expression. *Arthritis Rheum.* 52, 2386–2395.

15. De Croos, J.N., Dhaliwal, S.S., Grynpas, M.D., Pilliar, R.M., and Kandel, R.A. 2006. Cyclic compressive mechanical stimulation induces sequential catabolic and anabolic gene changes in chondrocytes resulting in increased extracellular matrix accumulation. *Matrix Biol.* 25, 323–331.

16. Madhavan, S., Anghelina, M., Rath-Deschner, B., Wypasek, E., John, A., Deschner, J., Piesco, N., and Agarwal, S. 2006. Biomechanical signals exert sustained attenuation of proinflammatory gene induction in articular chondrocytes. *Osteoarthritis Cartilage* 14, 1023–1032.

17. Chowdhury, T.T., Arghandawi, S., Brand, J., Akanji, O.O., Salter, D.M., Bader, D.L., and Lee, D.A. 2008. Dynamic compression counteracts IL-1β induced iNOS and COX-2 expression in chondrocyte/agarose constructs. *Arthritis Res. Ther.* 10, R35.

18. Chowdhury, T.T., Bader, D.L., and Lee, D.A. 2001. Dynamic compression inhibits the synthesis of nitric oxide and PGE_2 by IL-1β stimulated chondrocytes cultured in agarose constructs. *Biochem. Biophys. Res. Commun.* 285, 1168–1174.

19. Fitzgerald, J.B., Jin, M., Dean, D., Wood, D.J., Zheng, M.H., and Grodzinsky, A.J. 2004. Mechanical compression of cartilage explants induces multiple time-dependent gene expression patterns and involves intracellular calcium and cyclic AMP. *J. Biol. Chem.* 279, 19502–19511.

20. Mio, K., Saito, S., Tomatsu, T., and Toyama, Y. 2005. Intermittent compressive strain may reduce aggrecanase expression in cartilage: A study of chondrocytes in agarose gel. *Clin. Orthop. Relat. Res.* 433, 225–232.

21. Fanning, P.J., Emkey, G., Smith, R.J., Grodzinsky, A.J., Szasz, N., and Trippel, S.B. 2003. Mechanical regulation of MAPK signalling in articular cartilage. *J. Biol. Chem.* 278, 50940–50948.

22. Agarwal, S., Deschner, J., Long, P., Verma, A., Hofman, C., Evans, C.H., and Piesco, N. 2004. Role of NF-kappaB transcription factors in anti-inflammatory and pro-inflammatory actions of mechanical signals. *Arthritis Rheum.* 50, 3541–3548.

23. Akanji, O.O., Sakthithasan, P., Salter, D.M., and Chowdhury, T.T. 2010. Dynamic compression alters NFκB activation and IκB-α expression in IL 1β stimulated chondrocyte/agarose constructs. *Inflammation Res.* 59(1), 41–52.

24. Chowdhury, T.T, Bader, D.L, Shelton, J.C., and Lee, D.A .2003. Temporal regulation of chondrocyte metabolism in agarose constructs subjected to dynamic compression. *Arch. Biochem. Biophys.* 417, 105–111.

25. Bustin, S.A., Benes, V., Garson, J.A., Hellemans, J., Huggett, J., Kubista, M., Mueller, R. et al. 2009. The MIQ Eguidelines: Minimum information for publication of quantitative real-time PCR experiments. *Clin. Chem.* 55(4), 611–622. Epub Feb 26.

26. Lee, D.A. and Bader, D.L. 1997. Compressive strains at physiological frequencies influence the metabolism of chondrocytes seeded in agarose. *J. Orthop. Res.*, 15, 181–188.

27. Lee, D.A. and Knight, M.M. 2004. Mechanical loading of chondrocytes embedded in 3D constructs: *In vitro* methods for assessment of morphological and metabolic response to compressive strain. *Methods Mol. Med.* 100, 307–324.

28. Nolan, T., Hands, R.E., and Bustin, S.A. 2006. Quantification of mRNA using real-time RT-PCR. *Nat. Protoc.* 1, 1559–1582.

29. Pfaffl, M.W., Horgan, G.W., and Dempfle, L. 2002. Relative expression software tool (REST) for group wise comparison and statistical analysis of relative expression results in real time PCR. *Nucleic Acids Res.* 30(9), e36.

30. Livak, K.J. and Schmittgen, T.D. 2001. Analysis of relative gene expression data using real-time quantitative PCR and the 2-ΔΔCT method. *Methods* 25, 402–408.

31. Cook, P., Fu, C., Hickey, M., Han, E.S., and Miller, K.S. 2004. SAS programs for real-time RT-PCR having multiple independent samples. *Biotechniques* 37, 990–995.

32. Fundel, K., Haag, J., Gebhard, P.M., Zimmer, R., and Aigner, T. 2008. Normalization strategies for mRNA expression data in cartilage research. *Osteoarth. Cart.* 16, 947–955.

33. Karlen, Y., McNair, A., Perseguers, S., Mazza, C., and Mermod, N. 2007. Statistical significance of quantitative PCR. *BMC Bioinformat.* 8, 131.

34. Yuan, J.S., Wang, D., and Stewart, C.N. 2007. Statistical methods for efficiency adjusted real-time PCR quantification. *Biotechnol. J.* 3, 112–123.

35. Bustin, S.A. and Nolan, T. 2004. Pitfalls of quantitative real-time reverse-transcription polymerase chain reaction. *J. Biomol. Tech.* 15, 155–166.

Part VI

Applications

20 Enriching DNA Sequences with Nucleotide Variation by Thymidine Glycosylases Combined with Suppression PCR

Xinghua Pan, Alexander E. Urban, and Sherman M. Weissman

CONTENTS

20.1 INTRODUCTION

Understanding the associations between genetic and phenotypic variation has been a major task of modern molecular biology.[1–5] As an important part of this process a number of methods have been developed either for rapid detection of known variation in DNA sequence, or for detection of previously unappreciated sequence variation.[6–8] The latter category of approaches has been difficult to apply to large, complex DNA pools on a global scale until recently, when deep sequencing became available; yet even now deep sequencing a very complex genome such as the human genome at a sufficient depth is challenging.[9,10] *Escherichia coli* MutHLS proteins have been used to study variations via identity-by-descent (IBD) in *Saccharomyces cerevisiae*, mouse, and human genomes.[11–14] However, we and others have not obtained sufficient specificity with these enzymes to consistently detect single-nucleotide mutations. Another method, representative differential analysis, has been successfully used in scanning short- or long-fragment deletions,[15,16] but it is generally not sensitive to subtle nucleotide variation.

An ideal approach for screening nucleotide variations would fulfill the following criteria: (1) ability to detect all types of single-nucleotide variations, (2) high signal-to noise-ratio, (3) sufficient sensitivity to detect variants present at a low frequency in pooled samples, (4) ability to detect all of the variations in the DNA pool(s) studied in a single experiment, (5) no requirement for prior knowledge of the sequence and its variations, and (6) suitability for automation. The approach for global analysis of DNA allelic variations (GADAV) potentially meets all of these requirements. The most significant applications of GADAV would be in genetic mapping where it is necessary to distinguish the allelic variations within one DNA pool from that between two different pools, which may be combined with subsequent analysis with display gels, microarrays, and ultimately deep sequencing. In addition, GADAV has other applications such as screening for heterozygotic sites in an individual, or polymorphic sites in a population.

20.2   OUTLINE OF THE GADAV STRATEGY

Briefly, two pools of DNA are digested with a restriction endonuclease, ligated to a pair of specially designed adapters, and then mixed, denatured, and reannealed. The resulting mixture is subjected to TDG (thymine (or thymidine)-DNA glycosylase)-mediated enrichment of mismatch fragments (Mm) from perfectly matched fragments (Pm), followed by selective recovery of heterohybrids or homohybrids with suppression PCR (Figure 20.1). The perfect pool Pm can be subjected to a further round of enrichment of Mm fragments and selection of heterohybrids. The output can then be analyzed by gel display or hybridization to genomic or cDNA tiling sequence arrays, or library generation for deep sequencing. This procedure permits global screening of both DNA sequence variants in each pool, and of sequence variants that distinguish the sequences in one pool from those in the other. Combined with deep sequencing, GADAV would identify the nature and frequencies of the variations with a higher efficiency.

Our approach[17] uses a special group of mismatch repair enzymes, TDGs,[18–21] to separately enrich Mm and Pm DNA duplexes from a complex mixture. Although a number of other DNA glycosylases and mismatch recognizing enzymes for detection of DNA mutations have been described,[22–25] these strategies are not suitable for global scanning of DNA pools. The immobilized TDGs used here have two useful properties: First, they selectively remove bases from various internal unpaired nucleotides (when the human and a bacterial TDG are combined they cover all four types of single-nucleotide mismatches). Second, they bind to the abasic site thus generated in the presence of EDTA but release the DNA duplexes when magnesium is added.[17] Depending on the nature of the adapters that were ligated to the original DNA pools, either hetero- or homohybrids can be selectively recovered by suppression PCR.[17] Amplified Mm and Pm fragments can be compared by gel display, microarray, or identified at single-nucleotide resolution on deep sequencing. The outline of the strategy is described in Figure 20.1.

20.3   MATERIALS AND METHODS

20.3.1   Materials

20.3.1.1   Immobilized hTDG and mTDG

Human TDG (hTDG) and archaeon *Methanobacterium thermoautotropicum* DNA mismatch *N*-glycosylase (Mig.*Mth* or mTDG) were produced fused to glutathione *S*-transferase (GST),

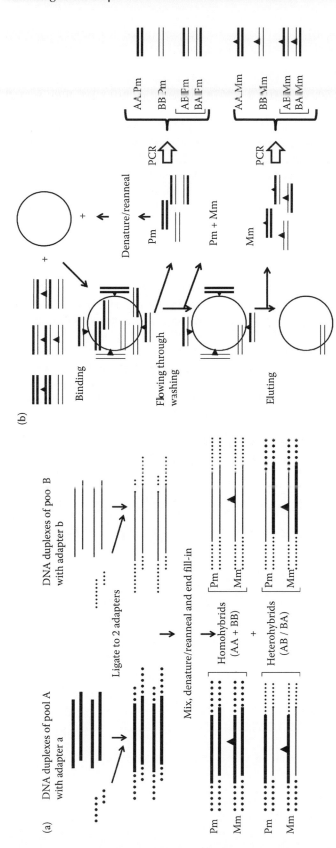

FIGURE 20.1 Outline of the strategy for global scanning of nucleotide variations between two DNA pools. AA and BB represent DNA duplexes in which both strands come from the same sources, or homohybrids. AB (each sequence potentially has two possible strand combinations: AB or BA, simplified as AB) represents DNA duplexes in which the two strands came from different pools, or heterohybrids. Pm represents perfectly matched DNA duplexes. Mm indicates DNA duplexes with one or more mismatched base pairs. ◄ indicates a mismatched base pair. ○ represents the TDG-immobilized on Gluatthione Sepharose 4B beads. (a) The construction of DNA templates and generation of the mixture of Pm and Mm fragments in both homohybrids and heterohybrids. The restriction enzyme-digested DNA pools "A" and "B" were ligated, respectively, to a set of long adapters (correspondingly "a" and "b") with different sequences as described in the Materials and Methods, denatured, reannealed, and filled in to create blunt ends with different sequences on the two ends of the heterohybrids (AB) and with an identical sequence on the two ends of the homohybrids (AA) and a different identical sequence on the two ends of the other homohybrids (BB). (b) The separation and enrichment in bulk of the Mm from the Pm fragments. The mixture generated above was loaded as the input onto the TDG-immobilized beads, bound in the presence of EDTA, washed with increasing concentrations of salt and eluted with magnesium. This process may be repeated for successive enrichments, in which the previous Pm output is subject to re-denaturation and annealing to deplete mismatch fragments. Then Mm and Pm fragment pools are obtained. Finally the heterohybrids (AB) and two homohybrids (AA or BB) can be selectively recovered by PCR (open arrow), and a total of six pools of fragments can be obtained. The adapter part of the fragments is not shown here.

immobilized on Glutathione Sepharose 4B beads (GE Healthcare), and stored in a buffer containing 50% glycerol at –80°C. They maintained their activities for up to 2 years.[17]

20.3.1.2   DNA Templates and Mm and Pm Oligonucleotide Duplexes

MspI-digested pBR322-DNA (New England Biolab) was 3′-^{32}P labeled with Taq polymerase at 72°C for 5 min in the presence of dGTP and α-^{32}P dCTP.

The mRNA pool was isolated with oligo(dT)-cellulose type 7 (GE Healthcare) from Trizol reagent (Invitrogen) extracted total RNA of the human lymphoblastoid cell line GM12729 (Coriell Cell Repositories). The cDNA pool was synthesized from this mRNA with the SuperScript choice system (Invitrogen). All steps were accomplished according to the manufacturers' instructions. Two 60 bp duplexes, one perfectly matched duplex (60Pm) and one containing a G/T mismatched base pair (60MmCG/GT), were made. The common oligonucleotide Co60 was 5′-P^{32} labeled with T4 nucleotide kinase (NEB) before being annealed with different complementary oligonucleotides, which were at excess molar (5:1) to generate the duplexes.[17]

20.3.1.3   Oligonucleotides for Adapters and Primers

A pair of heterohybrid-directing adapters (AdHeA, consisting of HeA1 and HeA2, and AdHeB consisting of HeB1 and HeB2) and their corresponding primers (HePa(HeP1), HePb(HeP2), HePaN(HeP1N), HePbN(HeP2N)) were made for selective recovery of the heterohybrids from a mixture of heterohybrids and homohybrids and for further subdivision of the amplicon[17] with suppression PCR based on the PCR suppression effect.[26,27]

When the selective nucleotide in the 3′ end of the subdivision primer (HePaN or HePbN) was "C," the third or fourth nucleotide from the 3′ end was modified so that it did not match the adapter sequence. Based on our observations and other reports,[26,28] this increases the specificity of the subdivision.

THE SEQUENCES FOR THE HETEROHYBRIDS-SELECTIVE ADAPTERS AND PRIMERS

Heterohybrids-Selective Adapters:

AdHeA
    HeA1: 5′-GTAATACGACTCACTATAGGGCTCGAGCGGCCGCCCGGGCAGGT
    HeA2: 5′-GATCACCTGCCC

AdHeB
    HeB1: 5′-TGTAGCGTGAAGACGACAGAAAGGGCGTGGTGCGGAGGGCGGT
    HeB2: 5′-GATCACCGCCCTCCG

Global Primers:

    HePa: 5′-GTAATACGACTCACTATAGGGC
    HePb: 5′-TGTAGCGTGAAGACGACAGAA

Subsetting Primers:

    HePaN: 5′-GCGGCCGCCCGGGCAGGTGATCN
    HePbN: 5′-CGTGGTGCGGAGGGCGGTGATCN
where the "N" is one nucleotide or a combination of two or three nucleotides.

When "N" in the subsetting primers above is "C", then
    HePaC: 5′-GCGGCCGCCCGGGCAGGTGTTCC
    HePbC: 5′-CGTGGTGCGGAGGGCGGTGTTCC

20.3.1.4 Other Reagents and Instruments

TDG binding buffer:
> 50 mM Tris–HCl, pH 8.3, 50 mM KCl, 5 mM EDTA, 0.2 mM ZnCl, 1 mM DTT, 0.25 mg/mL BSA.

TDG washing buffers:
> The same as TDG binding buffer, but with the separate addition of LiCl in concentrations of 100, 200, 300, 450, and 600 mM.

TDG elution buffers:
> 50 mM Tris–HCl pH 7.5, 50 mM KCl, 0.2 mM ZnCl, 1 mM DTT, 0.25 mg/mL BSA, and various amounts of MgCl$_2$ in concentrations of 30, 60, 100, 200, or 300 mM.

TDG perfect-match enrichment buffer:
> 50 mM Tris–HCl pH 7.5, 50 mM KCl, 20 mM EDTA, 0.2 mM ZnCl, 1 mM DTT, 0.25 mg/mL BSA.

Titanium Taq DNA polymerase (Clontech), T4 DNA ligase, and restriction endonucleases (New England Biolab), Wizard DNA clean-up system (Promega) or Centricon model 100 (Amicon), GeneAmp PCR system 9600 (Perkin Elmer), sequencing gel system model S2001 (Invitrogen), Liquid Scintillation Counter (Perkin Elmer), centrifuge, rotator, and isotope protection equipment were used in these experiments.

20.3.2 Experimental Protocol

20.3.2.1 Generation of Heterohybrids and Homohybrids from Two DNA Pools

In a pilot experiment, two cDNA pools "A" and "B," which were represented by two aliquots of a common human lymphoblastoid cDNA pool, were digested with Sau3AI in the provided buffer at 37°C for 3 h, purified with phenol–chloroform, and precipitated with 1/5 volume 3 M NaAc pH 5.0 and 2.5 volume of ethanol. The resulting DNA restriction fragment pools were separately ligated to the adapters AdHeA and AdHeB with T4 DNA ligase to form the constructs AdHeA–DNA–AdHeA and AdHeB–DNA–AdHeB. The ligation was stopped by adding 20 mM EDTA and heating at 65°C for 20 min. These two pools of constructs were then mixed together, extracted with phenol–chloroform, precipitated with ethanol, suspended in 4 μL 3 × EE buffer (30 mM EPPS [*N*-(2-hydroxyethyl) piperazine-*N'*-(3-propanesulfonic acid); pH 8.0] (Sigma-Aldrich) and 3 mM EDTA) and 1 μL of 5 M NaCl, carefully and thoroughly mixed by pipetting, overlaid with 1 drop of mineral oil, heated at 94°C for 5 min, quickly cooled to 80°C, then slowly cooled (over 5 h) to 67°C and held at 67°C for 20 h, and finally slowly cooled down to 60°C. This was executed in a thermocycler with a temperature profile of 94°C × 5 min, followed by cooling in one degree steps for 5 min per step from 80°C to 77°C, then 20 min per step from 76°C to 71°C, then 60 min per step from 70°C to 68°C, 20 h at 67°C, 60 min at 66°C, 65°C, and 64°C, and 20 min at each degree of temperature from 63°C to 60°C. The annealing time might be significantly shortened when a much less complex DNA pool such as a PCR-amplified subpool is tested. The sample was then diluted to 200 μL, and supplemented with PCR components including 1× buffer, each 100 μM 4 dNTPs, and 0.5× (1 μL) Titanium Taq DNA polymerase (Clonetech) without primers, and heated at 75°C for 10 min followed by 70°C for 5 min to fill in the DNA ends. The reaction was stopped by adding EDTA 20 mM, followed by purification with the Wizard DNA Clean Up system (Promega) or Centricon Model 100 (Amicon).

20.3.2.2 Enrichment and Separation of Mm Fragments from Pm Fragments

An 80 μL suspension of hTDG and/or mTDG beads was mixed with 500 μL TDG binding buffer, and 20–40 μg HaeIII-digested human genomic DNA (in more recent experiments, this was replaced with salmon sperm DNA (Invitrogen), sonicated to 100–600 bp in size) as a background-blocking carrier. The mixture was rotated gently at room temperature for 1 h. Depending on how much DNA is available and the complexity of the DNA sample, about 1–10 μg (usually 2–5 μg) of the mixture of DNA hybrids made above was loaded into this Eppendorf tube together with a monitoring DNA mixture (100 ng or less, as long as it provided a measurable isotope signal) composed of ^{32}P-labeled 60Pm and 60 MmCG/GT (1:1). The mixture was then rotated gently at room temperature between 4 h and overnight incubation. Two to four cycles of washing were carried out, each with 1000 μL of freshly prepared washing buffer with increasing LiCl (NaCl might also be used, but purification of the output is then more difficult) in concentrations from 100 to 600 mM until the ^{32}P isotope signal left in the Eppendorf was reduced to 5–10% of the total value estimated by Cerenkov radiation, measured in a liquid scintillation counter (PerkinElmer) without enhancer being added. The mixture was then equilibrated with the washing buffer, without EDTA, NaCl, or LiCl, at room temperature for 1 h. Finally, 2–4 cycles of elution were accomplished by successive application of 1000 μL fresh elution buffer with increasing $MgCl_2$ from 30 to 300 mM. Usually 95% of the remaining ^{32}P signal could be recovered. Each cycle of washing or elution was performed with incubation at room temperature for approximately 4 h, centrifugation at 1500 rpm for 2 min and removal of the supernatant. The eluates represented the Mm DNA duplex(es), and could be reloaded to the TDG beads for further enrichment when a higher specificity was desired.

For further enrichment of the Pm duplex(es), the initial flow-through portion (supernatant) obtained above before the first washing was phenol–chloroform–ethanol extracted, denatured, and reannealed as described above without filling-in, then mixed with 80 μL of a freshly prepared suspension of TDG beads equilibrated with 500 μL TDG perfect-match enrichment buffer and 20 μg blocking carrier DNA, rotated gently at room temperature between 4 h and overnight, followed by centrifugation at 1000 rpm for 2 min. The supernatant was collected as the Pm duplexes.

Fractions from different steps were phenol–chloroform extracted, ethanol precipitated, and dissolved in 20 μL of 10 mM Tris–HCl, pH 7.5, 1 mM EDTA for further analysis.

It is noteworthy that because a high concentration of salt was added, when phenol was added, the water (aqueous) phase and a phenol phase were often inverted. We divided the problem samples into two or more microfuge tubes, added an appropriate volume of chloroform to separate the aqueous phase from the phenol phase and then performed a standard phenol–chloroform extraction followed by an ethanol participation. In cases where a visible salt pellet was precipitated with DNA, redissolving the pellet with 200 μL of 0.2 M sodium acetate (pH 5.5) followed by ethanol participation could eliminate most of the salt.

20.3.2.3   Selective Amplification of Heterohybrids from a Mixture of DNA Fragments

The Mm and Pm output DNA samples were adjusted to an appropriate volume (10–100 μL) according to the effective DNA template concentration (the background-blocking DNA and the excess adapters were not taken into account), as evaluated by a trial PCR. From each sample, 2 μL was put into a PCR premixture including 1× buffer, 0.1 μM each of the inner primers HePa and HePb, 0.1 mM each of 4× dNTPs, and 1/250 volume (0.1 μL for 25 μL total PCR volume) Titanium Taq DNA polymerase (Clontech), and heated at 75°C for 10 min in a thermocycler. To recover heterohybrids, this solution was immediately connected to a thermal cycle profile for 14–24 cycles, consisting of 94°C for 40 s, 68°C for 50 s, and 70°C for 50 s. On a 1.5% agarose gel, most of the PCR products would be in the 150–800 bp range. Otherwise, the PCR mixture components, template concentration in the mixture or PCR cycle number should be adjusted. Because of the nature of PCR, too many cycles of amplification results in a shorter size range of fragments. However, if any key component (i.e., dNTPs, primer) was insufficient, too many cycles might result in very large fragments. When a single inner primer (HePa or HePb) is used together with appropriate concentration of other PCR components (esp. 0.5–1 μM primer, 0.25 mM 4 × dNTPs and 0.5 μL Taq polymerase) for 25–35 cycles of PCR, a given homohybrid (AA or BB) corresponding to the primer used will be produced. To reduce the complexity of a DNA pool for gel display, or to label and further amplify the amplicons (described below), the PCR product obtained was diluted 10–100-fold, and different combinations of primers (HePaN or HePa, and/or HePbN or HePb) were used in a second round of PCR. In this step, the PCR mixture contained 0.4 μM primers, 0.25 mM each of 4 × dNTPs, and 0.5 × -Taq polymerase, and 20–25 thermal cycles were performed.

20.3.2.4   Demonstration and Analysis of the Selected Fragments

Initially the DNA variation scanning and heterohybrid amplification results were demonstrated via gel display. For this purpose, the PCR was accomplished as described above except that one primer was labeled at the 5′ with ^{32}P using T4 nucleotide kinase and γ-^{32}P ATP. Alternatively, the product could be labeled by in-strand incorporation of α-^{32}PdCTP. The heterohybrids of Mm and Pm were separately amplified. Each 5 μL PCR product was mixed with 5 μL standard sequencing stop solution and heated at 94°C for 3.5 min. A volume of 3 μL of each product was loaded on a 6–8% polyacrylamide/urea gel.

The candidate mismatch fragment bands were recovered by needle puncture and amplified with the same primers used to generate the sample for gel display. The PCR products were digested with Sau3AI and cloned into pBluescript II SK(−) (Stratagene/Agilent) or TOPO TA Cloning vector (Invitrogen). Inserts of single clones were PCR-amplified with the flanking primers of the vectors and transferred to duplicate nylon membranes (Hybond-N+ , from Amersham Pharmacia Biotech). The pools of Mm and Pm output cDNA fragments from the column were ^{32}P labeled via PCR and separately hybridized to the membranes. When necessary, the candidate sequences containing an internal mismatch were used to design primers and to confirm the existence of the nucleotide variation in the original DNA pools via PCR–SSCP or direct sequencing.

For high-throughput identification of the Mm and Pm output fragments, hybridization was performed to a genomic tiling microarray consisting of PCR amplicons representing the nonrepetitive sequence of human chromosome 22q.[29] Deep sequencing of the pools was also applied to detect the detailed nature of the DNA variations.

20.4   RESULTS

20.4.1   Enrichment of a MM DNA Duplex from Digested PBR322 PM Fragments

It was previously demonstrated that large DNA fragments showed some nonspecific affinity for the beads. To test the effectiveness of the enrichment of Mm fragments from the background of large Pm fragments, the labeled MspI-digested pBR322 (5 μg), containing fragments up to 622 bp in length, was mixed with labeled 60 MmCG/GT and applied to the immobilized mTDG beads. After one cycle, the mismatch-enriched eluate showed 10–30-fold enrichment for the 60 MmG/T compared to larger plasmid fragments in the input DNA mixture, gel displayed and measured in a PhosphoImager (Molecular Dynamics). Conversely, this Mm fragment was relatively depleted (10–30-fold less) from the initial flow-through portion even without further enriching Pm fragments (Figure 20.2). Therefore, the ratio of Mm/Pm in the eluate to the ratio of Pm/Mm in the flow-through portion was 100- to 900-fold. The addition of a 25-fold excess unlabeled digested genomic DNA as a carrier improved the specificity.

20.4.2   Scanning the Polymorphic Fragments in a Human cDNA Pool

A human cDNA pool was analyzed for allelic variations in the coding sequences. Five micrograms of cDNA was digested with Sau3AI and heterohybrid-selecting adapters were added as described above. After Mm enrichment with immobilized hTDG, the output pools of Mm and Pm fragments were amplified with a radioactive DNA primer and the resulting heterohybrid fragments fractionated by electrophoresis on an 8% acrylamide sequencing gel (Figure 20.3, panel a1). The complexity of the mixture of fragments was also reduced in test runs by using DNA primers whose 3′ end extended beyond the Sau3AI recognition sequence by one or two specifically chosen bases (Figure 20.3, panel a2). Because Mm and Pm DNA pools were amplified to the same final yield of total product, while there were fewer Mm than Pm fragments, the relatively high intensity of bands in the Mm lanes did not reflect the relative depletion or enrichment of these fragments.

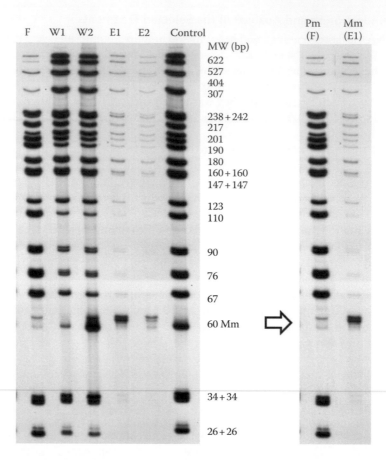

FIGURE 20.2 Enrichment of a 60 bp mismatched DNA duplex from MspI-restricted pBR322 fragments. The DNAs were labeled, mixed, and loaded onto the mTDG beads. The right panel demonstrated an immediate in-parallel comparison of candidate output of Mm to Pm. The Mm enrichment procedure was as described in Materials and Methods. Control, the input mixture control; F, initial flow through; W1 and W2 were successive washes and E1 and E2 were successive eluates. The right panel highlights the side-by-side comparison of the two key portions of output, in which the 60 bp Mm (open arrow) was depleted in portion F and enriched in E1.

Approximately 40 bands that appeared to be significantly enriched in the Mm portion were recovered from the gels by PCR. Because the homohybrid fragments had symmetric adapters, and the DNA recovered from one band probably contained multiple sequences of DNA fragment, it could not be sequenced directly but had to be subcloned. On average, four clones from each band were sequenced. However, the PCR products prepared with different selective primers with subdividing nucleotide(s) at their ends could be sequenced directly, because in this case, the complexity of each PCR product was significantly reduced, and we could often get single DNA species from one gel band. The resulting fragments included DNA sequences from the MHC Class I and Class II alleles, a number of ribosomal proteins, mitochondrial proteins, ubiquitins, and actins (70%), and from a variety of single copy genes (30%). Several different types of fragments were often recovered from a single band, both because of the complexity of the cDNA mixtures and the impurity of the gel excised bands. To analyze the enrichment of mismatch fragments, individual cloned fragments were amplified by PCR and subject to Southern blotting with probes prepared from either the total Mm or total Pm match DNA pools (Figure 20.3b). The result showed that a large fraction of the fragments were enriched in the mismatch pool, although the degree of enrichment varied over a fairly wide range.

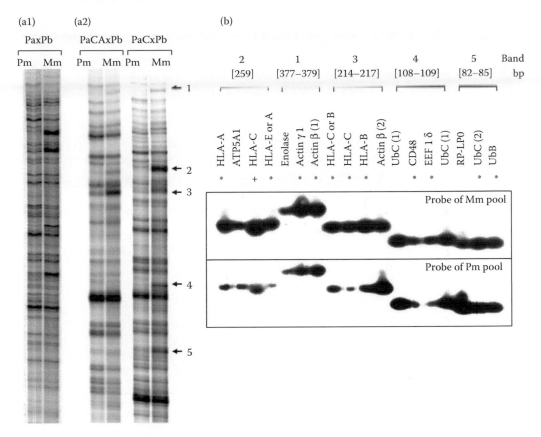

FIGURE 20.3 Gel display and Southern blot analysis of Mm fragments of a human cDNA pool. Following an immobilized hTDG-mediated procedure as described in Materials and Methods, the resulting pools of Mm and Pm were subjected to selective heterohybrid amplification. (a1) The Mm and Pm outputs recovered by PCR with primers HePa x HePb (20 cycles). (a2) Two subsets with decreasing coverage. The products of the first PCR with HePa × HePb (14 cycles) were diluted 100-fold, 2 μL of the diluted product was taken as the input for a second PCR amplification with subdividing primers HePaC × HePb or HePaCA × HePb (20 cycles). The numbers 1, 2, 3, 4, and 5 were different candidate Mm bands. (b) PCR-Southern blot check of the enriched DNA fragments amplified from the bands indicated in (a2). The marker "*" showed the clone's hybridization signal was significantly stronger on the membrane hybridized with Mm probes than that with Pm probes. The "+" refers to the clone derived (217 bp) from band 3. The "(1)" and "(2)" after actin β and UbC indicated two different fragments of one gene.

20.4.3 ANALYZING THE ENRICHED DNA POOLS ON A GENOMIC TILING ARRAY

To increase resolution and throughput, the enriched pools of Mm and Pm were labeled separately with Cys3 and Cys5 and hybridized to a chromosome 22 (Ch22q) tiling array.[29,30] The DNA tiling microarray comprising nearly all of the nonrepetitive sequences of human Chromosome 22 was constructed by producing approximately 20,000 PCR fragments, ranging in size from 0.3 to 1.4 kb (mean size = 720 bp), thus covering the entire nonrepetitive sequence of chromosome 22q.[29] The PCR fragments are printed over three glass slides, with each slide also containing control spots.

The scatter plot (Figure 20.4) shows the results for one of the Chromosome 22 microarray slides (from the set of three slides). There were 8448 chromosome 22 probes on this slide, 38 of which were detected with significantly enriched signal for Mm pool comparing to Pm pool,

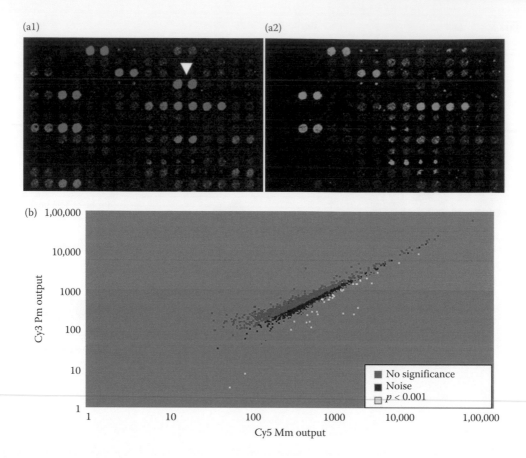

FIGURE 20.4 **(See color insert.)** Ch22 tiling microarray analysis of Mm fragments of the human cDNA pool. The same Mm and Pm outputs recovered in Figure 20.3a1 were labeled using Klenow fragment (New England Biolabs) for introducing amino-allyl moieties. The reactions were primed with random decamers in the presence of an amino-allyl-modified cytosine. The Mm and Pm pools were ethanol precipitated and resuspended in 0.1 M $NaHCO_3$ to facilitate coupling of the Cy5 and Cy3, respectively, monoamine dyes (Amersham) to the amino-allyl functional group. After the coupling reaction, the labeled pools were separated from unincorporated Cy5 mono-amine dye by ethanol precipitation and resuspended in hybridization buffer ($5\times$ SSC [$1\times$ SSC is 0.15 M NaCl plus 0.015 M sodium citrate], 25% formamide, 0.05% sodium dodecyl sulfate [SDS]). The microarray slides were prehybridized in $5\times$ SSC–25% formamide–$10\times$ Denhardt's solution–0.1% bovine serum albumin in HybChambers (GeneMachines, San Carlos, California) at 42°C for 1 h. The probe was applied to the microarray. Microarrays were hybridized at 42°C for 14–15 h and then washed twice in $2\times$ SSC–0.05% SDS for 2 min, once in $0.1\times$ SSC–0.05% SDS for 5 min, and 4 times in $0.1\times$ SSC for 1 min. Microarrays were scanned with an Axon 4000A scanner, and images were analyzed with GenePix Pro3.0 software. Microarray signals were normalized with the ExpressYourself analysis program (http://bioinfo.mbb.yale.edu/ExpressYourself).[31] Positively hybridizing fragments were identified as those with logged Cy5/Cy3 ratios more than 2.0 standard deviations above the mean for fragments with similar total intensities. (a) A representative area of the slides for Mm pool, with 635 nm Cy5 (a1) and for Pm pool, with 532 nm Cy3 (a2). The white arrow shows a significantly enriched Mm signal (with duplicate probes). (b) The scatter plot of the whole Ch22 tiling microarray.

indicating that these probes represented sequences containing nucleotide mismatch(es) in their respective positions on Chromosome 22q. Considering that only one restriction endonuclease, Sau3A1, was used for the fragmentation of the cDNA pool, most of the sequences that are too long or too short were not covered. Additional restriction digestion endonuclease(s) would give a better coverage.

20.5 DISCUSSION

20.5.1 Future Developments and Potential Applications

We have developed a relatively quick procedure for screening an entire human cDNA pool for single-nucleotide variations, with the potential for high throughput. This method can be applied to any DNA pool or DNA pool pair, whether derived from a single individual or a population or a pair of pools. In addition, all the identical DNA fragments shared by individuals within a pool can potentially be obtained. The strategy for separating heterohybrids from homohybrids is relatively simple and has satisfactory selectivity. A similar strategy was demonstrated later to be able to selectively amplify the homohybrids under certain conditions.[32]

We first used a gel display method to demonstrate the feasibility of the method. Further, we showed that tiling arrays of unique genomic fragments increased the throughput for large-scale applications, although a cDNA tiling array would be more appropriate for analyzing a cDNA library. This approach was subsequently employed by a different operator using microarrays to detect successfully a number of polymorphisms and mutations in the coding sequences from melanoma.[33]

A remaining challenge for GADAV is the formation of DNA duplexes by strands from different members of a family of closely related sequences, or orthologous loci, such as the MHC Class I genes, or actin genes. This could be partially overcome by hybridization to genomic or cDNA tiling arrays. This mismatch between loci could also be distinguished from the allelic sequence variation when self-annealed DNA is used as a control if the DNA sample is from a haploid genome. In addition, mostly because of the pitfalls above, the procedure is not sufficient to enumerate directly mutations or variations in large genomes; nevertheless, it does provide important enrichment, which may be followed up with additional verifications. Also, when this approach is applied to human or other mammalian genomic DNA, reducing the complexity by a prior isolation of subsets of the DNA by subdividing PCR would be desirable when gel display is followed for sequence identification[17] using restriction enzyme Sau3A1 digestion. However, this should not be necessary when microarray or deep sequencing is applied. To increase the Mm to Pm ratio, endonuclease V and VII and other glycosylases might be useful to combine with TDGs to deplete Mm fragments from the pool. With these enzymes, other subtle DNA variations such as small deletions/insertions and bi-, tri-, and other multiple nucleotide substitutions might also be covered. Biotinylated hydroxylamine,[24] an abasic endonuclease or DNA polymerase beta might also be introduced to remove the random abasic sites generated during DNA manipulation and further improve the signal-to-noise ratio.

This approach can be applied for exhaustive screening of single-nucleotide variations for a whole simple or medium complexity DNA pool, a subpool, or a given chromosome region. It is immediately applicable for DNA variation scanning in microorganisms such as bacterial or yeast genomes, where the genomes are haploid and quite simple. Again it might contribute to the discovery of new SNPs (single-nucleotide polymorphisms).

In complex genomic DNA scanning, this approach could detect areas of loss of heterozygosity, which would appear as regions without detectable polymorphism, as well as somatic single base mutations arising in malignancies, and mutations arising by large-scale mutagenesis in inbred model organisms.

The method suggested here may be applicable to the study of certain human genetic variants, including sporadic dominant mutations that prevent fertility, as well as familial dominant mutations and recessive mutations. Combined with quantitative microarray hybridization and DNA pooling strategies, which can be used for allele-frequency estimation, this approach could be used to screen the nucleotide variations in terms of allele frequency distortion for any phenotype.

Next-generation sequencing technology, provides a single-nucleotide resolution and high-throughput sequencing, while the cost of sequencing is dropping fast, but a routine sequencing of large numbers of whole eukaryotic genomes is currently not yet feasible. GADAV potentially dramatically improves the sequencing efficiency by reducing the complexity from a whole genome to a narrow candidate pool of target sequences with nucleotide difference. This target-enrichment strategy gives a new dimension beyond array-capturing and PCR-based multiplexed sequencing.[34] On the other hand, combining the next-generation sequencing with GADAV could also resolve the biggest challenge in GADAV, which is the background coming in the output that is actually derived from different members of a family of closely related sequences (orthologous loci). Overall, GADAV potentially provides a chance to allow a pool of multiple, even thousands of samples, to be sequenced in one flow-cell lane for the target regions with variations. In this case, the representation of allele frequencies can be predicted, because in the output pools the fragments collected in GADAV are proportional to the frequencies of the fragments with nucleotide variation.

20.5.2 SUMMARY

Efficient global scanning of nucleotide variations in DNA sequences between related, complex DNA samples remains a challenge. Deep sequencing a whole genome gives an ultimate resolution, but it is still cost inefficient to do so for many research scenarios, especially for large sets of samples. In the current report, we present GADAV as an approach to this problem. We have employed two immobilized thymidine DNA glycosylases, which in combination would cover all of the four groups of single-nucleotide mismatches, to capture and enrich DNA fragments containing internal mismatched base pairs and separate these fragments as a pool from perfectly base-paired fragments as another pool. Enrichments of up to several hundred-fold were obtained with one cycle of treatment. We have employed a strategy, suppression PCR, for selective amplification of heterohybrids, which can also be used for selective amplification of homohybrids. By combining these methods together, the single-nucleotide variations either between two DNA pools or within one DNA pool can be obtained in one process. When a pool is pre-enriched for the sequences with mismatched nucleotides, deep sequencing would be significantly more efficient. This approach has been applied to total cDNA from normal or cancer human cell line, and combined with gel display and microarray analysis, and has several potential applications in global scanning of nucleotide variations or polymorphisms in a moderately complex DNA pool, such as microorganismal genomes, and in mapping simple or complex genetic traits in complex genomes such as those of human and mouse. The next-generation sequencing resolves the background pitfall, increases the throughput with accuracy, and brings a new life and window for a wider application of GADAV.

ACKNOWLEDGMENTS

We thank Dr. R. Lasken (Molecular Staging, Inc.) for management throughout this project, and Dr. Michael Snyder (Stanford University) for support in microarray analysis. This work received financial support from the National Cancer Institute Grants R21 CA088326 and R33 CA88326.

REFERENCES

1. Wang, W.Y., Barratt, B.J., Clayton, D.G., and Todd, J.A. Genome-wide association studies: Theoretical and practical concerns. *Nat Rev Genet* **6**, 109–18, 2005.
2. Hardy, J. and Singleton, A. Genomewide association studies and human disease. *N Engl J Med* **360**, 1759–68, 2009.
3. Ku, C.S., Loy, E.Y., Pawitan, Y., and Chia, K.S. The pursuit of genome-wide association studies: Where are we now? *J Hum Genet* **55**, 195–206, 2010.
4. Jonsson, J.J. and Weissman, S.M. From mutation mapping to phenotype cloning. *Proc Natl Acad Sci USA* **92**, 83–5, 1995.
5. Gray, I.C., Campbell, D.A., and Spurr, N.K. Single nucleotide polymorphisms as tools in human genetics. *Hum Mol Genet* **9**, 2403–8, 2000.
6. Gunderson, K.L. et al. Whole-genome genotyping. *Methods Enzymol* **410**, 359–76, 2006.

7. Kristensen, V.N., Kelefiotis, D., Kristensen, T., and Borresen-Dale, A.L. High-throughput methods for detection of genetic variation. *Biotechniques* **30**, 318–22, 324, 326 passim, 2001.

8. Syvanen, A.C. Accessing genetic variation: Genotyping single nucleotide polymorphisms. *Nat Rev Genet* **2**, 930–42, 2001.

9. Chan, E.Y. Next-generation sequencing methods: Impact of sequencing accuracy on SNP discovery. *Methods Mol Biol* **578**, 95–111, 2009.

10. Kim, J.I. et al. A highly annotated whole-genome sequence of a Korean individual. *Nature* **460**, 1011–5, 2009.

11. Beaulieu, M., Larson, G.P., Geller, L., Flanagan, S.D., and Krontiris, T.G. PCR candidate region mismatch scanning: Adaptation to quantitative, high-throughput genotyping. *Nucleic Acids Res* **29**, 1114–24, 2001.

12. Cheung, V.G. et al. Linkage-disequilibrium mapping without genotyping. *Nat Genet* **18**, 225–30, 1998.

13. McAllister, L., Penland, L., and Brown, P.O. Enrichment for loci identical-by-descent between pairs of mouse or human genomes by genomic mismatch scanning. *Genomics* **47**, 7–11, 1998.

14. Nelson, S.F. et al. Genomic mismatch scanning: A new approach to genetic linkage mapping. *Nat Genet* **4**, 11–8, 1993.

15. Lisitsyn, N., Lisitsyn, N., and Wigler, M. Cloning the differences between two complex genomes. *Science* **259**, 946–51, 1993.

16. Wallrapp, C. and Gress, T.M. Isolation of differentially expressed genes by representational difference analysis. *Methods Mol Biol* **175**, 279–94, 2001.

17. Pan, X. and Weissman, S.M. An approach for global scanning of single nucleotide variations. *Proc Natl Acad Sci USA* **99**, 9346–51, 2002.

18. Hardeland, U. et al. Thymine DNA glycosylase. *Prog Nucleic Acid Res Mol Biol* **68**, 235–53, 2001.

19. Horst, J.P. and Fritz, H.J. Counteracting the mutagenic effect of hydrolytic deamination of DNA 5-methylcytosine residues at high temperature: DNA mismatch *N*-glycosylase Mig.Mth of the thermophilic archaeon *Methanobacterium thermoautotrophicum* THF. *EMBO J* **15**, 5459–69, 1996.

20. Neddermann, P. et al. Cloning and expression of human G/T mismatch-specific thymine-DNA glycosylase. *J Biol Chem* **271**, 12767–74, 1996.

21. Waters, T.R., Gallinari, P., Jiricny, J., and Swann, P.F. Human thymine DNA glycosylase binds to apurinic sites in DNA but is displaced by human apurinic endonuclease 1. *J Biol Chem* **274**, 67–74, 1999.

22. Vaughan, P. and McCarthy, T.V. Glycosylase mediated polymorphism detection (GMPD)—A novel process for genetic analysis. *Genet Anal* **14**, 169–75, 1999.

23. Bazar, L.S. et al. Mutation identification DNA analysis system (MIDAS) for detection of known mutations. *Electrophoresis* **20**, 1141–8, 1999.

24. Chakrabarti, S. et al. Highly selective isolation of unknown mutations in diverse DNA fragments: Toward new multiplex screening in cancer. *Cancer Res* **60**, 3732–7, 2000.

25. Zhang, Y., Kaur, M., Price, B.D., Tetradis, S., and Makrigiorgos, G.M. An amplification and ligation-based method to scan for unknown mutations in DNA. *Hum Mutat* **20**, 139–47, 2002.

26. Matz, M., Usman, N., Shagin, D., Bogdanova, E., and Lukyanov, S. Ordered differential display: A simple method for systematic comparison of gene expression profiles. *Nucleic Acids Res* **25**, 2541–2, 1997.

27. Diatchenko, L. et al. Suppression subtractive hybridization: A method for generating differentially regulated or tissue-specific cDNA probes and libraries. *Proc Natl Acad Sci USA* **93**, 6025–30, 1996.

28. Lizardi, P.M. et al. Mutation detection and single-molecule counting using isothermal rolling-circle amplification. *Nat Genet* **19**, 225–32, 1998.

29. Rinn, J.L. et al. The transcriptional activity of human Chromosome 22. *Genes Dev* **17**, 529–40, 2003.

30. Euskirchen, G. et al. CREB binds to multiple loci on human chromosome 22. *Mol Cell Biol* **24**, 3804–14, 2004.

31. Luscombe, N.M. et al. ExpressYourself: A modular platform for processing and visualizing microarray data. *Nucleic Acids Res* **31**, 3477–82, 2003.

32. Makrigiorgos, G.M., Chakrabarti, S., Zhang, Y., Kaur, M., and Price, B.D. A PCR-based amplification method retaining the quantitative difference between two complex genomes. *Nat Biotechnol* **20**, 936–9, 2002.

33. Liu, M.M., Weissman, S.M., and Tang, L. Identification of coding single nucleotide polymorphisms and mutations by combination of genome tiling arrays and enrichment/depletion of mismatch cDNAs. *Anal Biochem* **356**, 117–24, 2006.

34. Mamanova, L. et al. Target-enrichment strategies for next-generation sequencing. *Nat Methods* **7**, 111–8, 2010.

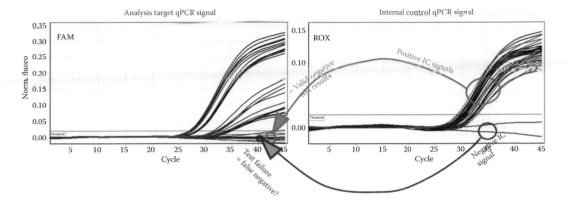

FIGURE 2.1 Use of an internal control for quality assurance of a hepatitis C virus real-time RT-PCR assay. Red amplification curves: invalid HCV results disclosed by the negative internal control signal; green curves: valid negative HCV results confirmed by the positive internal control signal; blue curves: HCV positive samples.

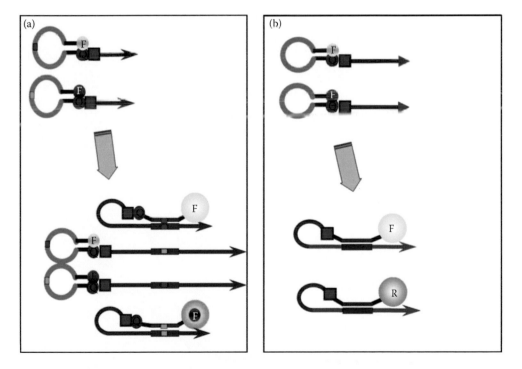

FIGURE 6.1 Two formats for genotyping by Scorpions. (a) Allele-specific hybridization method in which the difference between the alleles is detected by the probe element. When the Scorpion is incorporated into a mismatched amplicon, the stem-loop of the primer construct is favored over the mismatched hybrid yielding improved discrimination. (b) Allele specific priming format. In this format, discrimination depends upon differential primer extension. The extended products are all then targets for the probe element.

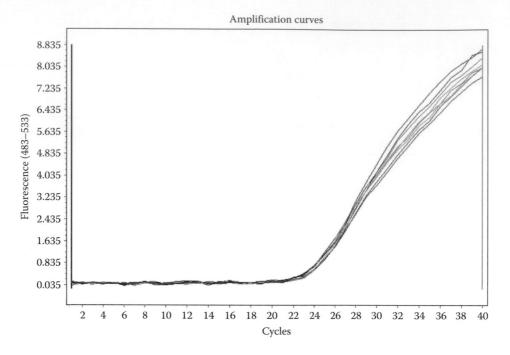

FIGURE 12.4 Human PTENP1 pseudogene assay for hPTEN—results for four primer combinations around a single probe. The red primer pair was selected in this case.

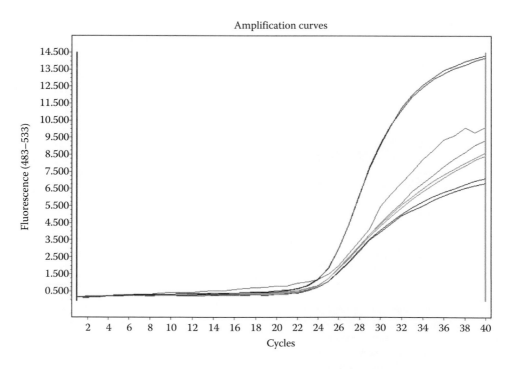

FIGURE 12.5 Human PTEN transcript assay—results for four primer combinations around a single probe. The purple primer pair is clearly the best for this assay.

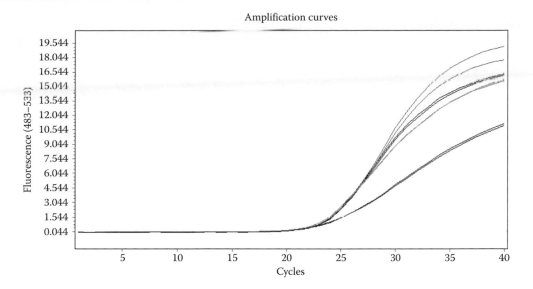

FIGURE 12.6 *Macaca* assay for Collagen 1A1—results for four primer combinations for an SYBR Green I assay using diluted PCR product as the template. The blue primer pair was selected in this case as giving the best amplification curve although the Cq values were not the lowest.

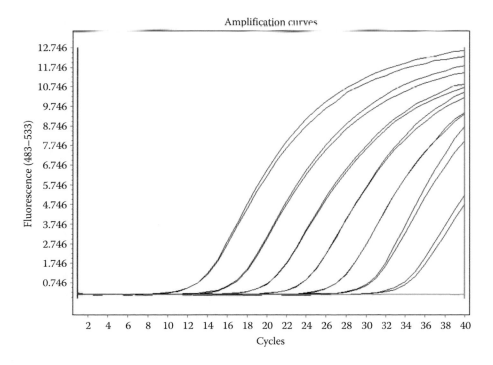

FIGURE 12.7 7-Log standard curve for the hPTENP1 assay. The PCR efficiency is 94%, the slope of −3.47 with a *y*-intercept of 38.42. The green line is the NTC.

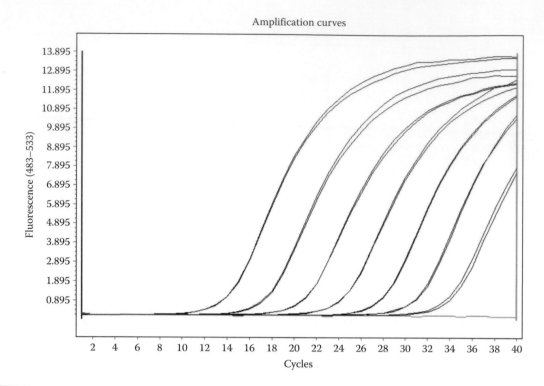

FIGURE 12.8 7-Log standard curve for the hPTEN assay. The PCR efficiency is 98%, the slope of −3.37 with a *y*-intercept of 37.53. The green line is the NTC.

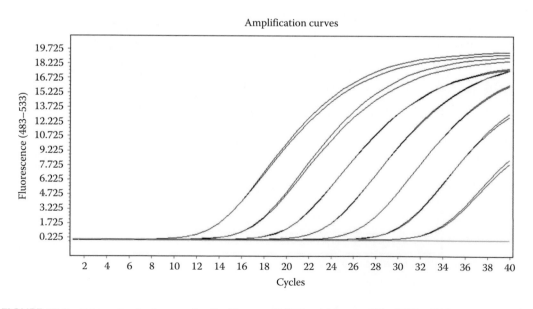

FIGURE 12.9 7-Log standard curve for the *Macaca* Collagen 1A1 assay. The PCR efficiency is 99%, the slope of −3.31 with a *y*-intercept of 37.69. The detection used SYBR Green I. The green line is the NTC.

(b)

3D line plot

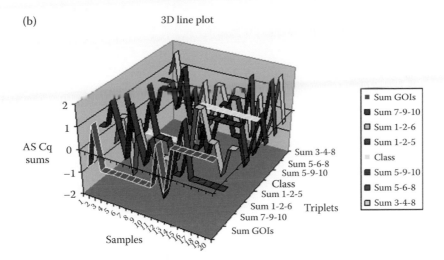

FIGURE 16.2   (b) A 3D line plot of a selection of the triplet sum combinations, including the ones with lowest *p*-value. The phenotypic indicator represented by the Class 3D line. We observe the apparent good correlation between the genes of interest's triplet sum combination and the phenotypic indicator.

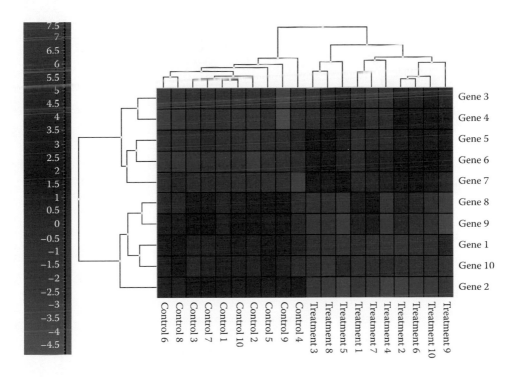

FIGURE 18.2   Heatmap analysis of a set of 10 significantly regulated genes between 10 untreated control animals and 10 animals, treated with steroid hormones.

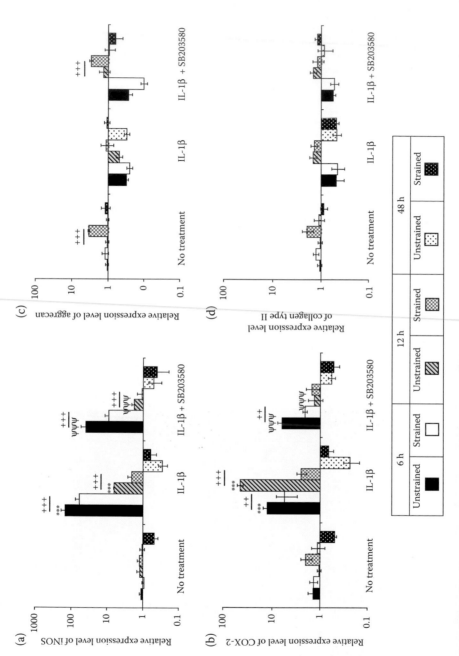

FIGURE 19.3 Effect of dynamic compression (15%, 1 Hz) on iNOS (a), COX-2 (b), aggrecan (c), and collagen type II (d) expression in unstrained and strained constructs cultured under no-treatment conditions or with 10 ng/mL IL-1β and/or 10 μM SB203580 for 6, 12, and 48 h. Ratios for the target were calibrated to the mean unstrained (untreated) control value and normalized to GAPDH. Ratios were log transformed and expressed on a logarithmic scale (arbitrary units). Bars represent the mean and SEM of 16–18 replicates from four separate experiments. Two-way ANOVA with post hoc Bonferroni-corrected t-tests was used to compare data. Significant differences are indicated as follows: + indicates comparisons for unstrained versus strained; * indicates comparisons in unstrained samples for no treatment versus IL-1β; ψ indicates comparisons in unstrained samples for IL-1β versus IL-1β + SB203580.

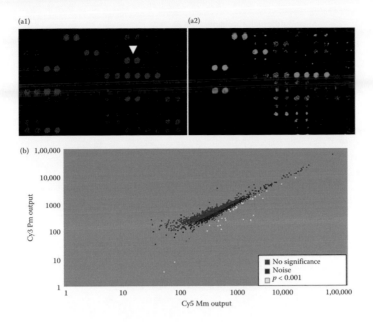

FIGURE 20.4 Ch22 tiling microarray analysis of Mm fragments of the human cDNA pool.

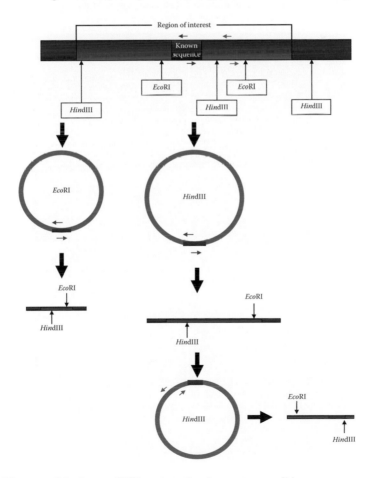

FIGURE 21.1 Diagram of the inverse PCR strategy for chromosome walking.

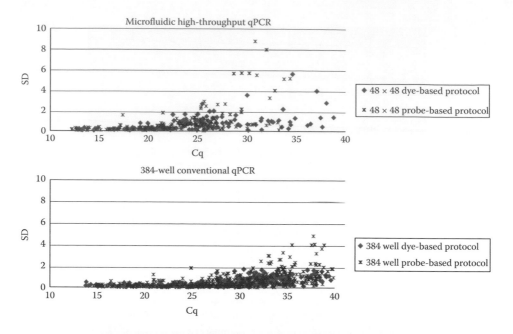

FIGURE 23.1 Dependence of standard deviation on Cq in conventional and microfluidic qPCR systems.

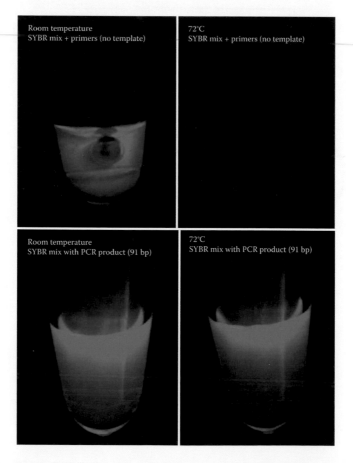

FIGURE 23.2 Dye fluorescence in conventional qPCR at different temperatures.

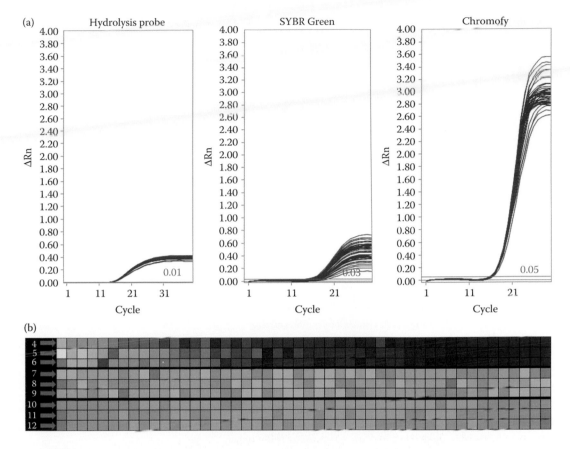

FIGURE 23.6 Amplification curves (a) and heatmap produced with the BioMark software (b). Blue squares represent high Cqs, yellow low Cqs. Lines 4–6 show SYBR Green signal, lines 7–9 show Chromofy, and lines 10–12 show probe signal.

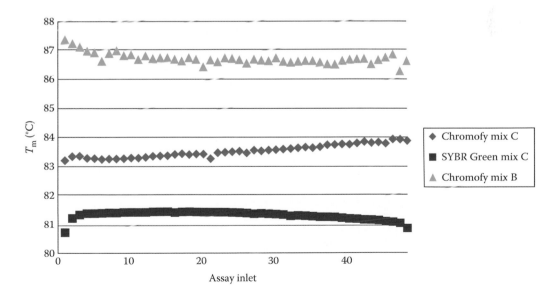

FIGURE 23.7 T_m uniformity across the 48 × 48 array. Mix B = KAPA Fast Universal, mix C = Finnzymes SYBR Flash mix.

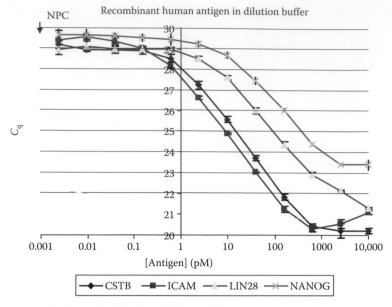

Antigen, sub-cellular localization and size:

ICAM-1, membrane bound, 57.8 kDa

CSTB, cytoplasmic protein, 11.1/22.2 kDa (monomer, can form dimer)

LIN28, nuclear localized, 22.7 kDa

NANOG, nuclear localized, 34.6 kDa

FIGURE 27.3 Homogenous TaqMan protein expression data: Recombinant human protein. Recombinant human antigens were diluted in a fourfold series over a 6-log range.

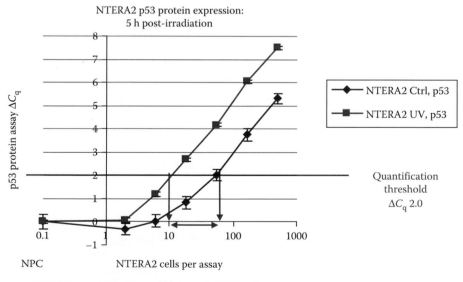

• p53 Relative quantification cell-intercept: 56/10 cells
• p53 protein: 5.6-fold UV induction

FIGURE 27.5 Relative quantification, ΔC_q squared method. The human embryonal carcinoma cell line NTERA2 was subjected to genotoxic stress.

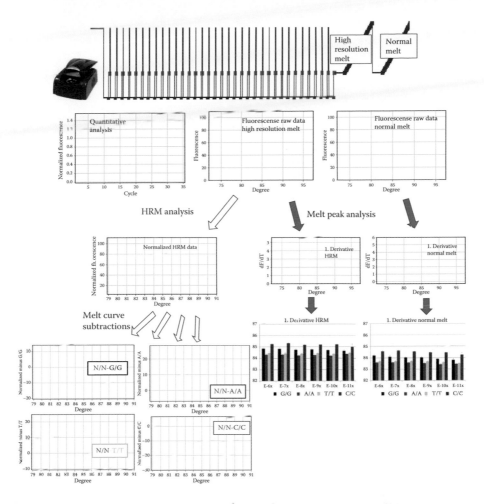

FIGURE 28.6 The progress of qPCR and melt analysis in Corbett's Rotor-Gene 6500 HRM instrument.

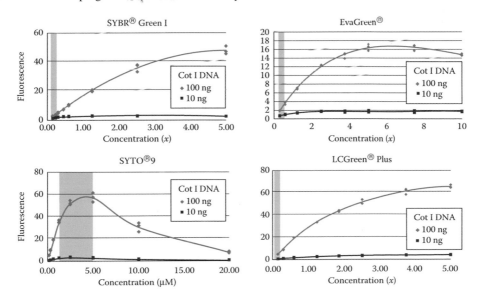

FIGURE 28.7 The saturating/nonsaturating ability of four dsDNA-specific fluorescent dyes.

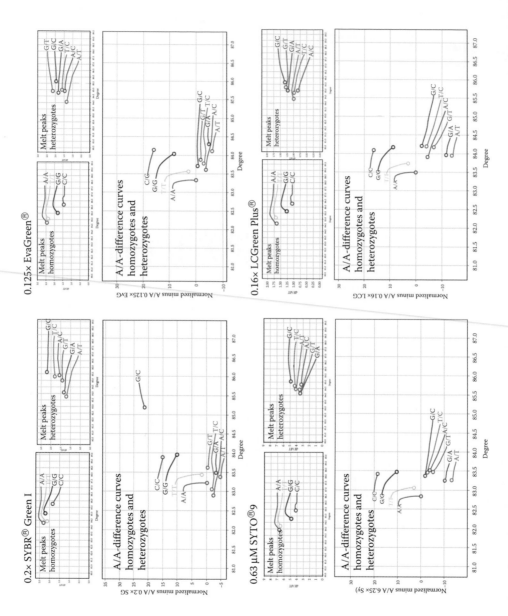

FIGURE 28.8 Comparison of the SNP homozygote and heterozygote genotyping ability of SYBR Green I, LCGreen Plus, EvaGreen, and SYTO9. The homozygote and heterozygote melt peak analyses are shown separately for visualization purposes. By the preparation of the difference curves the homozygous A/A-melt curve was used as reference since this was the least stable homozygote genotype, and thereby separated the heterozygotes from homozygotes most efficiently.

21 Chromosome Walking by Inverse PCR

Michael Gotesman, Selwyn A. Williams,
Jorge A. Garcés, and Ray H. Gavin

CONTENTS

21.1 INTRODUCTION

Inverse polymerase chain reaction (*i*PCR) is a powerful tool that can be used for the sequence analysis of DNA when only one end of a DNA sequence is known. *i*PCR[1–3] amplifies DNA from a circular template using primers with their 3′ ends directed away from each other (Figure 21.1). Cloning and sequencing PCR products are facilitated by the use of vectors, for example, pCR®2.1 (Invitrogen), which contain a thymine overhang. *i*PCR is very useful for supplementing conventional PCR screens that often yield only a substantial fragment of the gene of interest. An *i*PCR strategy used in conjunction with commercial cloning vectors and modern sequencing technology can rapidly yield accurate DNA sequencing results for several kilobases (kB) and allows one to "walk" both upstream and downstream of the known DNA sequence to obtain additional sequence. The methodology facilitates amplification of unknown DNA flanking sequences without the labor involved in constructing and screening libraries. Identification of flanking sequences on either side of a known sequence can be challenging. Other PCR-based methods for gene isolation, including random priming or the use of adapter sequences (anchored and rapid amplification of cDNA ends (RACE) PCR), employ a single sequence-specific primer, consensus or degenerate primer, or randomly prime off the template. These approaches frequently yield a low signal-to-noise ratio. *i*PCR requires two sequence-specific primers and generally yields a high signal-to-noise ratio. In addition, the *i*PCR primer pair can be used to amplify multiple circularized gene fragments that contain overlapping restriction sites. However, the *i*PCR technique requires circularization of DNA fragments within an optimum size range (see Section 21.4).

    *i*PCR has been useful in studies of genomic DNA from a variety of models, including humans,[1–3] *Lactobacillus*,[4] the red seaweed *Grateloupia*,[5] *Vibrio*,[6] *Tetrahymena*,[7–9] the marine yeast *Williopsis*

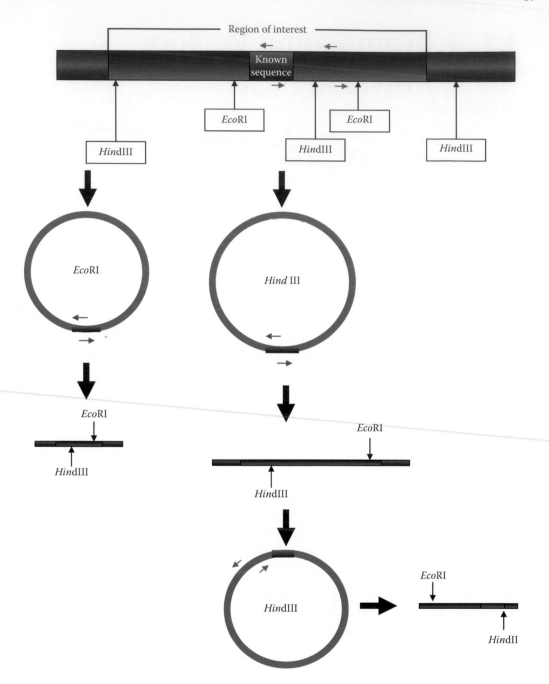

FIGURE 21.1 (**See color insert.**) Diagram of the inverse PCR strategy for chromosome walking.

saturnus,[10] and environmental metagenomes.[11] *i*PCR is amenable to studies of protein analysis,[12–14] DNA regulatory sequences,[15,16] RNAi,[17] and insertional mutagenesis.[18–21] Other recent use of this technique includes the detection of chromosome rearrangement in cancer patients,[22] sex determination in *Bubalus bubalis* (cattle),[23] and locating sites of integration in mice using Cre/loxp recombination system[24] and in silkworm using the *piggyBac* transposon system.[25] In this chapter, we describe protocols for using chromosome walking with *i*PCR to clone and sequence contiguous DNA upstream and downstream of a known site.

21.2 MATERIALS AND METHODS

The main steps in *i*PCR (Figure 21.1) are as follows:

1. Purify and cut the genomic DNA sample with an appropriate restriction enzyme (*Eco*RI or *Hin*dIII in this example).
2. Ligate the fragments into circular templates.
3. Perform *i*PCRs using primer sequences within the region of known sequence.
4. Analyze *i*PCR products with appropriate restriction endonucleases (*Eco*RI and *Hin*dIII in this example).
5. Ligate *i*PCR product into pCR2.1 and use the ligated vector to transform *Escherichia coli*.
6. Use blue/white screening to search for candidate screening colonies.
7. (Optional) Reanalyze plasmid with appropriate restriction endonucleases (*Eco*RI and *Hin*dIII in this example).
8. Use M13F and M13R to sequence plasmids.
9. After the *i*PCR product has been cloned and sequenced, a full-length contiguous fragment may be amplified by conventional PCR using primer sequences at or adjacent to the restriction sites from a genomic DNA template.
10. Use sequencing results to search for new primers 5′ and 3′ of the starting fragment to yield more sequence of interest.
11. Cut the genomic DNA sample with the second restriction enzyme. This creates additional fragments that will serve as template for the *i*PCR reaction and facilitates amplification of novel sequences both upstream and downstream of the known sequence (Figure 21.1; second set of *Hin*dIII sites and primers denoted by red arrows).
12. Obtain additional clones by *i*PCR. The overlap between the new and old clones allows for proper orientation of the *i*PCR fragments and prevents the creation of gaps in a contiguous sequence. PCR primers are designed from regions of known sequence as depicted in Figure 21.1.
13. Amplify full-length contiguous DNA fragment by conventional PCR using primer sequences at or adjacent to the outermost restriction sites. Using this strategy, between 5 and 10 kb of novel contiguous DNA sequence information can be obtained after only two rounds of *i*PCR. Further *i*PCR cycles using fragments left and right of the starting fragment will yield more sequence. Each cycle of the strategy allows the user to "walk" both upstream and downstream of a known DNA sequence.

21.2.1 PRIMERS

We have used a variety of primers for the amplification of ciliated protozoan sequences. The primers were designed to amplify from circular templates and therefore have their 3′ ends directed away from each other. These primers were generally 21- to 24-mers and 50% GC-rich with a T_m between 60°C and 68°C. A stock solution at 100 μM was prepared by resuspending the primers in sterile distilled water or TE buffer (1×).

21.2.2 DNA TEMPLATES

Tetrahymena genomic DNA was prepared by using a commercially available DNA Extraction Kit (Stratagene) with a final ethanol precipitation step. DNA was digested in small tubes, each containing 30 μg of genomic DNA, 30 units of *Eco*RI or *Hin*dIII, enzyme buffer (supplier's directions), and distilled water to 50 μL. The tubes were incubated at 37°C for 1 h. Restriction digest products were purified using a Chromospin column (Clontech). Alternatively, DNA digestion products were purified by gel electrophoresis, which was also useful for the identification of components of optimum size for circularization (see Section 21.4). For gel purification, the band of interest was excised with a clean, sharp razor blade and purified using a Gel Extraction Kit (Qiagen).

21.2.3 Circularizing DNA Templates

Serial dilutions of the digested, purified DNA were prepared. A range of ligation reactions with varying concentrations (10 ng to 1.0 µg) of the purified DNA digests was set up in small tubes. Each reaction mixture contained the appropriate amount of diluted DNA and six Weiss units of T4 ligase in ligation buffer consisting of 66 mM Tris–HCl (pH 7.6), 6.6 mM $MgCl_2$, 0.1 mM ATP, 0.1 mM spermidine, 10 mM DTT, and stabilizers. The tubes were incubated at 16°C for 60 min. The circularized fragments were purified using a CHROMASPIN column.

21.2.4 Inverse PCR

Serial dilutions of the circularized DNA described in Section 21.2.3 were prepared, and PCRs were set up using varying concentrations of circularized DNA. Each PCR contained

- 5 µL 10× Advantage 2 PCR[1] Buffer (Clontech Laboratories)*
- 1 µL 50× dNTP mix[†]
- 1 µL (each) primers (0.5 µM each)
- 5 ng to 0.5 µg circularized DNA
- 1 µL 50× Advantage 2 Polymerase Mix[3] (Clontech Laboratories)[‡]
- PCR Grade water to 50 µL

The PCR program consisted of

- Pre-PCR hold at 94°C for 10–15 s
- 18 rounds of denaturation at 94°C for 12 s
- Annealing/extension at 65°C for 3–5 min (1 min/kb)
- 12 cycles of denaturation at 94°C for 12 s
- Annealing/extension at 65°C for 3–5 min (1 min/kb, adding time increments of 12 s per cycle)
- Final extension at 72°C for 10–15 min to add 3′ "A" overhangs

21.2.5 Analysis of PCR Products

PCR products were analyzed on 1% agarose gels in Tris-acetate-ethylenediaminetetraacetic acid (TAE) buffer using standard electrophoresis protocols. To confirm that the PCR product was the result of authentic amplification from a circular template, an aliquot of the PCR product was digested with the same enzyme (*Eco*RI or *Hin*dIII) that generated the original template DNA, and the digestion products were analyzed by gel electrophoresis.

21.2.6 Cloning and Sequencing the PCR Products

The PCR-amplified products were cloned into pCR2.1 by ligating 3 µL of PCR product to 1 µL of pCR2.1 in a solution that consisted of 1 µL of T4 ligase, 1 µL of 10× ligase buffer, and 4 µL of sterile water incubated overnight at 4°C. The ligation product was used to transform *E. coli* that were subsequently plated on *Luria-Bertani* agar supplemented with 100 µg/mL ampicillin and X-Gal.

---

* 10× Advantage 2 PCR Buffer contains: 400 mM Tricine–KOH (pH 8.7 at 25°C), 150 mM KOAc, 35 mM Mg(OAc)$_2$, 37.5 µg BSA, 0.05% Tween-20, 0.05% Nonidet-P40.

† 50× dNTP mix contains: 10 mM each of dATP, dCTP, dGTP, and dTTP, final concentration of 0.2 mM each nucleotide.

‡ 50× Advantage 2 Polymerase Mix contains Titanium *Taq* DNA polymerase, a nuclease-deficient, N-terminal deletion of *Taq* DNA polymerase plus *Taq*Start Antibody to provide automatic hotstart PCR, and a minor amount of a proofreading polymerase.

Plasmids were extracted from white *E. coli* colonies and reanalyzed on 1% agarose gels as previously described. M13F and M13R primers were used to sequence the plasmid.

21.2.7 CHROMOSOME WALKING USING SEQUENCING RESULTS

Sequencing results can be used to design new primers 3′ and 5′ of the original primer site for repeated rounds of *i*PCR. Typical DNA sequencing reactions yield about 1000 bases; therefore, new primers that are about 100 bases from the end of original sequencing results should be used to extend/walk up the gene.

21.3 RESULTS

Circularized *Eco*RI- or *Hin*dIII-generated genomic DNA fragments yielded sequence information upstream and downstream of the original known DNA sequence (Figure 21.1). PCR amplification of sequences in circularized *Eco*RI-digested genomic DNA yielded sequences predominantly downstream of the PCR priming site (Figure 21.1). Amplification of circularized *Hin*dIII-digested fragments produced sequences upstream and downstream of the PCR priming site (Figure 21.1). Digestion of an aliquot of the amplified product with either *Eco*RI or *Hin*dIII produced two fragments as expected for authentic amplification from a circular template (Figure 21.1).

21.4 DISCUSSION

In this chapter, we have described a general method that can be used to apply the *i*PCR technique to clone contiguous sequences upstream and downstream of a known DNA sequence. Although the technique has been described for the amplification of *Tetrahymena* sequences, it is readily adaptable for use with any genomic DNA template.

A suitable restriction enzyme should generate fragments that are 2–3 kB in length and contain a four base overhang to facilitate ligation. Select an enzyme that has a recognition site predicted to occur frequently within the template DNA based on GC and AT content. For example, in *Tetrahymena*, the AT content of genomic DNA is high, and enzymes such as *Eco*RI and *Hin*dIII, which cut at AT-rich sites, yield fragments in the desired size range. In contrast, enzymes such as *Bam*H1, which cut at GC-rich sites, yield fragments that are too long for efficient use in *i*PCR. Hybridization blotting can be used to confirm the identity of the restriction fragments.

The T_m of the primers should not differ by more than 3–5°C from each other. If the T_m of the primers varies by 5–10°C, use a two-step PCR from the highest to the lowest T_m value. For example, in a two-step PCR reaction in which the T_m of one primer is 68°C and the T_m of the second primer is 62°C start with a cycle consisting of a 94° denaturation step followed by a 68°C combined annealing/extension step. Subtract 0.2°C from the annealing/extension step in each subsequent PCR cycle. A 5-s time increment should be added to the annealing/extension step of each subsequent cycle to compensate for a slight decrease in the DNA polymerase's rate of nucleotide incorporation as the reaction progresses. The last of 30 cycles consist of a 94°C denaturation step followed by a 62°C combined annealing/extension step.

Taq polymerase is ideal for the *i*PCR technique described in this chapter. *Taq* polymerase can be induced to add an extra adenine 3′ overhang to replicated DNA transcripts and therefore can be used to clone the transcript directly into pCR2.1 or other commercial cloning vectors that are prelinearized with 3′ thymine overhangs. The multiple cloning site of pCR2.1 is in the *lacZ* gene, and supplementation of media with X-Gal helps researchers use blue/white screening to locate plasmids containing inserts. Cloning a DNA transcript into pCR2.1 allows use of M13R and M13F primer sites within pCR2.1 to sequence the vector insert and circumvents the need to create new sequencing primers.

Many of the problems encountered with *i*PCR can often be traced to characteristics of the template DNA. In creating the circular DNA template, there is competition between concatamer formation and circularization of DNA fragments. The optimum DNA concentration that promotes circularization varies with the length of the DNA to be circularized and must be determined empirically for each template. In general, the optimum size range for efficient circularization is between 2 and 3 kb. Fragments outside this range fail to circularize efficiently. To lower the high melting temperatures required for denaturation of closed circular DNA, the DNA can be linearized at a site between primers. Alternatively, a PCR buffer containing dimethyl sulfoxide (DMSO) and glycerol (5%) as cosolvents can be used to overcome difficult secondary structure restrictions.

Although *i*PCR products are derived from monomeric circular templates, the presence of concatamers, even at low DNA concentrations, can yield unwanted amplification of noncontiguous sequences. A quick and effective way of distinguishing between an *i*PCR product and a concatamer-based amplification is to digest an aliquot of the amplified product with the same restriction enzyme used to generate the genomic template fragments. An authentic *i*PCR product contains a single internal restriction site corresponding to the original template-generating enzyme. Restriction digestion using this enzyme would therefore cleave the PCR product into two subfragments that, in most cases, are readily resolvable by agarose gel electrophoresis. However, if the restriction site is equidistant from both primers, both subfragments will have the same molecular size. The sum of the molecular sizes of the two subfragments should be equal to the size of the original *i*PCR product. A restriction digest of a concatamer-based amplification will always yield in excess of two subfragments due to the presence of multiple internal cleavage sites. If the *i*PCR generates multiple amplification products, each band should be excised and gel-purified separately. Aliquots of the gel purified DNA can then be used for restriction digestion analysis. This method provides a simple means of verifying an *i*PCR product prior to cloning and sequencing. Amplify3.1 (http:// engels.genetics.wisc.edu/amplify/) is a helpful algorithm in designing primers for amplification by quickly simulating PCRs, thereby helping researchers design primers that are least likely to target undesired regions.

ACKNOWLEDGMENT

The original research on the use of inverse PCR with *Tetrahymena* DNA was supported by grants MCB 9808301, MCB 0130624, and 0517083 from National Science Foundation to R.H.G.

REFERENCES

1. Tsuei, D.J., Chen, P.J., Lai, M.Y., Chen, D.S., Yang, C.S., Chen, J.Y., and Hsu, T.Y. 1994. Inverse polymerase chain reaction for cloning cellular sequences adjacent to integrated hepatitis B virus DNA in hepatocellular carcinomas. *J. Virol. Meth.* 49:269–284.
2. Le, H., Singh, S., Shih, S.J., Du, N., Schnyder, S., Loredo, G.A., Bien, C., Michaelis, L., Toor, A., Diaz, M.O., and Vaughan, A.T. 2009. Rearrangements of the MLL gene are influenced by DNA secondary structure, potentially mediated by Topoisomerase II binding genes, *Chromosomes Cancer* 48:806–815.
3. Singh, S., Le, H., Shih, S.J., and Vaughan, A.T. 2010. Suberoylanilide hydroxyamic acid modification of chromatin architecture affects DNA break formation and repair. *Int. J. Radiat. Oncol. Biol. Phys.* 76:566–573.
4. Ito, M., Kim, Y.G., Tsuji, H., Kiwaki, M., Nomoto, K., Tanaka, R., Okada, N., and Danbara, H. 2010. A practical random mutagenesis system for probiotic *Lactobacillus casei* using Tn5 transposition complexes. *J. Appl. Microbiol.* 109:657–666.
5. García-Jiménez, P., García–Maroto, F., Garrido-Cárdenasb, J.A., Ferrandizc, C., and Robainaa, R.R. 2009. Differential expression of the ornithine decarboxylase gene during carposporogenesis in the thallus of the red seaweed *Grateloupia imbricata* (Halymeniaceae). *J. Plant Physiol.* 166:1745–1754.
6. Han, F., Gong, Q.H., Song, K., Li, J.B., and Yu, W.G. 2004. Cloning, sequence analysis and expression of gene alyVI encoding alginate lyase from marine bacterium *Vibrio* sp. *QY101. DNA Seq.* 15:344–350.

7. Yao, M.C., and Yao, C.H. 1994. Detection of circular excised DNA deletion elements in *Tetrahymena thermophila* during development. *Nucleic Acids Res.* 22:5702–5708.

8. Garcés, J., and Gavin, R.H., 1998. A PCR screen identifies a novel, unconventional myosin heavy chain gene (MYO1) in *Tetrahymena thermophila*, *J. Euk. Microbiol.* 45:252–259.

9. Liu, X., and Gorovsky, M.A., 1996. Cloning and characterization of the major histone H2A genes completes the cloning and sequencing of known histone genes of *Tetrahymena thermophila*. *Nucleic Acids Res.* 24:3023–3030.

10. Peng, Y., Liu, G.L., Yu, X.J., Wang, X.H., Jing, L., and Chi, Z.M. 2010. Cloning of Exo-β-1,3-glucanase gene from a marine yeast *Williopsis saturnus* and its overexpression in *Yarrowia lipolytica*. *Mar. Biotechnol.* 13:193–204.

11. Yamada, K., Terahara, T., Kurata, S., Yokomaku, T., Tsuneda, S., and Harayama, S. 2008. Retrieval of entire genes from environmental DNA by inverse PCR with pre-amplification of target genes using primers containing locked nucleic acids. *Environ. Microbiol.* 10:978–987.

12. Amouric, A., Quéméneur, M., Grossi, V., Liebgott, P.P., Aurial, R., and Casalot, L. 2009. Identification of different alkane hydroxylase systems in *Rhodococcus ruber* strain SP2B, an hexane-degrading actinomycete. *J. Appl. Microbiol.* 108:1903–1916.

13. Topp, S.H., Rasmussen, S.K., Mibus, H., and Sander, L. 2009. A search for growth related genes in *Kalanchoe blossfeldiana*. *Plant Physiol. Biochem.* 47:1024–1030.

14. Wang, F., Hao, J., Yang, C., and Sun, M. 2010. Cloning, expression, and identification of a novel extracellular cold-adapted alkaline protease gene of the marine bacterium strain YS-80-122. *Appl. Biochem. Biotechnol.* 162:1497–1505.

15. Digeon, J.F., Guiderdoni, E., Alary, R., Michaux-Ferrière, N., Joudrier, P., and Gautier, M.F. 1999. Cloning of a wheat puroindoline gene promoter by IPCR and analysis of promoter regions required for tissue-specific expression in transgenic rice seeds. *Plant Mol. Biol.* 39:1101–1112.

16. Chen, L., Tu, Z., Cong, J.H.L., Yan, Y., Jin, L., Yang, G., and He, G. 2009. Isolation and heterologous transformation analysis of a pollen-specific promoter from wheat (*Triticum aestivum* L.). *Mol. Biol. Rep.* 37:737–744.

17. Kurttia, T.J., Mattilaa, J.T., Herrona, M.J., Felsheima, R.F., Baldridgea, G.D., Burkhardta, N.Y., Blazarb, B.R., Hackettc, P.B., Meyerd, J.M., and Munderloha, U.G. 2008. Transgene expression and silencing in a tick cell line: A model system for functional tick genomics Insect. *Biochem. Mol. Biol.* 38:963–968.

18. Sakurai, Y., Komatsu, K., Agematsu, K., and Matsuoka, M. 2009. DNA double strand break repair enzymes function at multiple steps in retroviral infection. *Retrovirology* 6:114.

19. Robert, V.J., and Bessereau, J.L. 2010. Manipulating the *Caenorhabditis elegans* genome using mariner transposons. *Genetica* 138:541–549.

20. Yu, J.G., Lee, G.H., Kim, J.S., Shim, E.J., and Park, Y.D. 2010. An insertional mutagenesis system for analyzing the Chinese cabbage genome using *Agrobacterium* T-DNA. *Mol. Cells* 3:267–275.

21. Kawazu, Y., Fujiyama, R., Noguchi, Y., Kubota, M., Ito, H., and Fukuoka, H. 2010. Detailed characterization of *Mirafiori lettuce virus*-resistant transgenic lettuce. *Transgenic Res.* 19:211–220.

22. Braekeleer, E.D., Meyer, C., Douet-Guilbert, N., Morel, F., Le Bris, M.J., Berthou, C., Arnaud, B., Marschalek, R., Férec, C., and Braekeleer, M.D. 2010. Complex and cryptic chromosomal rearrangements involving the MLL gene in acute leukemia: A study of 7 patients and review of the literature. *Blood Cells Mol. Dis.* 44:268–274.

23. Alves, B.C.A., Hossepian de Lima, V.F.M., and Moreira-Filho, C.A., 2009. Development of Y-chromosome-specific SCAR markers conserved in *Taurine*, *Zebu* and *Bubaline* cattle. *Reprod. Dom. Anim.* 45:1047–1051.

24. Colombo, S., Kumasaka, M., Lobe, C., and Larue, L. 2010. Genomic localization of the Z/EG transgene in the mouse genome. *Genesis* 48:96–100

25. Yue, Z., Xi, L., Guang Li, C., RenYu, X., and Cheng Liang, G. 2009. Expression of hIGF-I in the silk glands of transgenic silkworms and in transformed silkworm cells. *Sci. China Ser. C-Life Sci.* 52:1131–1139.

22 Expression Profiling of MicroRNAs by Quantitative Real-Time PCR

The Good, the Bad, and the Ugly

Mirco Castoldi, Paul Collier, Tania Nolan, and Vladimir Benes

CONTENTS

22.1 OVERVIEW

During the last decade, quantitative real-time PCR (qPCR) has reached its full potential as an essential tool for use in the quantitative analysis of nucleic acids. qPCR is the method of choice for validating results obtained from whole-genome screening (e.g., with microarray or Next Generation Sequencing) and in the screening for a clinically significant, pre-defined subset of genes to be evaluated in patients as predictive[1-3] biomarkers. The intrinsic nature of microRNAs (miRNAs) and their growing importance as diagnostic markers has required the development of dedicated technologies to perform detailed and reproducible analysis, starting with RNA extracted from various sources and with a different range of methods. Currently, biomedical researchers aim to analyze miRNA expression routinely in archival collections of formalin-fixed paraffin embedded (FFPE) material,[4,5] in RNA extracted from laser-capture microdissected (LCM) samples[6,7] as well as from single-cells.[8,9] Additionally, novel approaches have been developed to analyze serum and plasma circulating nucleic acids.[10-14] With the increased demand to achieve complete expression-profiles of miRNAs from a minute amount of material and the technical challenges associated with each methodology, miRNA profiling is slowly reaching the limits of established qPCR techniques.

Therefore, this chapter compares and describes available qPCR methodologies to detect and profile the expression of mature miRNA. Through the chapter, we have introduced novel methodological approaches and compared them to existing techniques. Importantly, in contrast to presenting a stepwise description of different platforms, we discuss the expression profiling of mature miRNAs by qPCR in five main sections: (1) RNA isolation, (2) cDNA synthesis, (3) primer design, (4) detection of amplified products, and (5) data analysis. Within each section, we address the technical challenges associated with each of the described approaches and outline the possible solutions.

22.2   INTRODUCTION

Although the first miRNA (lin-4) was discovered in 1993 by Ambros and colleagues,[15] it was not until the seminal miRNA publications in 2001[16] that the global extent of miRNA-regulatory activity on the eukaryotic transcriptome emerged. Currently, sequences of over 1500 human miRNAs are deposited in the miRBase database (Version 18, November 2011,[17] http://www.mirbase.org/). From bioinformatic predictions, it is suggested that miRNAs may regulate up to 30% of protein-coding genes.[18] Importantly, recent experimental findings corroborate these *in silica* predictions.[19] Significantly, it was shown that by altering the expression of a single miRNA, the expression levels of hundreds, if not thousands, of genes are directly (or indirectly) regulated.[20] Therefore, alteration of miRNA expression in a disease compared to a healthy state and/or correlation of miRNA expression with clinical parameters (i.e., disease progression or therapy response) may indicate that miRNA profiles can serve as clinically relevant biomarkers (see Chapter 19 for a further discussion on miRNA biomarkers). Furthermore, gaining information with regard to differential miRNA expression is an important first step toward further functional characterization that may allow determination of which disease-causing genes are specifically regulated by miRNAs. Whatever the question being addressed, accurate information on miRNA expression in a specific cell type or tissue is often considered an important first step. Currently, a range of quantitative polymerase chain reaction (qPCR)-based approaches are available for the profiling of miRNAs and in this chapter, the reader will find insights into the application of qPCR to assay the expression of miRNA. Specifically the purpose of this chapter is to provide the technical know-how, allowing for the selection of the method most suitable to the experimental setup.

22.3   WHAT ARE miRNAS?

miRNAs are an abundant class of short-, single-stranded, noncoding RNAs (~22 nts) that regulate gene expression at the posttranscriptional level.[21] miRNA activity has been associated with the control of a wide range of processes, such as development,[22,23] differentiation,[24,25] and metabolism.[26] miRNAs control the expression of messenger RNAs by binding to partial complementary target sequences within the 3′ untranslated regions (UTRs) of the protein-coding transcript.[27] The interaction between any given miRNA and its target results in either translation inhibition[28] or mRNA degradation,[29] effectively reducing the transcriptional output of a target gene, without affecting its transcription rate. miRNA biogenesis consists of three sequential steps[30,31]: miRNA genes are transcribed (generally by RNA polymerase II[32]) as long primary transcripts that can be up to several kilobases in length (e.g., the miR-122 primary transcript is over 5 kb both in human and in mouse[33]) that may encode for either a single (i.e., monocistronic) or a large number (e.g., miR-17-92 miRNA polycistron[34]) of miRNAs. The generation of mature miRNA requires that the miRNA primary transcripts undergo two sequential cleavage steps that are carried out by ribonucleoproteins (RNPs) with RNase III-like activities named Drosha[35,36] and Dicer.[37] The first cleavage that takes place in the nucleus (carried out by Drosha) generates the miRNA precursors (pre-miRNAs, 60–100 nts). These intermediate miRNA forms are exported to the cytoplasm (by Exportin 5[38,39]) where a second RNase III-like complex (Dicer) cleaves the pre-miRNAs to generate mature miRNAs, which are then loaded into the RNA-induced silencing complex (RISC[40,41]). The activated miRNA-RISC (often indicated as miRISC) recognizes and binds to miRNA-specific sequences within the 3′UTR of target mRNAs.[27]

Note: For simplicity, in the remaining part of this chapter, "mature-miRNAs" will be referred to simply as "miRNA."

22.4 WHY IS miRNA QUANTIFICATION OF INTEREST?

Why is it important to identify and validate changes in miRNA expression? The accurate profiling of miRNA expression represents an important tool for the investigation of the physiological and pathological states of organisms. Current knowledge indicates that miRNAs participate in the regulation of the transcriptional output of a given cell or a tissue. In many cases, detection of differential expression of miRNAs has established the basis for miRNA functional analysis and specific miRNA expression patterns can provide valuable diagnostic and prognostic indications, for example, in the context of human malignancies.[42–44] Furthermore, miRNAs are important for the replication of pathogenic viruses,[45,46] and small RNAs are also encoded by the genomes of several viruses.[47,48] More recently, the discovery that circulating miRNAs[49] (i.e., as opposed to cellular miRNAs) are present in the blood of humans and other vertebrates has enabled new approaches for the search of novel noninvasive biomarkers.[50] Specifically, profiling of circulating miRNAs is gaining increasing importance as a potential diagnostic and prognostic tool for risk stratification of important diagnostic patient subsets.

22.5 CHALLENGES TO miRNA EXPRESSION PROFILING

The expression profiling of mature miRNAs is technically demanding and profiling miRNA expression with high specificity and sensitivity requires careful evaluation of the miRNA-specific features, as listed in Table 22.1. Currently, miRNA levels can be assessed by various methodological

TABLE 22.1
Challenges Associated with miRNAs Expression Profiling by qPCR

Challenges	Reason
1. Pri-, pre-, and mature miRNAs	miRNA biogenesis is characterized by three maturation steps resulting in the presence of three miRNA forms in the same RNA. Our method of choice should be able to either exclude the unwanted targets from the cDNA. synthesis or should be able to discriminate among them during qPCR.
2. Single strand, ~22 nts long	As the mature miRNA are short and single stranded, it is a challenge to design sensitive and specific RT primers.
3. Lack of common sequences	Mature miRNAs are not characterized by the presence of common sequences such as the poly(A) tail that we find associated to mRNA transcripts. Hence, it is not possible to directly purify or universally reverse transcribe mature miRNAs.
4. Predicted T_m varies between 40°C and 80°C	Mature miRNAs are characterized by variable GC content (33–89%) resulting in a range of predicted T_m (42–91°C). Hence, it is challenging to design qPCR primers that offer both specificity and sensitivity.
5. Closely related family members	Mature miRNA are organized in families, which might be characterized by the presence of homologous family members (e.g., Let-7 family). It is very challenging to design miRNA-specific primers that are able to discriminate among closely related sequences, which might differ by a single nucleotide (e.g., Let-7a and Let-7e).

approaches, including northern blotting,[51] oligonucleotide macroarrays, qPCR-based detection, single-molecule detection in liquid phase,[52] oligonucleotide microarrays,[53] *in situ* hybridization (ISH),[54] as well as massively parallel sequencing.[55,56] Accordingly, the method of choice for analysis will dictate the requirements for RNA isolation, miRNA enrichment, miRNA labeling, detection, and quantification. Any one of the techniques listed above has its own advantages and disadvantages, which must be carefully assessed. However, this chapter will focus exclusively on qPCR and on those factors that, if not correctly addressed, will have a negative impact on expression profiling of miRNAs using qPCR.

22.6   RNA ISOLATION AND QUALITY CONTROL

Over the last few years, our laboratory has tested, as well as developed and optimized dedicated protocols for the analysis and quantification of miRNA (miCHIP[57,58] and miqPCR). Furthermore, we have shown that the reliability and reproducibility of miRNA profiling are a function of the quality of the total RNA used as input material.[59] Thus, a robust method for RNA isolation and quality control (QC) is essential.[60] Generally, two different approaches can be used for the extraction of total RNA from tissues or cells. The commonly adopted techniques include phenol–chloroform-based methods followed by alcohol precipitation or phenol-free column-based approaches (for a comparison of methods, see Ref. 61). Historically, column-based extraction kits (e.g., Qiagen RNeasy) have been designed as total RNA extraction kits with a cutoff to eliminate the small RNA fraction selectively from the total RNA (less than 100 nts). Hence, unless specifically stated as appropriate for miRNA extraction, these approaches are not suitable for retaining RNAs smaller than 100 nts. To obviate this problem, several companies have introduced column-based RNA extraction kits, which can retain the total RNA also containing the small RNA fraction (e.g., Qiagen miRneasy, Exiqon miRcury, Origen Vintage RNA, or Norgen). In general, the purification of RNA through columns allows for the faster isolation of nucleic acids and more efficient removal of water-soluble chemicals that might inhibit the subsequent polymerase activity (e.g., ethanol or phenol). On the other hand, the test conducted in our laboratory suggests that RNA extracted using phenol–alcohol procedures might have a more complete representation of the initial miRNA population and higher yields compared to column-based methods (personal communication). Finally, it should be considered that, if required (i.e., due to the limited amount of RNA available), it is also possible to separate the small and the larger RNA into two fractions (Ambion miRVana, Qiagen mini-elute, Sigma mirPremier, or Ambion Purelink) allowing the same samples to be separately employed for both miRNA and transcriptomic analysis.

As a note, in our laboratory, we routinely isolate RNA using phenol–chloroform-based methods (e.g., Trizol from Invitrogen, Qiazol from Qiagen, or Tri Reagent from Sigma) and RNAs are concentrated through isopropanol or ethanol precipitation. However, there must be careful consideration as to how to extract and isolate RNA for the chosen downstream analysis of small RNA targets.[62]

The size of miRNA molecules makes it difficult to gain a complete and accurate representation of the content from the total RNA and losses through the extraction process can lead to false representation of some miRNA species. Hence, we must be aware that the RNA extraction method we selected must be tailored to suit different aspects of the experimental setup, and methods that are suitable for certain experimental conditions might fail in others. For example, our method of choice for RNA extraction is phenol–chloroform-based method. However, it has been previously shown that there can be less than optimal recovery rates of miRNAs with low guanine cytosine (GC) content when using phenol–chloroform extractions specifically for samples derived from a small number of cells.[6] Furthermore, losses due to factors such as structure and base composition may also be seen when using column-based extraction, with a degree of loss differing between various experimental setups. The loss of accurate representation of these miRNAs is probably caused not just by their structure and sequence but also by the carrier effect of longer cellular RNA and so, a critical factor in small RNA yield after extraction is the amount of total RNA input in the extraction process. Therefore, comparison of miRNA populations from RNAs extracted from identical samples but with different extraction kits, will definitely lead to false-positive and false-negative results. Hence, we strongly advise to select the most suitable extraction method and to carry out the complete study with the same approach.

Once RNAs are extracted, the quantity and the quality of the RNA must be assessed. The most common approach used to assess RNA integrity is to run an aliquot of the sample on an agarose gel and to visualize the RNA after ethidium bromide staining or by fluorimetric visualization. The intact total RNA will have sharp 28S and 18S rRNA bands. As the guiding principle, the 28S rRNA band should be approximately twice as intense as the 18S rRNA band. A drawback of this method is that to visualize the RNA, several hundred nanograms of material must be loaded onto the gel. A viable alternative for RNA analysis is provided by the various microfluidic instruments available (e.g., Agilent Bioanalyzer or BioRad Experion). An advantage of these instruments is that 1–5 ng of total RNA is the maximum amount needed to carry out the analysis of RNA integrity. Furthermore, this approach allows the estimation of quantity and quality of RNA samples in the same assay. Additionally, Agilent has developed chips to be used with its Bioanalyzer instruments that are specifically for analysis of samples consisting of small RNA. The use of these chips allows for the analysis of small RNA populations. As a general comment, assessing the RNA quality based on the degradation of the rRNA might not always be optimal, as partial degradation of mRNA or miRNA transcripts might occur even though the rRNAs are not degraded. However, we have compared the miRNA profiles in identical samples after control degradation using heat and we observed that when there is evidence of sample degradation, as determined by rRNA, the miRNA profile determined after microarray or qPCR differs from the nondegraded control sample.

22.7 ALTERNATIVE APPROACHES TO cDNA SYNTHESIS

The first step in any qPCR is the accurate conversion of RNA into its complementary DNA (cDNA). There are major challenges to converting miRNAs into cDNA (see Table 22.1), such as the limited length of the template (on average 22 nts) and the lack of common miRNA sequences that can be used for enrichment and amplification. In the past, it was thought that the enrichment for small RNA was an essential step to ensure sensitive and specific miRNA analysis. This is because the "target" sequence (i.e., the mature miRNA) is contained in the three different miRNA forms (i.e., mature, pre-, and pri-miRNAs). Hence, fractionation was used to "subtract" at least one of the miRNA forms (the pri-transcript) from the cDNA synthesis. However, more recently, size selection as a factor to increase the specificity of qPCR has been replaced by more accurate cDNA synthesis methods that allow discrimination of the different miRNA forms contained in the total RNA. Two different approaches to reverse-transcribe miRNAs have been employed. In the first approach, miRNAs are reverse transcribed individually by using miRNA-specific reverse-transcription (RT) primers. In the second approach, miRNAs are first extended with the addition of a common sequence and then reverse transcribed by using a universal reverse primer. Although the design of miRNA-specific RT primers can be time-consuming, this type of approach might help one to decrease background signals. Universal RT is useful if several, different miRNAs need to be analyzed. However, in certain cases, it might be impossible to discriminate between mature miRNAs and their precursor forms when using this approach.

22.7.1 cDNA Synthesis Using miRNA-Specific RT Primers

Despite the (mature) miRNAs length being in the same range as the standard polymerase chain reaction (PCR) primer (around 22 nts), it is possible to design miRNA-specific RT primers to prime RT and, at the same time, to elongate the miRNA with a universal sequence. The biggest constraint of this approach is that, to avoid artifacts during the qPCR run, the RT primers should match only a small portion (i.e., not more than 6–8 nts) of the target sequence at the 5'-end of the miRNA. Furthermore, to allow the annealing between RT primers and miRNA and to avoid unwanted annealing of the RT primers to the pre- and the pri-miRNAs forms that are characteristically folded to form stem-loop structures, the annealing and the initial steps of the extension are carried out at low temperatures (usually 16°C). The resulting cDNA is then used as a substrate for a qPCR reaction with a forward miRNA-specific primer (miSP) and a universal reverse primer. There are two different approaches that can be used to design miRNA-specific RT primers. While the 3'-end of the RT primer must be complementary to the miRNA, there are two different approaches for designing the 5'-end of the RT primer. The non-miRNA-specific portion of the RT primer is either folded to resemble a stem loop[9,63] (Figure 22.1a) or to have a simpler linear structure[64] (Figure 22.1b). Stem-loop RT primers are designed to have a short-, single-stranded part that is complementary to the 5'-end of miRNA and a double-stranded part (the stem) and the loop that contains the universal reverse primer-binding sequence. Stem-loop primers are more difficult to design but their structure effectively reduces the chances of annealing between the primer and the unwanted pre- and pri-miRNAs forms, helping to increase the specificity of the assay.

Similarly, the 3'-end of the linear primers is designed to be complementary to the target miRNA and therefore to initiate RT, while the 5'-end of the primer carries a universal sequence that will be used during subsequent qPCR amplification (the universal reverse primer). Although linear RT primers are simpler to design than stem-loop primers, it may not be possible to completely avoid the binding of these RT primers to the different miRNA forms (i.e., pre- and pri-miRNAs) during the cDNA synthesis. This error will eventually carry on through the qPCR resulting in artifacts. Therefore, the annealing and RT conditions must be carefully optimized to avoid any interaction between the RT primers and the pre- and pri-miRNAs (such as by using low-temperature annealing and elongation steps).

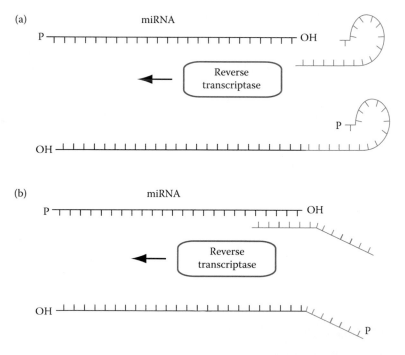

FIGURE 22.1 Schematic representations of cDNA synthesis with miRNA-specific RT primers. RT of miR-NAs by using (a) stem-loop or (b) linear miRNA-specific reverse transcription primers (MSPs). The MSPs are used to prime RT and contain universal sequences to prime amplification. (For detailed explanation, see the text.)

22.7.2 cDNA Synthesis by Tailing RNAs

An alternative approach to synthesizing cDNA from individual miRNAs is to extend the total population of miRNAs through the addition of a universal common sequence. In this approach, enzymes are used to elongate the 3′-end of miRNAs. To achieve this objective, currently available commercial platforms exploit the properties of the *Escherichia coli*-derived poly (A) polymerase (PAP).[65] The enzymatic activity of PAP catalyzes, in a template-independent fashion, the addition of adenosine to the 3′-end of RNA. Therefore, PAP allows the efficient polyadenylation of the 3′-end of all RNAs (including miRNAs) that are contained in the sample. The binding of an oligo(dT) primer to the poly(A) tail is used to prime RT. Importantly, in addition to the poly-dT stretch, the RT primer carries a universal sequence at the 5′-end that will be used during qPCR amplification. To increase the specificity of this assay, the stretch of "dTs" comprised between the 3′end of the miRNA and the 5′end of the universal sequence is precisely determined by anchoring the 5′end of the RT primer to the 3′-end of the miRNA; for this purpose, the "dTs" contain a number of degenerated oligonucleotides (in general 3 nts) followed by a predefined stretch of oligo(dT) and by the universal sequence (Figure 22.2a). The major drawback of this approach is that all the

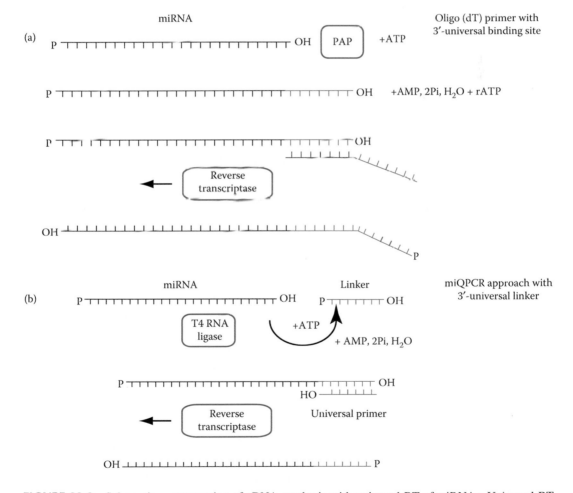

FIGURE 22.2 Schematic representation of cDNA synthesis with universal RT of miRNAs. Universal RT of all the mature miRNAs contained in the RNA sample by enzymatic tailing of the miRNAs by using (a) poly(A) polymerase or (b) T4 RNA ligase. miRNAs are first tailed by tailing and then reverse transcribed by using (a) oligo(dT) or (b) universal primers. (For detailed explanation, see the text.)

RNA species contained in the sample are a substrate of PAP; hence, they are polyadenylated and reverse transcribed in the process. As a consequence, the cDNA will also contain all the three forms for any miRNA contained in the RNA. Furthermore, this protocol does not allow for flexible primer design, and does not compensate for differences in annealing temperature associated with differences in CG content. In addition, standard DNA oligos applied to this approach do not allow the discrimination of the different miRNA forms during the qPCR.

More recently, Exiqon has introduced a novel approach for miRNA analysis by qPCR by combining the PAP-mediated universal RT of RNA with the power of locked-nucleic acids (LNAs)[66,67] when used for sequence discrimination. The poly-adenylation and cDNA-synthesis steps included in this assay are similar to methodological approaches marketed by other companies (i.e., Qiagen and Invitrogen). However, LNA primers are used during the amplification of target sequences and through the use of these modified primers, it is possible to employ two LNA-based mature miRNA-specific primers. This approach greatly reduces the chance of detection of the pre- and pri-miRNA forms.

22.8   NONCOMMERCIAL ALTERNATIVES FOR THE EXPRESSION PROFILING OF miRNAS: miQPCR ASSAY

Recently, our group has invented a novel approach to synthesize cDNA from miRNA that we have named miQPCR (Patent number EP 09 002 587.5, unpublished method. The full details of the miQPCR method are available after signing an MTA with EMBL patent office (Emblem)). miQPCR exploits a novel approach to elongate the 3′-end of mature miRNAs with a common linker adaptor. In this way, all the RNA species in the sample gain a universal 3′ sequence. The extended miRNAs are then converted into cDNA through standard RT using the linker as the RT primer site (Figure 22.2b). The advantage of miQPCR over existing universal elongation approaches using polyadenylation is that the addition of the linker extends all the RNAs within the sample with the same specific sequence. This allows both an increased efficiency for cDNA synthesis, which can now be performed at an optimal temperature range for the RT enzymes (i.e., 42–48°C instead of 16°C), as well as a greater flexibility of the miSP design. Our innovative approach allows the design of T_m-adjusted miSPs to such an extent that we can adjust the predicted T_m of about 98% of all miRNA sequences contained in the miRbase (>60,000 miRNAs) to fall between a predicted T_m of 55–60°C (that is the optimal T_m range for qPCR primers), without being necessary to include modifications such as LNAs or zip-nucleic acids (ZNAs).[68]

22.8.1   CHALLENGE OF miRNA-SPECIFIC PRIMER DESIGN

The specificity and sensitivity of qPCR assays rely on primer design. The design of miRNA-specific primers is particularly challenging. To date, over 60,000 miRNAs have been identified across several genomes (miRBase v18). The human genome alone encodes for more than 1500 individual miRNAs. miRNAs have highly heterogeneous GC content ranging from 33% of GC in miR-144 (T_m 42°C) to 89% of GC in miR-4665-3p (T_m 91°C), differences that generate a large spectrum of predicted T_ms. As qPCR is generally carried out at 60°C, primers designed to target miRNA with high GC content (and therefore with a predicted high T_m) might lack specificity, whereas primers designed to amplify miRNA with low GC content (with a predicted low T_m) might lack sensitivity. This problem needs to be carefully evaluated. Suitable strategies to overcome the technical issues presented by this large range of T_m are directly linked to (i) the type of cDNA we intend to synthesize and (ii) to the method that we will employ to detect the amplicon as it accumulates.

In general, a good starting point for the primer design is to access the miRBase database[17] and to analyze the sequence of the miRNA that is the target for amplification. The important factors to take into account are (i) the GC content and (ii) the presence of closely related, homologous family members. The CG content will influence the T_m of the DNA duplexes. If the predicted T_m is high (>65°C), then it may be necessary to increase specificity by decreasing the primer length. However, if the predicted T_m is low (<55°C), then sensitivity may become an issue. How can we efficiently increase the T_m? One way, which is independent of the type of synthesized cDNA, is to increase the DNA duplex T_m through the use of modified nucleic acid (e.g., LNAs[66,67] or ZNAs[68]). LNA is a synthetic RNA analog that confers increased thermostability to nucleic-acid duplexes. It has been shown that each incorporated LNA monomer increases the T_m of a DNA/DNA + LNA hybrid by up to 5°C. ZNA is an oligo modified by the addition of a polyspermine tail. It has been reported that the use of ZNA moiety might be helpful in increasing the affinity of oligonucleotides for their target by decreasing the electrostatic repulsions due to the polyanionic nature of nucleic acids.[68] Therefore, to achieve the adjustment of the predicted primer T_m to a uniform value, both LNA- and ZNA-modified nucleotides can be added to the primer sequences. However, the drawbacks of LNA and ZNA nucleotides are their cost; LNA and ZNA oligos are significantly more expensive than DNA nucleotides. Furthermore, it might be difficult to position the LNA/ZNA moiety correctly and these modified nucleotides cannot be freely used, as they must be acquired via patent-/license-holding companies (Exiqon and Sigma Aldrich, respectively).

The miQPCR approach that we have developed is proposed as a viable alternative to the use of either LNA or ZNA modification as a solution for T_m adjustment of miRNA-specific primers. The major benefit of the miQPCR method is that all RNA molecules contained in the sample have been identically modified at the 3′ end and this adapter can be utilized in the design of the assay. Given the ability to adjust the predicted T_m of miSP primers over this modification, the miQPCR approach is both more specific and more sensitive when compared to universal approaches based on poly-adenylation tailing. To adjust the T_m to an approximately uniform value, we simply extend the design of the miSPs to match the linker sequence (Figure 22.3), changing the predicted T_m to equal those of the miSPs for other miRNA targets. It is by having flexibility of designing miSPs using the extended sequence of the linker that miQPCR solves the problem of variable T_m within miRNA targets.

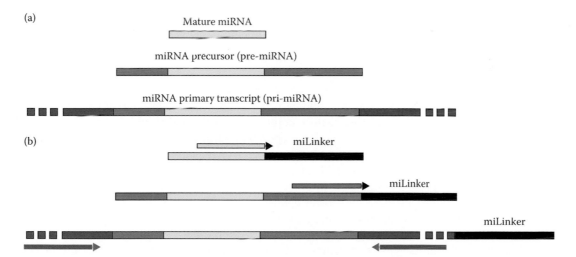

FIGURE 22.3   miQPCR approach enables flexible primer design. miQPCR technique embeds flexibility in the primer design, such that the T_m of miRNA-specific primers can be modified to fall within the desired range for robust and selective qPCR. (a) The target sequence (i.e., mature miRNA) is contained in the three different miRNA forms. (b) Following RNA elongation and RT, miRNA-specific primers are partially overlapping with nucleotides present at the 5′-end of the linker. By using this approach, we are able to adjust both the predicted T_m of the primers to the optimal T_m range as well as to design primers that are able to discriminate among the different miRNA forms.

22.9   HOW TO DISCRIMINATE AMONG MATURE, PRE-, AND PRI-ᴍɪRNAS

One of the challenges to miRNA detection is that there are up to three different RNA forms containing the target sequence (i.e., mature, pre-, and pri-miRNA). Therefore, if the preferred real-time quantitative polymerase chain reaction (RT-qPCR) assay protocol is unable to select the correct sequence to reverse transcribe or to discriminate among these three miRNA forms with specific primer design, the resulting data would be inconclusive due to inaccurate quantification. Importantly, the use of columns to size-fractionate RNA would not completely solve the problem, as both mature and pre-miRNA would be enriched in the small RNA fraction. RT-qPCR assays that perform cDNA synthesis on individual miRNAs (TaqMan® and the original Exiqon approach) perform the selection of the mature miRNA forms to be reverse transcribed during the primer hybridization step of the cDNA synthesis. RNA molecules of precursors and primary miRNA transcripts are folded into a stem-loop secondary structure. Therefore, if RT is carried out at low temperatures (around 16°C), the target sequences within the pre- and pri-miRNAs would not be accessible to the RT primer and effectively only mature miRNA should be reverse transcribed. Conversely, for qPCR assays that require a universal RT of miRNA, all the RNA molecules contained in the sample are reverse transcribed. Thus, following this approach, the PCR primers must be able to discriminate between the different miRNA forms. On the basis of our hands-on experience, among the different commercial platforms using PAP-based RT methods, only the Exiqon approach, thanks to LNA, is able to preferentially amplify the mature miRNAs instead of the pre- and pri-miRNAs. In this platform, all the RNAs in the sample are elongated through the addition of a poly(A) tail (Figure 22.2). The selection of the mature miRNA versus the precursor (and primary) form is controlled by using two LNA-based miRNA-specific qPCR primers. The forward primers comprise most of the miRNA sequence, whereas the reverse primer comprises the universal sequence, which has been introduced during RT, and a number of LNA moiety matching the 3′-end of the target miRNA.

Another method that enables discrimination between the three different miRNA forms is the miQPCR. When using miQPCR, all the RNAs contained in the sample are extended by a specific linker sequence. The presence of the linker sequence enables the design of miRNA-specific primers that are capable of discriminating between the different miRNA forms (Figure 22.3).

22.10   DETECTION OF AMPLIFIED PRODUCTS

The principle of qPCR is based on the detection, in real time, of a fluorescent reporter molecule whose signal intensity correlates with the amount of DNA present at each cycle of amplification. Increases in the reporter dye are plotted against amplification cycles and provide a measure of how much PCR product is generated. Therefore, it is important to identify which detector system is the most suitable to monitor the accumulation of PCR product. Although a number of fluorescent-detection technologies exist for performing qPCR against gDNA or cDNA from mRNA (SYBR Green I, hydrolysis [TaqMan] probes, Molecular Beacons, Light Upon eXtension [LUX], and HybProbes [LightCycler®]), only two of these have been applied to miRNA detection, SYBR Green I dye (Figure 22.4a), and hydrolysis probes (Figure 22.4b).

Detection of DNA duplexes using SYBR Green I dye is the simpler of the two options. SYBR Green I dye is a DNA-binding dye that fluoresces when binding to double-stranded DNA (dsDNA). This property is used to detect amplification products as they accumulate during the PCR run. However, when using SYBR Green I dye, it is not possible to discriminate between different PCR products. Since SYBR Green I dye will bind to any dsDNA, including nonspecific products such as primer–dimers, the final PCR product detected is a mix of specific and nonspecific amplicons. Obviously, this interferes with the accurate quantification of the target sequence, and it is important to evaluate the quality of the assay. To this end, when using SYBR Green I dye, a "melting point"

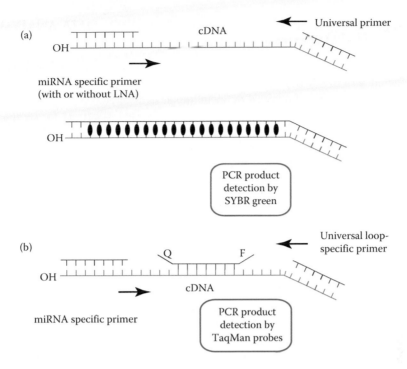

FIGURE 22.4 Amplicon detection by using fluorescent molecules: (a) detection of the amplicon by using SYBR Green I or (b) TaqMan probes. (For detailed explanation, see the text.)

analysis (also referred to as the "melting curve") should be performed. During the melting-curve analysis, the fluorescence emitted by the dye bound to PCR products is recorded at different temperatures (65–95°C). The rising temperature denatures the dsDNA with a consequent reduction of the fluorescent signal because the dye dissociates from single-stranded DNAs. Since T_m is specific for each PCR product, the number of peaks in the dissociation curve will indicate the number of PCR products generated within each well. A good dissociation curve has a single peak (i.e., one PCR product), whereas the presence of multiple peaks indicates the presence of a multiple amplification products instead of a single amplification product.

Detection of the miRNA target amplification product may be achieved through the hydrolysis probe principle. A master mix prepared according to this principle contains a universal and an miRNA-specific primer as well as a third oligonucleotide (the hydrolysis probe) that is designed to hybridize to the central region of the amplified PCR product (i.e., not overlapping with the primers). Hydrolysis probes are modified with a fluorescent reporter at the 5′- and a quencher at the 3′-end. The close proximity of the fluorescent reporter to its quencher molecule prevents fluorescence at the emission wavelength of the reporter. Importantly, the *Taq* DNA polymerase possesses a 5′–3′ exonuclease activity. Therefore, as the polymerase extends the primer and amplifies the template, it will reach the hydrolysis probe. To proceed, the exonuclease function of the enzyme is used to degrade the probe, separating the fluorescent reporter from the quencher such that the reporter fluorescence is released. The emission of fluorescence is proportional to the number of PCR products generated and this allows for accurate quantification of the amplified target. The use of hydrolysis probes increases detection specificity because, even if there are primer–dimers or other nonspecific amplification products, these will not be recognized by the hydrolysis probe and so, they do not contribute to the signal (although it will negatively influence the efficiency of the reaction).

22.11   ASSESSMENT OF SPECIFICITY AND SENSITIVITY, DATA ANALYSIS, AND NORMALIZATION

Comparison of miRNA expression across different samples requires ways to standardize and normalize the efficiency of qPCR runs. Similar to mRNA analysis, the reliability and reproducibility of miRNA experiments can be improved by including several reference genes in the experiment. The readout across the samples from the reference genes is used to correct for sample-to-sample as well as run-to-run variations. Ideally, we would like to identify "housekeeping-like" miRNAs, comparable to traditional reference genes such as analogs to glyceraldehyde-3-phosphate dehydrogenase (GAPDH), β-actin (ACTB), or α-tubulin (TUBA1), which could be used as references to normalize miRNA expression. Unfortunately, to date, no miRNA universal reference gene has been identified. Therefore, the selection of reference genes to be used as a normalizer system is still rather empirical. For example, in the experimental setup that we have adopted, genome-wide approaches (i.e., microarrays) are used to identify differentially regulated as well as invariant miRNAs. qPCR is then used for validation and we can then randomly select three small nucleolar RNAs (snoRNAs) as reference genes and, in addition, we can include at least two invariant miRNAs (i.e., miRNAs with constant expression across the samples as determined by microarray) and use these miRNAs as endogenous controls along with snoRNAs. Once the qPCR run is completed, the relative quantity of each miRNA is estimated by the $\Delta\Delta C_q$ method.[69] Dedicated qPCR software, such as Relative Expression Software Tool (REST)[70] or qBASE[71] is used for these calculations.

An alternative approach is to perform normalization without reference genes. Recently, Vandesompele et al.[72] described a different approach to normalize qPCR experiments involving miRNA quantification. In view of all the difficulties in identifying reliable reference genes, they have opted to exclude reference genes. In their work, Vandesompele's group has normalized qPCR–miRNA profiles against the mean total miRNA expression value in a system comparable to conventional microarray normalization. Although this method generates stable results, possibly beyond the quality achievable when using snoRNAs as normalizers, this approach is only applicable when performing the analysis of a large enough number of miRNAs (i.e., in the hundreds), and so is more appropriate for those using approaches resulting in capture of all miRNAs as cDNA in a multiplexed form (e.g., Exiqon and miQPCR systems).

22.12   WHAT IS NEW ON THE HORIZON? EXPRESSION PROFILING OF CIRCULATING miRNAS

It was recently discovered that cell-free miRNAs circulate in the blood. Circulating miRNAs are either included in lipid or lipoprotein complexes, such as microvesicles, exosomes,[73] or complexes within the RISC complex.[74] Therefore, although ribonucleases are present in the plasma and serum, these miRNAs are highly stable and are readily detected using qPCR. The existence of a population of circulating miRNAs in the blood of healthy, as well as diseased individuals[75] has raised the possibility that changes in miRNA levels may serve as a novel biomarker for diagnosis, prognosis, and monitoring of patient response to treatment.[76] Additionally, the secretory mechanism, biological function, as well as the reason for the existence of extracellular miRNAs remains unknown, therefore, opening the way to novel branches of miRNA research. In view of the great interest in detecting circulating miRNAs, dedicated serum, and plasma miRNA PCR panels have been added to the portfolio of major Biotech companies (e.g., Exiqon). There are three major technical challenges associated with the analysis of circulating miRNA using qPCR: (i) plasma and serum contain polymerase inhibitors, (ii) circulating nucleic acid cannot be detected using spectrophotometers or fluorospectrometers (such as Nanodrop, Bioanalyzer, or Qubit®), and (iii) so far, no standardized reference genes are available to serve as normalization systems. Technical approaches have been developed to solve the problems associated with the presence of inhibitors. Furthermore, current protocols that are in use for the analysis of circulating miRNAs adopt normalization to the total volume of serum or plasma from which miRNA was extracted and used in the reaction. Furthermore, the addition of artificial miRNA spikes, prior to RNA isolation allows for the assessment of, and correction for, differences in RNA recovery, cDNA synthesis, as well as qPCR efficiency. Hence, the remaining challenge associated with the analysis of circulating miRNAs is the lack of reliable normalization, which until the introduction of further technological developments, must be empirically determined.

22.13   CONCLUSION

It is important to consider what can be learned from the analysis of miRNA expression profiles. Alteration of miRNA expression in a disease, compared to a healthy state and/or correlation of miRNA expression with clinical parameters such as disease progression or therapy response may indicate that miRNA profiles can serve as clinically relevant biomarkers. In addition, the information regarding differential miRNA expression in cellular processes such as differentiation, proliferation, or apoptosis is an important first step for further functional characterization that may allow determination of which disease-causing genes are specifically regulated by miRNAs. However, in most cases, miRNA target genes are unknown; hence, the determination of changes in miRNA expression can only be considered as the starting point. This information needs to be complemented with bioinformatic analysis to reveal target-gene predictions and functional validation of putative miRNA targets *in vitro* and *in vivo*. Only in this way, will we be able to draw conclusions about the relevance of changes in miRNA expression profiles in physiological as well as in a diseased state.

REFERENCES

1. Ranade, A.R., Cherba, D., Sridhar, S., Richardson, P., Webb, C., Paripati, A., Bowles, B., and Weiss, G.J. 2010. MicroRNA 92a-2*: A biomarker predictive for chemoresistance and prognostic for survival in patients with small cell lung cancer. *J Thorac Oncol*, 5, 1273–1278.
2. Yang, N., Kaur, S., Volinia, S., Greshock, J., Lassus, H., Hasegawa, K., Liang, S., Leminen, A., Deng, S., Smith, L. et al. 2008. MicroRNA microarray identifies Let-7i as a novel biomarker and therapeutic target in human epithelial ovarian cancer. *Cancer Res*, 68, 10307–10314.
3. Mattie, M.D., Benz, C.C., Bowers, J., Sundinger, K., Wong, L., Scott, G.K., Fedele, V., Ginzinger, D., Getts, R., and Haqq, C. 2006. Optimized high-throughput microRNA expression profiling provides novel biomarker assessment of clinical prostate and breast cancer biopsies. *Mol Cancer*, 5, 24.
4. De Preter, K., Mestdagh, P., Vermeulen, J., Zeka, F., Naranjo, A., Bray, I., Castel, V., Chen, C., Drozynska, E., Eggert, A. et al. 2011. miRNA expression profiling enables risk stratification in archived and fresh neuroblastoma tumor samples. *Clin Cancer Res*, 17, 7684–7692.
5. Goswami, R.S., Waldron, L., Machado, J., Cervigne, N.K., Xu, W., Reis, P.P., Bailey, D.J., Jurisica, I., Crump, M.R., and Kamel-Reid, S. 2010. Optimization and analysis of a quantitative real-time PCR-based technique to determine microRNA expression in formalin-fixed paraffin-embedded samples. *BMC Biotechnol*, 10, 47.
6. Nonn, L., Vaishnav, A., Gallagher, L., and Gann, P.H. 2010. mRNA and micro-RNA expression analysis in laser-capture microdissected prostate biopsies: Valuable tool for risk assessment and prevention trials. *Experimental and molecular pathology*, 88, 45–51.
7. Wang, S., Wang, L., Zhu, T., Gao, X., Li, J., Wu, Y., and Zhu, H. 2010. Improvement of tissue preparation for laser capture microdissection: Application for cell type-specific miRNA expression profiling in colorectal tumors. *BMC Genomics*, 11, 163.
8. White, A.K., VanInsberghe, M., Petriv, O.I., Hamidi, M., Sikorski, D., Marra, M.A., Piret, J., Aparicio, S., and Hansen, C.L. 2011. High-throughput microfluidic single-cell RT-qPCR. *Proc Natl Acad Sci U S A*, 108, 13999–14004.
9. Tang, F., Hajkova, P., Barton, S.C., Lao, K., and Surani, M.A. 2006. MicroRNA expression profiling of single whole embryonic stem cells. *Nucleic Acids Res*, 34, e9.
10. Zhou, X., Marian, C., Makambi, K.H., Kosti, O., Kallakury, B.V., Loffredo, C.A., and Zheng, Y.L. 2012. MicroRNA-9 as potential biomarker for breast cancer local recurrence and tumor estrogen receptor status. *PLoS One*, 7, e39011.

11. Ding, X., Ding, J., Ning, J., Yi, F., Chen, J., Zhao, D., Zheng, J., Liang, Z., Hu, Z., and Du, Q. 2012. Circulating microRNA-122 as a potential biomarker for liver injury. *Molecular Medicine Reports*, 5, 1428–1432.

12. Liu, H., Zhu, L., Liu, B., Yang, L., Meng, X., Zhang, W., Ma, Y., and Xiao, H. 2012. Genome-wide microRNA profiles identify miR-378 as a serum biomarker for early detection of gastric cancer. *Cancer Lett*, 316, 196–203.

13. Tomimaru, Y., Eguchi, H., Nagano, H., Wada, H., Kobayashi, S., Marubashi, S., Tanemura, M., Tomokuni, A., Takemasa, I., Umeshita, K. et al. 2012. Circulating microRNA-21 as a novel biomarker for hepatocellular carcinoma. *J Hepatol*, 56, 167–175.

14. Kosaka, N., Iguchi, H., and Ochiya, T. 2010. Circulating microRNA in body fluid: A new potential biomarker for cancer diagnosis and prognosis. *Cancer Sci*, 101, 2087–2092.

15. Lee, R.C., Feinbaum, R.L., and Ambros, V. 1993. The *C. elegans* heterochronic gene lin-4 encodes small RNAs with antisense complementarity to lin-14. *Cell*, 75, 843–854.

16. Ambros, V. 2001. microRNAs: Tiny regulators with great potential. *Cell*, 107, 823–826.

17. Griffiths-Jones, S., Saini, H.K., van Dongen, S., and Enright, A.J. 2008. miRBase: Tools for microRNA genomics. *Nucleic Acids Res*, 36, D154–158.

18. Lewis, B.P., Burge, C.B., and Bartel, D.P. 2005. Conserved seed pairing, often flanked by adenosines, indicates that thousands of human genes are microRNA targets. *Cell*, 120, 15–20.

19. Vinther, J., Hedegaard, M.M., Gardner, P.P., Andersen, J.S., and Arctander, P. 2006. Identification of miRNA targets with stable isotope labeling by amino acids in cell culture. *Nucleic Acids Res*, 34, e107.

20. Lim, L.P., Lau, N.C., Garrett-Engele, P., Grimson, A., Schelter, J.M., Castle, J., Bartel, D.P., Linsley, P.S., and Johnson, J.M. 2005. Microarray analysis shows that some microRNAs downregulate large numbers of target mRNAs. *Nature*, 433, 769–773.

21. Chen, P.Y. and Meister, G. 2005. microRNA-guided posttranscriptional gene regulation. *Biol Chem*, 386, 1205–1218.

22. Morton, S.U., Scherz, P.J., Cordes, K.R., Ivey, K.N., Stainier, D.Y., and Srivastava, D. 2008. microRNA-138 modulates cardiac patterning during embryonic development. *Proc Natl Acad Sci U S A*, 105, 17830–17835.

23. Visvanathan, J., Lee, S., Lee, B., Lee, J.W., and Lee, S.K. 2007. The microRNA miR-124 antagonizes the anti-neural REST/SCP1 pathway during embryonic CNS development. *Genes Dev*, 21, 744–749.

24. Shivdasani, R.A. 2006. MicroRNAs: Regulators of gene expression and cell differentiation. *Blood*, 108(12), 3646–3653.

25. Zhang, J., Jima, D.D., Jacobs, C., Fischer, R., Gottwein, E., Huang, G., Lugar, P.L., Lagoo, A.S., Rizzieri, D.A., Friedman, D.R. et al. 2009. Patterns of microRNA expression characterize stages of human B-cell differentiation. *Blood*, 113, 4586–4594.

26. Krutzfeldt, J. and Stoffel, M. 2006. MicroRNAs: A new class of regulatory genes affecting metabolism. *Cell Metab*, 4, 9–12.

27. de Moor, C.H., Meijer, H., and Lissenden, S. 2005. Mechanisms of translational control by the 3' UTR in development and differentiation. *Semin Cell Dev Biol*, 16, 49–58.

28. Meister, G. 2007. miRNAs Get an Early Start on Translational Silencing. *Cell*, 131, 25–28.

29. Wu, L., Fan, J., and Belasco, J.G. 2006. MicroRNAs direct rapid deadenylation of mRNA. *Proc Natl Acad Sci U S A*, 103, 4034–4039.

30. Bartel, D.P. 2004. MicroRNAs: genomics, biogenesis, mechanism, and function. *Cell*, 116, 281–297.

31. Kim, V.N. 2005. MicroRNA biogenesis: Coordinated cropping and dicing. *Nat Rev Mol Cell Biol*, 6, 376–385.

32. Faller, M. and Guo, F. 2008. MicroRNA biogenesis: There's more than one way to skin a cat. *Biochim Biophys Acta*, 1779, 663–667.

33. Gatfield, D., Le Martelot, G., Vejnar, C.E., Gerlach, D., Schaad, O., Fleury-Olela, F., Ruskeepaa, A.L., Oresic, M., Esau, C.C., Zdobnov, E.M. et al. 2009. Integration of microRNA miR-122 in hepatic circadian gene expression. *Genes Dev*, 23, 1313–1326.

34. Hayashita, Y., Osada, H., Tatematsu, Y., Yamada, H., Yanagisawa, K., Tomida, S., Yatabe, Y., Kawahara, K., Sekido, Y., and Takahashi, T. 2005. A polycistronic microRNA cluster, miR-17–92, is overexpressed in human lung cancers and enhances cell proliferation. *Cancer Res*, 65, 9628–9632.

35. Davis, B.N., Hilyard, A.C., Nguyen, P.H., Lagna, G., and Hata, A. 2010. Smad proteins bind a conserved RNA sequence to promote microRNA maturation by Drosha. *Mol Cell*, 39, 373–384.

36. Zeng, Y., Yi, R., and Cullen, B.R. 2005. Recognition and cleavage of primary microRNA precursors by the nuclear processing enzyme Drosha. *Embo J*, 24, 138–148.

37. Bernstein, E., Caudy, A.A., Hammond, S.M., and Hannon, G.J. 2001. Role for a bidentate ribonuclease in the initiation step of RNA interference. *Nature*, 409, 363–366.

38. Bohnsack, M.T., Czaplinski, K., and Gorlich, D. 2004. Exportin 5 is a RanGTP-dependent dsRNA-binding protein that mediates nuclear export of pre-miRNAs. *Rna*, 10, 185–191.

39. Kohler, A. and Hurt, E. 2007. Exporting RNA from the nucleus to the cytoplasm. *Nat Rev Mol Cell Biol*, 8, 761–773.

40. Zamore, P.D., Tuschl, T., Sharp, P.A., and Bartel, D.P. 2000. RNAi: Double-stranded RNA directs the ATP dependent cleavage of mRNA at 21 to 23 nucleotide intervals. *Cell*, 101, 25–33.

41. Gregory, R.I., Chendrimada, T.P., Cooch, N., and Shiekhattar, R. 2005. Human RISC couples microRNA biogenesis and posttranscriptional gene silencing. *Cell*, 123, 631–640.

42. Ladeiro, Y., Couchy, G., Balabaud, C., Bioulac-Sage, P., Pelletier, L., Rebouissou, S., and Zucman-Rossi, J. 2008. MicroRNA profiling in hepatocellular tumors is associated with clinical features and oncogene/tumor suppressor gene mutations. *Hepatology*, 47, 1955–1963.

43. Asangani, I.A., Rasheed, S.A., Nikolova, D.A., Leupold, J.H., Colburn, N.H., Post, S., and Allgayer, H. 2007. MicroRNA-21 (miR-21) post-transcriptionally downregulates tumor suppressor Pdcd4 and stimulates invasion, intravasation and metastasis in colorectal cancer. *Oncogene*, 27, 2128–2136.

44. Calin, G.A., Sevignani, C., Dumitru, C.D., Hyslop, T., Noch, E., Yendamuri, S., Shimizu, M., Rattan, S., Bullrich, F., Negrini, M. et al. 2004. Human microRNA genes are frequently located at fragile sites and genomic regions involved in cancers. *Proc Natl Acad Sci U S A*, 101, 2999–3004.

45. Chang, J., Guo, J.T., Jiang, D., Guo, H., Taylor, J.M., and Block, T.M. 2008. Liver-specific microRNA miR-122 enhances the replication of hepatitis C virus in nonhepatic cells. *J Virol*, 82, 8215–8223.

46. Ura, S., Honda, M., Yamashita, T., Ueda, T., Takatori, H., Nishino, R., Sunakozaka, H., Sakai, Y., Horimoto, K., and Kaneko, S. 2009. Differential microRNA expression between hepatitis B and hepatitis C leading disease progression to hepatocellular carcinoma. *Hepatology*, 49, 1098–1112.

47. Bennasser, Y., Le, S.Y., Yeung, M.L., and Jeang, K.T. 2004. HIV-1 encoded candidate micro-RNAs and their cellular targets. *Retrovirology*, 1, 43.

48. Cullen, B.R. 2006. Viruses and microRNAs. *Nat Genet*, 38 Suppl, S25–30.

49. Schwarzenbach, H., Hoon, D.S., and Pantel, K. 2011. Cell free nucleic acids as biomarkers in cancer patients. *Nat Rev Cancer*, 11, 426–437.

50. Brase, J.C., Wuttig, D., Kuner, R., and Sultmann, H. 2010. Serum microRNAs as non-invasive biomarkers for cancer. *Mol Cancer*, 9, 306.

51. Valoczi, A., Hornyik, C., Varga, N., Burgyan, J., Kauppinen, S., and Havelda, Z. 2004. Sensitive and specific detection of microRNAs by northern blot analysis using LNA modified oligonucleotide probes. *Nucleic Acids Res*, 32, e175.

52. Neely, L.A., Patel, S., Garver, J., Gallo, M., Hackett, M., McLaughlin, S., Nadel, M., Harris, J., Gullans, S., and Rooke, J. 2006. A single-molecule method for the quantitation of microRNA gene expression. *Nat Methods*, 3, 41–46.

53. Krichevsky, A.M., King, K.S., Donahue, C.P., Khrapko, K., and Kosik, K.S. 2003. A microRNA array reveals extensive regulation of microRNAs during brain development. *RNA*, 9, 1274–1281.

54. Kloosterman, W.P., Wienholds, E., de Bruijn, E., Kauppinen, S., and Plasterk, R.H. 2006. In situ detection of miRNAs in animal embryos using LNA-modified oligonucleotide probes. *Nat Methods*, 3, 27–29.

55. Buermans, H.P., Ariyurek, Y., van Ommen, G., den Dunnen, J.T., and t Hoen, P.A. 2010. New methods for next generation sequencing based microRNA expression profiling. *BMC Genomics*, 11, 716.

56. Hackenberg, M., Sturm, M., Langenberger, D., Falcon-Perez, J.M., and Aransay, A.M. 2009. miRanalyzer: a microRNA detection and analysis tool for next-generation sequencing experiments. *Nucleic Acids Res*, 37, W68–76.

57. Castoldi, M., Benes, V., Hentze, M.W., and Muckenthaler, M.U. 2007. miChip: A microarray platform for expression profiling of microRNAs based on locked nucleic acid (LNA) oligonucleotide capture probes. *Methods*, 43, 146–152.

58. Castoldi, M., Schmidt, S., Benes, V., Hentze, M.W., and Muckenthaler, M.U. 2008. miChip: An array-based method for microRNA expression profiling using locked nucleic acid capture probes. *Nat Protoc*, 3, 321–329.

59. Ibberson, D., Benes, V., Muckenthaler, M.U., and Castoldi, M. 2009. RNA degradation compromises the reliability of microRNA expression profiling. *BMC Biotechnol*, 9, 102.

60. Becker, C., Hammerle-Fickinger, A., Riedmaier, I., and Pfaffl, M.W. 2010. mRNA and microRNA quality control for RT-qPCR analysis. *Methods*, 50, 237–243.

61. Eldh, M., Lotvall, J., Malmhall, C., and Ekstrom, K. 2012. Importance of RNA isolation methods for analysis of exosomal RNA: Evaluation of different methods. *Mol Immunol*, 50, 278–286.

62. Kim, Y.K., Yeo, J., Kim, B., Ha, M., and Kim, V.N. 2012. Short structured RNAs with low GC content are selectively lost during extraction from a small number of cells. *Mol Cell*, 46, 893–895.

63. Chen, C., Ridzon, D.A., Broomer, A.J., Zhou, Z., Lee, D.H., Nguyen, J.T., Barbisin, M., Xu, N.L., Mahuvakar, V.R., Andersen, M.R. et al. 2005. Real-time quantification of microRNAs by stem-loop RT-PCR. *Nucleic Acids Res*, 33, e179.

64. Raymond, C.K., Roberts, B.S., Garrett-Engele, P., Lim, L.P., and Johnson, J.M. 2005. Simple, quantitative primer-extension PCR assay for direct monitoring of microRNAs and short-interfering RNAs. *RNA*, 11, 1737–1744.

65. Martin, G. and Keller, W. 1998. Tailing and 3'-end labeling of RNA with yeast poly(A) polymerase and various nucleotides. *RNA*, 4, 226–230.

66. Kauppinen, S., Vester, B., and Wengel, J. 2006. Locked nucleic acid: High-affinity targeting of complementary RNA for RNomics. *Handb Exp Pharmacol*, 173, 405–422.

67. Wengel, J., Petersen, M., Nielsen, K.E., Jensen, G.A., Hakansson, A.E., Kumar, R., Sorensen, M.D., Rajwanshi, V.K., Bryld, T., and Jacobsen, J.P. 2001. LNA (locked nucleic acid) and the diastereoisomeric alpha-L-LNA: Conformational tuning and high-affinity recognition of DNA/RNA targets. *Nucleosides Nucleotides Nucleic Acids*, 20, 389–396.

68. Moreau, V., Voirin, E., Paris, C., Kotera, M., Nothisen, M., Remy, J.S., Behr, J.P., Erbacher, P., and Lenne-Samuel, N. 2009. Zip Nucleic Acids: New high affinity oligonucleotides as potent primers for PCR and reverse transcription. *Nucleic Acids Res*, 37, e130.

69. Livak, K.J. and Schmittgen, T.D. 2001. Analysis of relative gene expression data using real-time quantitative PCR and the 2(-Delta Delta C(T)) Method. *Methods*, 25, 402–408.

70. Pfaffl, M.W., Horgan, G.W., and Dempfle, L. 2002. Relative expression software tool (REST) for group-wise comparison and statistical analysis of relative expression results in real-time PCR. *Nucleic Acids Res*, 30, e36.

71. Hellemans, J., Mortier, G., De Paepe, A., Speleman, F., and Vandesompele, J. 2007. qBase relative quantification framework and software for management and automated analysis of real-time quantitative PCR data. *Genome Biol*, 8, R19.

72. Vandesompele, J., De Preter, K., Pattyn, F., Poppe, B., Van Roy, N., De Paepe, A., and Speleman, F. 2002. Accurate normalization of real-time quantitative RT-PCR data by geometric averaging of multiple internal control genes. *Genome Biol*, 3, RESEARCH0034.

73. Lasser, C., Eldh, M., and Lotvall, J. 2012. Isolation and characterization of RNA-containing exosomes. *Journal of Visualized Experiments: JoVE*, 59, e3037.

74. Turchinovich, A., Weiz, L., Langheinz, A., and Burwinkel, B. 2011. Characterization of extracellular circulating microRNA. *Nucleic Acids Res*, 39, 7223–7233.

75. Arroyo, J.D., Chevillet, J.R., Kroh, E.M., Ruf, I.K., Pritchard, C.C., Gibson, D.F., Mitchell, P.S., Bennett, C.F., Pogosova-Agadjanyan, E.L., Stirewalt, D.L. et al. 2011. Argonaute2 complexes carry a population of circulating microRNAs independent of vesicles in human plasma. *Proc Natl Acad Sci U S A*, 108, 5003–5008.

76. Kroh, E.M., Parkin, R.K., Mitchell, P.S., and Tewari, M. 2010. Analysis of circulating microRNA biomarkers in plasma and serum using quantitative reverse transcription-PCR (qRT-PCR). *Methods*, 50, 298–301.

23 Dye-Based High-Throughput qPCR in Microfluidic Platform BioMark™

David Svec, Vendula Rusnakova, Vlasta Korenkova, and Mikael Kubista

CONTENTS

23.1 INTRODUCTION

In biological and medical studies, large-scale gene expression and transcriptome analyses are essential tools.[1] Based on the goals of the investigator, different methods such as RNA microarrays, massive parallel sequencing, or high-throughput real-time quantitative polymerase chain reaction (qPCR) can be used. There are currently four platforms available for high-throughput qPCR: The BioMark™ from Fluidigm, the OpenArray from Life Technologies, the LightCycler® 1536 from Roche, and the Smartchip™ from Wafergen (Table 23.1). These instruments offer the broadest dynamic range, high versatility of experimental setup, low cost per sample, and the lowest demands on the amount of sample material, allowing analysis of minute sample amounts, including single cells.[2,3] One focus of our center is gene expression profiling of single cells and we have used the BioMark as well as the OpenArray system. When we acquired the BioMark, most of our assay libraries were based on nonspecific dyes and so we explored using these established dye-based assays in microfluidic qPCR and extensively optimized the experimental protocol. In this report, we focus on the use of the BioMark system, which uses microfluidic chips called "dynamic arrays" for gene expression profiling, "digital arrays" for digital PCR, and reusable genotyping arrays for single nucleotide polymorphism (SNP) applications.

When using the dynamic array, each sample is mixed with each assay combinatorially in the integrated fluidic circuits (IFC) resulting in $48 \times 48 = 2304$ or $96 \times 96 = 9216$ qPCR reactions. The complete workflow takes 3–4 h and requires approximately 1 h of hands-on time. When compared to conventional qPCR, the high-throughput platform requires 24 or 96 times less effort and time (considering 384- or 96-well blocks). Furthermore, when using the dynamic array, all the reactions are run in parallel under identical thermal and optical conditions; hence, no interplate calibration is needed for studies of up to 96 samples and less bias is introduced. The automatic mixing in the

TABLE 23.1

High-Throughput qPCR Platforms Available as of Autumn 2011[b-e]

	Fluidigm 48.48 Dynamic Array (+Access Array[a])	Fluidigm 96.96 Dynamic Array	Fluidigm FR 48.48 (Genotyping)	Fluidigm 192.24 (Genotyping)	Fluidigm 12.765 Digital Array	Fluidigm 48.770 Digital Array	Life Tech. Openarray	Wafergen Smartchip	Roche LightCycler® 1536
Priming time	11 min	20 min	11 min	10 min	6 min	30 min	—	—	—
Loading time	60 min	95 min	60 min	30 min	40 min	40 min	30 min	>10 min	10–20 min
Number of samples	48	96	48 (reusable)	192	12	48	12–48 (3×)	1–384	1–384
Assays per sample	48	96	48 (reusable)	24	Single/multiplex	Single/multiplex	18–224 (3×)	1–1296	1–1536
qPCR reactions	2304[a]	9216	2304	4608	9180	36,960	3072 (3×)	5184	1536
Minimum input/ sample	5 µL	5 µL	5 µL	4 µL	8 µL	4 µL	3–5 µL	100–400 nL	500–2000 nL
Reaction volume	10 nL	6.75 nL	8 nL	8 nL	6 nL	0.85 nL	33 nL	100 nL	500–2000 nL
Detection	Probe/dye	Probe/dye	Probe	Probe	Probe/dye	Probe/dye	Probe/dye	Probe/dye	Probe/dye
Loader	MX	HX	WX	RX	MX	MX	AccuFill System	Nanodispenser	Innovadyne™
Launched	Spring 2007	Fall 2008	Fall 2010	May 2011	Fall 2006	Spring 2009	Spring 2009	Spring 2010	Summer 2009

a The Access array is for downstream sequencing.

b BioMark™ HD System.

c OpenArray® Real-Time PCR Platform.

d The SmartChip Cycler.

e LightCycler® 1536 Real-Time PCR System.

loading station (IFC controllers with type names MX, HX, WX, and RX, specific for particular arrays) dramatically reduces manual liquid handling. Only 48 + 48 pipetting steps are needed to prepare 2304 qPCRs and 96 + 96 to prepare 9216 qPCRs. Pipetting error can be further reduced by using automatic dispensers and multichannel pipettes, as the arrays are compatible with standard footprints. The cost saving on reagents is substantial, since the BioMark consumes 200 times less reagents and sample material than conventional microtiter plate-based instruments running 20 μL reactions. The volume of the sample mix, composed of mastermix, including dsDNA binding dye + loading buffer + cDNA + ROX, pipetted in an array inlet is 5 μL, which is distributed into 48 or 96 reaction chambers. The reaction volume is 10 nL in the 48 × 48 dynamic array and 6.75 nL in the 96 × 96 dynamic array (Table 23.1).

Nanoliter-scale qPCR requires concentrated samples with a sufficient number of template molecules to allow homogeneous distribution into the reaction chambers. The reaction volumes vary somewhat among the microfluidic arrays, but are typically 2000 times smaller than conventional 20 μL qPCRs. In practice, this usually requires that the samples are preamplified.[4] Preamplification is required because, for example, if 2 μL of cDNA containing 30 copies or more of a target gene is used in a conventional 20 μL qPCR, the reproducibility of technical replicates will be satisfactory. But if the same sample is used in a 48 × 48 array, each chamber will contain an average of 0.015 copies. Only a few chambers will contain target, while most will be empty. Hence, the data would not be reproducible.[5] Poisson variability is higher in the BioMark than in conventional systems due to low template concentration yielding high C_q values (Figure 23.1).[6,7]

When the first BioMark dynamic array was launched in 2007,[8] hydrolysis probes were exclusively used for gene expression analysis. The cost of a probe assay is typically 5–10 times that of primers only, which can add a substantial amount to the cost of entire profiling experiments. Testing and validating assays requires substantial resources when using high-throughput profiling. The assay

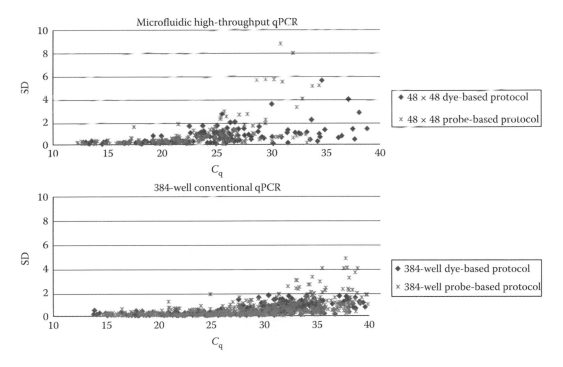

FIGURE 23.1 **(See color insert.)** Dependence of standard deviation on C_q in conventional and microfluidic qPCR systems. In the conventional system, 10 μL qPCR were run in a 384-well plate. The microfluidic system was a 48 × 48 array with 10 nL qPCRs. The same samples and assays were analyzed, although for the BioMark the material was preamplified through 14 cycles.

must be tested to determine efficiency, sensitivity, and specificity.[5] Assays in which primer dimers are formed should be excluded as they will result in unreliable data due to poor sensitivity and compromised reproducibility. This factor is frequently ignored by those using probes for template detection, arguing that nonspecific products do not generate signal. However, parallel amplifications compete for reagents, which leads to an underestimation of the specific target.[9] When using dye-based assays, melt profiles of PCR products can be recorded after each run with no additional cost and these can be used to reveal the presence of competing reactions. Studies of SNPs and methylation patterns can be done at minimal cost by using a similar strategy but by applying postamplification, high-resolution melt (HRM) analysis. If sequencing of the PCR product is required, all PCR products can be harvested as one pool from the 48 × 48 access array ready to be sequenced, instead of microinjecting the nanoliter-scale volumes one by one.

There are some important technical differences between conventional systems using microtiter plates and the microfluidic platforms. The BioMark optical system uses a xenon lamp, optical filters (by default; three out of four filter positions are mounted with filters that are compatible with detection of ROX [6-carboxy-X-rhodamine], FAM [5-carboxyfluorescein], and VIC/JOE [6-carboxy-4′,5′-dichloro-2′,7′-dimethoxyfluorescein]), a charged coupled device (CCD) 9M pixel camera, and a cooling system.[4] The passive reference dye, ROX, is always used in the BioMark platform to identify the chambers and to normalize the signal (Figure 23.3). Before the start of cycling, the CCD camera is focused onto the wells. Then, the autoexposure function sets an exposure time for each fluorescence channel based on the background signal. The objective is to set a baseline such that even a minute increase in fluorescence can be detected with high sensitivity by the CCD chip. The autoexposure calibration is performed by collecting several images with different exposure times. An optimum exposure time is calculated and is then used for the acquisition of qPCR signals. With increasing background, the exposure time to register fluorescence during qPCR is shortened. If the exposure time during acquisition is too short, images are underexposed, which leads to poor-quality data and, in the worst-case scenario, amplification which is not detected above the background.

The calibration setting protocol dates back to the time when the BioMark was introduced in November 2006,[15] and only probe-based assays were used. This included using a calibration temperature of 20°C, and initially the calibration temperature was not adjustable. The calibration settings for Fluidigm's EvaGreen protocol based on the use of Taqman Gene Expression Master Mix® (TGEMM, Life Technologies) use a different approach to that presented here. This requires that a new detector profile is artificially created in the data collection software to reprogram the sensitivity of the CCD. This setting is only optimum for the combination of EvaGreen and TGEMM and can lead to under- or overexposure of the CCD if used for other combinations of dye and mastermix. Here, we present a new calibration approach of the CCD in the BioMark that does not depend on the dye and master mix used, and include the use of the novel dye Chromofy™.

23.2 MATERIALS AND METHODS

Intensity of fluorescence, loading uniformity, and the performance of reporter dyes in the microfluidic qPCR assays using the 48 × 48 dynamic arrays were tested. A GAPDH qPCR assay developed within the SPIDIA project (www.spidia.eu) with efficiency >90% and no primer dimer formation in conventional qPCR was used for tests on the BioMark platform with dye-based and probe-based detection and using several master mixes. The same primer pair was used for all studies: Human GAPDH assay targeting NM_002046.3, 91 bp amplicon, Fwd primer: CCTCCACCTTTGACGCT, Rev primer: TTGCTGTAGCCAAATTCGTT, probe FAM-BHQ1: AGCTTGACAAAGTGGTCGTTGAGGGCAATG. Primers (MWG Operon) were desalted and the hydrolysis probe was HPLC purified (MWG Operon). Final concentrations in the reaction chamber were 400 nM of each primer, 200 nM hydrolysis probe, and recommended amount of 2× DA Assay

Loading Reagent (Fluidigm). The selection of mastermixes used with the probe were ABI Taqman Gene Expression MasterMix (TGEMM) according to Fluidigm's protocol. Reporter dyes were used in KAPA SYBR Fast Universal Mix™ and Finnzymes SYBR Flash Mix™. These mixes were supplied either with SYBR Green I or without SYBR Green I to which we added Chromofy or EvaGreen. Chromofy 5000× (TATAA Biocenter, 6.85 mM) was diluted with nuclease-free water to a 400× working stock. The final concentration in the reaction chamber was 6×. EvaGreen 20× (Biotium) was used in the final 1× concentration in qPCR. Tests were performed using the Universal human cDNA (BioChain Institute Inc.) in different concentrations or after preamplification: Preamplfication of 20 genes (SPIDIA) was carried out in Biorad IQ Supermix™ and contained 5 μL of universal cDNA in 50 μL reaction volume, 50 nM each primer final concentration, and used a amplification protocol consisting of 3 min 95°C followed by 18 cycles of 20 s 95°C, 1 min 58°C, and 30 s 72°C. Test reactions consisting of 5 μL of sample mix contained 0.5 μL of 20× DA sample loading reagent (Fluidigm) and 2.5 μL of mastermix. Concentrations of ROX reference dye (50× Invitrogen, stock 25 μM) in test arrays varied from 25, 50, to 125 nM in qPCR. Prior to loading of samples and assay reagents into the inlets, the array was primed in the NanoFlex™ 4-IFC Controller (Fluidigm), 5 μL of sample mix was pipetted into sample inlets, and 5 μL of assay mix into the assay inlets. Technical replicates were performed on the assay as well as on the sample level to evaluate repeatability. All dynamic arrays were thermocycled using the BioMark Real-Time PCR System (Fluidigm) under the following conditions: Calibration temperature was manually set to 20°C or 72°C made possible with a change in the Windows registry using a *.reg file provided by Fluidigm. The ROX/FAM-MGB detection channel was selected. The temperature protocol was 600 s 95°C activation, 40 cycles followed by either a two-step probe protocol: 95°C 20 s, 60°C 60 s (signal acquisition at 60°C), or a three-step dye protocol: 95°C 20 s, 58°C 20 s, and 72°C for 30 s (signal acquisition at 72°C). Cycling was followed by melt curve analysis with amplicon incubated from 65°C to 95°C at 0.5°C increments and 1 s wait between each increment. Data were collected using the BioMark Data Collection software with separate protocols for the qPCR and for the melting analysis. Data were analyzed using the Fluidigm Real-Time PCR Analysis v3.0.2. and the Fluidigm Melting Curve Analysis v.3.0.2. Linear baseline correction was used and threshold was manually adjusted to be as close as possible to the background. A set of validated assays that had been designed for the EU project SPIDIA (www.spidia.eu) were used for comparison of the BioMark data with measurements performed on a conventional qPCR platform (Roche LightCycler® 384). Images of PCR tubes with GAPDH PCR product were taken using stereomicroscope with green fluorescent protein filter. The melting temperature of GAPDH product in KAPA SYBR mix is 81°C.

23.3 RESULTS AND DISCUSSION

When the BioMark system was installed in our laboratory, several high-throughput projects were planned based on panels of established dye-based assays. The first choice of dye was SYBR Green™, which is the most popular qPCR reporter dye. However, SYBR Green I turned out to perform very poorly on the Biomark (Figure 23.3). After extensive screening of other dyes, we identified Chromofy as a strong, alternative candidate. The performance of the assays (hydrolysis probes, Chromofy) was evaluated on standard human cDNA at different concentrations. The uniformity of the loading process was tested using 48 assay replicates (GAPDH) on cDNA at different concentrations analyzed in sample triplicates.

23.3.1 Calibration of Background Signal

While DNA binding dyes usually generate higher-plateau fluorescent signal than probes and a larger ΔRn value in conventional qPCR systems,[10–12] we found the opposite to occur in the initial runs with SYBR Green on the BioMark. The poor signal had a major impact on data quality in terms of low reproducibility and poor sensitivity. We found that this was caused by inappropriate calibration settings of the CCD camera, which was optimized for probe-based detection. The calibration temperature was set to 20°C by default. The background fluorescence of free probes does not change appreciably with temperature. However, in a reaction mix containing free dye, there is negligible fluorescence, although the reaction mix contains primers and possibly other residual DNA to which the dye may bind. This bound dye fluoresces and its intensity decreases with temperature (Figure 23.2).[10,13]

The strong background signal of the dye at 20°C during calibration results in short exposure time during acquisition, which leads to underexposure during qPCR. To improve the quality of measured data in dye-based reads, we performed background calibration at elevated temperature. For the dyes, we found calibration at 72°C to perform well (Figure 23.3). We were not able to test higher calibration temperature because this is limited in the BioMark Data Collection software.

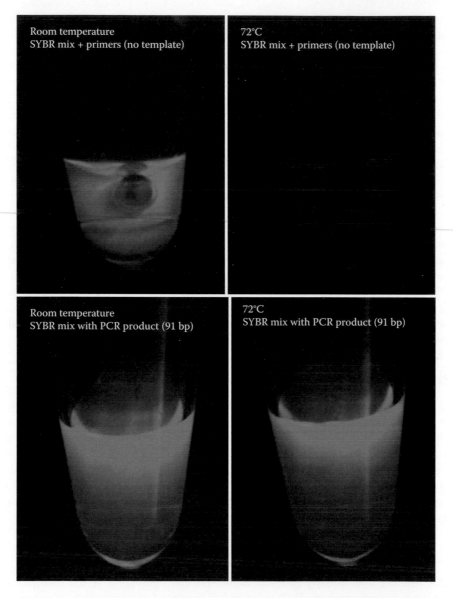

FIGURE 23.2   (See color insert.) Dye fluorescence in conventional qPCR at different temperatures. Left: Two images show a tube with 10 μL SYBR Green master mix (KAPA Universal SYBR mix) in the presence of primers (400 nM per GAPDH primer) before cycling at room temperature and at 72°C. Right: Two images show the tube after cycling in the presence of PCR product at room temperature and at 72°C.

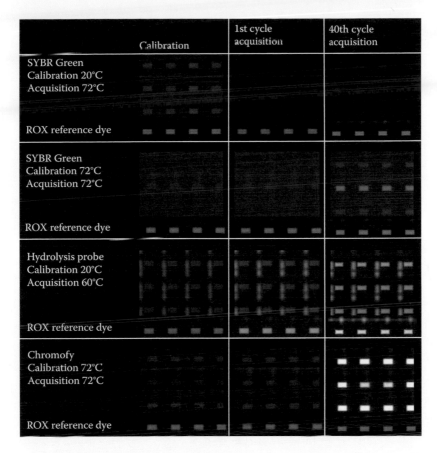

FIGURE 23.3 Camera images showing 12 qPCR chambers in a 48×48 dynamic array. Top row: 20°C calibration. Dye signal is strong during calibration and exposure time is set to 0.39 s. During acquisition at 72°C, no signal is detected after one cycle, and even after 40 cycles, the signal is very weak. Second row: 72°C calibration. Dye signal is weak during calibration and exposure time is set to 3.6 s. Already after one cycle, chambers are distinguishable, and after 40 cycles, strong signals develops. Third row: Using hydrolysis probes, fluorescence at calibration temperature is low and exposure time is set to 2.15 s. Bottom: Chromofy signal calibrated at 72°C, which results in 1.94 s exposure time. ROX reference dye signal shows that fluorescence is independent of calibration temperature. First to fourth row concentration and exposure time: 125 nM, 0.33 s; 125 nM, 0.23 s; 25 nM, 2.15 s; 125 nM, 0.119 s.

This limit depends on the thermal protocol and for some reason that is unclear, this limit is set to the temperature at which signal is acquired. In most three-step PCR protocols, this is at 72°C, while for two-step probe-based PCR, acquisition is typically at 60°C.

To demonstrate the effect of calibration temperature, three dynamic arrays with identical sample and assay setup were used, but signal was generated using either hydrolysis probes, SYBR Green I or Chromofy (Table 23.2). For each array, autoexposure at 20°C was performed. The run was then aborted and new autoexposure was measured at 72°C followed by running qPCR. We observed that the exposure time was almost 10 times longer after calibrating at 72°C compared to calibration at 20°C. There was a slight decrease of exposure times when using hydrolysis probes and including ROX, which suggests that there was only a minor increase of background fluorescence from these labels with decreased temperature. ΔFAM is the difference between initial raw fluorescence signal and the signal measured after the last cycle. Raw fluorescence data measured in the reporter channel are normalized by dividing by the raw data of the passive ROX reference channel and are denoted Rn. Rn values and raw fluorescence data were accessed using the real-time PCR analysis software

TABLE 23.2

Effect of Calibration Temperature on Detection Chemistry

	Hydrolysis Probe—Mix A					Eva Green—Mix A					SYBR Green—Mix B					Chromofy—Mix B				
	ROX (s)	FAM (s)	ΔFAM Raw Signal	Max ΔRn		ROX (s)	FAM (s)	ΔFAM Raw Signal	Max ΔRn		ROX (s)	FAM (s)	ΔFAM Raw Signal	Max ΔRn		ROX (s)	FAM (s)	ΔFAM Raw Signal	Max ΔRn
20°C	1.11	0.63	x	x		1.13	0.22	x	x		1.21	0.391	x	x		1.01	0.53	x	x
72°C	0.88[a]	0.37[a]	4000[a]	0.3[a]		0.84	3	16,000	0.6		0.9	3.15	1600	0.12		0.79	2.73	24,000	2.6

Note: Identical setup of assays and samples are compared. Mix A = TGEMM, Mix B = KAPA Fast Universal.

[a] 60°C.

TABLE 23.3
Effect of Calibration Temperature for Different qPCR Chemistries

Calibration Temperature (°C)	ROX (s)	FAM (s)	Probe/Dye	Master Mix	Array	ΔFAM Raw Signal	Max ΔRn
20	2.159	1.893	Probe	A	48 × 48	25,000	0.4
60	1.516	1.635	Probe	A	48 × 48	19,000	0.9
20	2.21	0.391	SYBR Green	C	48 × 48	1100	0.14
60	1.87	3.15	SYBR Green	C	48 × 48	6000	0.8
72	1.75	3.35	SYBR Green	C	48 × 48	5500	0.7
72	0.34	2.33	Chromofy	C	48 × 48	36,000	2
72	0.328	1.386	Chromofy	C	96 × 96	46,000	2.4

Note: Mix A = TGEMM, Mix C = Finnzymes SYBR Flash Mix.

TABLE 23.4
Sensitivity and Reproducibility

	20°C Calibration	60°C Calibration	72°C Calibration
Mean SD of triplicates	1.14	0.65	0.26
Missing values (%)	15.28	34.30	4.17

module (Fluidigm). The maximum values in initial and terminal cycles within the array were read. (+)Rn represents fluorescence values for a reaction that contain all components, including template, while (−)Rn is the background reading for a negative sample. For positive reactions, the (−)Rn signal is obtained from the early cycles before fluorescence starts to increase. The normalized signal $\Delta Rn = (-)Rn + (+)Rn$ is the fluorescence signal generated using particular ROX concentration and run settings (exposure times).

The qPCR was completed using calibration readings at 72°C and data are presented in Table 23.2. To demonstrate higher sensitivity with calibration at lower temperature, we performed a second study using a different set of dynamic arrays (Table 23.3).

One sample was analyzed using SYBR Green I as the reporter for a set of assays using different calibration temperatures in the qPCR (Table 23.4). Samples were analyzed in triplicate using 48 assays. This generated a total of 144 data points. The assays were loaded in all chips in the same order, which minimized the uniformity bias experienced with SYBR Green I (see below). We observed lower C_q values, reduced standard deviations of technical replicates, and higher call rates after calibration at 72°C than with calibration at 20°C temperature.

23.3.2 Data Uniformity

Sample mixes are conventionally loaded into the IFCs from one edge of the array to fill all the reaction chambers. The surface of the polydimethylsiloxane (PDMS)[14] that contacts the sample mix is much greater than the surface of the polypropylene in conventional microtiter plates. To validate the dye-based protocol, a uniformity test was performed, with the identical GAPDH assay loaded into all 48 assay inlets. Finnzymes Dynamo master mix was used to test the performance of SYBR Green I in first array (1 μL of 20× diluted universal cDNA per 5 μL sample mix) and Chromofy performance was tested on a second array (1 μL of 40× diluted universal cDNA per 5 μL sample mix). Both arrays were run using autoexposure at 72°C. Figure 23.4 shows how C_q varies from the

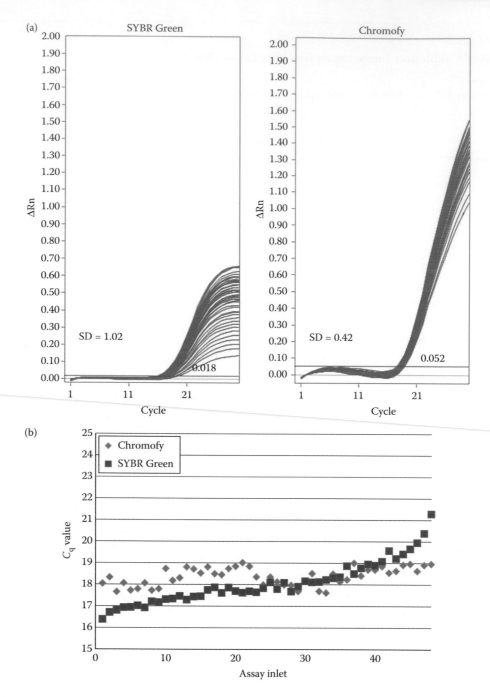

FIGURE 23.4 Amplification curves (a) and C_q values (b) showing uniformity of SYBR and Chromofy assays in 48×48 arrays. All 48 assay inlets were loaded with GAPDH primers.

loading influx. For the SYBR Green I, reporter C_q increases with five cycles from the inlet to the last-filled reaction chamber. For Chromofy, the increase is much more modest (Table 23.5). This drift in C_q probably reflects adsorption of the reporter dye to the PDMS of the IFCs (Figure 23.5).[14]

After optimizing the calibration of the CCD and reagent conditions of the Chromofy protocol, we compared the performance with the recommended probe protocol (Figure 23.6, Table 23.6).

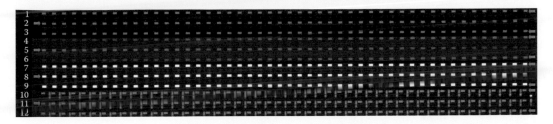

FIGURE 23.5   CCD images showing selected qPCR chambers in ROX and FAM channels. Lines 1–3 are ROX signals, and lines 4–6 show SYBR Green signal. Lines 7–9 show Chromofy signal, and lines 10–12 hydrolysis probe signal.

TABLE 23.5

Uniformity of SYBR and Chromofy Assays Run in a BioMark 48 × 48 Channel

	Mean C_q (48 Replicates)	SD (48 Replicates)	ΔC_q (Max–Min)	ΔC_q (Mean of Five Left Chambers–Mean of Five Right Chambers)
SYBR Green	18	1.01	4.9	3.38
Chromofy	18.3	0.4	1.3	0.9

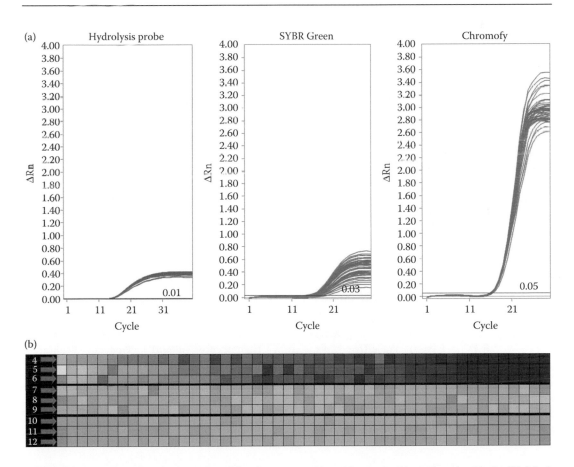

FIGURE 23.6   **(See color insert.)** Amplification curves (a) and heatmap (b) produced with the BioMark software. Blue squares represent high C_qs, yellow low C_qs. Lines 4–6 show SYBR Green signal, lines 7–9 show Chromofy, and lines 10–12 show probe signal.

TABLE 23.6

Reproducibility and Loading Uniformity of Probe- and Dye-Based Assays

	Mean C_q (48 Replicates)	SD (48 Replicates)	ΔC_q (Max–Min)	ΔC_q (Mean of Five Left Chambers–Mean of Five Right Chambers)
Probe	14.64	0.18	0.80	0.09
SYBR Green	16.70	1.00	4.96	3.44
Chromofy	16.11	0.14	0.75	0.30

The Chromofy assay shows comparable reproducibility and loading uniformity to the probe-based assay. The optimized conditions were for the probe assay: 0.625 µL of universal cDNA per 5 µL of sample mix, for the SYBR Green I assay: 0.05 µL cDNA per 5 µL sample mix, and for the Chromofy assay: 0.2 µL cDNA per 5 µL sample mix.

23.3.3 MELTING CURVE ANALYSIS

Optimized calibration temperature also improves the readings of fluorescence during melt curve analysis. The BioMark data collection software has a surprising limit for the maximum calibration temperature that can be used. This is the temperature at which acquisition of the HRM signal starts. For example, if we run a melt curve analysis from 65°C to 95°C, the highest possible calibration temperature is 65°C. The performance of HRM analysis under various conditions is summarized in Table 23.7.

The melting temperature (T_m) is fairly homogeneous across the array when using SYBR Green I or Chromofy reporter (Figure 23.7), demonstrating that it is not appreciably affected by the progressive dye adsorption revealed above. Poor reproducibility is only observed at the edges of the array and is most likely due to evaporation through the partially permeable MSDS caused by extensive heating during thermal cycling.

PCR products can be categorized based on the melting profiles when using version 2.6.× (and higher) of the Fluidigm Melting Curve Analysis. This is useful to identify qPCRs with nonspecific amplification products such as primer dimers. An example of this application on single-cell material is shown in Figure 23.8.

TABLE 23.7

Effect of Calibration Temperature on Melt Curve Analysis

Calibration Temperature (°C)	ROX (s)	FAM (s)	Probe/Dye	Master Mix	Array	Decrease of FAM Fluorescence	–dRn/dT	Average T_m (°C) GAPDH Product	SD (48 Replicates)
20	2.65	1.76	SYBR Green	C	48 × 48	9000	0.02	N/A	N/A
60	1.38	3.64	SYBR Green	C	48 × 48	20,000	0.03	N/A	N/A
65	1.58	3.13	SYBR Green	C	48 × 48	21,000	0.03	81.31	0.16
65	0.3	0.87	Chromofy	C	48 × 48	18,000	0.03	83.52	0.21
65	0.1	0.33	Chromofy	B	48 × 48	21,000	0.03	86.71	0.19
65	0.35	0.22	Chromofy	C	96 × 96	17,000	0.03	85.05[a]	0.14[a]

Note: Mix A = TGEMM, Mix B = KAPA Fast Universal, Mix C = Finnzymes SYBR Flash Mix.

[a] 96 replicates.

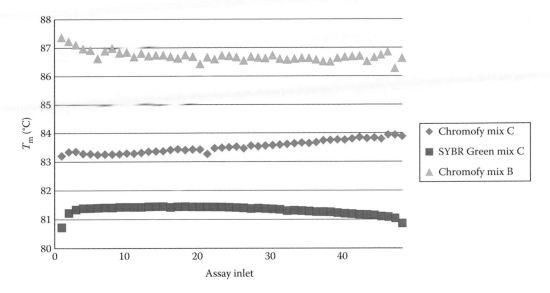

FIGURE 23.7 **(See color insert.)** T_m uniformity across the 48×48 array. Mix B = KAPA Fast Universal, mix C = Finnzymes SYBR Flash Mix.

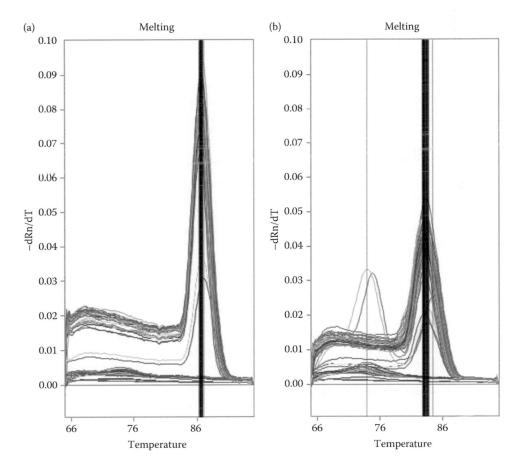

FIGURE 23.8 High-throughput melting analysis of PCR products. (a) Only specific PCR products formed. (b) Some reactions have formed aberrant PCR products characterized by lower T_m.

23.4 CONCLUSIONS

In the BioMark IFC, the fluorescence reporter signal collected during qPCR is normalized to the signal from the passive reference dye ROX based on the background signals collected before the onset of the experiment. In the standard BioMark protocol, the background signal is collected at ambient temperature. ROX is a negatively/uncharged dye at neutral pH that does not interact with DNA. It has a rigid structure and constant, very high fluorescence quantum yield that is essentially independent of temperature. Hydrolysis probes are most frequently based on fluorescein (FAM) reporter and rhodamine quencher (TAMRA) dyes, which have similar properties. Therefore, when using hydrolysis probes, there is essentially no dependence of fluorescence intensity on temperature. Since neither ROX nor hydrolysis probe fluorescence depends on temperature, it is possible to record background signal at ambient temperature even though fluorescence during PCR is recorded at substantially higher temperature.

Most free DNA-binding dyes used as reporters in qPCR are cationic asymmetric cyanines that bind preferentially to double-stranded DNA but show residual binding to single-stranded DNA.[13] The strong fluorescence induced upon binding is induced by the dye being held in a tight conformation, usually in the groove of the double helix or squeezed in between bases in intercalative mode. The tight binding restricts internal motion within the dye that dissipates excitation energy through nonradiative processes, giving rise to intensive fluorescence. The tightness of the binding is temperature-dependent. At elevated temperatures, binding is looser, thus increasing nonradiative decay and thereby reducing fluorescence emission. The background fluorescence of asymmetric cyanine dyes such as SYBR Green I and Chromofy at ambient temperature is therefore much lower than their fluorescence at the temperatures used to record fluorescence during qPCR. As demonstrated in this report, calibration of the CCD at ambient temperature when using dye reporters results in poor quality of the fluorescence signals measured during the qPCR, and the performance can be dramatically improved by calibrating the CCD at high temperature. We also show that the common reporter dye SYBR Green I adsorbs to the PDMS of the ICS resulting in a large drift of C_q values with the distance from the inlet. The adsorption is much less with dye Chromofy.

ACKNOWLEDGMENTS

This work was supported in part by the European Community Seventh Research Framework Programme project SPIDIA (http://www.spidia.eu) Grant Agreement No. 222916 and to the Biotechnology Institute of the Czech Academy of Sciences and the TATAA Biocenter and in part by AV0Z50520701 granted by Ministry of Youth, Education and Sports of the Czech Republic. David Svec is in part supported by the PhD fellowship from Charles University.

REFERENCES

1. Tang, F., Lao, K., and Surani, M.A. Development and applications of single-cell transcriptome analysis. *Nat Methods* **8**, S6–S11, 2011.
2. Stahlberg, A. et al. Defining cell populations with single-cell gene expression profiling: Correlations and identification of astrocyte subpopulations. *Nucleic Acids Res* **39**, e24, 2011.
3. Stahlberg, A. and Bengtsson, M. Single-cell gene expression profiling using reverse transcription quantitative real-time PCR. *Methods* **50**, 282–288, 2010.
4. Mengual, L., Burset, M., Marin-Aguilera, M., Ribal, M.J., and Alcaraz, A. Multiplex preamplification of specific cDNA targets prior to gene expression analysis by TaqMan arrays. *BMC Res Notes* **1**, 21, 2008.
5. Bustin, S.A. et al. The MIQE guidelines: Minimum information for publication of quantitative real-time PCR experiments. *Clin Chem* **55**, 611–622, 2009.
6. Rutledge, R.G. and Stewart, D. Assessing the performance capabilities of LRE-based assays for absolute quantitative real-time PCR. *PLoS One* **5**, e9731, 2010.
7. Wang Z, Spadoro J. Determination of target copy number of quantitative standards used in PCR-based diagnostic assays, in *Gene Quantification* (ed. Ferre F) pp. 31–43 (Birkhäuser, Boston, 1998).

8. Heid, C.A., Stevens, J., Livak, K.J., and Williams, P.M. Real time quantitative PCR. *Genome Res* **6**, 986–994, 1996.

9. Lind, K., Stahlberg, A., Zoric, N., and Kubista, M. Combining sequence-specific probes and DNA binding dyes in real-time PCR for specific nucleic acid quantification and melting curve analysis. *Biotechniques* **40**, 315–319, 2006.

10. Wittwer, C.T., Herrmann, M.G., Moss, A.A., and Rasmussen, R.P. Continuous fluorescence monitoring of rapid cycle DNA amplification. *Biotechniques* **22**, 130–131, 134–138, 1997.

11. Newby, D.T., Hadfield, T.L., and Roberto, F.F. Real-time PCR detection of Brucella abortus: A comparative study of SYBR green I, 5'-exonuclease, and hybridization probe assays. *Appl Environ Microbiol* **69**, 4753–4759, 2003.

12. Arikawa, E. et al. Cross-platform comparison of SYBR Green real-time PCR with TaqMan PCR, microarrays and other gene expression measurement technologies evaluated in the MicroArray Quality Control (MAQC) study. *BMC Genomics* **9**, 328, 2008.

13. Nygren, J., Svanvik, N., and Kubista, M. The interactions between the fluorescent dye thiazole orange and DNA. *Biopolymers* **46**, 39–51, 1998.

14. Perkel, J.M. Microfluidics—Bringing new things to life science. *Science Magazine* (AAAS OPMS, 2008), DOI: 10.1126/science.opms.p0800029.

15. Fluidigm's BioMark™ System Sales Surpass the Century Mark.

24 Surface Plasmon Resonance-Based Biosensor Technology for Real-Time Detection of PCR Products

Giordana Feriotto and Roberto Gambari

CONTENTS

24.1 INTRODUCTION

Surface plasmon resonance (SPR) and biosensor technologies for biospecific interaction analysis (BIA)[1] enable monitoring of DNA:DNA and DNA:RNA hybridization in real time.[2] In the BIAcore biosensor system (Figure 24.1a), plane-polarized light is reflected from the gold-coated sensor chip (Figure 24.1b), where the molecular interactions take place.[2] SPR in the gold layer results in the extinction of the reflected light at a specific angle. This angle (SPR angle) varies with the refractive

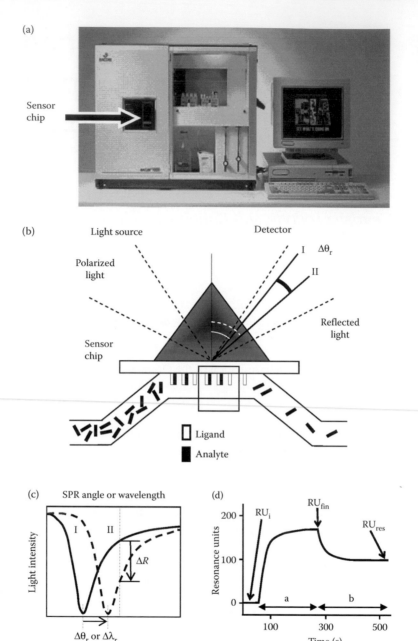

FIGURE 24.1 (a) BIAcore™ instrument. (b) Schematic view of the BIAcore flow-based biosensor. In the BIAcore biosensor systems, the light source provides a beam of light, which is introduced to the SPR coupler (prism) to excite a surface plasmon. The plane-polarized light is reflected from the gold-coated sensor chip, where the molecular interactions take place. The detector records the angular spectrum of reflected light. When molecules bind to the chip, the refractive index of the solution close to the sensor chip and the SPR angle change ($\Delta\theta_r$). The change in SPR angle is measured in resonance units (RU). (c) Typical angular or wavelength spectra of light coupled with a surface plasmon. Depending on which modulation approach is used in the SPR sensor, the refractive index change results in the shift in the SPR angle $\Delta\theta_r$, resonant wavelength $\Delta\lambda_r$, or the change in the amplitude of reflected light ΔR. (d) Typical sensorgram in which biospecific interactions are measured in resonance units (RU). RU_i, initial RU; RU_{fin}, final RU, obtained after the hybridization step; RU_{res}, residual RU, obtained after the washing step. Segments "a" of the graph: Injection of the probe, association phase. Segments "b": Injection of HBS-EP, dissociation phase.

index of the solution close to the other side of the sensor chip (Figure 24.1b, I and II). When molecules (ligands) bind to the chip, the refractive index changes, and with it, the SPR angle (Figure 24.1c, $\Delta\theta_r$). Other SPR sensors detect changes in refractive index by measuring the modulation of the wavelength or intensity of light (Figure 24.1c, $\Delta\lambda_r$ and ΔR). The changes of the SPR angle, wavelength, or light intensity are measured in resonance units (RU).[2] After ligand immobilization, the injection of analyte(s) results in a further increase in RU, only when molecular interactions between the ligand and the analyte occur (Figure 24.1d, segment "a" of the graph). Moreover, injection of binding/running buffer allows for determination of whether this interaction is stable or not (Figure 24.1d, segment "b" of the graph). Sensor chips can be regenerated by removing all the bound analytes by short pulses with suitable buffers.

The change in refractive index is proportional to the change in absorbed mass, permitting the quantitative monitoring of ligand interaction, such as nucleotide hybridization, in real time.[1,2] Therefore, SPR-based BIA has been proposed as a tool to identify and characterize PCR products in a number of biological applications,[3–8] including detection of HIV-1 infection,[3] genetic mutations involved in cystic fibrosis (CF, ΔF508, and W1282X),[5–7] and in thalassemia ($\beta°39$, $\beta°IVSI$-1, β^+IVSI-6, β^+IVSI-110)[8,9] and sequences of genetically modified organisms (Roundup-Ready™ soybean,[10] Bt-176 Maximizer maize).[11,12] The general setup of SPR-based BIA for DNA:DNA hybridization is shown in Figure 24.1b. Biotinylated target oligonucleotides (ligands), complementary to the nucleotide sequences to be studied, are immobilized on streptavidin-coated flow cells (sensor chip SA5) and molecular probes (analytes) injected.[5,6] For SPR-based characterization of PCR products of diagnostic relevance, two different approaches are usually considered. In the first, target PCR products are immobilized on the chip, and the probe (oligodeoxyribonucleotide [ODN], peptide nucleic acid [PNA],[13,14] RNA, or PCR products) is injected. In the second, the molecular probes (usually ODNs, PNAs, or PCR products) are immobilized on the chip, and the PCR products to be analyzed are injected.

Following the injection of ligand, the sensorgram reproducibly shows a gradual increase of RU (starting from initial RU, RU_i) up to the final RU values (RU_{fin}) at the end of the injection (Figure 24.1d). The amount of bound probe can, therefore, be calculated by subtracting the RU_i values from the RU_{fin} values ($RU_{fin} - RU_i$). After washing with HEPES-buffered Saline-EP (HBS-EP), the RU may drop or remain constant, depending on the stability of the DNA:probe complex. The RU remaining after this step is referred to as RU_{res} (Figures 24.1d and 24.2). In some cases, the informative values are $RU_{res} - RU_i$. Sensor chip flow cells are regenerated following a short injection of NaOH.

With respect to molecular probes, some reports indicate that PNAs[13] could be of great interest.[13–16] PNAs are indeed DNA mimics able to hybridize to complementary DNA with high efficiency, since in these molecules, the negatively charged sugar–phosphate backbone is replaced by neutral N-(2-aminoethyl)glycine units.[13] The absence of electrostatic repulsion during DNA:PNA hybrid formation allows even short PNAs to hybridize efficiently to target DNA.[14] In addition, unlike oligonucleotides, PNAs are expected to bind with high efficiency to single-stranded PCR products since their binding could be independent of the secondary structure of target DNA.[14] Nevertheless, the stability of PNA:DNA hybrids is greatly affected by the presence of a single base mismatch.[15]

24.2 MATERIALS AND METHODS

24.2.1 Primer, Molecular Probes, and Target Oligonucleotides

Synthetic HPLC-purified oligonucleotides were purchased from a variety of molecular biology companies such as Sigma-Genosys (Cambridge, U.K.). PNAs were obtained from Professor Rosangela Marchelli (Department of Organic and Industrial Chemistry, Parma University, Italy), but they can also be purchased from several biotechnology companies such as PRIMM (S. Raffaele Biomedical Science Park, Milan, Italy). All primers, ODN and PNA probes, and biotinylated target oligonucleotides were resuspended in sterile, distilled water to obtain 100-μM stock solutions.[5–11]

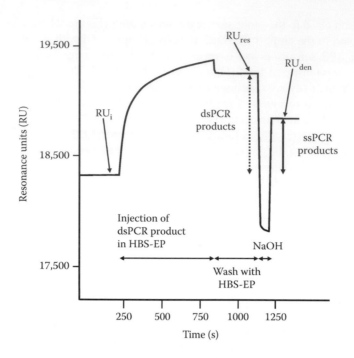

FIGURE 24.2   Increase in resonance units following the injection of 60 µL of 3.5-µM biotinylated double-stranded PCR products on a streptavidin-coated sensor chip. PCR products were injected and washed in HBS-EP and denaturated with 50 mM NaOH. At the end of the procedure, single-stranded PCR products are stably immobilized on the sensor chip (RU$_{den}$). (Modified from Feriotto, G. et al., *Hum. Mutat.*, 18, 70, 2001; Feriotto, G. et al., *J. Agric. Food Chem.*, 51, 4640, 2003. Copyright 2003 American Chemical Society.)

24.2.2   DNA TEMPLATES AND PCR CONDITIONS

Genomic DNA was prepared by lysing white blood cells with 600 µg/mL of Proteinase K (Eurobio, Les Ulis Cedex, France) in 50 mM Tris–HCl, pH 8, 100 mM EDTA, 100 mM NaCl, 1% SDS for 12 h at 55°C and purified by phenol extraction and ethanol precipitation using standard methods.[7] For the extraction of soybean and maize DNA, the Wizard Magnetic DNA Purification System for food (Promega Corporation, Madison, Wisconsin) was used.[10–12] Sequences of PCR primers and DNA or PNA probes used in the experiments reviewed in this work for identification of gag HIV-1 sequences,[3] cystic fibrosis W1282X mutation,[6,7] β-thalassemia mutations,[8,9] Roundup-Ready soybean,[10,12] and Bt-176 Maximizer maize[11,12] sequences are shown in Table 24.1. PCR experiments were performed using a PE Applied Biosystems GeneAmp 9600 thermal cycler. Each PCR was performed in a final volume of 50–100 µL, containing 50–500 ng of target genomic DNA, 50 mM KCl, 10 mM Tris–HCl, pH 8.8, 1.5 mM MgCl$_2$, 33 µM dNTPs, 0.33 µM PCR primers, and 2 U/reaction of *Taq* DNA polymerase (Finnzymes Oy, Espoo, Finland). Thirty to forty PCR cycles were usually performed, and conditions for the denaturation steps were 30 s, 95°C while specific experimental conditions for annealing and extension are reported in Table 24.1. Biotinylated PCR products were generated using an excess (twofold) of the unbiotinylated primer. Asymmetric PCR products were obtained using between 10:1 and 100:1 primer ratio.[5–12]

24.2.3   SPR-BASED BIA

BIAcore 1000™ analytical system (BIAcore AB, Uppsala, Sweden) was used in all experiments. Sensor chips SA5 were precoated with streptavidin, and the running buffer was HEPES buffered

TABLE 24.1

SPR-Based BIA of PCR Products: Sequences of Used ODN and PNA Molecules

Detection of HIV1 Gene Sequences

SK145 (PCR primer)	5′-AGTGGGGGGACATCAAGCAGCCATGCAAAT-3′
SK39 (PCR primer)	5′-TTTGGTCCTTGTCTTATGTCCAGAATGC-3′
SK150 (DNA probe)	5′-TGCTATGTCACTTCCCCTTGGTTCTCTC-3′

PCR conditions: annealing, 1 min, 55°C; elongation, 1 min, 72°C
Asymmetric PCR: SK145:SK39 (10:1) 250 ng/25 ng

Detection of CF W1282X Mutation

CF3 (5′biotinylated PCR primer)	5′-AAGGAGAAATCCAGATCGA-3′
CF2 (PCR primer)	5′-GCTCACCTGTGGTATCACT-3′
N-CF-12 (normal biotinylated target DNA)	5′-Biot-CAGTGGAGGAAA-3′
M-CF-12 (mutated biotinylated target DNA)	5′-Biot-CAGTGAAGGAAA-3′
N-CF-12/10/9[a] (DNA probe)	5′-TT*TCCTCCACTG*-3′
M-CF-12/10/9[a] (DNA probe)	5′-TT*TCCTTCACTG*-3′
N-PNA-CF-9 (PNA probe)	H-TCCTCCACT-NH2
M-PNA-CF-9 (PNA probe)	H-TCCTTCACT-NH2

PCR conditions: annealing, 30 s, 58°C; elongation, 20 s, 72°C

Detection of β-Thalassemia Mutations

Biot-beta-glob-F (PCR primer)	5′-biot-AGTTGGTGGTGAGGCCCTG-3′
Beta-glob-R (PCR primer)	5′-CCCATAACAGCATCAGGAGTGG-3′
N-beta39 (DNA probe)	5′-CTCTGGGTCCAA-3′
M-beta39 (DNA probe)	5′-CTCTAGGTCCAA-3′
N-beta(I)1 (DNA probe)	5′-CCAACCTGCCC-3′
M-beta(I)1 (DNA probe)	5′-CCAATCTGCCC-3′
N-beta(I)6 (DNA probe)	5′ CCTTGATACCA-3′
M-beta(I)6 (DNA probe)	5′-CCTTGATGCCA-3′
N-beta(I)110 (DNA probe)	5′-AGACCAATAGGC-3′
M-beta(I)110 (DNA probe)	5′ AGACTAATAGGC-3′

PCR conditions: annealing, 20 s, 65°C; elongation, 20 s, 72°C

Detection of Roundup-Ready Soybean Transgene

RupR1 (PCR primer)	5′-TGTATCCCTTGAGCCATGTTG-3′
RupR2 (5′biotinylated or not, PCR primer)	5′-CGCACAATCCCACTATCCTTC-3′
RupR-15 (DNA probe)	5′-CTAGAGTCAGCTTGT-3′
Lec1 (PCR primer)	5′-ATGGGCTTGCCTTCTTTCTC-3′
Lec2 (5′biotinylated or not, PCR primer)	5′-CCGATGTGTGGATTTGGTG-3′
Lec-15 (DNA probe)	5′-TCAAGTCGTCGCTGT-3′

PCR conditions: annealing, 30 s, 62°C; elongation, 10 s, 72°C
Asymmetric PCR: RupR1:RupR2 or Lec1:Lec2 (50:1) 0.5 µM/0.01 µM

Detection and Quantification of Bt-176 Maize Transgene

Bt-F (5′biotinylated PCR primer)	5′-AGCCTGTTCCCCAACTACGAC-3′
Bt-R (PCR primer)	5′-TGGTGTAAATCTCGCGGGTC-3′
Bt-p (DNA probe)	5′-GGTGCGGATGGGGTAG-3′
ZM-F (5′biotinylated PCR primer)	5′-TGCAGCAACTGTTGGCCTTAC-3′
ZM-R (PCR primer)	5′-TGTTAGGCGTCATCATCTGTGG-3′
ZM-p (DNA probe)	5′-ATCATCACTGGCATCG-3′

PCR conditions: 30 cycles; annealing, 30 s, 65°C; elongation, 15 s, 72°C

[a] 9- and 10-mer ODN probe sequences are underlined and italicized, respectively.

saline-EP (HBS-EP), containing 10 mM HEPES, pH 7.4, 0.15 M NaCl, 3 mM EDTA, 0.005% (v/v) Surfactant P20, and all were from BIAcore AB (Uppsala, Sweden).

24.2.3.1 Protocol A: Determination of the Optimal Conditions for Hybridization, Including Detection of Point Mutations

For protocol A, the target oligonucleotide is immobilized on a flow cell of the sensor chip, and molecular probes are injected. This protocol is routinely employed to determine experimental conditions for optimal hybridization. When a biotinylated target oligonucleotide is immobilized on a SA5 sensor chip, the binding kinetics are fast and DNA remains bound, even after NaOH washing.[6] When designing molecular probes for the detection of point mutations, great attention should be paid to the length of the synthetic oligonucleotides employed. Under standard BIA conditions (HBS-EP, 25°C), oligonucleotides longer than 14 bp cannot usually be used to discriminate efficiently between full-matched and mismatched biotinylated target oligonucleotides that are immobilized onto the sensor chip.

24.2.3.2 Protocol B: Immobilization of Target PCR Products and Injection of Molecular Probes

The first step of protocol B is to produce a biotinylated, double-stranded PCR product. In general, it is advisable to employ an excess of nonbiotinylated over biotinylated primer for the production of target PCR products. This minimizes the amount of unincorporated biotinylated primer in the product mixture. In the second step, the final PCR product is further purified using Microcon-30 (Millipore Corporation, Bedford, Massachusetts). The third step is the injection of the biotinylated PCR product onto flow cells of an SA5 sensor chip. In this case, the binding kinetic is slow (Figure 24.2) and repeated injections of biotinylated PCR products are usually necessary to reach saturation. The denaturation step with 50-mM NaOH induces a decrease of the RU value (RU_{den}) to about 1/2 of the previously immobilized dsPCR products [$RU_{res} - RU_i$]. The [$RU_{den} - RU_i$] value is dependent on the amount of single-stranded (ss) PCR products stably immobilized onto the sensor chip by biotin–streptavidin interactions.[7,11] After these steps, molecular probes are injected.

24.2.3.3 Protocol C: Immobilization of Molecular Probes and Injection of Target PCR Products

The first step of protocol C uses biotinylated ODNs, PNAs, or PCR products as molecular probes that are immobilized onto sensor chip flow cells. With this experimental approach, the target DNA molecules to be analyzed are usually produced by asymmetric PCR and are directly injected or are injected after a heat treatment with denaturing agents.

Table 24.2 summarizes reports present in the recent literature on the use of the described SPR-BIA protocols for detection and characterization of PCR products.[3–12,17–20]

24.3 RESULTS

24.3.1 Choice of Primers and Probes

In SPR-based analysis of PCR products, the choice of the primers is relevant, at least for some of the experimental approaches described. For protocol C, immobilized probes are employed and asymmetric PCR products are injected. Therefore, primers should be designed for optimal yields of single-stranded asymmetric PCR products. In our experience, this is an important parameter. In addition, single-stranded PCR products (either injected as for protocol C or immobilized on the chip as for protocol B) should exhibit low levels of secondary structure in the BIA experimental conditions, to maximize the hybridization efficiency to molecular probes. Accordingly, it is advisable to study possible secondary structures with the help of dedicated software. For instance, the *mfold* software (version 3.0) developed by ZU.K.er and Mathews[9,21] can be used to determine secondary

TABLE 24.2

Examples of SPR-Based Biosensor Technology for Real-Time Detection of Polymerase Chain Reaction (PCR) Products

Organism	Gene/Mutation	Probe	Protocol[a]	Reference
HIV-1		ODNs	C	3
E. coli	Verotoxin 2 subunit A	ODNs	C	17
		PNAs	C	18
Human	p53 gene mutations	ODNs	B, C	4
	CF ΔF508 mutation	ODNs	A, C	5
	CF W1282X mutation	ODNs	A, B	6
		PNAs	A, B	7
	Thal β°39 mutation	ODNs	A, B	8,9
	Thal β°IVSI-1 mutation	ODNs	B	9
	Thal β+IVSI-6 mutation	ODNs	B	9
	Thal β+IVSI-110 mutation	ODNs	B	9
GMO,[b] soybean	Roundup-Ready	ODNs	A, B, C	10,12
GMO,[b] common markers	P35S promoter, terminator TNOS	ODNs	A, C	19
GMO,[b] maize	Bt-176	ODNs	B	11,12
Fusarium culmorum		ODNs	C	20

[a] See text for details.

[b] Genetically modified organism.

structures of single-stranded PCR products. The analysis should be performed with parameters identical to the BIA experiment, such as the standard conditions 25°C and 0.15 M NaCl.

The length of the probes is also very important for the detection of point mutations with BIA. For example, the presence of a single-nucleotide mismatch does not affect the efficiency of hybridization of injected normal or mutated W1282X CF-17-mer probes to complementary CF-21-mer oligonucleotide immobilized on the sensor chip.[7] Therefore, shorter oligonucleotides should be used for the detection of the CF-associated single-nucleotide mutation W1282X. For instance, while both normal and mutated W1282X CF-12-mer probes hybridize to the target M-CF-21 DNA, the generated DNA:DNA hybrids exhibit different stabilities.[7] Consequently, point mutations can be identified by looking at the difference between residual and initial resonance units ($RU_{res} - RU_i$). Finally, shorter oligonucleotide probes (CF-9-mer) allow for the detection of single-nucleotide mismatches during the association phase; in this case, the difference between final and initial resonance units ($RU_{fin} - RU_i$) is informative. Despite the results obtained with either the CF-12 or CF-9-mer, both probes are informative, and the use of longer probes generates higher RU levels and, therefore, greater sensitivity. This is a very important parameter when these probes are targeted against immobilized PCR products.

24.3.2   CHARACTERIZATION OF PCR PRODUCTS IMMOBILIZED ON FLOW CELLS

In a typical experiment, PCR products are immobilized and molecular probes are injected to detect the presence or the absence of a sequence of interest. This approach has been applied to the detection of Roundup-Ready and lectin soybean gene sequences by injecting RupR-15 or Lec-15 probes (see Table 24.1) to flow cells carrying Roundup-Ready or lectin PCR products. These oligonucleotides generate stable hybrids with the target sequence and high RU_{fin} values.[10] No cross-hybridization of Roundup-Ready probes with immobilized lectin PCR products, or of lectin probes with immobilized Roundup-Ready PCR products, is detectable.[10,12]

In another example, the identification of the CF-associated point mutation W1282X is shown in Figure 24.3. Here, immobilized PCR products were used from normal, W1282X heterozygous, and W1282X homozygous samples. It was clear that 12-mers are useful to detect W1282X CF mutations when the stability of the generated hybrids is analyzed ($RU_{res} - RU_i$ values). The cystic fibrosis index (CF index) was determined as the value $(RU_{res} - RU_i)(N)/(RU_{res} - RU_i)(M)$, where $(RU_{res} - RU_i)(N)$

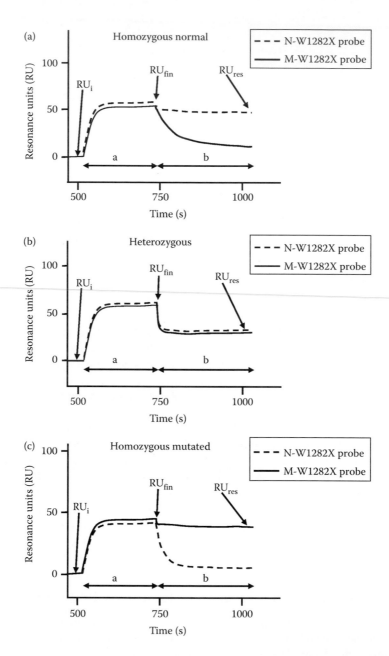

FIGURE 24.3 Sensorgrams obtained after injection of 25 μL containing 0.5 μg of normal (N-W1282X) and mutated (M-W1282X) CF 12-mer ODN probes onto flow cells carrying ssPCR products from normal genomic DNA (a), or heterozygous (b) and homozygous (c) W1282X samples. Segments "a" of the panels: Injection of the probes, association phase. Segments "b": Injection of HBS-EP, dissociation phase. (Modified from Feriotto, G. et al., *Hum. Mutat.*, 18, 70, 2001.)

are the values obtained with the "normal" N-CF-12 ODN probe and the $(RU_{res}-RU_i)(M)$ values are those obtained with the "mutated" M-CF-12 ODN probe. The CF index was found to be high (4.3 ± 0.8) when PCR products from normal subjects were employed (see the representative sensorgrams shown in Figure 24.3a). In contrast, this value approached 1 (1.05 ± 0.35) when PCR products from heterozygous W1282X subjects were employed (see Figure 24.3b). Finally, the CF index was always found to be lower than 0.5 (0.28 ± 0.1) when PCR products from homozygous W1282X samples were immobilized on the SA5 sensor chip (see Figure 24.3c). Thus, SPR-based BIA allows for the detection of point mutations and discrimination between heterozygous and homozygous carriers.[6]

The same protocol was also applied to the analysis of multiple mutations of the human beta-globin gene. To this aim, large-target PCR products were immobilized on sensor chips and then probes designed to detect the four most frequently detected mutations in the Italian population were sequentially injected. The results demonstrate that injections of normal and mutated probes to a large PCR product containing the sites for beta°39, beta°IVSI-1, beta+IVSI-6, and beta+IVSI-110 thalassemia mutations give informative results for all the mutations when the $(RU_{res} - RU_i)$ values are analyzed. Figure 24.4 shows a representative example of the sensorgrams obtained by sequential

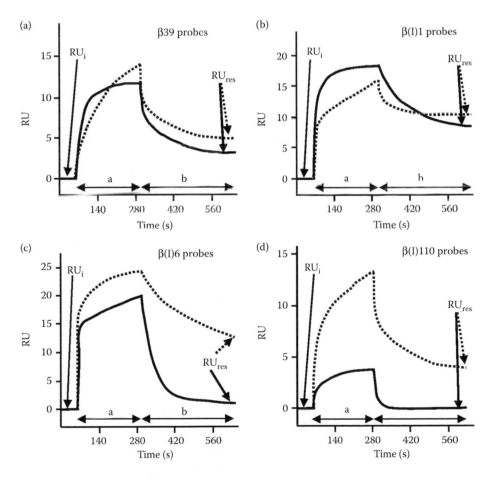

FIGURE 24.4 Sensorgrams obtained after injection of 20 μL containing 0.2 μg of normal (dotted lines) and mutated (solid lines) ODN probes, identifying the β°39 (a), β°IVSI-1 (b), β+IVSI-6 (c), and β+IVSI-110 (d) thalassemia mutations, onto flow cells carrying ssPCR products from a double-heterozygous β°39/β°IVSI-1 patient. Segments "a" of the panels: Injection of the probes, association phase. Segments "b" of the panels: Injection of HBS-EP, dissociation phase. (Modified from Feriotto, G. et al., *Lab. Invest.*, 84, 796, 2004.)

TABLE 24.3

Beta-THAL Index in Homozygous N/N, Heterozygous N/M, and Homozygous M/M Subjects

β-Thalassemia Mutation	Target	Subjects	Determinations	Beta-THAL Index $(RU_{res} - RU_i)N/(RU_{res} - RU_i)M$
β°39	N/N	5	5	8.90 ± 4.65
	N/M	4	3	1.38 ± 0.25
	M/M	1	3	0.42 ± 0.09
β°IVSI-1	N/N	7	18	15.23 ± 7.09
	N/M	2	5	1.37 ± 0.26
	M/M	1	7	0.30 ± 0.22
β⁺IVSI-6	N/N	8	17	6.71 ± 3.89
	N/M	1	6	0.59 ± 0.10
	M/M	1	6	0.098 ± 0.050
β⁺IVSI-110	N/N	8	3	5.33 ± 0.58
	N/M	1	3	1.30 ± 0.61
	M/M	1	3	0.16 ± 0.10

Source: Feriotto, G. et al., *Lab. Invest.*, 84, 796, 2004.

injections of the same four normal (dotted lines) and mutated (solid lines) ODN probes on immobilized PCR products from a double heterozygous beta°39/beta°IVSI-1 beta-thalassemia patient. In this case, the target PCR products hybridize stably (see $(RU_{res} - RU_i)$ values) with the following probes: M-beta39 and N-beta39 (Figure 24.4a), M-beta(I)1 and N-beta(I)1 (Figure 24.4b), N-beta(I)6 and N-beta(I)110 (Figure 24.4c-d, dotted line). Accordingly, M-beta(I)6 and M-beta(I)110 (Figure 24.4c and d, solid lines) probes are completely released from the hybridization complexes during the dissociation phase.

For reproducible molecular diagnosis of all the four thalassemia mutations analyzed, the "beta-thalassemia index" (beta-THAL index) has been calculated as the value $(RU_{res} - RU_i)(N)/(RU_{res} - RU_i)(M)$, where $(RU_{res} - RU_i)(N)$ are the data obtained with the normal N-beta probes, and $(RU_{res} - RU_i)(M)$ are those obtained with the M-beta probes, which are able to recognize the thalassemia mutations. The beta-THAL index of 10 normal subjects was always higher than 2, between 2 and 0.6 when PCR products from seven heterozygous subjects were used, and lower than 0.6 when PCR products from six homozygous thalassemia patients were immobilized on the sensor chip. From the results summarized in Table 24.3, it can be concluded that the discrimination between normal subjects, heterozygous patients, and homozygous patients is readily achieved for all four of the mutations analyzed.[9]

24.3.3 IMMOBILIZATION OF MULTIPLEX PCR PRODUCTS FOR QUANTITATIVE DETERMINATION OF GENETICALLY MODIFIED SEQUENCES

This protocol has been used for quantitative determinations of genetically modified Bt-176 Maximizer maize. For the specific and sensitive detection of genetically modified material in food, the PCR method has proved to be an invaluable tool. The detection of PCR products can be achieved by performing SPR-based BIA. Maize zein and Bt-176 sequences were amplified using a primer mix consisting of Bt-R, ZM-R, and biotinylated Bt-F and ZM-F primers (Table 24.1). Genomic DNA extracted from maize powders containing 0.5% and 2% Bt-176 GM maize was used as the PCR template. The two biotinylated multiplex PCR products were immobilized into independent flow cells of a sensor chip (Figure 24.5a). In a third flow cell, a multiplex PCR product obtained from conventional maize was immobilized. After immobilization, 16-mer oligonucleotide probes recognizing maize zein and Bt-176 sequences were sequentially injected (Figure 24.5b–d). As expected, in the control experiment, no hybridization was obtained following injection of the Bt-probe on the flow cell carrying multiplex PCR from conventional maize (Figure 24.5b, solid line). In contrast, an increase of $RU_{fin} - RU_i$ values was observed after the injection of the zein probe (Figure 24.5b, dotted line). Similar $RU_{fin} - RU_i$ values were obtained when zein probe was injected on flow cells carrying the multiplex PCR products amplified from maize powders containing 0.5% and 2% Bt-176 GM maize (Figure 24.5c and d). When the Bt-176 probe was injected, different $RU_{fin} - RU_i$ values were obtained, corresponding to the different concentrations of Bt-176 genetically modified organism (GMO) in template genomic DNA (compare solid lines of Figure 24.5c and d). The SPR BIA fold increase of the *Bt indices* $((RU_{fin} - RU_i)Bt\text{-}176/(RU_{fin} - RU_i)$ zein) from maize sample containing 0.5% Bt-176 sequences to maize sample containing 2% Bt-176 sequences was 3.14 (average of independent experiments).[11,12] Comparison of the fold increase of the GMO Bt-176 sequences obtained from SPR BIA (3.14) and from real-time quantitative PCR (4.53) demonstrated comparable efficiency of the two techniques in discriminating material containing different amounts of Bt-176 maize.

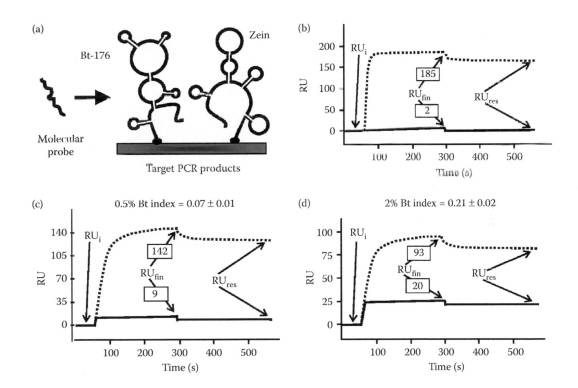

FIGURE 24.5 (a) Experimental strategy for GMO detection and quantification using SPR-based BIA and sensor chips carrying both Bt-176 and zein PCR products obtained by multiplex PCR. (b–d) Sensorgrams obtained after injection of 25 μL containing 40 pmol of Bt-176 (solid lines) and zein (dotted lines) probes on sensor chips carrying multiplex Bt-176 and zein PCR products obtained from maize containing 0% (panel b), 0.5% (panel c), and 2% (panel d) Bt-176 sequences. RU_{fin} values relative to each injection are boxed. 0.5% and 2% Bt index ($RU_{fin\ Bt\text{-}176}/RU_{fin\ zein}$), obtained from four independent experiment, are reported. (Reprinted with permission from Feriotto, G. et al., Quantitation of Bt-176 maize genomic sequences by surface plasmon resonance-based biospecific interaction analysis of multiplex polymerase chain reaction (PCR). *J. Agric. Food Chem.*, 51, 4640, 2003. Copyright 2003 American Chemical Society.)

The results obtained suggest that SPR-based BIA is an easy, speedy, and automated approach to quantify Bt-176 Maximizer sequences in maize.[11] To our knowledge, this is the first example of quantification of PCR products by SPR-based BIA.[11,12]

24.3.4 Use of Peptide Nucleic Acids as Molecular Probes

PNAs have proved to be excellent probes for identifying the W1282X CF mutation. In the experiment shown in Figure 24.6, the hybridization behavior of 9-mer PNA probes was compared to that of 9-mer ODN probes. Interestingly, both N-PNA-CF-9 and M-PNA-CF-9 hybridize to the relative full complementary target DNAs (Figure 24.6a, dotted line; Figure 24.6b, solid line).[7] In this case, hybridization is much more efficient than that of the 9-mer ODN probes and the generated PNA:DNA hybrids are much more stable than DNA/DNA hybrids (Figure 24.6c, dotted line). No hybridization occurs between mismatched PNA probes and target W1282X DNA (Figure 24.6a, solid line; Figure 24.6b, dotted line).[7] When 9-mer PNA and DNA probes were injected onto immobilized PCR products, it was found that only PNA probes were able to give informative results.[7]

24.3.5 Injection of Asymmetric PCR Products to Flow Cells Carrying Specific Probes

This protocol was demonstrated to be feasible for the detection of specific gene sequences, as in the case of HIV-1 infection,[3] and for the identification of GMOs.[10,12] In the case of HIV-1 detection, gag-specific probes recognizing a region of the HIV-1 genome were immobilized. The data obtained demonstrate that this technology is suitable for identifying HIV-1 sequences after direct injection

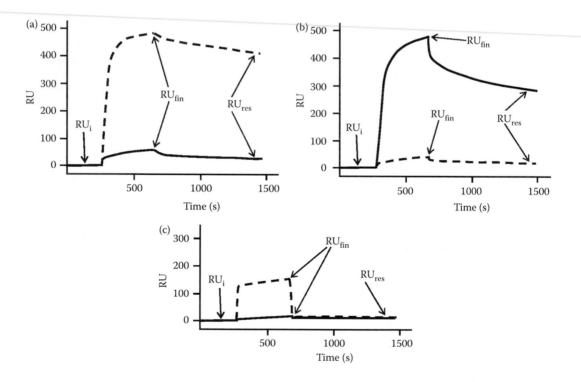

FIGURE 24.6 Sensograms obtained after injection of 25 μL containing 0.5 μg of normal (dotted lines) and mutated (solid line) W1282X CF 9-mer PNA (a, b) or ODN (c) probes to SA5 sensor chip flow cells carrying normal (a, c) or mutated (b) W1282X 12-mer target DNA. (Modified from Feriotto, G. et al., *Lab. Invest.*, 81, 1415, 2001.)

TABLE 24.4

Detection of Roundup and Lectin PCR Products Immobilized on Sensor Chip with ODN or PCR Probes

	Injected Probes			
	15-Mer ODNs		Asymmetric PCR Products	
Immobilized Target DNA	RupR	Lectin	RupR	Lectin
RupR PCR products	13	0	367	0
Lectin PCR products	0	32	0	772

Source: Modified from Gambari, R., Feriotto, G., *J. AOAC Int.*, 89, 893, 2006.

of either purified or unpurified asymmetric PCR products.[3] Table 24.4 shows the hybridization efficiency for flow cells carrying immobilized target PCR products that were used in a study of soybean Roundup-Ready GMO sequences.[12] The results obtained with asymmetric Roundup-Ready and lectin PCR product probes injected into flow cells demonstrate that, although the values of $RU_{fin} - RU_i$ were significantly different, both approaches lead to informative results.[12]

24.4   DISCUSSION

24.4.1   Advantages of SPR-Based BIA for Characterization of PCR Products

The major advantages of using SPR-based analysis employing optical-based biosensors are that (1) analysis is performed in a few minutes, (2) no label is required, and (3) in protocols employing immobilization of molecular probes on the flow cells, the sensor chip can be reused several times. The main limitation is the high cost for most of the commercially available instruments.

With respect to the use of $(RU_{fin} - RU_i)$ or $(RU_{res} - RU_i)$ values, the choice depends on the ability of the employed probes in discriminating target DNA during the association phase of the BIA analysis. In this case, the higher $(RU_{fin} - RU_i)$ values are preferably used for analytical determinations, with highly reproducible results (for instance, in the detection of GMO or HIV-1 sequences).[3,10–12] We cannot generalize on the use of $(RU_{fin} - RU_i)$ values for the detection of point mutations; this strongly depends on the length of the probes and the secondary structures of PCR-generated single-stranded target DNAs. In most instances, when $(RU_{fin} - RU_i)$ values are not informative for detecting point mutations, $(RU_{res} - RU_i)$ values should be taken into consideration (such as in the case shown in Figures 24.3 and 24.4).[7,9]

24.4.2   Problems and Possible Solutions

A partial list of problems and possible solutions is shown in Table 24.5. To maximize the hybridization efficiency in SPR-based analysis of PCR products, the choice of the probes is crucial.[21,22] It appears to be important to determine possible secondary structures of target PCR products; in fact, probes hybridizing with single-stranded PCR regions not involved in secondary structures are expected to be more efficient than probes recognizing double-stranded stretches of target PCR product.[6–9,21–23] The choice of the PCR primers, as well as the possible use of PNAs, is relevant, at least for some of the experimental approaches described.[7,9] In addition, when secondary structure is a problem, one of the possible solutions is to follow the approach described by Kai et al.,[17] producing two asymmetric, but complementary, PCR products of different lengths. After heat denaturation and gradual cooling, these anneal to generate double-stranded molecules with protruding ssDNA ends,

TABLE 24.5

Problems and Possible Solutions of SPR-Based BIA of PCR Products

Problems	Possible Solutions
Asymmetric PCR is not produced efficiently	Synthesize two asymmetric but complementary PCR products differing in length and exposing after annealing the sequence recognized by probes[17]
	Produce dsPCR products using a biotinylated PCR primer and purify single stranded target molecules[12,19]
	Produce dsPCR products using a chimeric 5′-RNA-DNA-3′ primer followed by degradation of the 5′ RNA stretch[22]
Heavy secondary structure of asymmetric PCR products	Change PCR primers
	Increase hybridization temperature[23]
	Change hybridization buffers[16,20,23]
	Use PNA probes[7,16]
	Synthesize two asymmetric but complementary PCR products differing in length and exposing after annealing the sequence recognized by probes[17]
	Produce dsPCR products using a chimeric 5′-RNA-DNA-3′ primer followed by degradation of the 5′ RNA stretch[22]
Hybridization to immobilized probes is not efficient	Change PCR primers in order to modify possible secondary structure of asymmetric PCR product[9]
	Increase the distance between biotin and probe sequences[2]
Probes do not discriminate between full-matched and mismatched target DNA	Decrease length of probes[6]
	Increase hybridization temperature
	Change hybridization buffer
High bulk effects of PCR products	Purify PCR products with Microcon-30[6–12]

exposing the probe-binding site.[17] If the yields of asymmetric PCR are low, a biotinylated primer and streptavidin-coated magnetic beads can be helpful to capture the PCR product strands.[12,19]

24.4.3 FUTURE PERSPECTIVES

Several major issues should be investigated in the near future. The first is the increase of sensitivity of SPR-based assays, enabling the analysis of the genotype and/or gene expression at the level of a single cell. The second is the development of SPR array biosensors, with the objective of parallel analysis of a large number of biomolecular interactions allowing, at least in principle, transcriptomic studies.

Recently published data have described SPR array biosensors based on multielement transduction systems combined with conventional SPR spectroscopy for monitoring binding events on macro- or micropatterned arrays created on disposable sensor chips.[24] These platforms allow for the analysis of several independent, biospecific binding events simultaneously.[25] Studies aimed at developing SPR-based array biosensors for applications in the field of DNA:DNA hybridization are expected to generate very interesting results in the near future.[26,27]

Moreover, SPR imaging (SPR-I)[27,28] has recently been applied to DNA analysis as an extremely versatile method for detecting interactions of biomolecules in a microarray format. SPR-I employs optical detectors for spatial monitoring of localized differences in the reflectivity of the incident light from an array of biomolecules linked to chemically modified gold surfaces.[28] Label-free and real-time analyses can be carried out with high-throughput and low sample consumption.

In SPR-I, DNA or PNA probes that are complementary to the target sequence to be analyzed are immobilized onto the surface. When a hybrid is produced with the target DNA, the SPRI signal is amplified with gold nanoparticles, allowing ultrasensitive detection of DNA at a 1-fm concentration and still presenting single-nucleotide mismatch recognition.[29]

These novel experimental strategies are expected to allow for easy analysis of DNA from a limited number of isolated cells (in theory even single cells).

24.5 SUMMARY

SPR biosensors have become the main tool for the study of biomolecular interactions of nucleic acids in medical diagnostics. In addition, they have been increasingly applied in the detection of biological substances for food safety and security. This chapter reviews the main principles of SPR biosensor technology for monitoring PCR hybridization in real time and discusses the applications of this technology for rapid, sensitive, and specific detection of PCR analytes or for single-base mutation detection.

ACKNOWLEDGMENTS

This research was supported by Telethon, by Fondazione Cassa di Risparmio di Padova e Rovigo, and by MUR-PRIN-2007.

REFERENCES

1. Johnsson, U. et al., Real-time biospecific interaction analysis using surface plasmon resonance and a sensor chip technology, *BioTechniques*, 11, 620, 1991.
2. Nilsson, P. et al., Real-time monitoring of DNA manipulations using biosensor technology, *Anal. Biochem.*, 224, 400, 1995.
3. Bianchi, N. et al., Biosensor technology and surface plasmon resonance for real-time detection of HIV-1 genomic sequences amplified by polymerase chain reaction, *Clin. Diagn. Virol.*, 8, 199, 1997.
4. Nilsson, P. et al., Detection of mutations in PCR products from clinical samples by surface plasmon resonance, *J. Mol. Recognit.*, 10, 7, 1997.
5. Feriotto, G. et al., Detection of ΔF508 (F508del) Mutation of the Cystic Fibrosis Gene by Surface plasmon Resonance and Biosensor Technology, *Hum. Mutat.*, 13, 390, 1999.
6. Feriotto, G. et al., Biosensor Technology for Real-Time Detection of the Cystic Fibrosis W1282X Mutation in CFTR, *Hum. Mutat.*, 18, 70, 2001.
7. Feriotto, G. et al., Peptide nucleic acids and biosensor technology for real-time detection of the cystic fibrosis W1282X mutation by surface plasmon resonance, *Lab. Invest.*, 81, 1415, 2001.
8. Feriotto, G. et al., Surface plasmon resonance and biosensor technology for real-time molecular diagnosis of beta° 39 thalassemia mutation. *Mol. Diagn.*, 8, 33, 2004.
9. Feriotto, G. et al., Real-time multiplex analysis of four beta-thalassemia mutations employing surface plasmon resonance and biosensor technology. *Lab. Invest.*, 84, 796, 2004.
10. Feriotto, G. et al., Biosensor Technology and Surface Plasmon Resonance for real-time detection of genetically modified Roundup Ready soybean gene sequences, *J. Agr. Food Chem.*, 50, 955, 2002.
11. Feriotto, G. et al., Quantitation of Bt-176 maize genomic sequences by surface plasmon resonance-based biospecific interaction analysis of multiplex polymerase chain reaction (PCR). *J. Agric. Food Chem.*, 51, 4640, 2003.
12. Gambari, R., Feriotto, G., Surface plasmon resonance for detection of genetically modified organisms in the food supply. *J AOAC Int.*, 89, 893, 2006.
13. Nielsen, P. E. et al., Sequence selective recognition of DNA by strand displacement with a thymine-substituted polyamide, *Science*, 254, 1497, 1991.
14. Wang, J., DNA biosensors based on Peptide Nucleic Acid (PNA) recognition layers, *Biosens. Bioelectron.*, 13, 757, 1998.
15. Wang, J. et. al., Detection of point mutation in the p53 gene using peptide nucleic acid biosensor, *Anal. Chim. Acta*, 344, 111, 1997.

16. Sawata, S. et al., Application of Peptide Nucleic Acid to the direct detection of deoxyribonucleic acid amplified by polymerase chain reaction, *Biosens. Bioelectron.*, 14, 397, 1999.
17. Kai, E. et al., Detection of PCR products in solution using surface plasmon resonance, *Anal. Chem.*, 71, 796, 1999.
18. Kai, E. et al., Detection of PCR products of Escherichia coli O157:H7 in human stool samples using surface plasmon resonance (SPR), *FEMS Immunol. Med. Microbiol.*, 29, 283, 2000.
19. Mariotti, E. et al., Surface plasmon resonance biosensor for genetically modified organisms detection, *Anal. Chim. Acta*, 453, 165, 2002.
20. Zezza, F. et al., Detection of Fusarium culmorum in wheat by a surface plasmon resonance-based DNA sensor. *J. Microbiol. Methods*, 66, 529, 2006.
21. Mathews, D. H. et al., Expanded sequence dependence of thermodynamic parameters improves prediction of RNA secondary structure, *J. Mol. Biol.*, 288, 911, 1999.
22. Miyachi, H. et al., Application of chimeric RNA–DNA oligonucleotides to the detection of patogenic microorganisms using surface plasmon resonance, *Anal. Chim. Acta*, 407, 1, 2000.
23. Giakoumaki E. et al., Combination of amplification and post-amplification strategies to improve optical DNA sensing, *Biosens Bioelectron.*, 19, 337, 2003.
24. O'Brien, M. J. II et al., A surface plasmon resonance array biosensor based on spectroscopic imaging, *Biosens. Bioelectron.*, 16, 97, 2001.
25. Nelson, B. P. et al., Surface plasmon resonance imaging measurements of DNA and RNA hybridization adsorption onto DNA microarrays, *Anal. Chem.*, 73, 1, 2001.
26. Brockman, J. M. et al., Surface plasmon resonance imaging measurements of ultrathin organic films, *Annu. Rev. Phys. Chem.*, 51, 41, 2000.
27. Lecaruyer, P. et al., Surface plasmon resonance imaging as a multidimensional surface characterization instrument–application to biochip genotyping, *Anal. Chim. Acta*, 573–574, 333, 2006.
28. Piliarik, M. et al., A new surface plasmon resonance sensor for high-throughput screening applications. *Biosens. Bioelectron.*, 20, 2104, 2005.
29. D'Agata, R. et al., Ultrasensitive detection of DNA by PNA and nanoparticle-enhanced surface plasmon resonance imaging. *Chem Bio Chem*, 9, 2067, 2008.

25 Current Innovations in the Development and Application of qPCR in a Resource-Limited Setting

Clare Watt and Jim Huggett

CONTENTS

25.1 INTRODUCTION

The application of a technologically advanced molecular tool such as quantitative polymerase chain reaction (qPCR) may seem incongruous in resource-limited settings where the advantages to both the researcher and the local community may initially appear doubtful. While disease burden is greatest in resource-limited settings (sub-Saharan Africa alone carries 25% of the world's disease burden[1]), should the highly technical molecular diagnostic tool of qPCR be applied to answering such questions where limited resources could well be spent addressing the causes of disease or improving treatment rather than the diagnosis? The benefit to the individual from rapid, accurate disease diagnosis, as long as diagnosis goes hand in hand with prompt and effective treatment, is clear.[2–4] The benefit to the wider community is also immediately obvious for diagnosis of infectious disease. But what are the possible benefits of using qPCR for such disease diagnosis? The use of any diagnostic or research tool should be driven by its usefulness in answering the clinical/scientific questions posed, regardless of the setting. It is clear that qPCR offers many advantages for disease diagnosis, such as speed, increased sensitivity of detection, accurate quantification of the target DNA, ease of data comparison, the ability to amplify and quantify multiple targets in the same assay, the use of a small starting volume of sample, and a completely closed system of analysis compared to conventional PCR and equivalent molecular tests such as lateral flow assays. Both commercial and in-house assays have been developed for many important diseases common in resource-poor settings, including malaria, HIV, and tuberculosis. The development of the assay is the first step to conducting qPCR in a resource-limited setting, accurately, appropriately, and sustainably for the benefit of all.

25.1.1 qPCR CURRENT TECHNOLOGIES: WHICH AND WHERE?

There exist many platforms (Part III of this book) and approaches for qPCR (Part II of this book), which have been outlined elsewhere in this book. In a resource-poor setting, where would qPCR outperform other approaches for infectious disease diagnosis? The level at which qPCR would be used in the diagnostic chain, point-of-care (POC), peripheral laboratory, or central laboratory, needs to be addressed. While the diagnostic questions at each level are often different, so are the constraints on implementation of a qPCR diagnostic and the degree of resource-poorness. In general, at all levels, it would be necessary to confirm

- A constant, reliable, direct current (achievable by the use of UPS [universal/uninterruptible power supply] batteries if electricity is supplied by a generator or less-than-reliable mains electricity)
- Well-established supply chains for reagents and consumables, including cold supply chains where necessary
- A clean working space
- Appropriate waste management
- A skilled work force[5]

The ultimate goal of any infectious disease diagnostic would be POC, where diagnosis could occur simply, quickly, and accurately, enabling treatment to commence immediately. This level of diagnosis has additional advantages in a resource-poor setting where communities are often geographically isolated, mobile, and extremely poor, so unlikely to see professional health care. Therefore, a near-patient qPCR test could be extremely beneficial in such environments. While the microfluidic revolution is ongoing, there exist a number of currently available mobile qPCR platforms, which integrate to a greater or lesser extent sample extraction; Cepheid's GeneXpert and Idaho Technologies Incorporated's RAPID. A minimal amount of sample processing prior to extraction and qPCR is required for the GeneXpert. A 10-min cell lysis and protein denaturation step is conducted using reagents and plasticware that are provided in the kit, prior to loading onto the platform where nucleic acid extraction and qPCR analysis are then conducted.[6] The upstream processing for the RAPID (ruggedized advanced pathogen identification device) requires the use of a centrifuge, a separate kit, and also more laboratory space and infrastructure. The biological targets currently amplified by these commercial platforms are dictated by the manufacturers' response to their potential market. GeneXpert does provide cartridges for amplification of over 16 infectious disease agents,[7] which may be present in resource-poor countries, but only two of which are likely to be high diagnostic priority in such settings. The RAPID platform is similar, although assays have been developed outside those commercially available for the emerging arthropod-borne viruses Chikungunya (CHIK) and O'nyong-nyong (ONN).[8] Both the RAPID and GeneXpert platforms are relatively compact, and both require a laptop computer to run the software for the analysis and a constant electricity supply. There does exist an extremely portable, hand-held LATE-PCR machine, the Bio Seeq Plus (Smiths Detection), which can be powered for up to 10 h by lithium batteries. The Bio Seeq Plus does not require a separate computer for data analysis. While an extremely useful platform and a good example of how technology can be miniaturized and mobilized, only six assays are available and these are for biological warfare agents. The machine does not incorporate a nucleic acid extraction option, with input into the cartridges limited to powders or concentrated aerosols.[9]

Both the GeneXpert and the RAPID platforms were optimized for their mobility, which is huge compared to conventional machines, yet it is unlikely that these kinds of platforms would ever be used for routine POC disease diagnosis in resource-poor settings in the current healthcare infrastructure. They are expensive, require external power sources, and their ruggedness in dusty, hot, humid environments is yet to be tested extensively.[3]

If it is indeed unlikely that an affordable POC qPCR platform for the resource-poor setting will be developed in the near future, the application of qPCR to infectious disease diagnosis could be improved by processing the biological samples in the field prior to transport to peripheral or central laboratories. A simple extraction of blood samples for HIV provirus testing has been developed for

POC sample processing.[10] Nucleic acid extraction can still pose a bottleneck for the application of mobile qPCR technology. It is of paramount importance that the biological sample extracted is not contaminated and is also extracted sufficiently to provide useable results. Nucleic acid extraction, however, still remains poorly researched for many biological samples and assumptions about the quality, stability, and level of inhibitors continue, rendering the quantitative nature of the qPCR analysis dubious, at best, on many occasions.[11] Certainly, pooling samples for qPCR analysis can increase the cost-effectiveness of detection,[12] but there are also increased risks of a contaminant entering the pooled samples and producing spurious results due to the extremely sensitive detection offered by qPCR.[8] The nucleic acid extraction for the GeneXpert machine is simple but extracted samples do need to be processed within 1 h of collection or else stored at −20°C.[13] This would necessitate the GeneXpert platform being within 1 h of the POC sample collection or transfer of unextracted biological samples or the patient to the peripheral or central laboratory. A recent study assessing the usefulness of the GeneXpert for diagnosis of multidrug-resistant tuberculosis infection in peripheral healthcare centers as an alternative to microscopy detected 90.3% of culture-positive patient samples compared to 67.1% by microscopy in a much shorter time frame and with less patient sampling.[3]

A novel approach to enable qPCR-based infectious disease diagnosis at POC is to bring the necessary infrastructure and technical skill to the isolated communities or populations using a mobile laboratory. In this approach, a laboratory equipped with routine laboratory diagnostics that could include qPCR platforms, water, electricity (by means of a mobile generator), and trained staff is fitted into the back of a lorry. Therefore, so long as the lorry can reach the community, laboratory investigations can take place. It may be that the diagnostics take a couple of hours to complete, but the likelihood of the patient not receiving their results due to absence is extremely reduced when compared to attendance at a peripheral laboratory. By moving the laboratory, other infrastructure constraints such as storage of consumables, cold chain supply, and maintenance of the equipment are also obviated, as these are based at the central laboratory from where the mobile laboratory is based. The mobile laboratory bridges the gap between highly technical advances in molecular diagnostics and simplification and modification of the same for POC, immediate diagnosis, which can take years. This could be invaluable for the health of isolated communities who are unlikely to reach treatment centers or, if they do, unlikely to return to peripheral health centers for the results of their diagnosis. Such mobile laboratories have worked well in reaching specific populations for HIV testing,[14] and a more integrated approach to disease diagnosis is currently being trialled in rural west Tanzania;[15] however, they are only suitable for chronic infections and not for acute medical emergencies often associated with diseases like malaria or sepsis.

In many resource-poor settings, central laboratories are often sufficiently supported by infrastructure to enable the incorporation of advanced molecular techniques such as qPCR. Peripheral laboratories are likely to be less well supported with infrastructure and would require some investment to enable qPCR to be routinely conducted. As discussed above, the self-contained platforms such as GeneXpert or RAPID do provide the most facile approach for a diagnostic laboratory to implement qPCR-based analysis of patient samples whatever be the setting. However, the monetary saving from these platforms, as they exclude the need to establish and maintain extensive well-equipped laboratories, are somewhat balanced out by the high cost of the initial purchase of such stand-alone machines. In addition, the cost of the diagnostic cartridges, even with the hefty reductions often offered by big biotech companies to such customers, do remain high and ultimately unaffordable. Regular servicing would also be essential to maintain standards of quality control, but is often the cost most likely to be sacrificed and least likely to be included in costings.

Aside from the physical infrastructure, the most important component of any laboratory is the staff. It may be that only 1.3% of the world's health workforce are found in sub-Saharan Africa[16] but the figures are considerably worse for laboratory technicians. To reach the millennium development goals (MDG) for diagnosis of tuberculosis by smear microscopy by 2015, an additional 23,000 laboratory technicians would be required globally.[17] Capacity-building through on-the-job training provides a valuable resource for both the laboratory scientist and the country as more technically skilled laboratory workers are trained. In the short term, however, while such shortfalls in skilled staff exist, those with the skills are a valuable commodity and as such are often difficult to retain for long-term projects.

25.1.2 qPCR Current Applications for Infectious Disease Diagnosis

Were infrastructure not limiting, qPCR could be applied to infectious disease diagnosis in any resource-poor setting, but for what kind of infectious disease diagnosis would a qPCR outperform other cheaper, less technically demanding diagnostics? There are a number of characteristics specific to qPCR, which potentially hold the key.

One of the advantages of qPCR over other conventional detection methods is the sensitivity of detection. qPCR is able to quantify accurately the target over a large dynamic range, normally at least seven orders of magnitude without the need to conduct dilutions of the sample under investigation. Indeed, a comparison of two Roche assays for HIV-1 RNA detection from plasma, the Amplicor (conventional PCR) and the Taqman (Applied BioSystems qPCR), recorded increased sensitivity of detection by the Taqman assay[18] with the additional benefit of removing the need to predict the viral load before applying the correct sensitivity of test. The detection of very low viral loads could be vital in the rapid diagnosis of infants born of HIV-positive mothers. The ability to describe easily in detail the disease burden of individuals can provide important information for clinicians regarding responses to antiretroviral treatment.

The ability to monitor the PCR in real time is another advantage of qPCR, especially where fast infectious disease diagnosis is paramount such as in emerging infectious diseases. After an outbreak of Ebola hemorrhagic fever in northern Uganda in 2000–2001, a one-step reverse transcriptase (RT) qPCR has been developed. This RT qPCR was used to measure Ebola viral load to allow correlation with patient survival retrospectively, providing important information on disease epidemiology.[19] The RT qPCR had higher detection sensitivities than the two-step (nested) RT-PCRs used in the field in the Ugandan outbreak, with less risk of contamination and faster time to detection and is envisaged to be used as the first line of detection in future Ebola outbreaks.

Molecular approaches to disease diagnosis also enable organisms from different genera, families, or even orders to be assayed in one test. A similar approach has been developed to diagnose the cause of genital ulcers in a rural population in Uganda.[20] Two multiplex qPCRs for *Haemophilus ducreyi*, *Treponema pallidum*, and *Herpes simplex* virus type 1 and type 2 were conducted in Rakai, Uganda, using an ABI 7900 HT real-time PCR instrument. Additional swabs were independently analyzed in the United States for quality control, with 96% agreement between the two laboratories. The multiplex qPCR assays improved the detection sensitivity of the previous PCR approach, decreased hands-on time, and could be used to provide important information to clinicians regarding the presence of active or latent infection, particularly in HIV-positive individuals.

Aside from direct use of qPCR for infectious disease diagnosis, the level of scrutiny afforded by the quantitative approach can play a vital role in describing disease epidemiology. This can be especially important where eradication is the goal, such as for trachoma. qPCR has helped to determine the burden of *Clamydia trachomatis* infection by age in populations with different levels of endemicity of trachoma,[21] which was impossible to do by clinical presentation and difficult with cell culture techniques. Such work is vital if trachoma is to be eradicated. Were it possible to conduct this population analysis prior to mass treatment, the use of the antibiotic azithromycin could be focused, reducing both cost and misappropriate treatment,[22] making the existing donations of antibiotic achieve far more than blanket mass treatment of whole regions.

Misdiagnosis, underdiagnosis, and overdiagnosis can be common where quality control checks are not in place, or are difficult to conduct due to the absence of any alternative method of detection. Molecular diagnosis is most valuable for pathogens where culture is currently impossible or very slow, such as differentiation of multidrug-resistant tuberculosis.[3] Additionally, molecular diagnosis can be valuable where differentiation by simple laboratory techniques, such as microscopy, is difficult. For example, the presence of the amoeba *Entamoeba histolytica* in stool from microscopically positive samples in Ethiopia was less than 1% when analyzed by qPCR.[23,24] Only the presence of the red blood cell-engulfing trophozoites differentiate *E. histolytica* from the noninvasive *E. dispar* when examined under the microscope. While in regions with declining malaria prevalence (>11%), such as the Gambia and Guinea Bissau, conventional PCR detection of *Plasmodium falciparum* infection was up to three times more sensitive than rapid detection tests and microscopy,[25] which form the mainstay of malaria diagnosis in highly endemic regions. As a platform for disease diagnosis, microscopes are incredibly versatile, have low cost of upkeep, and are relatively easy to use with minimal infrastructure. As with any diagnostic tool, human error cannot be prevented, but perhaps with microscopy these errors can have huge consequences. It is likely that the frequency with which the microscopist encounters the disease-causing agent can affect the frequency with which it is recorded; in nonendemic countries, malaria in returning travelers was only detected with microscopy in 89% of qPCR-positive samples.[26] A similar underdiagnosis of HIV using rapid diagnostic tests was observed in South Africa when compared to the PCR results, where suboptimal test use and poor levels of quality control were identified as the main causes.[27] The ability to automate and easily standardize means that such user errors related to disease prevalence and familiarity should be minimal with the qPCR approach, while errors relating to suboptimal test use should become apparent immediately upon analysis of the standards in a qPCR. It is vital that appropriate quality control and quality assurance standards are established alongside any qPCR diagnostic. In 2010, of 340 accredited laboratories in sub-Saharan Africa, only 28 were located outside of South Africa and almost all were private, parastatal, or donor-sponsored, not state-run.[28]

25.1.3   Where Should We Be Aiming?

As we have discussed above, qPCR can and is being used for infectious disease diagnosis in resource-poor settings. For such an analytical platform to become more commonplace, a number of changes need to occur outside of just the local physical infrastructure discussed in Section 25.1.1. Without functioning healthcare systems, the implementation of new diagnostic tests will only happen in isolation. The need for such systems is well-recognized[29,30] and well-funded infectious disease programs could provide the impetus and the method for change.[31] It was estimated that in mid-2006, the Global Fund contributed 18% of international antiretroviral MDG targets, 29% of directly observed therapy (DOTS) targets (for treatment of tuberculosis), and 9% of insecticide-treated mosquito nets (ITNs) (to prevent malaria) in sub-Saharan Africa,[32] with the MDG deadline set for 2015.

Alongside a functioning healthcare system, capacity-building should enable a research base to be developed and led by the communities affected. International partnerships will no doubt contribute toward this goal,[33] but it is vital that the ownership of the research is by those affected, and the capacity that is being built is needed by the same.[34] Specifically for qPCR, the philosophy behind adopting commercial assays or in-house assays can highlight the different approaches. Assays developed for commercial platforms by the manufacturers by their very nature are aimed at infectious diseases where a profit can be made. The philanthropic "at-cost" sales to resource-poor countries will only ever be made possible when the assay can be sold for profit outside of these regions; only diseases that affect the resource-rich are likely to be developed solely by the commercial platform manufacturers. Perhaps funding agencies and not-for-profit organizations such as the Foundation for Innovative New Diagnostics (FIND) can help bridge the gap in assay development and technology specifically tailored to the resource-poor setting.

In-house assays have the benefit of being developed by anybody with a basic molecular biology skill set, so long as a large enough databases exist of the nucleic acid sequences of the organism you wish to amplify. You need not have the skills to design the qPCR assay, as commercial manufacturers such as ABI and Sigma provide a primer and probe design service,[8] which provides the customer with details of the primer and probe sequences. However, often, in-house qPCR assays are poorly evaluated, difficult to quality control, have poorly described protocols in the literature, and are published in journals that are not free to access and so are unlikely to be read by researchers based in a resource-poor setting. The abundance of different in-house assays also makes interlaboratory comparison complex[35] and reduces any economies of scale that might be possible through using just one commercial platform.

In-house qPCR assays are also likely to be prone to far more user errors than the commercial extraction/qPCR platforms. These errors can somewhat be abated by the presence of good controls and standards, and good quality-control measures. However, in many instances, even when the in-house qPCR is vigorously controlled, the upstream processes are not, with no attempt made to examine for the presence of inhibitors in the extract or the general condition of the extracted nucleic acid.[36] In whatever system is employed, a bad input will absolutely yield a bad output.

When resources are limited, even though qPCR may provide a huge advantage in diagnostic capability, it may yet prove more cost-effective to only apply this tool after preliminary screening has occurred. Rapid testing for HIV in infants prior to confirmation by PCR reduced costs of diagnosis in Uganda from \$23.47 per infant to \$7.58, reducing total program costs by up to 40%.[37] Additionally, where the level of detection is anticipated to be low, pooling of samples prior to individual analysis could reduce costs drastically.[8] Such pooling of samples will inevitably not be appropriate for POC-based testing, but is a relevant approach where the existing healthcare infrastructure can accommodate centralized processing.

Whether in-house or commercial platform, the goal for qPCR in the resource-poor setting must be conducted by local personnel in local laboratories. The skill set required to conduct the assay and interpret the data for a commercial extraction/qPCR platform is much less than an in-house assay that would additionally require a larger laboratory infrastructure (separate areas for DNA extraction, master-mix preparation, and DNA addition). Currently, the ability to fragment these skills is often utilized in research papers, whereby qPCR is conducted in a resource-rich setting after DNA extraction in a resource-poor setting, even within the same country.[25] Of course, this is a sensible solution to resource limitation, but does raise the question of when skill division becomes akin to parachute research.[38] Do quality control and data analysis outside the resource-poor country just become parachute analysis, with control of data streams confined to the resource-rich? For qPCR, the ubiquity of the Internet now allows data to be transferred to experts almost instantly, which may be a practical solution in the short term, but are likely to build barriers to continued development in the long term.

25.2   CONCLUDING REMARKS

qPCR is an incredible tool for the diagnosis of infectious diseases common to many resource-poor settings. Where resources are limited, effective and quick diagnosis can help limited resources be more accurately focused to appropriate treatment, which would ultimately enable the cause of the disease to be tackled. As such, it can and should be used for both infectious disease diagnosis and research in resource-poor countries, with the caveat that the use be applied to only those settings and diseases where the advantages for the community are obvious and sustainable. The huge inequity between resource-rich and resource-poor countries makes ethical considerations intrinsically linked with any application of qPCR technology. Where it does exist, it will act as a stepping stone to future qPCR technologies that have advantages for the individual, the researcher, and the global community.

REFERENCES

1. Audit of Progress Against the 2005 Recommendations, Commission for Africa, accessed April 15, 2013. http://www.commissionforafrica.info/wp-content/uploads/2010/09/cfa-report-2010-chapter-audit.pdf.
2. Kallander K, Hildenwall H, Waiswa P, Galiwango E, Peterson S, Pariyo G. 2008. Delayed care seeking for fatal pneumonia in children aged under five years in Uganda: A case-series study. *Bulletin of the World Health Organization*;86:332–8.

3. Boehme CC, Nicol MP, Nabeta P, Michael JS, Gotuzzo E, Tahirli R et al. 2011. Feasibility, diagnostic accuracy, and effectiveness of decentralised use of the Xpert MTB/RIF test for diagnosis of tuberculosis and multidrug resistance: A multicentre implementation study. *Lancet*;377:1495–505.

4. Greenaway C, Menzies D, Fanning A, Grewal R, Yuan L, Fitzgerald MJ. 2001. Delay in diagnosis among hospitalised patients with active tuberculosis—Predictors and outcomes. *American Journal of Respiratory and Critical Care Medicine*;165:927–33.

5. Huggett JF, Green C, Zumla A. 2009. Nucleic acid detection and quantification in the developing world. *Biochemical Society Symposia*;37:419–23.

6. The New GeneXpert System, Cepheid, accessed April 15, 2013. http://www.cepheid.com/media/files/brochures/GeneXpert-Brochure.pdf.

7. Xpert MTB/RIF. Automated molecular detection of TB and MTB screening in peripheral laboratories. Introduction to the New Technology, Perkins., M., 41st Union World Conference on Lung Health, Nov 2010. http://www.finddiagnostics.org/resource-centre/presentations/iualtd_41st_union_conf_berlin_2010_FIND_Forum/index.html.

8. Smith DR, Lee JS, Jahrling J, Kulesh DA, Turell MJ, Groebner JL, O'Guinn ML. 2009. Development of field-based real-time reverse transcription-polymerase chain reaction assays for detection of chikungunya and O'nyong-nyong viruses in mosquitoes. *The American Journal of Tropical Medicine and Hygiene*;81:679–84.

9. Technology Information Summary:BioSeeq Plus, US EPA ORD NHSRC, July 2009. http://www.epa.gov/ordnhsrc/pubs/TISBioSeeqPLUS.pdf.

10. Jangam SR, Yamada DH, McFall SM, Kelso DM. 2009. Rapid, point-of-care extraction of human immunodeficiency virus type 1 proviral DNA from whole blood for detection by real-time PCR. *Journal of Clinical Microbiology*;47:2363–8.

11. Cannas A, Kalunga G, Green C, Calvo L, Katemangwe P, Reither K et al. 2009. Implications of storing urinary DNA from different populations for molecular analyses. *PLoS ONE*;4:e6985.

12. Abarshi MM, Mohammed IU, Wasswa P, Hillocks RJ, Holt J, Legg JP et al. 2010. Optimization of diagnostic RT-PCR protocols and sampling procedures for the reliable and cost-effective detection of Cassava brown streak virus. *Journal of Virological Methods*;163:353–9.

13. The New GeneXpert System, Cepheid, accessed April 15, 2013. http://www.cepheid.com/media/files/brochures/GeneXpert-Brochure.pdf.

14. van Schaik N, Kranzer K, Wood R, Bekker LG. 2010. Earlier HIV diagnosis: Are mobile services the answer? *South African Medical Journal*;100:671–4.

15. ADAT Study- Mobile Diagnostic and Training Centre (MDTC), NIMR-MMRP, accessed April 15, 2013. http://www.mmrp.org/projects/Interventions.html.

16. Commission for Africa. Our common interest: Report of the Commission for Africa, 2005. http://www.commissionforafrica.org.

17. Van Deyn A. 2008. Sustainability of laboratory scale-up: What does it entail? *International Journal of Tuberculosis and Lung Disease*;12:S24.

18. Oliver AR, Pereira SF, Clark DA. 2007. Comparative evaluation of the automated Roche TaqMan real-time quantitative human immunodeficiency virus type 1 RNA PCR assay and the Roche AMPLICOR version 1.5 conventional PCR assay. *Journal of Clinical Microbiology*;45:3616–9.

19. Towner JS, Rollin PE, Bausch DG, Sanchez A, Crary SM, Vincent M et al. 2004. Rapid diagnosis of Ebola hemorrhagic fever by reverse transcription-PCR in an outbreak setting and assessment of patient viral load as a predictor of outcome. *Journal of Virology*;78:4330–41.

20. Suntoke TR, Hardick A, Tobian AAR, Mpoza B, Laeyendecker O, Serwadda D et al. 2009. Evaluation of multiplex real-time PCR for detection of *Haemophilus ducreyi*, *Treponema pallidum*, herpes simplex virus type 1 and 2 in the diagnosis of genital ulcer disease in the Rakai District, Uganda. *Sexually Transmitted Infections*;85:97–101.

21. Solomon AW, Holland MJ, Burton MJ, West SK, Alexander NDE, Aguirre A et al. 2003. Strategies for control of trachoma: Observational study with quantitative PCR. *Lancet*;362:198–204.

22. Harding-Esch EM, Edwards T, Sillah A, Sarr I, Roberts CH, Snell P et al. 2009. Active trachoma and ocular *Chlamydia trachomatis* infection in two Gambian regions: On course for elimination by 2020? *PLoS Neglected Tropical Diseases*;3:e573.

23. Kebede A, Verweij JJ, Endeshaw T, Messele T, Tasew G, Petros B, Polderman AM. 2004. The use of real-time PCR to identify *Entamoeba histolytica* and *E-dispar* infections in prisoners and primary-school children in Ethiopia. *Annals of Tropical Medicine and Parasitology*;98:43–8.

24. Kebede A, Verweij JJ, Petros B, Polderman AM. 2004. Misleading microscopy in amoebiasis. *Tropical Medicine & International Health*;9:651–2.

25. Satoguina J, Walther B, Drakeley C, Nwakanma D, Oriero EC, Correa S et al. 2009. Comparison of surveillance methods applied to a situation of low malaria prevalence at rural sites in The Gambia and Guinea Bissau. *Malaria Journal*;8:12.

26. Khairnar K, Martin D, Lau R, Ralevski F, Pillai DR. 2009. Multiplex real-time quantitative PCR, microscopy and rapid diagnostic immuno-chromatographic tests for the detection of *Plasmodium* spp.: Performance, limit of detection analysis and quality assurance. *Malaria Journal*;8:17.

27. Wolpaw BJ, Mathews C, Chopra M, Hardie D, de Azevedo V, Jennings K, Lurie MN. 2010. The failure of routine rapid HIV testing: A case study of improving low sensitivity in the field. *BMC Health Services Research*;10:73.

28. Gershy-Damet G-M, Rotz P, Cross D, Belabbes EH, Cham F, Ndihokubwayo J-B et al. 2010. The World Health Organization Africa region laboratory accreditation process. *American Journal of Clinical Pathology*;134:393–400.

29. Bates I, Maitland K. 2006. Are laboratory services coming of age in sub-Saharan Africa? *Clinical Infectious Diseases*;42:383–4.

30. Petti CA, Polage CR, Quinn TC, Ronald AR, Sande MA. 2006. Laboratory medicine in Africa: A barrier to effective health care. *Clinical Infectious Diseases*;42:377–82.

31. Harries AD, Jensen PM, Zachariah R, Rusen D, Enarson DA. 2009. How health systems in sub-Saharan Africa can benefit from tuberculosis and other infectious disease programmes. *International Journal of Tuberculosis and Lung Disease*;13:1194–9.

32. Komatsu R, Low-Beer D, Schwartlander B. 2007. Global Fund-supported programmes' contribution to international targets and the Millennium Development Goals: An initial analysis. *Bulletin of the World Health Organization*;85:805–11.

33. Haines A, Dockrell H. 2009. Building research capacity through international partnerships. *International Health*;1:109–10.

34. Tedmanson, D. 2010. Whose capacity needs building? Reflections on capacity building initiatives in remote Australian Indigenous communities, in Prasad, A. (Ed.) *Against the Grain: Advances in Postcolonial Organization Studies*, Advance in Organization Studies Series, Liber & CBS Press, Copenhagen.

35. Rouet F, Menan H, Viljoen J, Ngo-Giang-Huong N, Mandaliya K, Valea D et al. 2008. In-house HIV-1 RNA real-time RT-PCR assays: Principle, available tests and usefulness in developing countries. *Expert Review of Molecular Diagnostics*;8:635–50.

36. Huggett J, Novak T, Garson J, Green C, Morris-Jones S, Miller R, Zumla A. 2008. Differential susceptibility of PCR reactions to inhibitors: An important and unrecognised phenomenon. *BMC Research Notes*;1:70.

37. Menzies NA, Homsy J, Pitter JYC, Pitter C, Mermin J, Downing R et al. 2009. Cost-effectiveness of routine rapid human immunodeficiency virus antibody testing before DNA-PCR testing for early diagnosis of infants in resource-limited settings. *The Pediatric Infectious Disease Journal*;28:819–25.

38. Costello A, Zumla A. 2000. Moving to research partnerships in developing countries. *BMJ*;321:827–9.

26 A Review of Isothermal Nucleic Acid Amplification Technologies

Gavin J. Nixon and Claire A. Bushell

CONTENTS

26.1 INTRODUCTION

Nucleic acid amplification (NAA) technologies are fundamental to modern day molecular diagnostics and can be found within a diverse range of applications and markets ranging from clinical testing to food analysis.[1–6] These approaches provide a uniquely powerful suite of technologies for the amplification of nucleic acid materials to analytically relevant concentrations.

Leading the molecular biology revolution in the early 1980s was a novel *in vitro* approach developed by Mullis and coworkers, utilizing DNA polymerase and oligonucleotide primers called the polymerase chain reaction (PCR).[7,8] PCR was quickly established as the dominant NAA technology within the research and commercial communities due to features such as relative ease of assay design and high levels of sensitivity and specificity. However, PCR-based amplification approaches are limited by a number of factors. These limitations include the requirement for complex thermal cycling instruments (routinely Peltier-based) with defined experimental conditions, susceptibility to well characterized PCR inhibitors,[9–12] complex licensing landscape and initial restriction to DNA templates. Hence, from the early 1990s, alternative NAA technologies were actively developed and commercialized.

Molecular biological processes found within microorganisms were investigated and adapted to generate a multitude of potential *in vitro* amplification technologies that did not require thermal cycling. These isothermal approaches[13–15] provided the means to circumvent the existing licensing landscape, potentially simplify the NAA process and broaden the range of diagnostic targets to include RNA. Early isothermal technologies (Table 26.1) such as nucleic acid sequence-based amplification (NASBA)[16] and transcription-mediated amplification (TMA) modeled the replication strategies of retroviruses,[17] while strand displacement amplification (SDA)[18] utilized template amplification/restriction processes found within bacteria.

Applications of these early isothermal NAA technologies were generally restricted to clinical diagnostics and incorporated into commercial systems such as the BD ProbeTec™ ET System (BD Diagnostics, Sparks, Maryland) and the APTIMA COMBO® Assay System (Gen-Probe Inc., San Diego, California) which focussed on the diagnosis of sexually transmitted infection (STI) and viral infection. Subsequently, a number of other isothermal technologies have emerged with a range of performance characteristics (detailed in Table 26.1) that have helped one to drive the uptake of isothermal approaches over the last 10 years (Figure 26.1).

In addition to dispensing with the need for thermal cycling instruments, isothermal approaches can also offer advantages such as rapid reaction time and reduced susceptibility to inhibitors. This has opened the possibility of applying molecular methods to applications that were previously difficult to achieve with PCR, such as near-patient diagnostics. This chapter contains an overview of technological developments and explores specific applications of isothermal NAA within the microbial diagnostic community.

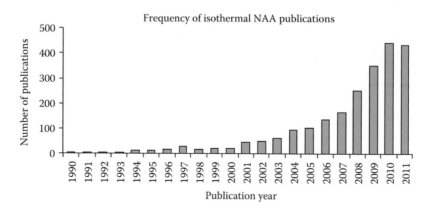

FIGURE 26.1 Time-based analysis of isothermal publications. Bibliographic search (January 1990–December 2011) of CAS databases using SciFinder® search tool (American Chemical Society, Washington) with search terms "nucleic acid amplification" and "isothermal" was used to generate the data set (2301 references). The data set was analyzed for the annual change in publication levels within the stated time interval. While this limited analysis of the isothermal NAA technological landscape may not capture all relevant publications, it should provide an indication of relative publication levels.

TABLE 26.1

Selected Isothermal Technologies Compared against PCR

Feature	Comparator	Isothermal Technology								
Name	Polymerase chain reaction	Nucleic acid sequence-based amplification	Single primer isothermal amplification	Strand displacement amplification	Rolling circle amplification	Loop-mediated isothermal amplification	Helicase dependent amplification	Cross priming amplification	Multiple displacement amplification	Recombinase polymerase amplification
Abbreviation	PCR	NASBA	SPIA	SDA	RCA	LAMP	HDA	CPA	MDA	RPA
Amplification target	DNA RNA (RT-PCR)	RNA	DNA RNA	DNA RNA	DNA RNA	DNA RNA	DNA RNA	DNA	DNA	DNA
Temperature	Thermal cycling	37–42°C	45°C, 50°C	37°C	30–60°C	60–65°C	37°C 60–65°C	60°C	30°C	37°C
Design complexity	Simple	Simple	Simple	Complex	Complex	Complex	Simple	Simple	Simple	Simple
Multiplex	Yes	Yes	No	No	Yes	Yes	Yes	Yes	No	Yes
Detection	Electrophoresis, real-time: fluorescence	Electrophoresis, real-time: fluorescence	Electrophoresis	Electrophoresis, real-time: fluorescence	Electrophoresis, real-time: fluorescence	Electrophoresis, real-time: fluorescence, turbidity	Electrophoresis, real-time: fluorescence	Electrophoresis, real-time: fluorescence	Electrophoresis	Electrophoresis, real-time: fluorescence
Denaturing agents	Heat	RNase H	RNase H	Restriction enzymes	Strand displacement polymerases	Strand displacement polymerases	Helicase	Strand displacement polymerases	Strand displacement polymerases	Strand displacement polymerases
Typical detection time	<60 min	<60 min	180 min	70 min	Various >30 min	<30 min	<30 min	90 min	Various <16 h	<30 min
Susceptibility to matrix components	Yes, for example, humic acid, haem	Yes, for example, enzymatic inhibitors	Unknown	Yes, for example, blood, bilirubin	Unknown	Yes, for example, calcium ions	Unknown	Unknown	Unknown	Unknown
Key Refs	[7,8]	[16]	[19,20]	[18]	[21]	[22]	[23]	[24]	[25]	[26]

26.2   ISOTHERMAL NAA TECHNOLOGIES

26.2.1   NUCLEIC ACID SEQUENCE-BASED AMPLIFICATION AND TRANSCRIPTION-MEDIATED AMPLIFICATION

NASBA (bioMérieux SA, Marcy l'Etoile, France)[16] and TMA (Gen-Probe Inc, San Diego, California) represent commercial developments of the self-sustained sequence replication (3SR) process[17,27] and are early examples of successful isothermal NAA technologies. Both approaches utilize T7 RNA polymerase to synthesize RNA and reverse transcriptase to generate DNA from the RNA templates. Later adaptations to the technologies have enabled the use of DNA as the starting template through the use of processes such as denaturation and enzymatic digestion.[28,29]

Owing to similarities between the two isothermal technologies (Figures 26.2 and 26.3), this section will focus on the mechanism of NASBA using RNA templates. The NASBA process utilizes an oligonucleotide primer containing a T7 RNA polymerase promoter sequence that initiates reverse transcription upon hybridization to the target RNA using avian myeloblastosis virus reverse transcriptase (AMV-RT) and creates a RNA:DNA heteroduplex. The RNA strand is degraded by RNAse H which releases the DNA strand and makes it available for the reverse primer to bind and synthesize a second strand of DNA. The resultant double-stranded DNA (dsDNA) molecule contains T7 RNA polymerase promoter sequences which also facilitates primer-independent transcription. Each newly synthesized RNA amplicon serves as a template for a new round of primer-driven replication, leading to an exponential expansion of the RNA target under isothermal conditions. The isothermal amplification process can be performed at temperatures between 37°C and 42°C.

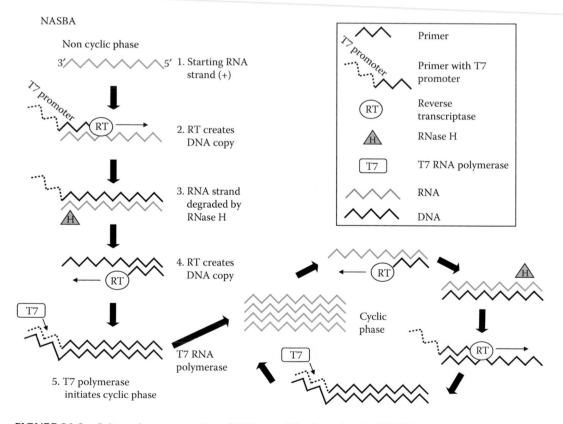

FIGURE 26.2   Schematic representation of RNA amplification using the NASBA process.

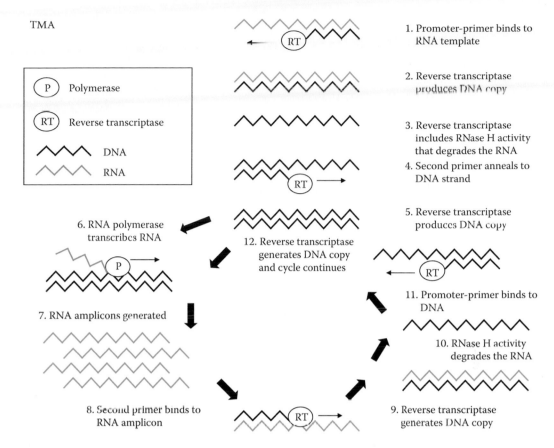

FIGURE 26.3 Schematic representation of RNA amplification using the transcription-mediated amplification (TMA) process.

TMA is similar to NASBA, but differs by the use of Moloney murine leukemia virus reverse transcriptase (MMLV-RT) instead of AMV-RT and does not require the supplementary addition of RNAse H. These reverse transcriptase enzymes have different specificities and efficiencies of incorporation.

NASBA is well established within the clinical diagnostics area for viral and bacterial applications[13,30–34] whereas TMA is generally restricted to diagnostic systems provided by Gen-Probe Inc.[35–40] targeting areas such as viral and STI diagnostics.

Successful amplification can be detected through the use of standard approaches such as electrophoresis or capture probes. Further developments of NASBA and TMA have incorporated fluorescent probe-based detection components such as Molecular Beacons[41] that allow the real-time monitoring of isothermal amplification processes for detection and quantification applications.[34,42] Other areas of research have focussed on their integration into miniaturized systems with near-patient applicability.[43,44]

26.2.2 SINGLE PRIMER ISOTHERMAL AMPLIFICATION

Single primer isothermal amplification (SPIA)® and Ribo-SPIA® are relatively recent proprietary isothermal technologies that were developed and commercialized by NuGEN Technologies Inc. (San Carlos, California) to target DNA and RNA templates respectively.[19,20] SPIA linear isothermal amplification generates many copies of single-stranded DNA (ssDNA) sequences that are complementary to the initial DNA template or RNA (in the case of Ribo-SPIA). The SPIA

Ribo-SPIA®

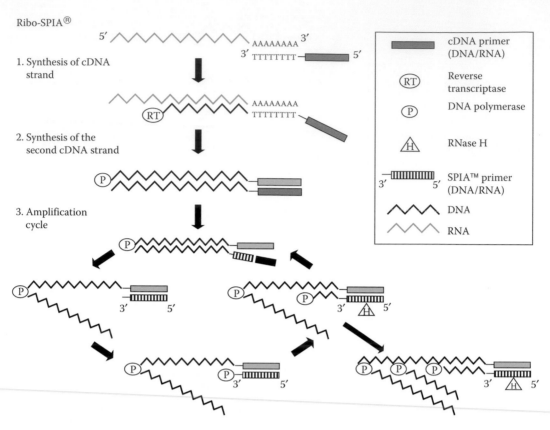

FIGURE 26.4 Schematic representation of RNA amplification using single primer isothermal amplification (Ribo-SPIA).

process (Figure 26.4) utilizes composite DNA/RNA sequence primers (DNA at the 3′ end and RNA at the 5′ end), which anneal to the DNA template. A DNA polymerase, possessing strand-displacement activity, extends the primer and RNase H is used to cleave the 5′ RNA section of the primer. This process frees the priming site to allow a new primer to anneal. Extension of the new primer by DNA polymerase displaces the product of the previous extension round and continuous cycles of this process result in efficient amplification at a single temperature (47–50°C). Ribo-SPIA involves a series of amplification steps prior to the SPIA process in order to generate ds cDNA from a mRNA template.[20]

The SPIA technique has been applied to gene-expression studies as a method for generating amplified pools of RNA material to facilitate the characterization of rare transcripts.[45,46] Amplification of entire transcriptomes is also possible by whole-transcript (WT)-Ribo-SPIA/WT-Ovation™ RNA Amplification System (NuGEN Technologies Inc), which involves the use of a random DNA sequence in the 3′ portion of the chimeric primer for single-strand cDNA synthesis.[20,47]

26.2.3 STRAND DISPLACEMENT AMPLIFICATION

SDA is a well-established isothermal amplification technique that forms the technological basis for the BD ProbeTec™ system (BD Diagnostics, Sparks, Maryland). SDA was originally described by Walker et al. in 1992[18,48] and involves a two-stage process comprising target generation and SDA reaction cycling (Figure 26.5). The process uses a restriction endonuclease enzyme and exonuclease-deficient polymerase to produce exponential target accumulation. After an initial heat denaturation of DNA template molecules at 95°C, the reaction proceeds at 37°C.

SDA target generation process

1. Target denatured and primers bound

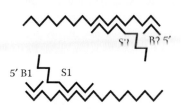

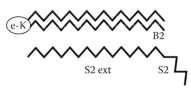

2. Primers extended and displaced by exo-Klenow (e-K)

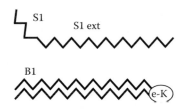

3. Opposite primers annealed to S1 ext and S2 ext

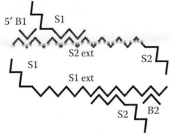

4. Primers extended and strand displacement initiated by exo-Klenow

5. Annealing of opposite primers and extension by exo-Klenow

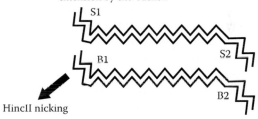

HincII nicking

SDA reaction cycle process

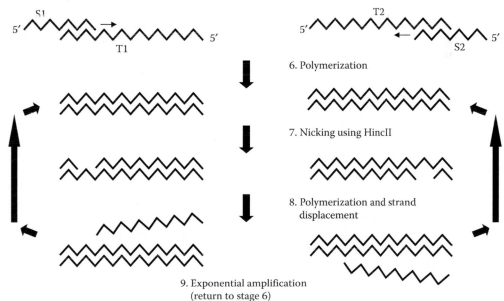

6. Polymerization

7. Nicking using HincII

8. Polymerization and strand displacement

9. Exponential amplification (return to stage 6)

FIGURE 26.5 SDA amplification process.

The first stage (Figure 26.5) uses two primer sets (S1, S2 and B1, B2) to create a template containing hemiphosphorothioate (thioate linkage between nucleosides) Hinc II restriction endonuclease recognition sites. The four primers are extended simultaneously using exo-Klenow and result in displacement of the S1 and S2 products from the original strand. The second stage exploits the activity of Hinc II to nick the Hinc II site, and the exo-Klenow extends the 3′-end at this point, both regenerating the restriction site for repeat digestion and displacing the downstream strand as it progresses. The displaced strand in turn acts as a template for the opposite SDA primer, producing a sustainable cycle of exponential amplification.[18]

Reaction times for the amplification stage were reduced from approximately 2 h in the infancy of the technology to 30 min[49] which made it a rapid alternative to PCR-based amplification. SDA is conducted at relatively low stringency (using 37°C as the incubation temperature) which means that the technology is susceptible to nonspecific amplification processes. This characteristic is especially problematic when nontarget DNA is present within a sample at large excess to specific target sequences. Approaches to minimize the generation of nonspecific amplification products include the use of co-solvents to increase stringency and improved experimental design.[18]

Modifications to the original SDA design have been made to both improve detection sensitivity and broaden its application. One such example is the addition of a fluorescent hairpin probe to the SDA reaction, which permits hybridization to an ssDNA target. Once annealed, the probe is able to open its stem and allow primer hybridization, which subsequently triggers a polymerization reaction.[50] This adaptation has been used for detection of DNA viruses such as parvoviridae.[50]

SDA was commercialized by Becton Dickinson and incorporated into their ProbeTec System™ for the high-throughput analysis of clinical specimens. This platform has been used for SNP detection using fluorescent probes incorporated into the SDA reaction process.[51]

Isothermal amplification technologies such as SDA are well suited for DNA biosensor applications as they typically require constant temperatures during the detection process. Recently, the detection of human cytomegalovirus (HCMV) using SDA associated with a piezoelectric biosensor has been reported.[52]

26.2.4 ROLLING CIRCLE AMPLIFICATION

Rolling circle amplification (RCA) is used extensively *in vivo* by viruses and plasmids to synthesize DNA sequences. This biomolecular process has been exploited to synthetically generate repeated template copies (concatemers) from a circular template.[21,53] In its simplest form, RCA operates with a circularized oligonucleotide as a template, to which a single primer anneals (Figure 26.6). A DNA polymerase such as phi29 DNA polymerase subsequently extends the primer to generate a complementary sequence, while the strand displacement activity of the polymerase ensures that the simultaneous unwinding of the newly created duplex occurs as amplification advances. The RCA reaction is typically conducted between 30°C and 37°C.

Further development of the technique by Lizardi and coworkers in 1998[54] combined phi29 DNA polymerase-based RCA with ligation-dependent circularizable oligonucleotides termed "padlock probes"[55,56] to detect point mutations in low-level human genomic DNA samples. Circularizable probes hybridize to the target sequence leaving a small gap that is filled through the binding/ligation of a small allelic-specific phosphorylated oligonucleotide or gap-filled using a DNA polymerase and ligase. The RCA phase utilizes a two-primer approach in which the first primer initiates the RCA reaction, and as the amplicon is generated, a second primer anneals to the concatemers. Following on from this, a sequence of strand displacement and creation of annealing sites for the initial primer occurs which yields hyperbranched RCA products.

Yields can also be increased by implementation of multiply primed RCA,[57] which combines phi29 DNA polymerase with a random priming approach to produce replication forks that enable exponential amplification. The use of random hexamer primers is reported to give rise to a 40-fold increase in the amplification rate when compared to linear RCA.[57]

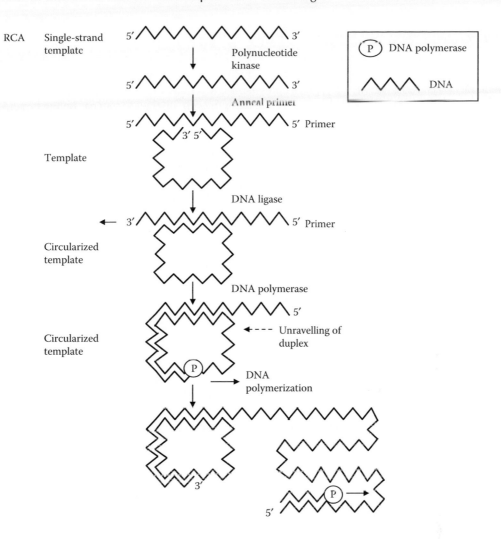

FIGURE 26.6 Scheme for RCA generation of a concatemer library.

The generation of nonspecific amplification products can be problematic when using RCA-based approaches due to potential priming at sites where primer dimers are formed.[58] Recently, a novel variant of RCA called primer-generation RCA (PG-RCA) has been described[59] which minimizes the formation of such products. The process utilizes a circular probe in conjunction with a nicking enzyme and the thermostable Vent (exo-) DNA polymerase to continually generate a new priming site throughout the reaction at elevated temperature (60°C incubation). This removes the need for design of external primers and reduces the potential for nonspecific artifact generation associated with the primers. The inclusion of ssDNA-binding proteins (SSB) has also shown potential for improving the efficiency of RCA reactions and reducing the generation of nonspecific products.[58]

RCA has a variety of applications that include viral DNA detection,[60–62] identification of fungal pathogens,[63] and bacterial subtypes using SNPs.[64] Studies have also looked into the potential for using RCA in the field of food authenticity, for example, the detection of genetically modified (GM) varieties of maize and soya.[65] Whole genome amplification (WGA) developments of the technology are also available commercially through GE Healthcare (Chalfont St. Giles, U.K.) as illustra™ GenomiPhi™ amplification kits.

26.2.5   Loop-Mediated Isothermal Amplification

Loop-mediated isothermal amplification (LAMP) isothermal NAA technology[22] was developed by the Eiken Chemical Corporation Ltd (Tokyo, Japan) and utilizes the auto-cycling, strand displacement, DNA synthesis activity of *Bst* DNA polymerase together with a set of four specially designed inner (FIP, BIP) and outer primers (F3, B3) that recognize six distinct sequences on the target DNA (Figure 26.7). The forward inner primer (FIP) containing sequences of the sense and antisense strands of the target DNA initiates the LAMP reaction. Strand displacement DNA synthesis primed by the F3 outer primer releases a ssDNA product that serves as the template for DNA synthesis primed by a second set of backward inner primer (BIP) and B3 outer primer, producing a stem-loop DNA structure. Subsequent rounds of LAMP cycling involve an inner primer hybridizing to the loop and initiating displacement DNA synthesis that generates the original stem-loop DNA and a new stem-loop DNA with a stem twice as long. The auto-cycling reaction proceeds and is capable of synthesizing 10^9 copies of the target in less than an hour.[22] A complex set of products is generated that comprises stem-loop DNAs with several inverted repeats of the target and "cauliflower-like" structures with multiple loops formed by annealing between alternately inverted repeats of the target in the same strand. Assay amplification efficiency and sensitivities are enhanced through the use of an additional pair of "loop" primers that bind internally.

LAMP represents a relatively open technology that has seen a large uptake within the research and commercial communities over the last 10 years. Key assay performance characteristics such as rapidity,[66–68] good sensitivity,[69–71] and excellent specificity (conferred by multiple primer binding)[72–74] have driven the popularity of this approach and resulted in LAMP being highly cited within scientific literature.

LAMP-based assays have quickly established a presence within a wide variety of testing sectors including clinical diagnostics such as viral[75–77] and tuberculosis (TB) testing,[78–80] and nonclinical areas such as food diagnostics (pathogen detection and authenticity).[81–83]

26.2.6   Helicase-Dependent Amplification

Helicase-dependent amplification (HDA) utilizes the natural ability of a helicase enzyme to unwind dsDNA and expose primer annealing sites for subsequent priming, without the requirement for heat denaturation.[23] HDA reactions may be performed at a single temperature and the overall process is relatively simple, in comparison to many other isothermal amplification techniques. In the original design, a DNA helicase from *Escherichia coli* (UvrD helicase) was selected to work in association with single-stranded binding proteins (SSBs) that prevented the separated strands from rebinding prior to primer annealing at 37°C. Once hybridization and extension of the initial primers occurred, the newly formed DNA duplex was then separated by the helicase and the cycle of amplification continued.[23] However, during further development of the technique it was identified that using the thermostable UvrD helicase from *Thermoanaerobacter tengcongensis* improved assay specificity by facilitating an increase of reaction temperature to 65°C and eliminating the need for accessory proteins (tHDA)[84] (Figure 26.8).

The application of HDA has been extended to include amplification of RNA by reverse transcription thermophilic HDA (RT-tHDA).[85] This method combines a reverse transcription step with the HDA assay, whereby the double-stranded products of the reverse transcription process are separated by helicase to generate single-stranded RNA and cDNA molecules. The RNA acts as a template for further reverse transcription reactions and the cDNA enters the main tHDA amplification cycle. The addition of SSBs to this assay also increases the speed of the reaction and improves its suitability for integrated near-patient applications.[85]

Another development of this technology facilitates the amplification of very long DNA fragments using circular HDA (cHDA).[86] This is based on the T7 bacteriophage replication system and uses a pair of specific primers to amplify a circular DNA template with high processivity.

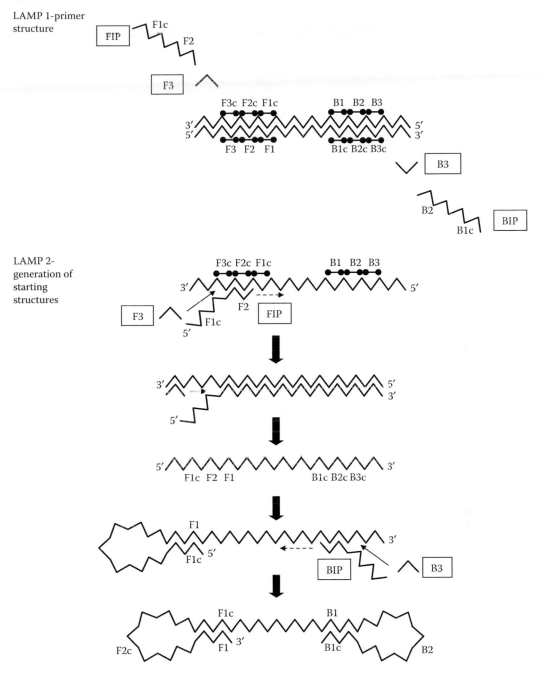

FIGURE 26.7 Schematic representation of the LAMP amplification process utilizing core F3, B3, FIP, and BIP primer set. The amplification stage follows on from step 2, for full details refer to Notomi and coworkers.[22]

Rapid diagnostic tests that may be utilized in a near-patient setting are of increasing demand, particularly in resource-limited environments and where speed is paramount, such as in the control of hospital infections. A diagnostic test has been developed by BioHelix Corporation (Beverly, Massachusetts) for the detection of *Staphylococcus aureus* using HDA.[87] The NAA technology has also been combined with hydrolysis probe-based detection to develop bio-defence assays for the rapid detection of pathogens such as *Bacillis anthracis*.[88] Additionally the technology has

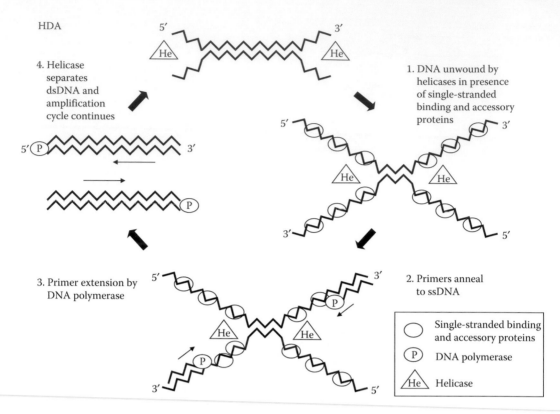

HDA

4. Helicase separates dsDNA and amplification cycle continues

1. DNA unwound by helicases in presence of single-stranded binding and accessory proteins

3. Primer extension by DNA polymerase

2. Primers anneal to ssDNA

○ Single-stranded binding and accessory proteins

Ⓟ DNA polymerase

He Helicase

FIGURE 26.8 Schematic representation of the HDA reaction cycle.

the potential to be integrated into microfluidic devices for applications such as bacterial species diagnostics[89] and human immunodeficiency virus type 1 (HIV-1) testing within resource-limited environments.[90]

26.2.7 CROSS PRIMING AMPLIFICATION

Cross priming amplification (CPA) is a relatively new isothermal NAA technology developed by Ustar Biotechnologies (Hangzhou) Ltd. (Hangzhou, China) that shows similarities with the LAMP approach. The technology originates from the development of diagnostic tools for TB detection in resource-limited environments.[24] CPA is based on multiple cross-linked primers, which amplify a DNA template under isothermal conditions (Figure 26.9). Six or eight primers are involved in the generation of cross priming sites and subsequent amplification of a target sequence at a constant temperature of 65°C.

Originally the performance of Ustar CPA isothermal amplification and detection kit was evaluated for *Mycobacterium tuberculosis* from sputum specimens, and showed an improvement in sensitivity compared to some of the established analysis methods.[24] The technology has also been used for the rapid detection of *Enterobacter sakazakii*, a bacterium which can be found in contaminated powdered infant milk formula, using CPA associated with immuno-blotting analysis.[91]

26.2.8 MULTIPLE DISPLACEMENT AMPLIFICATION

Multiple displacement amplification (MDA)[25,57] employs multiple primers to generate long strands of DNA that are suitable for applications such as WGA[25] and plasmid amplification.[57]

CPA

Generation of cross priming amplification sites

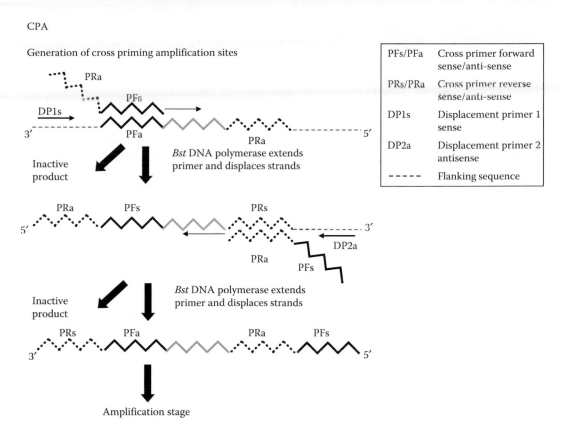

PFs/PFa	Cross primer forward sense/anti-sense
PRs/PRa	Cross primer reverse sense/anti-sense
DP1s	Displacement primer 1 sense
DP2a	Displacement primer 2 antisense
- - - - -	Flanking sequence

Amplification stage

FIGURE 26.9 Simplified schematic representation of the CPA amplification methodology.

MDA can be catalyzed by the phi29 DNA polymerase or by the large fragment of the *Bst* DNA polymerase. phi29 DNA polymerase is typically used in MDA reactions due to inherent high efficiency (using a highly processive enzyme) and low error rates[92] that, together with exonuclease-resistant primers, enable WGA.[25] The process occurs at a constant temperature of 30°C and usually takes between 4 and 6 h, although protocols containing incubations for as long as 18 h have been devised. During the reaction, random hexamers anneal to the template DNA which is subsequently extended by the polymerase. The extended strands are displaced from the starting template by upstream primer extension, and the displaced products act as template for secondary priming events (Figure 26.10). This process continues until a plateau is achieved (regardless of the quantity of starting material).

MDA

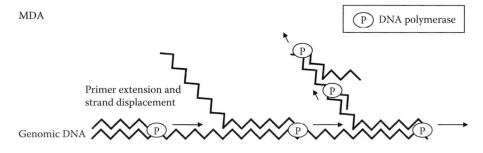

Primer extension and strand displacement

Genomic DNA

P DNA polymerase

FIGURE 26.10 Principle of MDA with a genomic DNA template.

It is important to consider an assessment of amplification bias when using WGA approaches and several studies have addressed this parameter.[25,93,94] Minimizing the nonspecific products generated during the amplification process (originating from contaminants or oligonucleotides) improves the quality of template for downstream analysis. In a recent study, the addition of trehalose to an optimized MDA reaction was found to reduce the occurrence of template-independent products, thus improving specificity.[95]

MDA can be performed on templates derived from a range of clinical sample types, including whole blood, buccal cells,[93] and single cells.[96] There are a number of MDA reagents that are commercially available, including REPLI-g® (Qiagen GmbH, Hilden, Germany), PicoPlex™ (Rubicon Genomics, Inc., Ann Arbor, Michigan) and RepliPHI™ (Epicentre Biotechnologies, Madison, Wisconsin). MDA-based approaches have been used in a recent study to transfer MDA to a digital format (dMDA) for the quantification of DNA fragment contamination. This method was found to be more sensitive than the equivalent PCR-based analysis for detection of microbial DNA fragments.[97]

26.2.9 RECOMBINASE POLYMERASE AMPLIFICATION

Recombinase polymerase amplification (RPA) is an isothermal technology that has been developed and commercialized by TwistDX Ltd (Babraham, U.K.). RPA utilizes a similar concept to PCR in which a pair of oligonucleotides binds to opposite strands of template DNA and generates complementary sequences using a DNA polymerase[26] (Figure 26.11). Unlike PCR, RPA utilizes recombinase–primer complexes to identify and initiate strand displacement, which negates the need for thermal template denaturation. The T4 uvsX recombinase strand displacement activity is stabilized by ssDNA-binding proteins (T4 gp32) and primer extension is catalyzed by the strand displacing DNA polymerase Pol I (*Bsu*). The process uses additional reaction additives (T4 uvsY, a recombinase loading factor and Carbowax 20M crowding agent) to establish optimal reaction conditions. Exponential amplification is achieved by the cyclic repetition of this process, which is typically completed within 30 min.[26,98]

The speed and sensitivity of the RPA technology lends itself to diagnostic applications such as near-patient care[99] and the digital isothermal-based quantification of nucleic acids.[98]

26.3 APPLICATIONS

26.3.1 VIRAL DIAGNOSTICS

26.3.1.1 RNA-Based Detection

Isothermal amplification technologies that are capable of detecting RNA sequences, such as NASBA[16] and TMA, are important to the viral diagnostics sector. Evidence of their suitability has been demonstrated by the development of a wide range of detection assays for RNA-based viruses. HIV infections were one of the first areas to be targeted by emerging isothermal techniques,[16,100] as they required rapid detection and represented a growing diagnostic need across the world. The resultant HIV NASBA assays demonstrated good diagnostic characteristics (e.g., sensitivity and specificity) that compared well with contemporary PCR-based[101] and antigen determination assays.[102]

Highly infectious diseases, such as influenza, which can spread rapidly among populations, necessitate rapid diagnosis in order to manage patients appropriately. Certain virulent strains of the virus, for example, Influenza A virus subtype H5N1 (avian influenza), pose a particular threat to human health and require early containment. RT-LAMP technology has been applied to the development of diagnostic tests for influenza viruses.[103–105] In addition, avian influenza outbreaks have the potential to seriously impact on the poultry industry and wild avian populations, which has led to the development of NASBA-based approaches for the detection of such strains.[106] The detection of other respiratory illnesses such as severe acute respiratory syndrome (SARS) coronavirus have also benefited from the application of isothermal techniques.[107,108]

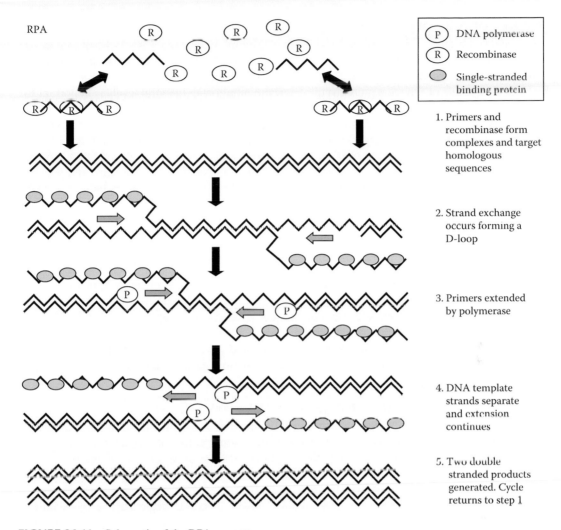

FIGURE 26.11 Schematic of the RPA process.

A major diagnostic challenge facing all nucleic acid-based detection methods for influenza viruses, however, is the high frequency of viral RNA mutation. Therefore, in order to maintain assays of sufficient sensitivity and of relevance for the currently circulating strains, primer redesign may often be necessary.[109]

A number of different isothermal amplification methods have been successfully applied to the detection of hepatitis A and C viruses within the environmental and clinical diagnostic sectors. Contamination of food by Hepatitis A virus (HAV) may cause acute gastroenteritis even though it is often present at diagnostically challenging concentration levels. The use of NASBA to detect HAV may offer improved sensitivity over RT-PCR detection of the viral RNA.[110] TMA has similarly been employed to identify Hepatitis C virus (HCV) in end-stage renal disease patients, and potentially improves the analytical accuracy over ELISA-based assays.[111]

NASBA assays have been reported to enable early detection of other RNA viruses including dengue virus[112] and West Nile virus.[113,114] There are a number of commercially available kits that use NASBA technology for detecting a variety of viruses, such as the NucliSENS range by bioMérieux SA (Marcy l'Etoile, France).

26.3.1.2 DNA-Based Detection

Isothermal techniques targeting DNA are well suited for viral diagnostics. LAMP-based approaches have widely been applied to viral DNA detection and currently dominate the majority of publications in this area. Diagnostic LAMP assays have been developed for a range of viruses, hepatitis B (HBV) being one of the most prominent,[115–117] where monitoring the viral levels is important for management of the infection.

NASBA was originally restricted to RNA templates but has been modified through a number of approaches such as the inclusion of a pre-amplification enzymatic digestion and denaturation step[29,118] to allow for the detection of DNA targets. Yates and coworkers employed DNA NASBA for HBV detection using a method that involved targeting a specific DNA sequence and generating RNA amplicons.[118] The good sensitivity and wide dynamic range of the assay make it a useful alternative tool to PCR.

LAMP assays have been developed for detection of other clinically important pathogens, including human papillomavirus,[119] human herpesvirus 6 (HHV-6),[120] and human herpesvirus 7 (HHV-7),[121] as well as herpes simplex virus type 2 (HSV-2).[122] Chen et al.[123] used a rapid LAMP-based method to detect porcine parvovirus DNA by targeting the VP2 gene and demonstrated better performance than comparable PCR-based assays. With amplification complete within 45 min, the simple and rapid nature of the analysis has a wide appeal in many diagnostic settings.

In addition to LAMP, other isothermal techniques have also been applied to viral DNA diagnostics such as RCA. Detection of porcine circovirus type 2 DNA was successfully achieved using a randomly primed RCA approach.[124] This method enabled direct detection of the viral DNA from clinical samples without the initial virus isolation stage, thus simplifying the analytical process.

26.3.1.3 Case Study: Human Cytomegalovirus Detection

CMV is a DNA virus, commonly occurring in humans, that is asymptomatic in the majority of cases.[125] However, for immunocompromised patients (including transplant recipients, AIDS patients and young babies),[126] CMV is associated with a risk of potentially severe complications.

Continuous monitoring of CMV viral load in high-risk groups is important for managing patient treatment and currently, the most widely used methods are based on qPCR analysis.[127] However, there is a significant time period between patient sample collection, molecular analysis within a specialized laboratory, and reporting of results back to clinicians. CMV viral load testing represents a diagnostic situation whereby the clinician requires timely access to patient viral load data so that treatment can either be quickly initiated or adapted in response to rising viral titers.

Alternative isothermal NAA such as LAMP assays have been applied to CMV diagnostics that target DNA rather than RNA.[128,129] A NASBA-based test has been used to detect viral RNA associated with human CMV,[130,131] and is available commercially (NucliSENS® CMV pp67) from bioMé (Marcy l'Etoile, France). Utilization of this assay has proved to be clinically valuable in monitoring the efficacy of antiviral therapy[132,133] and progression of CMV disease.[134] A quantitative approach targeting the immediate-early gene, UL123 (IE1), has been demonstrated to be useful in monitoring for subclinical infection.[135]

Isothermal NAA technologies such as LAMP and NASBA offer the potential to reduce the time lag between sampling and delivery of results, which would significantly improve the diagnostic process. The potential for streamlining the diagnostic process and miniaturizing methods like LAMP is an area of growing interest.[115,136] The transfer of LAMP onto microfluidic chip formats has been demonstrated for pathogen detection using a method called micro-LAMP (μLAMP).[136] The amplification stage of this viral DNA detection system, tested with Pseudorabies virus (PRV), takes less than 1 h to complete and the entire diagnostic process, from sample arrival to final result, is achievable within 2 h or less. LAMP shows strong potential for integration into rapid near patient devices for nucleic acid testing within areas such as CMV viral load monitoring (Figure 26.12).

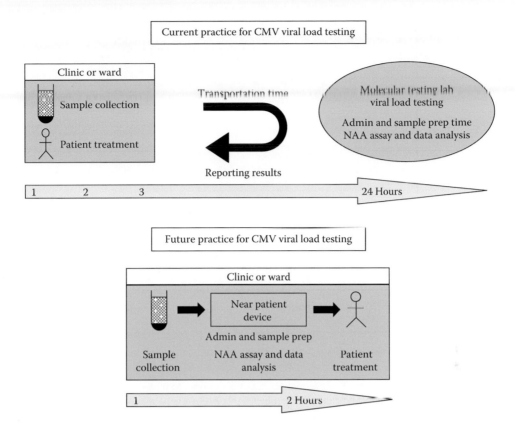

FIGURE 26.12 Comparing the typical processes involved in laboratory-based and near-patient cytomegalovirus viral load testing.

26.3.2 Bacterial Diagnostics

The bacterial diagnostic world encompasses a wide variety of commercially and clinically important organisms that have defined requirements for identification methods. Until the advent of NAA-based technologies, bacterial diagnostics was limited to classical microbiological and immunological approaches that suffered from issues such as lack of speed and sensitivity. The application of PCR[8] to bacterial diagnostics has revolutionized the detection and enumeration of bacterial targets.[5,6] The PCR technique exhibits high molecular sensitivity and specificity that reduces the requirement for classical microbiological analyses.

Alternative NAA technologies such as SDA[18] and TMA were also rapidly adopted by the diagnostic community (Figure 26.13) due to the commercial and clinical pressures for non-PCR approaches that avoided limitations associated with PCR, for example, licensing, target type and improved performance characteristics.

26.3.2.1 Clinical Bacterial Diagnostics

The earliest examples of isothermal NAA technologies can be found within diagnostic systems for STIs where they currently represent a routine analytical approach. A well-established commercial instrument system is represented by the BD ProbeTec™ system (BD Diagnostics, Sparks, Maryland), which was originally developed in the 1990s and utilizes SDA technology to detect and quantify key pathogenic organisms such as *C. trachomatis* and *N. gonorrhoeae*.[137–141] Additional commercial success was established with the DTS® and TIGRIS® DTS® Systems developed by Gen-Probe Inc (San Diego, California) that incorporate the TMA amplification technology to

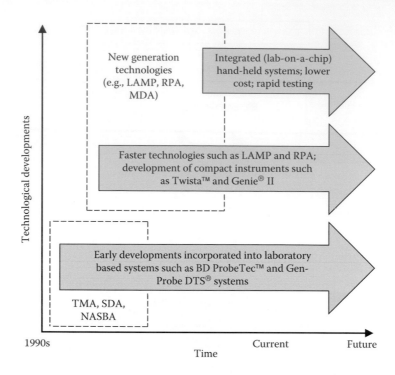

FIGURE 26.13 General overview of isothermal NAA developments. Manufacturer details: BD ProbeTec™ (BD Diagnostics, Sparks, Maryland) system, Gen-Probe DTS® system (Gen-Probe Inc., San Diego, California), Twista™ (TwistDX, Babraham, U.K.) and Genie® II (OptiGene Ltd., Horsham, U.K.).

deliver a similar range of testing capabilities as the BD ProbeTec™ system.[141–143] However, these systems (Figure 26.13) are laboratory-based and fundamentally unsuited to near-patient applications due to large instrument footprints and processing requirements.

Later developments (Figure 26.13) have exploited newer isothermal technologies such as LAMP[22] that offer improved performance characteristics (e.g., reduced detection times) and encouraged the development of near-patient diagnostic systems within the bacterial testing community.[144,145] These emerging technologies are suitable for integration into point-of-care/miniaturized systems as evidenced by the wide variety of work currently being conducted within the field.[89,99,136,146,147]

26.3.2.2 Food and Environmental Diagnostics

Food and environmental diagnostics represent active areas of commercial and research development due to the general applicability of NAA techniques. Diagnostic systems utilizing isothermal NAA technologies are increasingly providing viable alternatives to standard PCR-based approaches within key sectors such as food safety/microbiology,[3,82,148,149] environmental testing,[150,151] plant pathogen testing,[152,153] and testing for genetically modified organisms.[154,155] These isothermal technologies are amenable for point-of-testing scenarios and are being integrated into compact microfluidic-type devices[99,156] that offer the potential for robust field-based testing.

26.3.2.3 Case Study: Tuberculosis Diagnostics

TB bears an important public health priority within both the developed and developing world due to the increasing prevalence of new virulent forms of the organism that demonstrate resistance to front-line medications and an increasing association with HIV/TB coinfection.[157] The effective implementation of rapid diagnosis and drug susceptibility testing can facilitate effective treatment and limit the spread of TB.

Nucleic acid amplification tests (NAATs) are well established within the field of TB diagnostics[158–160] and incorporate a range of amplification technologies. Commercially available systems for TB NAAT generally comprise large analyzers such as the PCR-based COBAS® TaqMan® MTB Test (Roche Diagnostics Corp, Indianapolis, Indiana), TMA-based Gen-Probe Amplified MTD® system and SDA-based BD ProbeTec™ system. The Gen-Probe Amplified MTD system is a FDA-approved isothermal method for the detection of *M. tuberculosis* from respiratory specimens (smear-positive and smear-negative) with good levels of sensitivity and specificity.[161]

While large laboratory-based TB diagnostic systems provide high specificity and sensitivity performance levels,[162] simpler, cheaper, and more compact systems are required for usage outside reference laboratory centers or within resource-limited situations. Compact integrated instruments such as the Cepheid GeneXpert System (Cepheid, Sunnyvale, California) have been developed to address this diagnostic need[163] using PCR-based technologies.

In comparison, modern isothermal NAA methodologies such as LAMP offer the potential for rapid and low resource TB diagnostic systems. These features are driving their uptake by international healthcare initiatives such as the Foundation for Innovative New Diagnostics (FIND) (http://www.finddiagnostics.org) who are encouraging the development of TB tests for resource-limited scenarios. TB NAATs that use loop-mediated isothermal amplification with a simple visual colorimetric readout[79,164] are currently under evaluation. These low-cost alternatives to relatively expensive diagnostic systems potentially offer improved access to diagnosis within the developed (near-patient access) and developing world (resource-constrained scenario).

26.3.3 WHOLE-GENOME AMPLIFICATION

The development of molecular biological techniques such as PCR has provided a suite of tools that are capable of both sequence-specific target amplification and WGA. WGA was developed in 1992[165,166] to increase the quantity of nucleic acid template material for testing and is a valuable tool when template is limiting such as for single-cell genomics and forensic analysis (Figure 26.14). Initial WGA approaches utilized PCR with random or degenerative primers to amplify genomic regions—degenerate oligonucleotide PCR (DOP-PCR)[165] and primer extension preamplification (PEP).[166] Subsequent WGA developments focussed on the application of isothermal NAA technologies that demonstrated improved characteristics such as lower error rates, reduced sequence bias and longer amplified product lengths.

26.3.3.1 Isothermal WGA Technologies

As previously discussed, MDA[25,57] is an isothermal NAA technology that utilizes the strand displacement activity of phi29 DNA polymerase to generate a network of branched DNA structures. MDA is particularly suitable for WGA applications due to the utilization of phi29 DNA polymerase which does not dissociate from the genomic DNA template (unlike Taq polymerase) and is capable of generating DNA fragments up to 100 kb in size. The enzyme has a $3'-5'$ exonuclease proofreading activity and provides error rates 100 times lower than Taq DNA polymerase-based methods.[25] Other advantages comprise a longer than average product length (>10 kb) which allows for restriction fragment-length polymorphism (RFLP) analysis and the ability to generate microgram levels of amplified products.[96] It has been demonstrated that the technology is capable of generating reliable amplification from 1 to 10 copies of human genomic DNA, as well as amplification directly from biological samples including crude whole-blood and tissue culture cells.[25]

MDA-based WGA has been applied to a wide variety of areas, such as comparative genomic hybridization, that require high levels of unbiased genomic enrichment. Single-cell analysis is an increasingly popular study area that has benefited from WGA,[96,167,168] which has also been used to successfully diagnose aneuploidy in human embryos. In addition to clinical diagnostics, MDA has also been used to monitor microbial populations from uncultivated samples.[169] Commercial MDA kits are available from companies such as QIAGEN GmbH (Hilden, Germany) and Rubicon Genomics, Inc (Ann Arbor, Michigan) that, respectively, market the technology as REPLI-g® and PicoPlex™.

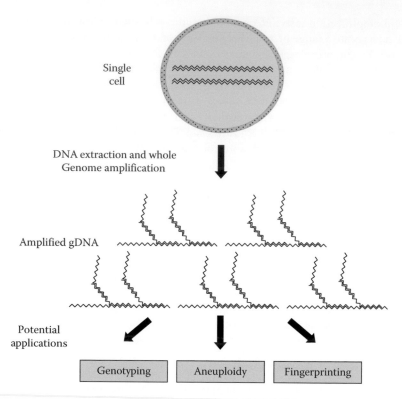

Single cell

DNA extraction and whole
Genome amplification

Amplified gDNA

Potential
applications

| Genotyping | Aneuploidy | Fingerprinting |

FIGURE 26.14   Typical whole genome amplification application—single-cell analysis.

WGA approaches that incorporate alternative isothermal NAA technologies such as SPIA[20] and RCA[170] have been developed and commercialized. One example is the Ovation WGA®

System (NuGEN Technologies, Inc., San Carlos, California) that is based on the SPIA process and is capable of amplifying DNA from low levels of starting material.

26.3.3.2   Case Study: Preimplantation Genetic Diagnostics

Preimplantation genetic diagnostics (PGD) enables the identification of genetic defects in embryos created through *in vitro* fertilization (IVF) prior to implantation. PGD can be used effectively to screen embryos for genetic defects from at-risk individuals using a single blastomere (containing approximately 6 pg genomic DNA). The field of PGD has benefited greatly from the use of WGA techniques,[96,171,172] which enable complex molecular tests to be performed. Diagnostic testing for Fragile X Syndrome was improved with the addition of an MDA step prior to fluorescent PCR analysis. This enabled a reduction in the number of embryos failing to return a diagnostic result from 45% without MDA to only 14% with the additional amplification step.[173]

Isothermal WGA technologies have also been applied to the development of genetic screening tests for aneuploidy[168,174,175] as a result of the failure of fluorescence *in situ* hybridization (FISH)-based methods to deliver the required clinical efficacy for the treatment of infertility. These technologies provide clinicians with the molecular tools to amplify small amounts of genetic material and perform previously challenging diagnostic analyses.

26.4   DISCUSSION

The *in vitro* application of biological nucleic acid replication processes has led to the development of modern molecular diagnostics and revolutionized the detection and quantitation of RNA and DNA materials. PCR-based technologies have been central to these developments and represent the bulk of current nucleic acid testing.

However, since the earliest days of PCR, alternative isothermal NAA strategies have been devised that offer enhanced performance characteristics such as rapid reaction time and reduced susceptibility to inhibitors. These isothermal amplification strategies are involved in a growing range of nucleic acid detection technologies, encompassing a wide variety of applications from environmental pathogen detection to PGD. Isothermal NAA technologies have moved beyond specialized clinical areas (STI and viral infection testing) to the wider diagnostic community. This expansion has been driven by the development of newer approaches such as LAMP.

Isothermal technologies employ single temperature amplification profiles and offer the potential for very rapid assay times. These technologies are characterized by greater flexibility in terms of enzyme choice/nucleic acid template and exploit alternative NAA mechanisms to heat denaturation utilized by PCR for separating DNA duplexes. The relative simplicity of the thermal dynamics opens up the techniques for use both in laboratory settings, as well as incorporation within portable or hand-held devices.

The ability to perform molecular analyses without the need for a thermocycler opens up the potential for the development of compact near patient devices. Technological advances within the fields of microfluidics and engineering should support these developments. The arrival of compact commercial isothermal NAA-based instrumentations such as the Twista™ (TwistDX, Babraham, U.K.) and Genie® II (OptiGene Ltd., Horsham, U.K.), together with ongoing research into areas such as hand-held detectors[176,177] and the growth of near-patient diagnostic markets, demonstrate the utility of isothermal NAA technologies in the modern diagnostic world.

Although there has been considerable development and refinement of isothermal detection methods in recent years, there are still areas where improvements would further enhance these technologies. Introducing control processes to regulate the starting point of isothermal reactions (such as a heat activation stage) has the potential to improve assay performance and quantitative potential. Potential regulatory systems include the application of heat-activated dNTPs[178] and photoactivated polymerases.[179] Further work toward reducing background signals and nonspecific priming in some isothermal assays would also be beneficial to the diagnostic potential of this field.

Characterizing and benchmarking the performance associated with isothermal systems is necessary for continued development within research and commercial sectors. This is particularly important when reliable quantitative measurements are required. The regulatory framework associated with clinical testing and diagnostic devices such as the European IVD Directive 98/79/EC[180] necessitates a good understanding of measurement issues and the appropriate application of strategies such as method validation and reference materials. A robust measurement infrastructure for isothermal NAAT technologies will support innovation and growth within this promising area.

It is evident from the wide range of applications highlighted in this chapter that isothermal technologies have an important and growing role to play in the future of molecular diagnostics. Future developments of improved and novel isothermal NAA technologies have the potential to strengthen the position of the sector as a viable alternative to PCR-based techniques, broaden the availability of diagnostic systems within resource-limited environments and expand their application within the general diagnostic community.

ACKNOWLEDGMENTS

The authors would like to thank Dr. Jim Huggett and Dr. Carole Foy for their help during the preparation of this book chapter. This work was funded by the U.K. National Measurement System.

REFERENCES

1. Cho WC. Molecular diagnostics for monitoring and predicting therapeutic effect in cancer. *Expert Rev Mol Diagn* 2011;11:9–12.

2. Waiblinger HU, Grohmann L, Mankertz J, Engelbert D, Pietsch K. A practical approach to screen for authorised and unauthorised genetically modified plants. *Anal Bioanal Chem* 2010;396:2065–72.

3. Shi X-M, Long F, Suo B. Molecular methods for the detection and characterization of foodborne pathogens. *Pure Appl Chem* 2010;82:69–79.

4. Pafundo S, Gulli M, Marmiroli N. Multiplex real-time PCR using SYBR® GreenER™ for the detection of DNA allergens in food. *Anal Bioanal Chem* 2010;396:1831–9.

5. Weile J, Knabbe C. Current applications and future trends of molecular diagnostics in clinical bacteriology. *Anal Bioanal Chem* 2009;394:731–42.

6. Muldrew KL. Molecular diagnostics of infectious diseases. *Curr Opin Pediatr* 2009;21:102–11.

7. Mullis K, Faloona F, Scharf S, Saiki R, Horn G, Erlich H. Specific enzymatic amplification of DNA *in vitro*: The polymerase chain reaction. *Cold Spring Harb Symp Quant Biol* 1986;51(Part 1):263–73.

8. Saiki RK, Scharf S, Faloona F, Mullis KB, Horn GT, Erlich HA, Arnheim N. Enzymatic amplification of beta-globin genomic sequences and restriction site analysis for diagnosis of sickle cell anemia. *Science* 1985;230:1350–4.

9. Khan G, Kangro HO, Coates PJ, Heath RB. Inhibitory effects of urine on the polymerase chain reaction for cytomegalovirus DNA. *J Clin Pathol* 1991;44:360–5.

10. Mahony J, Chong S, Jang D, Luinstra K, Faught M, Dalby D et al. Urine specimens from pregnant and nonpregnant women inhibitory to amplification of *Chlamydia trachomatis* nucleic acid by PCR, ligase chain reaction, and transcription-mediated amplification: Identification of urinary substances associated with inhibition and removal of inhibitory activity. *J Clin Microbiol* 1998;36:3122–6.

11. Al Soud WA, Radstrom P. Purification and characterization of PCR-inhibitory components in blood cells. *J Clin Microbiol* 2001;39:485–93.

12. Kontanis EJ, Reed FA. Evaluation of real-time PCR amplification efficiencies to detect PCR inhibitors. *J Forensic Sci* 2006;51:795–804.

13. Gill P, Ghaemi A. Nucleic acid isothermal amplification technologies: A review. *Nucleosides Nucleotides Nucleic Acids* 2008;27:224–43.

14. Kim J, Easley CJ. Isothermal DNA amplification in bioanalysis: Strategies and applications. *Bioanalysis* 2011;3:227–39.

15. Craw P, Balachandran W. Isothermal nucleic acid amplification technologies for point-of-care diagnostics: A critical review. *Lab Chip* 2012;12:2469–86.

16. Compton J. Nucleic acid sequence-based amplification. *Nature* 1991;350:91–2.

17. Guatelli JC, Whitfield KM, Kwoh DY, Barringer KJ, Richman DD, Gingeras TR. Isothermal, *in vitro* amplification of nucleic acids by a multienzyme reaction modeled after retroviral replication. *Proc Natl Acad Sci USA* 1990;87:1874–8.

18. Walker GT, Fraiser MS, Schram JL, Little MC, Nadeau JG, Malinowski DP. Strand displacement amplification—An isothermal, *in vitro* DNA amplification technique. *Nucleic Acids Res* 1992;20:1691–6.

19. Dafforn A, Chen P, Deng G, Herrler M, Iglehart D, Koritala S et al. Linear mRNA amplification from as little as 5 ng total RNA for global gene expression analysis. *Biotechniques* 2004;37:854–7.

20. Kurn N, Chen P, Heath JD, Kopf-Sill A, Stephens KM, Wang S. Novel isothermal, linear nucleic acid amplification systems for highly multiplexed applications. *Clin Chem* 2005;51:1973–81.

21. Fire A, Xu SQ. Rolling replication of short DNA circles. *Proc Natl Acad Sci USA* 1995;92:4641–5.

22. Notomi T, Okayama H, Masubuchi H, Yonekawa T, Watanabe K, Amino N, Hase T. Loop-mediated isothermal amplification of DNA. *Nucleic Acids Res* 2000;28:E63.

23. Vincent M, Xu Y, Kong H. Helicase-dependent isothermal DNA amplification. *EMBO Rep* 2004;5:795–800.

24. Fang R, Li X, Hu L, You Q, Li J, Wu J et al. Cross-priming amplification for rapid detection of *mycobacterium tuberculosis* in sputum specimens. *J Clin Microbiol* 2009;47:845–7.

25. Dean FB, Hosono S, Fang L, Wu X, Faruqi AF, Bray-Ward P et al. Comprehensive human genome amplification using multiple displacement amplification. *Proc Natl Acad Sci USA* 2002;99:5261–6.

26. Piepenburg O, Williams CH, Stemple DL, Armes NA. DNA detection using recombination proteins. *PLoSBiol* 2006;4:e204.

27. Fahy E, Kwoh DY, Gingeras TR. Self-sustained sequence replication (3SR): An isothermal transcription-based amplification system alternative to PCR. *PCR Methods Appl* 1991;1:25–33.

28. Berard C, Cazalis M-A, Leissner P, Mougin B. DNA nucleic acid sequence-based amplification-based genotyping for polymorphism analysis. *Bio Techniques* 2004;37:680–2, 4, 6.

29. Deiman B, Jay C, Zintilini C, Vermeer S, van SD, Venema F, van dWP. Efficient amplification with NASBA of hepatitis B virus, herpes simplex virus and methicillin resistant *Staphylococcus aureus* DNA. *J Virol Methods* 2008;151:283–93.

30. Gracias KS, McKillip JL. Nucleic acid sequence-based amplification (NASBA) in molecular bacteriology: A procedural guide. *J Rapid Methods Autom Microbiol* 2007;15:295–309.

31. Waldenmaier A. Molecular biological diagnostics of human papillomaviruses. *MTA Dialog* 2009;10:904–8.

32. Kao M-C, Durst RA. Detection of *Escherichia coli* using nucleic acid sequence-based amplification and oligonucleotide probes for 16S ribosomal RNA. *Anal Lett* 2010;43:1756–69.

33. Lau LT, Feng XY, Lam TY, Hui HK, Yu ACH. Development of multiplex nucleic acid sequence-based amplification for detection of human respiratory tract viruses. *J Virol Methods* 2010;168:251–4.

34. Sidoti F, Bergallo M, Terlizzi ME, Piasentin Alessio E, Astegiano S, Gasparini G, Cavallo R. Development of a quantitative real-time nucleic acid sequence-based amplification assay with an internal control using Molecular Beacon probes for selective and sensitive detection of human rhinovirus serotypes. *Mol Biotechnol* 2011;50:221–8.

35. Mikamo H, Tanaka K, Watanabe K. TMA (transcription mediated amplification) test for simultaneous detection of *Chlamydia trachomatis/Neisseria gonorrhoeae*. *Kensa to Gijutsu* 2006;34:1293–4.

36. Ross RS, Viazov S, Kpakiwa SS, Roggendorf M. Transcription-mediated amplification linked to line probe assay as a routine tool for HCV typing in clinical laboratories. *J Clin Lab Anal* 2007;21:340–7.

37. Nye MB, Schwebke JR, Body BA. Comparison of APTIMA *Trichomonas vaginalis* transcription-mediated amplification to wet mount microscopy, culture, and polymerase chain reaction for diagnosis of trichomoniasis in men and women. *Am J Obstet Gynecol* 2009;200:188.e1–7.

38. Bachmann LH, Johnson RE, Cheng H, Markowitz L, Papp JR, Palella FJ, Jr., Hook EW, 3rd. Nucleic acid amplification tests for diagnosis of *Neisseria gonorrhoeae* and *Chlamydia trachomatis* rectal infections. *J Clin Microbiol* 2010;48:1827–32.

39. Munson E, Napierala M, Basile J, Miller C, Burtch J, Hryciu.K. JE, Schell RF. *Trichomonas vaginalis* transcription-mediated amplification-based analyte-specific reagent and alternative target testing of primary clinical vaginal saline suspensions. *Diagn Microbiol Infect Dis* 2010;68:66–72.

40. Soni S, White JA. Self-screening for *Neisseria gonorrhoeae* and *Chlamydia trachomatis* in the human immunodeficiency virus clinic—High yields and high acceptability. *Sex Transm Dis* 2011;38:1107–9.

41. Tyagi S, Kramer FR. Molecular beacons: Probes that fluoresce upon hybridization. *Nat Biotechnol* 1996;14:303–8.

42. Leone G, Van SH, Van GB, Kramer FR, Schoen CD. Molecular Beacon probes combined with amplification by NASBA enable homogeneous, real-time detection of RNA. *Nucleic Acids Res* 1998;26:2150–5.

43. Asiello PJ, Baeumner AJ. Miniaturized isothermal nucleic acid amplification, a review. *Lab Chip* 2011;11:1420–30.

44. Gulliksen A, Keegan H, Martin C, O'Leary J, Solli LA, Falang IM et al. Towards a "Sample-In, Answer-Out" point-of-care platform for nucleic acid extraction and amplification: Using an HPV E6/E7 mRNA model system. *J Oncol* 2012;2012:12.

45. Tariq MA, Kim HJ, Jejelowo O, Pourmand N. Whole-transcriptome RNAseq analysis from minute amount of total RNA. *Nucleic Acids Res* 2011;39:e120.

46. Dafforn A, Chen P, Deng GY, Herrler M, Iglehart DM, Koritala S et al. Ribo-SPIA, a rapid isothermal RNA amplification method for gene expression analysis. *Drug Discov Ser* 2007;7:261–79.

47. Vermeulen J, Derveaux S, Lefever S, De Smet E, De Preter K, Yigit N et al. RNA pre-amplification enables large-scale RT-qPCR gene-expression studies on limiting sample amounts. *BMC Res Notes* 2009;2:235.

48. Walker GT, Little MC, Nadeau JG, Shank DD. Isothermal *in vitro* amplification of DNA by a restriction enzyme/DNA polymerase system. *Proc Natl Acad Sci USA* 1992;89:392–6.

49. Spears PA, Linn CP, Woodard DL, Walker GT. Simultaneous strand displacement amplification and fluorescence polarization detection of *Chlamydia trachomatis* DNA. *Anal Biochem* 1997;247:130–7.

50. Guo Q, Yang X, Wang K, Tan W, Li W, Tang H, Li H. Sensitive fluorescence detection of nucleic acids based on isothermal circular strand-displacement polymerization reaction. *Nucleic Acids Res* 2009;37:e20.

51. Wang S-S, Thornton K, Kuhn AM, Nadeau JG, Hellyer TJ. Homogeneous real-time detection of single-nucleotide polymorphisms by strand displacement amplification on the BD ProbeTec ET system. *Clin Chem* 2003;49:1599–607.

52. Chen Q, Bian Z, Chen M, Hua X, Yao C, Xia H et al. Real-time monitoring of the strand displacement amplification (SDA) of human cytomegalovirus by a new SDA-piezoelectric DNA sensor system. *Biosens Bioelectron* 2009;24:3412–8.

53. Liu D, Daubendiek SL, Zillman MA, Ryan K, Kool ET. Rolling circle DNA synthesis: Small circular oligonucleotides as efficient templates for DNA polymerases. *J Am Chem Soc* 1996;118:1587–94.

54. Lizardi PM, Huang X, Zhu Z, Bray-Ward P, Thomas DC, Ward DC. Mutation detection and single-molecule counting using isothermal rolling-circle amplification. *Nat Genet* 1998;19:225–32.

55. Nilsson M, Krejci K, Koch J, Kwiatkowski M, Gustavsson P, Landegren U. Padlock probes reveal single-nucleotide differences, parent of origin and *in situ* distribution of centromeric sequences in human chromosomes 13 and 21. *Nat Genet* 1997;16:252–5.

56. Nilsson M, Malmgren H, Samiotaki M, Kwiatkowski M, Chowdhary BP, Landegren U. Padlock probes: Circularizing oligonucleotides for localized DNA detection. *Science* 1994;265:2085–8.

57. Dean FBN, JR., Giesler, TL., Lasken, RS. Rapid amplification of plasmid and phage DNA using Phi 29 DNA polymerase and multiply-primed rolling circle amplification. *Genome Res* 2001;11:1095–9.

58. Inoue J, Shigemori Y, Mikawa T. Improvements of rolling circle amplification (RCA) efficiency and accuracy using *Thermus thermophilus* SSB mutant protein. *Nucleic Acids Res* 2006;34:e69.

59. Murakami T, Sumaoka J, Komiyama M. Sensitive isothermal detection of nucleic-acid sequence by primer generation-rolling circle amplification. *Nucleic Acids Res* 2009;37:e19.

60. Haible D, Kober S, Jeske H. Rolling circle amplification revolutionizes diagnosis and genomics of geminiviruses. *J Virol Methods* 2006;135:9–16.

61. Schubert J, Habekuss A, Kazmaier K, Jeske H. Surveying cereal-infecting geminiviruses in Germany—Diagnostics and direct sequencing using rolling circle amplification. *Virus Res* 2007;127:61–70.

62. Henriksson S, Blomstrom AL, Fuxler L, Fossum C, Berg M, Nilsson M. Development of an *in situ* assay for simultaneous detection of the genomic and replicative form of PCV2 using padlock probes and rolling circle amplification. *Virol J* 2011;8:37.

63. Kong F, Tong Z, Chen X, Sorrell T, Wang B, Wu Q et al. Rapid identification and differentiation of *Trichophyton* species, based on sequence polymorphisms of the ribosomal internal transcribed spacer regions, by rolling-circle amplification. *J Clin Microbiol* 2008;46:1192–9.

64. Tong Z, Kong F, Wang B, Zeng X, Gilbert GL. A practical method for subtyping of *Streptococcus agalactiae* serotype III, of human origin, using rolling circle amplification. *J Microbiol Methods* 2007;70:39–44.

65. Pang S, FQ, Shanahan D, Harris N. Investigation of the use of rolling circle amplification for the detection of GM food. *Eur Food Res Technol* 2007;225:59–66.

66. Horisaka T, Fujita K, Iwata T, Nakadai A, Okatani AT, Horikita T et al. Sensitive and specific detection of *Yersinia pseudotuberculosis* by loop-mediated isothermal amplification. *J Clin Microbiol* 2004;42:5349–52.

67. Wang Y, Chen P, Guo H, Chen Y, Liu H, He Q. Loop-mediated isothermal amplification targeting the apxIVA gene for detection of *Actinobacillus pleuropneumoniae*. *FEMS Microbiol Lett* 2009;300:83–9.

68. Polley SD, Mori Y, Watson J, Perkins MD, Gonzalez IJ, Notomi T et al. Mitochondrial DNA targets increase sensitivity of malaria detection using loop-mediated isothermal amplification. *J Clin Microbiol* 2010;48:2866–71.

69. Tsai S-M, Chan K-W, Hsu W-L, Chang T-J, Wong M-L, Wang C-Y. Development of a loop-mediated isothermal amplification for rapid detection of orf virus. *J Virol Methods* 2009;157:200–4.

70. Lee M-F, Chen Y-H, Peng C-F. Evaluation of reverse transcription loop-mediated isothermal amplification in conjunction with ELISA-hybridization assay for molecular detection of *Mycobacterium tuberculosis*. *J Microbiol Methods* 2009;76:174–80.

71. Nakao R, Stromdahl EY, Magona JW, Faburay B, Namangala B, Malele I et al. Development of loop-mediated isothermal amplification (LAMP) assays for rapid detection of *Ehrlichia ruminantium*. *BMC Microbiol* 2010;10:296.

72. Poon LLM, Wong BWY, Ma EHT, Chan KH, Chow LMC, Abeyewickreme W et al. Sensitive and inexpensive molecular test for falciparum malaria: Detecting *Plasmodium falciparum* DNA directly from heat-treated blood by loop-mediated isothermal amplification. *Clin Chem* 2006;52:303–6.

73. Lau YL, Meganathan P, Sonaimuthu P, Thiruvengadam G, Nissapatorn V, Chen Y. Specific, sensitive, and rapid diagnosis of active toxoplasmosis by a loop-mediated isothermal amplification method using blood samples from patients. *J Clin Microbiol* 2010;48:3698–702.

74. Francois P, Bento M, Hibbs J, Bonetti EJ, Boehme CC, Notomi T et al. Robustness of loop-mediated isothermal amplification reaction for diagnostic applications. *FEMS Immunol Med Microbiol* 2011;62:41–8.

75. Parida M, Horioke K, Ishida H, Dash PK, Saxena P, Jana AM et al. Rapid detection and differentiation of dengue virus serotypes by a real-time reverse transcription-loop-mediated isothermal amplification assay. *J Clin Microbiol* 2005;43:2895–903.

76. Boldbaatar B, Inoue S, Sugiura N, Noguchi A, Orbina JRC, Demetria C et al. Rapid detection of rabies virus by reverse transcription loop-mediated isothermal amplification. *Jpn J Infect Dis* 2009;62:187–91.

77. Yaqing H, Wenping Z, Zhiyi Y, Xionghu W, Shouyi Y, Hong Y et al. Detection of human Enterovirus 71 reverse transcription loop-mediated isothermal amplification (RT-LAMP). *Lett Appl Microbiol* 2012;54:233–9.

78. Enosawa M, Kageyama S, Sawai K, Watanabe K, Notomi T, Onoe S et al. Use of loop-mediated isothermal amplification of the IS900 sequence for rapid detection of cultured *Mycobacterium avium* subsp. *paratuberculosis*. *J Clin Microbiol* 2003;41:4359–65.

79. Boehme CC, Nabeta P, Henostroza G, Raqib R, Rahim Z, Gerhardt M et al. Operational feasibility of using loop-mediated isothermal amplification for diagnosis of pulmonary tuberculosis in microscopy centers of developing countries. *J Clin Microbiol* 2007;45:1936–40.

80. Geojith G, Dhanasekaran S, Chandran SP, Kenneth J. Efficacy of loop mediated isothermal amplification (LAMP) assay for the laboratory identification of *Mycobacterium tuberculosis* isolates in a resource limited setting. *J Microbiol Methods* 2011;84:71–3.

81. Ahmed MU, Hasan Q, Mosharraf HM, Saito M, Tamiya E. Meat species identification based on the loop mediated isothermal amplification and electrochemical DNA sensor. *Food Control* 2010;21:599–605.

82. Ye Y, Wang B, Huang F, Song Y, Yan H, Alam MJ et al. Application of *in situ* loop-mediated isothermal amplification method for detection of *Salmonella* in foods. *Food Control* 2011;22:438–44.

83. Zhang G, Brown EW, Gonzalez-Escalona N. Comparison of real-time PCR, reverse transcriptase real-time PCR, loop-mediated isothermal amplification, and the FDA conventional microbiological method for the detection of *Salmonella* spp. in produce. *Appl Environ Microbiol* 2011;77:6495–501.

84. An L, Tang W, Ranalli TA, Kim HJ, Wytiaz J, Kong H. Characterization of a thermostable UvrD helicase and its participation in helicase-dependent amplification. *J Biol Chem* 2005;280:28952–8.

85. Goldmeyer J, Kong H, Tang W. Development of a novel one-tube isothermal reverse transcription thermophilic helicase-dependent amplification platform for rapid RNA detection. *J Mol Diagn* 2007;9:639–44.

86. Xu Y KH, Kays A, Rice J, Kong H. Simultaneous amplification and screening of whole plasmids using the T7 bacteriophage replisome. *Nucleic Acids Res* 2006;34:e98.

87. Goldmeyer J, Li H, McCormac M, Cook S, Stratton C, Lemieux B et al. Identification of *Staphylococcus aureus* and determination of methicillin resistance directly from positive blood cultures by isothermal amplification and a disposable detection device. *J Clin Microbiol* 2008;46:1534–6.

88. Tong Y, Tang W, Kim H-J, Pan X, Ranalli TA, Kong H. Development of isothermal TaqMan assays for detection of biothreat organisms. *BioTechniques* 2008;45:543–4, 6, 8, 50, 52–53, 55, 57.

89. Mahalanabis M, Do J, Al MH, Zhang JY, Klapperich CM. An integrated disposable device for DNA extraction and helicase dependent amplification. *Biomed Microdevices* 2010;12:353–9.

90. Tang W, Chow WH, Li Y, Kong H, Tang YW, Lemieux B. Nucleic acid assay system for tier II laboratories and moderately complex clinics to detect HIV in low resource settings. *J Infect Dis* 2010;201 Suppl 1:S46–51.

91. Yulong Z XZ, Hongwei Z, Wei L, Wenjie Z, Xitai H. Rapid and sensitive detection of *Enterobacter sakazakii* by cross-priming amplification combined with immuno-blotting analysis. *Mol Cell Probes* 2010;24:396–400.

92. Spits C, Le Caignec C, De Rycke M, Van Haute L, Van Steirteghem A, Liebaers I, Sermon K. Optimization and evaluation of single-cell whole-genome multiple displacement amplification. *Hum Mutat* 2006;27:496–503.

93. Hosono S, Faruqi AF, Dean FB, Du Y, Sun Z, Wu X et al. Unbiased whole-genome amplification directly from clinical samples. *Genome Res* 2003;13:954–64.

94. Dickson PA, Montgomery GW, Henders A, Campbell MJ, Martin NG, James MR. Evaluation of multiple displacement amplification in a 5 cM STR genome-wide scan. *Nucleic Acids Res* 2005;33:e119.

95. Pan X, Urban AE, Palejev D, Schulz V, Grubert F, Hu Y et al. A procedure for highly specific, sensitive, and unbiased whole-genome amplification. *Proc Natl Acad Sci USA* 2008;105:15499–504.

96. Hellani A, Coskun S, Benkhalifa M, Tbakhi A, Sakati N, Al-Odaib A, Ozand P. Multiple displacement amplification on single cell and possible PGD applications. *Mol Hum Reprod* 2004;10:847–52.

97. Blainey PC, Quake SR. Digital MDA for enumeration of total nucleic acid contamination. *Nucleic Acids Res* 2011;39:e19.

98. Shen F, Davydova EK, Du W, Kreutz JE, Piepenburg O, Ismagilov RF. Digital isothermal quantification of nucleic acids via simultaneous chemical initiation of recombinase polymerase amplification reactions on SlipChip. *Anal Chem* 2011;83:3533–40.

99. Lutz S, Weber P, Focke M, Faltin B, Hoffmann J, Mueller C et al. Microfluidic lab-on-a-foil for nucleic acid analysis based on isothermal recombinase polymerase amplification (RPA). *Lab Chip* 2010;10:887–93.

100. Romano JW, Shurtliff RN, Sarngadharan MG, Pal R. Detection of HIV-1 infection *in vitro* using NASBA: An isothermal RNA amplification technique. *J Virol Methods* 1995;54:109–19.

101. Vandamme A-M, Van DS, Kok W, Goubau P, Fransen K, Kievits T et al. Detection of HIV-1 RNA in plasma and serum samples using the NASBA amplification system compared to RNA-PCR. *J Virol Methods* 1995;52:121–32.

102. van der Vliet GM, SchU.K.kink RA, van Gemen B, Schepers P, Klatser PR. Nucleic acid sequence-based amplification (NASBA) for the identification of mycobacteria. *J Gen Microbiol* 1993;139:2423–9.

103. Poon LLM, Leung CSW, Chan KH, Lee JHC, Yuen KY, Guan Y, Peiris JSM. Detection of human influenza A viruses by loop-mediated isothermal amplification. *J Clin Microbiol* 2005;43:427–30.

104. Imai M, Ninomiya A, Minekawa H, Notomi T, Ishizaki T, Tu PV et al. Rapid diagnosis of H5N1 avian influenza virus infection by newly developed influenza H5 hemagglutinin gene-specific loop-mediated isothermal amplification method. *J Virol Methods* 2007;141:173–80.

105. Kubo J. Development of a reverse transcription-loop-mediated isothermal amplification assay for detection of pandemic (H1N1) 2009 virus as a novel molecular method for diagnosis of pandemic influenza in resource-limited settings. *J Clin Microbiol* 2010;48:728–35.

106. Chantratita W, SU.Kasem C, Kaewpongsri S, Srichunrusami C, Pairoj W, Thitithanyanont A et al. Qualitative detection of avian influenza A (H5N1) viruses: A comparative evaluation of four real-time nucleic acid amplification methods. *Mol Cell Probes* 2008;22:287–93.

107. Poon LLM, Leung CSW, Tashiro M, Chan KH, Wong BWY, Yuen KY et al. Rapid detection of the severe acute respiratory syndrome (SARS) coronavirus by a loop-mediated isothermal amplification assay. *Clin Chem* 2004;50:1050–2.

108. Keightley MC, Sillekens P, Schippers W, Rinaldo C, St GK. Real-time NASBA detection of SARS-associated coronavirus and comparison with real-time reverse transcription-PCR. *J Med Virol* 2005;77:602–8.

109. Pasick J. Advances in the molecular based techniques for the diagnosis and characterization of avian influenza virus infections. *Transboundary Emerging Dis* 2008;55:329–38.

110. Jean J, Blais B, Darveau A, Fliss I. Detection of hepatitis A virus by the nucleic acid sequence-based amplification technique and comparison with reverse transcription-PCR. *Appl Environ Microbiol* 2001;67:5593–600.

111. Rigopoulou EI, Stefanidis I, Liaskos C, Zervou EK, Rizos C, Mina P et al. HCV-RNA qualitative assay based on transcription mediated amplification improves the detection of hepatitis C virus infection in patients on hemodialysis: Results from five hemodialysis units in central Greece. *J Clin Virol* 2005;34:81–5.

112. Wu S-JL, Lee EM, Putvatana R, Shurtliff RN, Porter KR, Suharyono W et al. Detection of dengue viral RNA using a nucleic acid sequence-based amplification assay. *J Clin Microbiol* 2001;39:2794–8.

113. Lanciotti RS, Kerst AJ. Nucleic acid sequence-based amplification assays for rapid detection of West Nile and St. Louis encephalitis viruses. *J Clin Microbiol* 2001;39:4506–13.

114. Parida M, Posadas G, Inoue S, Hasebe F, Morita K. Real-time reverse transcription loop-mediated isothermal amplification for rapid detection of West Nile virus. *J Clin Microbiol* 2004;42:257–63.

115. Lee S-Y, Lee C-N, Mark H, Meldrum DR, Lin C-W. Efficient, specific, compact hepatitis B diagnostic device: Optical detection of the hepatitis B virus by isothermal amplification. *Sens Actuators, B: Chemical* 2007;127:598–605.

116. Cai T, Lou G, Yang J, Xu D, Meng Z. Development and evaluation of real-time loop-mediated isothermal amplification for hepatitis B virus DNA quantification: A new tool for HBV management. *J Clin Virol* 2008;41:270–6.

117. Cai Z, Lou G, Cai T, Yang J, Wu N. Development of a novel genotype-specific loop-mediated isothermal amplification technique for Hepatitis B virus genotypes B and C genotyping and quantification. *J Clin Virol* 2011;52:288–94.

118. Yates S, Penning M, Goudsmit J, Frantzen I, van DWB, Van SD, Van GB. Quantitative detection of hepatitis B virus DNA by real-time nucleic acid sequence-based amplification with molecular beacon detection. *J Clin Microbiol* 2001;39:3656–65.

119. Hagiwara M, Sasaki H, Matsuo K, Honda M, Kawase M, Nakagawa H. Loop-mediated isothermal amplification method for detection of human papillomavirus type 6, 11, 16, and 18. *J Med Virol* 2007;79:605–15.

120. Ihira M, Yoshikawa T, Enomoto Y, Akimoto S, Ohashi M, Suga S et al. Rapid diagnosis of human herpesvirus 6 infection by a novel DNA amplification method, loop-mediated isothermal amplification. *J Clin Microbiol* 2004;42:140–5.

121. Yoshikawa T, Ihira M, Akimoto S, Usui C, Miyake F, Suga S et al. Detection of human herpesvirus 7 DNA by loop-mediated isothermal amplification. *J Clin Microbiol* 2004;42:1348–52.

122. Kaneko H, Iida T, Aoki K, Ohno S, Suzutani T. Sensitive and rapid detection of herpes simplex virus and varicella-zoster virus DNA by loop-mediated isothermal amplification. *J Clin Microbiol* 2005;43:3290–6.

123. Chen H-t, Zhang J, Yang S-h, Ma L-n, Ma Y-p, Liu X-t et al. Rapid detection of porcine parvovirus DNA by sensitive loop-mediated isothermal amplification. *J Virol Methods* 2009;158:100–3.

124. Navidad PD, Li H, Mankertz A, Meehan B. Rolling-circle amplification for the detection of active porcine circovirus type 2 DNA replication *in vitro*. *J Virol Methods* 2008;152:112–6.

125. Onorato IM, Morens DM, Martone WJ, Stansfield SK. Epidemiology of cytomegaloviral infections: Recommendations for prevention and control. *Rev Infect Dis* 1985;7:479–97.

126. Griffiths P. Prophylaxis against CMV infection in transplant patients. *J Antimicrob Chemother* 1997;39:299–301.

127. Sun H. Preemptive therapy for cytomegalovirus based on real-time measurement of viral load in liver transplant recipients. *Transpl Immunol* 2010;23:166–9.

128. SuzU.K.i R, Yoshikawa T, Ihira M, Enomoto Y, Inagaki S, Matsumoto K et al. Development of the loop-mediated isothermal amplification method for rapid detection of cytomegalovirus DNA. *J Virol Methods* 2006;132:216–21.

129. SuzU.K.i R. Heat denaturation increases the sensitivity of the cytomegalovirus loop-mediated isothermal amplification method. *Microbiol Immunol* 2010;54:466–70.

130. Goossens VJ, Blok MJ, Christiaans MH, van HJP, Sillekens P, Hockerstedt K et al. Diagnostic value of nucleic-acid-sequence-based amplification for the detection of cytomegalovirus infection in renal and liver transplant recipients. *Intervirology* 1999;42:373–81.

131. Merlino C, Bergallo M, Gregori G, Sinesi F, Bollero C, Cavallo R. HCMV pp67-mRNA detection by nucleic acid sequence-based amplification in renal transplant patients. *Microbiologica* 2002;25:1–8.

132. Hebart H, Rudolph T, Loeffler J, J M, T L, G J, Einscle H. Evaluation of the NucliSens CMV pp67 assay for detection and monitoring of human cytomegalovirus infection after allogeneic stem cell transplantation. *Bone Marrow Transplant* 2002;30:181–7.

133. Renoult E, Clermont M-J, Phan V, Buteau C, Alfieri C, Tapiero B. Prevention of CMV disease in pediatric kidney transplant recipients: Evaluation of pp67 NASBA-based pre-emptive ganciclovir therapy combined with CMV hyperimmune globulin prophylaxis in high-risk patients. *Pediatr Transplant* 2008;12:420–5.

134. Caliendo AM, St George K, Allega J, Bullotta AC, Gilbane L, Rinaldo CR. Distinguishing cytomegalovirus (CMV) infection and disease with CMV nucleic acid assays. *J Clin Microbiol* 2002;40:1581–6.

135. Greijer AE, Verschuuren EA, Harmsen MC, Dekkers CA, Adriaanse HM, The TH, Middeldorp JM. Direct quantification of human cytomegalovirus immediate-early and late mRNA levels in blood of lung transplant recipients by competitive nucleic acid sequence-based amplification. *J Clin Microbiol* 2001;39:251–9.

136. Fang X, Liu Y, Kong J, Jiang X. Loop-mediated isothermal amplification integrated on microfluidic chips for point-of-care quantitative detection of pathogens. *Anal Chem* 2010;82:3002–6.

137. Little MC, Andrews J, Moore R, Bustos S, Jones L, Embres C et al. Strand displacement amplification and homogeneous real-time detection incorporated in a second-generation DNA probe system, BDProbeTecET. *Clin Chem* 1999;45:777–84.

138. Templeton K, Roberts J, Jeffries D, Forster G, Aitken C. The detection of *Chlamydia trachomatis* by DNA amplification methods in urine samples from men with urethritis. *Int J STD AIDS* 2001;12:793–6.

139. McHugh TD, Pope CF, Ling CL, Patel S, Billington OJ, Gosling RD et al. Prospective evaluation of BDProbeTec strand displacement amplification (SDA) system for diagnosis of tuberculosis in non-respiratory and respiratory samples. *J Med Microbiol* 2004;53.1215–9.

140. Bachmann LH, Johnson RE, Cheng H, Markowitz LE, Papp JR, Hook FW, III. Nucleic acid amplification tests for diagnosis of *Neisseria gonorrhoeae* oropharyngeal infections. *J Clin Microbiol* 2009;47:902–7.

141. Hardwick R, Gopal RG, Mallinson H. Confirmation of BD ProbeTec Neisseria gonorrhoeae reactive samples by Gen-Probe APTIMA assays and culture. *Sex Transm Infect* 2009;85:24–6.

142. Boyadzhyan B, Yashina T, Yatabe JH, Patnaik M, Hill CS. Comparison of the APTIMA CT and GC assays with the APTIMA combo 2 assay, the Abbott LCx assay, and direct fluorescent-antibody and culture assays for detection of *Chlamydia trachomatis* and *Neisseria gonorrhoeae*. *J Clin Microbiol* 2004;42:3089–93.

143. Moncada J, Schachter J, Liska S, Shayevich C, Klausner JD. Evaluation of self-collected glans and rectal swabs from men who have sex with men for detection of *Chlamydia trachomatis* and *Neisseria gonorrhoeae* by use of nucleic acid amplification tests. *J Clin Microbiol* 2009;47:1657–62.

144. Zhang X, Liao M, Jiao P, Luo K, Zhang H, Ren T et al. Development of a loop-mediated isothermal amplification assay for rapid detection of subgroup J avian leU.K.osis virus. *J Clin Microbiol* 2010;48:2116–21.

145. Bista BR, Ishwad C, Wadowsky RM, Manna P, Randhawa PS, Gupta G et al. Development of a loop-mediated isothermal amplification assay for rapid detection of BK virus. *J Clin Microbiol* 2007;45:1581–7.

146. Mori Y, Notomi T. Loop-mediated isothermal amplification (LAMP): A rapid, accurate, and cost-effective diagnostic method for infectious diseases. *J Infect Chemother* 2009;15:62–9.

147. Andresen D. Helicase-dependent amplification: Use in OnChip amplification and potential for point-of-care diagnostics. *Expert Rev Mol Diagn* 2009;9:645–50.

148. Yamazaki W, Taguchi M, Kawai T, Kawatsu K, Sakata J, Inoue K, Misawa N. Comparison of loop-mediated isothermal amplification assay and conventional culture methods for detection of *Campylobacter*

jejuni and *Campylobacter coli* in naturally contaminated chicken meat samples. *Appl Environ Microbiol* 2009;75:1597–603.

149. Nadal A, Coll A, Cook N, Pla M. A molecular beacon-based real time NASBA assay for detection of *Listeria monocytogenes* in food products: Role of target mRNA secondary structure on NASBA design. *J Microbiol Methods* 2007;68:623–32.

150. Du F, Zhang Q, Yu Q, Hu M, Zhou Y, Zhao J. Soil contamination of *Toxoplasma gondii* oocysts in pig farms in central China. *Vet Parasitol* 2012;187:53–6.

151. Girones R, Ferrus MA, Alonso JL, Rodriguez-Manzano J, Calgua B, Correa Ade A et al. Molecular detection of pathogens in water—The pros and cons of molecular techniques. *Water Res* 2010;44:4325–39.

152. Tomlinson JA, Dickinson MJ, Boonham N. Detection of *Botrytis cinerea* by loop-mediated isothermal amplification. *Lett Appl Microbiol* 2010;51:650–7.

153. Le DT, Netsu O, Uehara-Ichiki T, Shimizu T, Choi IR, Omura T, Sasaya T. Molecular detection of nine rice viruses by a reverse-transcription loop-mediated isothermal amplification assay. *J Virol Methods* 2010;170:90–3.

154. Liu M, Luo Y, Tao R, He R, Jiang K, Wang B, Wang L. Sensitive and rapid detection of genetic modified soybean (Roundup Ready) by loop-mediated isothermal amplification. *Biosci, Biotechnol, Biochem* 2009;73:2365–9.

155. Lee D, La MM, Allnutt TR, Powell W. Detection of genetically modified organisms (GMOs) using isothermal amplification of target DNA sequences. *BMC Biotechnol* 2009;9:7.

156. Sato K, Tachihara A, Renberg B, Mawatari K, Sato K, Tanaka Y et al. Microbead-based rolling circle amplification in a microchip for sensitive DNA detection. *Lab Chip* 2010;10:1262–6.

157. Wells CD, Cegielski JP, Nelson LJ, Laserson KF, Holtz TH, Finlay A et al. HIV infection and multidrug-resistant tuberculosis: The perfect storm. *J Infect Dis* 2007;196 Suppl 1:S86–107.

158. Huggett JF, McHugh TD, Zumla A. Tuberculosis: Amplification-based clinical diagnostic techniques. *Int J Biochem Cell Biol* 2003;35:1407–12.

159. Nyendak MR, Lewinsohn DA, Lewinsohn DM. New diagnostic methods for tuberculosis. *Curr Opin Infect Dis* 2009;22:174–82.

160. Rachow A, Zumla A, Heinrich N, Rojas-Ponce G, Mtafya B, Reither K et al. Rapid and accurate detection of *Mycobacterium tuberculosis* in sputum samples by cepheid Xpert MTB/RIF assay-A clinical validation study. *PLoS One* 2011;6:e20458.

161. Coll P, Garrigo M, Moreno C, Marti N. Routine use of gen-probe amplified *Mycobacterium tuberculosis* direct (MTD) test for detection of *Mycobacterium tuberculosis* with smear-positive and smear-negative specimens. *Int J Tuberc Lung Dis* 2003;7:886–91.

162. Greco S, Girardi E, Navarra A, Saltini C. Current evidence on diagnostic accuracy of commercially based nucleic acid amplification tests for the diagnosis of pulmonary tuberculosis. *Thorax* 2006;61:783–90.

163. Helb D, Jones M, Story E, Boehme C, Wallace E, Ho K et al. Rapid detection of *Mycobacterium tuberculosis* and rifampin resistance by use of on-demand, near-patient technology. *J Clin Microbiol* 2010;48:229–37.

164. Zhu R-Y, Zhang K-X, Zhao M-Q, Liu Y-H, Xu Y-Y, Ju C-M et al. Use of visual loop-mediated isothermal amplification of rimM sequence for rapid detection of *Mycobacterium tuberculosis* and *Mycobacterium bovis*. *J Microbiol Methods* 2009;78:339–43.

165. Telenius H, Carter NP, Bebb CE, Nordenskjold M, Ponder BA, Tunnacliffe A. Degenerate oligonucleotide-primed PCR: General amplification of target DNA by a single degenerate primer. *Genomics* 1992;13:718–25.

166. Zhang L, Cui X, Schmitt K, Hubert R, Navidi W, Arnheim N. Whole genome amplification from a single cell: Implications for genetic analysis. *Proc Natl Acad Sci USA* 1992;89:5847–51.

167. Spits C, Le Caignec C, De Rycke M, Van Haute L, Van Steirteghem A, Liebaers I, Sermon K. Whole-genome multiple displacement amplification from single cells. *Nat Protoc* 2006;1:1965–70.

168. Treff NR, Su J, Tao X, Northrop LE, Scott RT, Jr. Single cell whole genome amplification technique impacts the accuracy of SNP microarray based genotyping and copy number analyses. *Mol Hum Reprod* 2011;17:335–43.

169. Chang H-W, Sung Y, Kim K-H, Nam Y-D, Roh SW, Kim M-S et al. Development of microbial genome-probing microarrays using digital multiple displacement amplification of uncultivated microbial single cells. *Environ Sci Technol* 2008;42:6058–64.

170. Wang G, Maher E, Brennan C, Chin L, Leo C, Kaur M et al. DNA amplification method tolerant to sample degradation. *Genome Res* 2004;14:2357–66.

171. Handyside AH, Robinson MD, Simpson RJ, Omar MB, Shaw MA, Grudzinskas JG, Rutherford A. Isothermal whole genome amplification from single and small numbers of cells. A new era for preimplantation genetic diagnosis of inherited disease. *Mol Hum Reprod* 2004;10:767–72.

172. Zheng YM, Wang N, Li L, Jin F. Whole genome amplification in preimplantation genetic diagnosis. *J Zhejiang Univ Sci B* 2011;12:1–11.

173. Burlet P, Frydman N, Gigarel N, Kerbrat V, Tachdjian G, Feyereisen E et al. Multiple displacement amplification improves PGD for fragile X syndrome. *Mol Hum Reprod* 2006;12:647–52.

174. Johnson DS, Gemelos G, Baner J, Ryan A, Cinnioglu C, Banjevic M et al. Preclinical validation of a microarray method for full molecular karyotyping of blastomeres in a 24-h protocol. *Hum Reprod* 2010;25:1066–75.

175. Treff NR, Su J, Tao X, Levy B, Scott RT, Jr. Accurate single cell 24 chromosome aneuploidy screening using whole genome amplification and single nucleotide polymorphism microarrays. *Fertil Steril* 2010;94:2017–21.

176. Smith MC, Steimle G, Ivanov S, Holly M, Fries DP. An integrated portable hand-held analyser for real-time isothermal nucleic acid amplification. *Anal Chim Acta* 2007;598:286–94.

177. Jenkins DM, Kubota R, Dong J, Li Y, Higashiguchi D. Handheld device for real-time, quantitative, LAMP-based detection of *Salmonella enterica* using assimilating probes. *Biosens Bioelectron* 2011;30:255–60.

178. KoU.K.hareva I, Haoqiang H, Yee J, Shum J, Paul N, Hogrefe RI, Lebedev AV. Heat activatable 3′-modified dNTPs: Synthesis and application for hot start PCR. *Nucleic Acids Symp Ser (Oxf)* 2008:259–60.

179. Chou C, Young DD, Deiters A. A light-activated DNA polymerase. *Angew Chem Int Ed Engl* 2009; 48:5950–3.

180. The European Parliament, Council of the European Union. Directive 98/79/Ec of the European Parliament and of the Council of 27 October 1998 on *in vitro* diagnostic medical devices. *Official Journal of the European Communities*, 1998.

27 Proximity Ligation Assay for Protein Quantification

*David W. Ruff, Mark E. Shannon, Shiaw-Min Chen,
Simon Fredriksson, and Ulf Landegren*

CONTENTS

27.1 INTRODUCTION

It goes without saying that DNA is an attractive storage medium for digital information, both in nature and in the laboratory. DNA-based information can be recorded, reposited, replicated, and read-out with little or no error. Sequencing or chemical synthesis technologies for reading and writing DNA sequences are currently in a phase where the rate of improvement far outpaces the canonical Moore's law, which predicts that the capacity for electronic information storage doubles roughly every 2 years.[1] This means that the cost for, and effort of, DNA sequencing is decreasing very rapidly and reagent costs for synthesis of large sets of probe molecules are trivial. Other DNA-based methods, such as real-time PCR, provide high precision and vast dynamic ranges for quantitative measurements of specific nucleic acid molecules at the levels of DNA or RNA.

The diversity of protein structure and function makes proteins much more difficult to analyze. Since these molecules execute most of the functions in the cell, they are of obvious interest for a better understanding of cell biology, and they are prime targets for both diagnostics and therapy. However, in stark contrast to the situation for nucleic acids, there is currently no way of designing reagents that can predictably bind a specific protein molecule along the lines of the simple DNA complementarity rules. Instead, it is necessary to resort to screening large sets of molecules for suitable binders, either by employing the immune systems of host organisms to isolate antibodies or by using *in vitro* selection techniques to identify affinity reagents from extensive molecular libraries.[2]

It has been particularly difficult to develop efficient tools for specific protein detection and quantification because analysis of this class of molecules presents a number of unique challenges.

First, concentration differences of 10 or more orders of magnitude exist among proteins known to be present in biological samples such as blood. For example, proteins from tissues captured in blood flow samples from patients in normal or disease state groups are likely to be present at ultra-low concentrations. Therefore, detection of marker proteins that could signal disease processes could easily require assay sensitivities and dynamic ranges that are far greater than those that have historically been possible using conventional protein analysis techniques.

Second, proteins can exist in many similar forms, but potentially with widely different functional roles. Alternative splicing of transcripts can give rise to protein variants with distinct biological properties. For example, the human calcitonin gene *CALCA* produces distinct transcripts that encode two unique peptide hormones (calcitonin and calcitonin gene-related peptide) that function in calcium regulation.[3] Further functional diversity is generated from this gene through the posttranslational cleavage of an inactive precursor, procalcitonin, to yield two functionally distinct mature species, calcitonin and katacalcin. Along the same lines of logic, there is a need to detect hundreds of different posttranslational modifications that can trigger new activities by protein molecules, as exemplified by phosphorylation of receptor proteins that transmit cellular stimuli along signaling cascades.

Third, proteins do not generally execute their actions in isolation, but as members of dynamic complexes of cooperating molecules that assemble or disassemble in response to initiating events, such as receptor–ligand binding. This means that techniques are required that can reveal, not only individual proteins but also complexes.

In the light of these challenges and the fact that a single antibody typically binds a few thousand square Ångström of a protein surface, the equivalent of six or so amino acids, it is quite apparent that single binding reagents are unable to sense the global state of an interrogated protein. Moreover, antibodies frequently lack the specificity to distinguish the protein of interest from others that exhibit similar epitopes, particularly when these proteins are present at much higher concentration. The problem of detection specificity can be alleviated, just as with PCR, by designing experiments such that positive detection signals require simultaneous recognition of two epitopes, as is done in sandwich immune assays. However, the advantages of dual recognition tend to be eroded when many proteins are being investigated in parallel, as the opportunities for cross reactions grow rapidly with multiplexing. This has been a serious impediment to specific and highly multiplexed protein assays using conventional dual antibody systems.[4]

To a limited extent, the power of DNA analyses has been enlisted for detection of immune reactions by attaching reporter DNA molecules to antibodies. In immuno-PCR, for example, antibodies that have become trapped on solid supports via antigen binding can be detected with very high sensitivity by PCR.[5] Similarly, for immunodetection coupled with rolling circle amplification (RCA), a localized DNA amplification reaction is elicited after binding of an oligonucleotide-tagged antibody to a target protein.[6] These two methods lack the full power that is typical of amplified DNA detection. However, protein detection specificity is completely dependent on the ability of the single detection antibody to distinguish between correct target molecules on the one hand and cross-reactive targets and nonspecific binding on the other. Accordingly, the methods amplify both signal and background in parallel, thereby limiting the gain in sensitivity.

The proximity ligation assay (PLA™) has been developed to address many of the above limitations of current approaches for protein analysis. This method requires target recognition by two or more antibodies or other affinity reagents, ensuring specificity of detection, as well as the ability also to detect very complex targets.[7] The affinity reagents are modified by covalent or noncovalent attachment of short DNA strands. When several affinity reagents are brought into proximity by virtue of binding the same target molecule or molecular complex, their attached DNA strands can undergo DNA ligation reactions. The DNA strands can then be amplified by methods like PCR, and in particular real-time PCR, for detection of molecules in solution phase over very wide concentration ranges. Alternatively, ligation reactions can be designed to result in circular DNA strands that are then locally amplified via RCA.[8] This form of the assay serves to image the distribution of the detected molecules and molecular complexes by microscopy.

The DNA ligation step allows the assay to be tuned to the desired degree of specificity by requiring two or more reagents to participate in the ligation event when they are found in close proximity. This is analogous to the use of nested PCR to enhance specificity and sensitivity when amplifying rare DNA sequences. Ligation also offers an important opportunity to restrict the reactions so that only the intended pairs of reagents can give rise to detectable amplification products in multiplex reactions, limiting cross reactions. This property means that highly parallel and specific detection reactions can be achieved for comprehensive measurements of protein levels or for protein imaging.

27.2 PROXIMITY LIGATION ASSAY OVERVIEW

As with other immunoassay procedures, selecting the appropriate affinity reagent is the most important step when setting up a PLA. A range of binder types have been used successfully, including DNA-based aptamers generated by *in vitro* selection methods and antibodies raised by immunizations. Aptamer-based PLA probes were first used for the detection of homodimeric PDGF-BB.[7] Homodimeric PDGF-BB detection required one specific aptamer affinity binder sequence probe extended by about 40 oligonucleotides to make each PLA probe. These studies demonstrated a lower limit of detection of ~24,000 PDGF-BB molecules and over a 1000-fold concentration range of linear detection. This approach detected ~1000-fold fewer molecules than required by a sandwich enzyme-linked immunosorbent assay (ELISA). In principle, aptamers have the advantage of being composed entirely of nucleic acid and they thereby enable convenient linking of the protein-binding elements to the ligatable reporter DNA sequences. These reagents work well; however, the sparse availability of suitable aptamer pairs limits the implementation of this strategy. In contrast, antibody-based reagents are much more widely available, making them, as a class, more amenable for generating a wide repertoire of assays. In addition, with well-established procedures for linking nucleic acids to antibodies now in place, antibodies are currently the most suitable affinity reagent for building a PLA.

As the performance of PLA depends directly on the affinity of the binders used, high-affinity antibodies are very desirable.[9] The antibodies can be a pair of matched monoclonal antibodies (mAbs), monoclonal/polyclonal pair (e.g., ELISA pair), or a single batch of an affinity-purified polyclonal antibody (pAb) raised against the whole native, or recombinant, antigen. An assay is generated from a single pAb by simply splitting it into two portions and coupling each portion to one of the two oligonucleotides required for the assay. In the assay, these reagents will bind randomly to the target analyte at several sites to provide dual recognition. The pAbs usually have sufficient affinity (often <1 nM) to enable a sensitive assay and they are more commonly available for target proteins than matched monoclonal antibody pairs.

27.3 PROXIMITY LIGATION ASSAY PROCESS

27.3.1 OVERVIEW OF FORMATS

PLA can be performed in solution and detected using qPCR or *in situ* with RCA monitoring as mentioned above. The former approach has the advantages that it enables analysis of many samples in a high-throughput fashion with a short time to results, as well as enabling a wide variety of sample types. Solution PLA is ideally suited for detection of protein targets that are free in a solution such as in serum samples, cell lysates, or tissue homogenates. Proteins from these samples can be detected either by using a homogenous assay with no washing steps or by first immobilizing the target protein on a solid surface using a capture antibody. These two assay formats are called "homogenous in solution PLA" and "solid phase in solution PLA," respectively. Homogenous in solution PLA is the most convenient format because no wash steps are used and only two affinity reagents are needed. It also has the advantage of requiring low sample input (1–2 µL and typically 1–10 cells, 0.1–1 ng cellular protein), which is most beneficial in cases where samples are limited, such as in analysis of biobanked samples. However, the solid-phase version can provide greater specificity because three binding events are required, a unique feature among immunoassays in general. Moreover, because this format incorporates washing steps and can be used with larger sample volumes (up to 100 µL cell lysate or serum/plasma, or 5 mg protein), it enables broader dynamic ranges and lower limits of detection, and interference from sample components is minimized. Homogenous in solution PLA kits are commercially available under the TaqMan® Protein Assays brand name from Life Technologies (Carlsbad, California).

The *in situ* PLA approach enables specific localization of a protein(s) within the histocytological context of a tissue or cell. The data from *in situ* PLA experiments can give both a visual account of a protein's location and a numeric measurement of its abundance in a sample.[10] *In situ* PLA makes use of an isothermal RCA to produce a locally confined signal at the original site of detection by the antibodies. The assay is particularly useful for detecting protein–protein interactions or posttranslational modifications in fixed cells and tissues. This provides the opportunity to digitally record detection signals from individual RCA products in cell signaling studies. *In situ* PLA is commercially available under the Duolink™ brand name from Olink Bioscience (Uppsala, Sweden). The Duolink kit contains generic secondary antiprimary antibody reagents enabling simple adoption of the technology for user-specific needs. The pair of secondary reagents carries the different oligonucleotides capable of templating the proximity-dependent hybridization event leading to the RCA. The kit also contains all the components required for ligation, amplification, fluorescence detection, and finally mounting for microscope imaging.

Both these assay formats can also be performed in multiplex by encoding the identity of the target molecules as a tag sequence in the nucleic acid reporters generated in PLA. Thereby, the different protein analytes are converted into uniquely detectable sequence elements. The ability to multiplex assays is an important feature of PLA. In traditional sandwich immunoassays, antibody cross-reactive binding events limit the multiplexing capability of both bead-based and planar arrays.[4] However, when using PLA, tests can be designed so that amplifiable signals only result when the correct dual recognition elements are brought into proximity through paired antibody binding. This feature enables higher multiplexing capability with PLA as binding of noncognate antibody pairs are not reported as a signal.[11]

27.3.2　HOMOGENOUS ASSAY COMPONENTS

The central component of PLA is the antibody–oligonucleotide conjugate, referred to as the PLA probe. PLA requires the construction of two PLA probes, one possessing a 5′-free oligonucleotide and the other a 3′-free oligonucleotide, covalently linked to appropriate antibodies (Figure 27.1). This attachment can be achieved via direct conjugation or indirectly through the tight association of a biotin group on an antibody and a preformed streptavidin–oligonucleotide (SO) conjugate. A notable advantage of using preformed SO conjugates is that there are many commercial suppliers of high-quality biotinylated antibodies. Also, many kits are available for antibody biotinylation. SO conjugates can therefore serve as "universal" reagents for building a PLA.

Chemical conjugation of the 5′-free and 3′-free oligonucleotides to streptavidin can be conveniently performed using common chemistries such as the maleimide process.[9] These "universal" SO compounds with 5′-free and 3′-free oligonucleotides can be mixed in separate reactions with

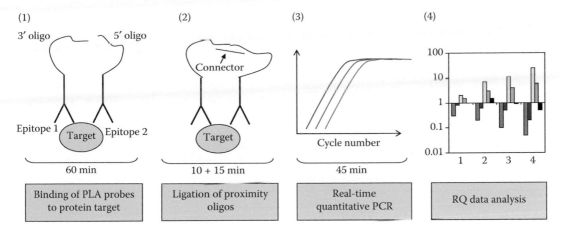

FIGURE 27.1 Proximity ligation: Homogenous assay overview. The homogenous proximity ligation assay uses two antibodies to bind to separate epitopes on a target. The antibodies are conjugated with two different DNA oligonucleotides—one has a 3′-free end and the other a 5′-free end (phosphorylated) (1). When the two conjugated antibodies bind and are in close proximity, the oligonucleotides can be ligated (2). The ligation product then serves as the template for real-time PCR amplification and quantification (3). The C_q points are transformed into normalized plots to calculate relative quantification values between different test samples (4).

equimolar ratios of biotinylated antibody to create a working PLA probe set. For certain applications, such as detection of proteins in serum and plasma, direct conjugation of the oligonucleotides to the antibodies may be advantageous to avoid background resulting from the SO reagents.[9]

The DNA oligonucleotides used for PLA probes contain the essential elements that form a functional amplicon upon specific DNA ligation; terminal sequences that are complementary to the connector oligonucleotide guiding ligation; target sequences for real-time PCR forward and reverse primers; and a target sequence for a fluorogenic probe. Since one oligonucleotide side contains both a primer and probe sequence, typical DNA strand design lengths for the probe pairs are approximately 60-mer and 40-mer.

The assay is commonly developed using pAbs after affinity purification using the immunogen. The requirement for purification stems from the fact that only a minority of antibodies in a crude antiserum or purified immunoglobulin fractions are directed against the immunogen after immunization. The presence of antibodies with irrelevant specificity can result in increased nonspecific background from PLA probes that are to be used in a homogenous assay format. The use of full-length or partial recombinant proteins as immunogens as opposed to short peptides increases the chances of generating superior "affinity pairs" from a single pAb preparation, as this increases the likelihood that several distinct epitopes can be recognized by the antibodies. Successful assays have been built using antibodies raised in goat, sheep, rabbit, and chicken.

mAbs can also be used in homogenous PLA. mAbs raised against distinct epitopes and with demonstrated performance in a functional paired binding assay format such as sandwich ELISA are ideal candidates for use in PLA. Typically, these mAbs are isolated from the purified IgG fraction produced by the hybridomas. In certain instances, a single mAb can be used to form a pair of binders (e.g., for detection of viral coat proteins, other surface markers, or homo-multimeric protein). Also, pAbs can be paired with mAbs in PLA as is commonly done in ELISA.

As noted above, other types of affinity reagents have been exploited in PLA systems. In principle, suitable binders include all classes of proteins and other molecules that form stable complexes with targets (e.g., via protein–protein interactions or small molecules that bind to specific epitopes on a target). Also, self-assembling monomers capable of forming multimeric complexes such as prions may be employed as PLA probes (one example being the detection of aggregate formation using the yeast Sup35 *in vitro* prion model system).[12]

Building a functional protein quantification assay requires three important target-specific ingredients: high-quality antibodies to make the PLA probes, a positive control sample, and a negative control sample. The positive control sample(s) should include both a recombinant protein and a cell lysate known to express the biologically active form of the protein. Unfortunately, recombinant proteins are not readily available for all target proteins. Even when they are available, recombinant proteins can be unstable or assume a conformation that is not reflective of the biologically active state. Thus, using actual biological samples known to express the target is preferable. A serial dilution curve of the sample should be created to ascertain assay characteristics such as dynamic range (the concentration range over which targets can be quantified) and limit of detection. As a minimum, the negative control can be the assay dilution buffer, but to properly address sample matrix effects, it is best to use dilutions of a sample such as a cell lysate, or serum and plasma, known not to include the investigated proteins.

27.3.3   Homogenous Assay Workflow

PLA is commonly performed on unfractionated cell and tissue lysates as well as diluted serum and plasma samples. For cell and tissue lysates, sample preparation buffers containing nonionic detergents such as Triton X-100 and NP-40 are compatible with the assay process. Commonly used protein sample preparation buffers formulated for immunoprecipitation assay workflows yield satisfactory PLA results. When testing macerated solid tissue lysates, removal of insoluble tissue debris by centrifugation is recommended. Protease and phosphatase inhibitor cocktails are recommended additives to the lysis reagent. Denaturing lysis buffers and sample treatments are not necessary for most applications. Denaturants such as urea and guanidinium must be diluted significantly to avoid inhibition of DNA ligation or qPCR. The anionic detergent sodium dodecyl sulfate is strongly inhibitory to Taq polymerase and must be avoided during the sample preparation process. Dilution of lysates in PBS containing blocking agents such as BSA provide superior performance.

The first step for the assay process involves the transfer of 2 μL of lysate sample containing at least 1–10 cells, or 0.1–1 ng protein, to a well in a standard 96-well PCR plate (see Figure 27.2). An equal volume of 2 μL PLA probes is then transferred. Working PLA probe concentrations are typically in the 35–250 nM range. PLA probes function optimally in this range because high-quality antibodies bind effectively to antigen epitopes and background ligation events are minimal in this range. The use of lower PLA probe concentrations may sharply reduce antibody binding rates. To prevent evaporation, the plate is sealed with an adhesive cover. After gentle mixing and a brief centrifugation, the PLA probe-binding step is initiated. The plate is incubated at 37°C on a thermal cycler with a heated cover. For many assays, 1 h is sufficient for complete binding. In some cases, an overnight incubation at 4°C promotes better antibody–antigen binding. After the binding incubation step is finished, the reactions are ready for the ligation step. A reaction buffer solution containing the DNA connector 18-mer oligonucleotide and a DNA ligase is added to the binding reaction. Ligation is completed within 10 min at 37°C. The ligase enzyme is then deactivated by the addition of a heat-labile protease. Subsequent denaturation of the protease is accomplished by a brief 95°C treatment. The ligation products are transferred to a real-time PCR plate for qPCR detection. Expedited cycling parameters using Fast Master Mix are recommended to reduce the time to results during a 40 cycle qPCR run (95°C denaturation 1 s, 60°C anneal/extension for 20 s each cycle). However, standard cycling conditions can also be used with equivalent results. For reliable relative quantification results, it is necessary to assay a series of 4–6 two- to fourfold dilutions of each sample along with a no protein control (NPC) to obtain an assay response curve. Assay curves are subsequently used to calculate fold changes in protein concentration between calibrator and experimental samples (see Section 27.5).

27.3.4   Solid-Phase Format

PLA can be performed by first immobilizing the target analyte onto a solid phase, typically using a capture antibody. The capture antibody may be attached to the inner wall of a PCR tube or to the

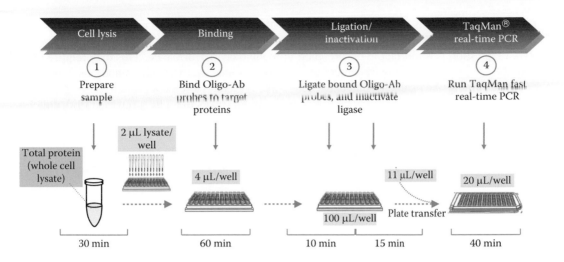

FIGURE 27.2 Homogenous proximity ligation assay workflow for cells. (1) Cell samples are lysed with a detergent containing sample preparation buffer. The initial cell concentrations should be in the range of 50–250 cells per μL or 5–25 ng tissue protein per μL. More concentrated samples may be required if the protein target has low abundance. To ensure linear range readouts, this initial concentrated lysate should be serially diluted by 1:2 to 1:4 for 3 or more dilution points. (2) A 2 μL volume of lysate is transferred to a 96-well PCR thermalcycler plate. The 3′-free and 5′-free PLA probes are mixed together immediately prior to distribution to the lysate wells. A 2 μL volume of PLA probes is used in each well. The plate is gently agitated to facilitate mixing of the solutions. The binding procedure is typically done at 37°C for 60 min. (3) A 96 μL ligation reaction buffer is added to the binding solution. This ligation mix contains DNA ligase and the connector oligonucleotide. This process takes 10 min at 37°C. The ligase enzymatic activity is quenched by addition of 2 μL of inactivation reagent. This step takes 15 min. (4) The ligation products (9 μL) are transferred to a qPCR plate and a 20 μL universal TaqMan assay reaction and cycling conditions applied to generate real-time PCR amplification results. Typical time-to-results are about 3.5 h, of which hands-on time is roughly 1.5 h. The workflow is most conveniently performed using a 12-tip multichannel pipet and using a 96-well PCR thermacycler plate with adhesive cover. To prevent evaporation, the assay incubations should be conducted in a temperature control device with a heated cover such as a PCR thermalcycler.

surface of magnetic beads. After addition of the sample to the capture tube and binding the target, the solid surface is washed with a buffer to remove sample matrix. The solid surface is then available for incubation in a solution containing the two PLA probes. The probe-binding step is followed by a second round of washing steps that serve to remove unbound PLA probes. The subsequent ligation and qPCR steps are carried out in a manner similar to that described for the homogenous PLA format. As mentioned above, the additional steps required for solid-phase PLA have the potential benefit of greater assay specificity, sensitivity, and dynamic range when compared to homogenous PLA.[13] As with homogenous PLA, the three antibodies needed for solid-phase PLA can be matched monoclonals or aliquots of a single pAb.

27.3.5 SPECIAL CONSIDERATIONS WHEN QUANTIFYING PROTEINS IN SERUM OR PLASMA SAMPLES

Achieving analysis of proteins with sufficient specificity in complex samples such as serum and plasma is challenging for many immunoassays. There are a few unique aspects to be considered when using homogenous PLA with these samples.[14] For example, PLA incubation buffers contain EDTA to remove potential nuclease activity that could degrade the oligonucleotide component of the probes. In addition, the enzymes used in the assay (i.e., ligase) require reducing conditions provided by small amounts of dithiothreitol (DTT) to operate fully. Heparin-treated plasma should be avoided since heparin is a well-known inhibitor of DNA polymerases. Plasma diluents containing carrier-blocking proteins used prior to PLA incubations can be chosen to reduce possible nonspecific signals to improve assay specificity. Assessment of protein recovery for the particular sample matrix in question is important to ensure good immunoassay performance. This may require dilution of the sample and/or normalization of the data using spiked-in exogenous standards, especially when multiplexing. An example of an exogenous spike is green fluorescent protein (GFP). C_q values acquired after detection of GFP may be compared between spikes into test serum and PLA dilution buffer. If the serum test sample has no matrix effect on PLA efficiency, the GFP C_q values will be identical between the buffer spike control and the test sample. Also, the GFP PLA signal may be used to normalize target PLA signals in studies that compare multiple subject serum samples. In this case, the difference in C_q values between the GFP and target are directly compared.

27.4 PROXIMITY LIGATION ASSAY RESULTS

The PLA data output reflects the sum of all of the properties associated with the biochemical processes that comprise the assay workflow. These include sample preparation, immunoassay binding, oligonucleotide ligation, and real-time PCR amplification and detection. It is important that sample lysis and dilution buffers are compatible with immunobinding as well as enzymatic detection. For instance, buffered lysis reagents containing nonionic detergents are suitable for cell lysis and are compatible with antibody binding, ligation, and qPCR reactions. Since the sample matrix is not removed by the washing steps in homogenous PLA, sample components remain present during ligation and qPCR detection and may inhibit these processes. In such cases, sample dilution prior to setting up the assay can often reduce or eliminate the inhibition. The quality of antibody affinity to the target protein dictates the assay specificity, sensitivity, detection limit, and the shape of the assay response curve. The PLA probe and connector oligonucleotide concentrations, as well as the DNA ligase enzyme activity, contribute to the level of background ligation signal.

As previously indicated, for each sample, a series of dilutions are assayed to generate C_q data curves (plotted as C_q vs. sample concentration) having a defined linear range. Using recombinant antigen, PLA sensitivity and dynamic detection range is often equal to or better than that of the ELISA.

Each sample generates a slope with a linear range enabling relative quantification determination. A typical recombinant antigen dilution series of a 5- to 6-log range of sample input produces a sigmoidal-shaped C_q plot (see Figure 27.3). When target concentrations reach high saturation inputs, the assay response reaches a plateau at a lowest C_q value. At the highest antigen input, the curve plateaus and C_q values may even start increasing again, thus resembling a slight "hook" (see ICAM on Figure 27.3). The hook-shaped curve is generated when the stoichiometric ratio of antigen to antibody–PLA probe is greatly exceeded. In this situation, the PLA probes are fully bound to antigen target leaving no unbound probes. If the antigen amounts exceeds this 1:1 ratio, this may cause an "antiproximity" effect that reduces the opportunities for PLA probe ligation.

Relative quantification determinations can be made on serial dilutions of lysates in PLA compatible buffers. An example of a cellular protein measurement experiment is displayed in Figure 27.4. Cultured cells are usually diluted to a 1 to 500 cell input range per binding reaction. Solid tissue lysate inputs are readily normalized by protein mass quantity. The micro-BCA kit is one commonly used chemistry kit system for determining protein concentration.[15] Typical input ranges are 0.1–100 ng (a convenient conversion tool is an estimation that an average mammalian cell contains roughly 100 pg of protein). Homogenous assays can often be used to quantify specific proteins in as few as 5–10 cells. The comparable detection limit for Western blots typically requires a sample input of about three orders of magnitude.

It is must be noted that a certain level of background ligations are generated by virtue of connector-mediated ligation of free unbound PLA probes. This background qPCR signal remains constant at sample input levels below the detection sensitivity threshold, and is designated the NPC. The NPC C_q values are useful in each experiment as the reference normalizer for assay run performance

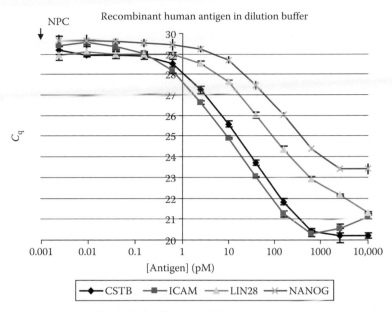

Antigen, sub-cellular localization and size:

ICAM-1, membrane bound, 57.8 kDa

CSTB, cytoplasmic protein, 11.1/22.2 kDa (monomer, can form dimer)

LIN28, nuclear localized, 22.7 kDa

NANOG, nuclear localized, 34.6 kDa

FIGURE 27.3 **(See color insert.)** Homogenous TaqMan protein expression data: Recombinant human protein. Recombinant human antigens were diluted in a fourfold series over a 6-log range. The most dilute sample point at 0.002 represents the NPC value. To minimize protein stability concerns, the lyophilized recombinant proteins were solubilized and immediately serially diluted and PLA commenced. Typical assay working ranges are 0.001 pM to 10 nM, linear dynamic ranges are 2–3 logs, lower limit of detection in the 0.5–5 pM range (0.17–65 pg/mL), and upper limit of detection in the 100–10,000 pM (1.7–650 ng/mL).

parameters, including PLA probe concentration, ligase activity, qPCR master mix and reaction parameters, and qPCR instrument-to-instrument variability.

The difference in C_q values between each sample dilution point and the NPC values are termed ΔC_q (see Figure 27.4b). The ΔC_q calculation minimizes day-to-day fluctuations in the performance of assays and qPCR instruments in much the same way as an ELISA blank assay sample does for OD background signal normalization. The ΔC_q value for the NPC to maximum assay response can be used as a quality control term for assay robustness. A convenient rule of thumb is that assays with a maximum ΔC_q of 5 or greater typically have a 1–2 order of magnitude linear response range. Each assay has a characteristic ΔC_q slope that is the product of antibody avidity and affinity, antigen structural complexity, as well as sample matrix influences. Antigens that are involved in protein–protein interactions have alternative conformations or have structural modifications such as glycosylation, and may have a variable number of antigenic sites available for the PLA probe binding. Under such circumstances, this may contribute to variable assay slopes and linear ranges for cell samples. This phenomenon, commonly referred to as nonparallelism, is also found in other immunoassay formats.[16,17]

The limit of detection of the assay can be defined using the ΔC_q plots. One approach is to classify signals with a ΔC_q of 1 or less as background. At the ΔC_q of 1, the number of ligations registered is twice that of the background NPC ligations. With background NPC values typically around a C_q of 30, the limit of detection requires roughly 128–256 target-PLA probe-mediated ligation events. If the

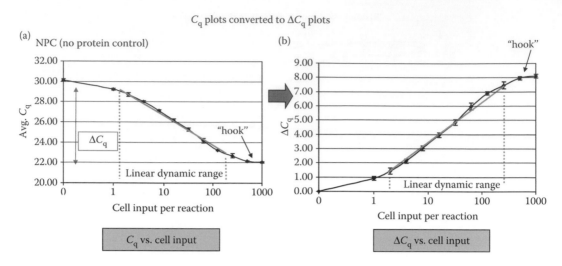

C_q plots converted to ΔC_q plots

(a) NPC (no protein control)

(b)   "hook"

C_q vs. cell input

ΔC_q vs. cell input

FIGURE 27.4 Homogenous TaqMan protein expression data: Cell lysate. (a) The human B-cell lymphoma cell line Raji was used to produce a 500 cell per μL lysate. A twofold serial dilution series was carried out to produce a standard curve using the cystatin B (CSTB) PLA system. Each dilution point had quadruplicate samples that are averaged with standard deviation error bars. The last sample point at 0 is denoted the no protein control (NPC). The NPC is important because it gives the background ligation C_q values for the reaction plate. This standard curve has a linear range from 2 to 125 cell input. The "hook" effect is caused by sample overtitration of available PLA probes. (b) C_q values are transformed into the ΔC_q scale. This normalizes the C_q signals against assay background ligations, enabling comparison of data sets from one qPCR run and instrument to another. The C_q's derived from protein quantification in the sample are subtracted from those derived from the NPC. The one cell input shows a ΔC_q of about 1. The full range of this assay response is 8 ΔC_q's.

proximity ligation process is 100% efficient, then the assay sensitivity is exquisitely high. However, the overall assay is predicted to be less than 100% efficient with the current homogenous assay configurations, so the limit of detection is likely to be in the low thousands of target molecules. Ongoing improvements in PLA chemistry technology are expected to further improve the sensitivity.

The specificity of the PLA process is facilitated by the requirement for two independent binding events to take place in proximity. Background cross-reactivity of the assay can be readily confirmed by spike matrix experiments. A lysate of a cellular sample that has been confirmed to express a spectrum of irrelevant protein targets is used in a dilution series for spiking known quantities of recombinant target protein. The ΔC_q values should remain constant in these spike matrix validation samples. The percentage recovery of the spike-in should be near 100% for all dilution points. Also, a simple approach to investigate whether unfamiliar sample types contain elements that contribute to nonspecific ligation signals is to use the mixed-probe method. The mixed probe procedure is used to detect any sample matrix promotion of PLA probe association using probes to two independent protein targets. A 5′-free oligonucleotide PLA probe specific for the first target is paired with a 3′-free oligonucleotide PLA probe specific for the second target. Addition of these pairs to a binding reaction with a test sample matrix should only result in an NPC background level of ligations. The scheme assumes that neither target has molecular association in the sample matrix. This is a useful test to verify the specificity of the ligation process with solid tissue samples, in particular when inputs require higher mass (>100 ng).

27.5 DATA ANALYSIS

One of the powerful features of PLA is the ability to quantify minute amounts of protein that are present in complex biological samples. However, specific data analysis protocols have been required that are suitable for the examination of the results derived from PLA experiments. The widely used real-time PCR $\Delta\Delta C_q$ method is not directly applicable to the analysis of PLA data.[18] The $\Delta\Delta C_q$ method relies on well-defined and similar assay slopes among the endogenous controls and test assays. Ideally, the slope of the plots from the serial dilutions of all assays should demonstrate 100% efficiency values. In conventional qPCR protocols for the analysis of target concentrations of genomic DNA and cDNA templates, C_q values derived from the serial dilution of the sample are plotted against the $\log_{10}$ of input sample. When the assay has a 100% efficient amplification of the amplicon, the gradient of these slopes are characteristically −3.3. PLA sample dilution input points, plotted on a $\log_{10}$ scale input versus C_q values, have slopes that are usually in the range of 2.5–3.1, and vary considerably for each different assay (Figures 27.3 and 27.4). To conduct relative quantity (RQ) calculations, two innovative approaches have been devised. These are based on comparisons of standard curves generated from designated reference and test samples. These mathematical schemes are used to convert the assay C_q values into ΔC_q numbers using the background ligation C_q point or alternatively into absolute ligation values using a real-time PCR C_q-to-copy number transformation. Both methods require a calibrator sample and a serial dilution curve.

27.5.1 ΔC_q Squared Method

The ΔC_q squared method is commonly used for the determination of relative protein expression in cell lysates. This method requires accurate cell counts or input protein mass and uses the NPC input point to normalize the sample input C_q values (see Figure 27.5). The first data manipulation is to subtract the sample input C_q's from the NPC to form the ΔC_q plot. Next, the linear portions of the data points are selected to form a linear plot. This process is repeated with the serial dilution

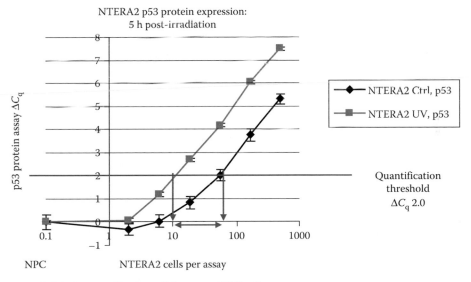

• p53 Relative quantification cell-intercept: 56/10 cells
• p53 protein: 5.6-fold UV induction

FIGURE 27.5 **(See color insert.)** Relative quantification, ΔC_q squared method. The human embryonal carcinoma cell line NTERA2 was subjected to genotoxic stress. A monolayer of cells was exposed to 50 J of UV irradiation and then cultured for 5 h to allow for the apoptotic response pathway to initiate. Control cells received no UV exposure. Protein lysates were prepared and serially diluted threefold. Using a ΔC_q threshold of 2.0, the UV irradiated standard curve crosses the linear curve at 10 cell input. The no-UV-treated cell curve required 56 cells to cross the ΔC_q 2.0 threshold. The cell input ratios are 5.6; therefore, the UV-irradiated cells have a 5.6-fold higher p53 protein level than the control cells.

curve from each test sample. A linear trend line equation is calculated through each curve. The user designates a ΔC_q threshold that is typically between 1.5 and 2.0. The cell input values at each linear trend line ΔC_q crossover point are then calculated. The ratio of cell input between these crossover points becomes the fold difference in target expression between the various samples.

27.5.2 LIGATION CONVERSION METHOD

The ligation conversion method is suitable for the determination of relative protein expression when using serum and plasma sample types. Traditional immunoassays result in data that are typically reported on a linear scale rather than on a $\log_2$ scale such as C_q values from real-time PCR. A convenient method to convert C_q's to linear values is to simply take 2^{C_q} for all raw data points prior to analysis thereby generating a more familiar type of relative protein quantification data. This method is based on the assignment that a single ligation template amplicon yields a C_q of 37 in a 20 μL qPCR and that the qPCR amplification efficiency of ligated template is 100%. Overall PLA slopes are less than −3.3 because of the antibody–antigen binding and subsequent ligation nonlinearity. Using this approach, a sample exhibiting a C_q readout of 33 is estimated to have 16 ligation events ($37–33 = \Delta C_q$ of 4, and $2^4 = 16$). Standard curves of serially diluted recombinant antigens may also be plotted on a graph of ligation number versus concentration scale. Such standard curves may then be used for absolute quantification by simply generating a trend line equation for assessment of antigen concentration in an unknown sample.

27.6   APPLICATIONS

PLA has been used to quantify proteins in diverse sample types and applications. Several reports describe the development of assays for measuring serum and plasma levels of cancer biomarkers, avian influenza viruses, stem cell proteins, protein–protein complexes, and bacterial markers (see Table 27.1).[14,19–21] The classes of protein targets measured include growth factors, transcription factors, cell surface receptors, and receptor ligands. One advantage of this assay is that sample purification procedures are not required to perform the detection. Cell specimens are homogenized in lysis buffer and unfractionated lysates are sampled directly in the PLA probe-binding step. Plasma and serum can be diluted in carrier protein solutions and assayed without further preparation.

27.6.1 STEM CELL MARKERS FOR PLURIPOTENCY EVALUATION

Human embryonic stem cells possess the ability to differentiate into any of the major cell lineages. This characteristic, called pluripotency, is crucial to the biological role that stem cells play in development. The presence of certain biomarkers is used to classify stem cells and their differentiation potential. PLA has been developed for a set of four key human stem cell pluripotency markers, LIN28, NANOG, OCT3/4, and SOX2. These assays have been used as a tool to characterize the pluripotency status in cell lines and reprogrammed cells. The assay specificity was evaluated on cell lysates produced from the pluripotent embryonal carcinoma cell line NTERA2 and the non-pluripotent Raji B-cell lymphoma cell line. Typical assay results for 500 NTERA2 cells give ΔC_q's for LIN28 (~6), NANOG (~3), OCT3/4 (~6), and SOX2 (~7), or a range of about 100–500 pg/mL.[21] Western blot detection of the same targets requires about a thousand-fold more NTERA2 cell material. Raji cell lysates do not generate ΔC_q's above the background noise level of 1.0. These data provide a clear demonstration that PLA can be used as a sensitive tool for evaluation of protein markers.

27.6.2 CORRELATION OF PROTEIN EXPRESSION TO MRNA EXPRESSION

The measurement of mRNA levels is often used as a surrogate for the estimation of protein levels. RNA sample preparation and real-time PCR assay quantification methods are now routinely used

TABLE 27.1

Examples of PLA Applications (Not Including *In Situ* PLA)

	Analyte	Comment	References
Detection of cytokines and growth factors	IL-2, IL-4, VEGF	Utility of PLA for detecting cytokines in small samples using antibody-based PLA probes	[9]
Mouse conditional knock-down verification	PDGF-BB	Antigen levels proven to be decreased in minute mouse glomeruli samples in knock-down verification	[29]
Microbial pathogens	Lawsonia intracellularis, porcine parvo virus	PLA detects bacteria and virus particles at sensitivities equal to qPCR detection of the organisms DNA	[30]
DNA–protein interactions	P53-DNA	Transcription factor binding analysis in samples containing very few cells	[31]
Multiplex protein detection	VEGF, IL-4, IL-10, IL-1α, TNFα, IL-7	PLA is highly suitable for multiplex assay development	[11]
Plasma biomarker research	Multiple analytes	Multiplex PLA is highly suitable for biomarker research in biobanked plasma	[14,27]
Protein complexes	PSA–API in plasma	PLA eliminated nonspecific background enabling complex measurement in plasma	[19]
Protein ligand–receptor interactions	VEGF and its receptor interactions	Use of PLA to screen for inhibitors of ligand–receptor interactions	[32]
Stem cell pluripotency determination	LIN28, NANOG, OCT4, SOX2	Detection of these markers is key to verifying pluripotentcy status of cells	[21]
Protein–mRNA correlation	LIN28, NANOG, OCT4, SOX2	Great advantage of using qPCR platform for both analyte readouts	[21]

in molecular and cellular biology applications. However, several published reports show discordant relationships between mRNA and protein levels in cell models.[22,23] Some reports place the discordance at about 50% of the mRNA–protein pairs studied in certain cancer cell systems, while other cell systems show better concordance between mRNA–protein pairs. In light of these observations, distinct advantages lie in the direct determination of protein levels in cellular response systems. A technical concern for these studies is the disparate sample preparation and analytical detection platforms that are required to measure the expression levels of each macromolecule. In several such studies, microarray signals record the mRNA prevalence and mass spectrometry the protein. Consequently, when comparing mRNA–protein pair expression profiles, the use of a single analytical platform such as real-time PCR is advantageous. One such recent study examining early differentiation of neuronal cells showed good concordance among several selected mRNA–protein pairs.[21] Cell culture lysates were split into two sample workflows, one for PLA and the other used to purify total RNA for one-step RT-qPCR assays. Pluripotency maintenance biomarkers LIN28, NANOG, OCT3/4, and SOX2 exhibited similar levels of down-regulation at the mRNA and protein levels. Neuronal differentiation biomarkers ALCAM and NCAM1 mRNA and protein expression levels were coordinately up-regulated. This sample preparation and assay integration approach greatly facilitates the analysis of mRNA–protein correlations in biological responses.

27.6.3 Correlation of miRNA Expression and Protein Signatures

MicroRNAs (miRNAs) are a class of cellular, small RNA molecules ~22 nucleotides in length that have been demonstrated to have roles in posttranscriptional regulation of gene expression. miRNAs are thought to exhibit their effect through destabilizing mRNA and reducing translational efficiency. It has been difficult to apply conventional proteomic approaches for the identification and quantification of miRNA influence on protein signatures. Such studies have usually been limited to reporter assays such as luciferase readouts and immunoblotting.[24] Other reports have employed quantitative mass spectrometry for resolving protein output.[25] Mass spectrometry approaches require large sample inputs and generate large data sets that are challenging to reconcile to miRNA quantification. Proximity ligation assay is an alternative proteomic tool for these studies. PLA is more sensitive and requires smaller sample inputs, and the data values can be directly correlated to C_q data obtained for mRNA and miRNA expression levels from the same sample source.[21]

27.6.4 PROTEIN–PROTEIN INTERACTIONS

Cellular processes are reliant on specific interactions between proteins and an array of macromolecules, including other proteins, genomic DNA, and RNA. Understanding the architecture and dynamics of these interactions is an important step for charting regulatory networks in cells. PLA can be harnessed for the identification, mapping, and quantification of these protein–protein interactions in cells. In one example, the interactions of prostate-specific antigen (PSA) with alpha-1-protease inhibitor (API) were characterized using PLA.[19] This study utilized three mAbs for a solid-phase assay approach: an anti-PSA capture mAb, anti-PSA mAb, and anti-API mAb detection PLA probes. PCR tube wells were coated with the anti-PSA capture mAb. Serum test samples were transferred and incubated for 1 h ambient temperature binding. After washing, the two mAb detection PLA probes were added for the PSA–API protein complex detection. This PLA process eliminated background issues from the vast excess levels of circulating free API and enabled accurate measurements of the PSA–API complex. Investigations employing quantitative PLA have a great potential to contribute to a better understanding of the cellular interactome.

27.6.5 BIOMARKERS IN SERUM

Homogenous PLA has been used in a few small-scale biomarker studies in the multiplexed format.[14,26] By assaying a number of case and control samples and performing multivariate statistics, increased diagnostic performance has been shown for cancer detection. Multiplexed PLA is especially suited for such studies as it can (1) utilize a broad range of available affinity reagents in the form of polyclonal antibodies, (2) achieve rapid assay development even in multiplex as individual assays do not interact during detection,[11] (3) sample consumption is low, enabling access to precious, limited biobanks, and (4) the sensitivity allows the discovery of novel, low-abundant biomarkers that are difficult to detect with other technologies. A high-throughput homogenous PLA process has been developed using four 24-plex panels containing 74 biomarkers for the detection of colorectal cancer signatures in human plasma samples.[27] These results demonstrate sub-pM sensitivity and plasma sample input requirement of only 1 μL per panel. PLA offers a valuable tool for biomarker applications using biobanked human serum and plasma samples.

27.6.6 PROTEIN DETECTION FROM DIVERSE SAMPLE TYPES

PLA detection has been used to detect proteins in samples that include bacteria, viral suspensions in animal tissue and fluid samples, protein aggregates, and cell surface markers on intact cells. Intact cells can be suspended and diluted in an isotonic buffer such as PBS and then added directly into the binding reaction with PLA probes. Examples of cell surface markers detected on intact mouse cell line P19 are ADAM9, CSTB, and ICAM1. As expected, internally localized proteins such as LIN28 are not detected on the cell surface. RNA aptamers have also been adopted as PLA probes for detecting a cell surface tumor antigen, the prostate-specific membrane antigen.[28] RNA aptamers have the ability to form more diverse structures than their DNA counterparts. RNA aptamers are readily produced *in vitro* using synthetic DNA constructs possessing the T7 RNA polymerase promoter motif. PLA systems are also being developed to monitor the phosphorylation status of pathway response proteins. One PLA probe is specific for the phospho-peptide motif while the other PLA probe against the total protein. Proximity ligation signals are recorded when the phosphorylated form of the protein is present in the binding reaction. In one model system, intracellular Akt Ser-473 phosphorylation is induced by the addition of PDGF to cell culture media. Detection of Akt phosphorylation by PLA required only 50 cells as opposed to some 6000 cells for a sandwich ELISA.

27.7 CONCLUDING REMARKS

The demonstrated utility of combining immunoassay specificity and dual molecular recognition with highly sensitive qPCR amplification is driving the expansion of PLA in many directions. PLA is being applied in new and dynamic ways to the study of normal and aberrant cellular processes associated with homeostasis, growth, and differentiation. With its simple and rapid workflow akin to traditional qPCR techniques and its capacity for multiplexing, PLA has enabled a broad range of studies that were previously either not possible or not practical, such as high-throughput cancer biomarker profiling, parallel protein and miRNA/mRNA correlations using the same analytical platform, and digital readouts for proteins.

The arrival of PLA has ushered in a new era in measuring proteins and has created future opportunities for ultrasensitive protein detection down to the level of the single cell. Zooming in to this level of granularity will enable an unprecedented look at the functional biology of individual cells as well as defined cell populations. With this achievement will come a more refined view of the modularity of cellular networks and the interplay of genotype, phenotype, and environmental factors on cellular processes. By the addition of PLA technology to the real-time PCR assay repertoire, biological systems can now be studied using one sample workflow to map the interactive activities of the "central dogma" key components, from genomic DNA output to mRNA, miRNA, and ncRNA to protein output and pathway responses.

ACKNOWLEDGMENTS AND NOTES

Ulf Landegren's research is supported by the Knut and Alice Wallenberg Foundation, the Swedish Research Council, Vinnova/Uppsala Bio, Uppsala Berzelii Centre and by the European Community's 6th and 7th Framework Programs.

The development work on PLA at Olink Bioscience is supported by the EU 7th Framework Program.

The products described in this chapter are *for research use only*. They are not intended for any animal or human therapeutic or diagnostic use.

REFERENCES

1. Carr, P.A. and Church, G.M. 2009. Genome engineering. *Nat Biotechnol*, 27, 1151.
2. Nygren, P.A. 2008. Alternative binding proteins: Affibody binding proteins developed from a small three-helix bundle scaffold. *FEBS J*, 275, 2668.
3. Auboeuf, D. et al. 2004. Differential recruitment of nuclear receptor coactivators may determine alternative RNA splice site choice in target genes. *Proc Natl Acad Sci USA*, 101, 2270.
4. Ellington, A.A. et al. 2010. Antibody-based protein multiplex platforms: Technical and operational challenges. *Clin Chem*, 56, 186.
5. Sano, T. et al. 1992. Immuno-PCR: Very sensitive antigen detection by means of specific antibody-DNA conjugates. *Science*, 258, 120.
6. Haab, B.B. and Lizardi, P.M. 2006. RCA-enhanced protein detection arrays. *Methods Mol Biol*, 328, 15.
7. Fredriksson, S. et al. 2002. Protein detection using proximity dependent DNA ligation assays. *Nat Biotechnol*, 20, 473.
8. Söderberg, O. et al. 2006. Direct observation of individual endogenous protein complexes *in situ* by proximity ligation. *Nat Methods*, 3, 995.
9. Gullberg, M. et al. 2004. Cytokine detection by antibody based proximity ligation. *Proc Natl Acad Sci USA*, 101, 8420.

10. Söderberg, O. et al. 2007. Proximity ligation: A specific and versatile tool for the proteomic era. *Genet Eng (NY)*, 28, 85.

11. Fredriksson, S. et al. 2007. Multiplexed protein detection by proximity ligation for cancer biomarker validation. *Nat Methods*, 4, 327.

12. Ruff, D.W. et al. 2008. Methods, compositions, and kits for detecting protein aggregrates, US Patent Application #20080003604.

13. Darmanis, S. et al. 2010. Sensitive plasma protein analysis by microparticle-based proximity ligation assays. *Mol Cell Proteomics*, 9, 327.

14. Fredriksson, S. et al. 2008. Multiplexed proximity ligation assays to profile putative plasma biomarkers relevant to pancreatic and ovarian cancer. *Clin Chem*, 54, 582.

15. Smith, P.K. et al. 1985. Measurement of protein using bicinchoninic acid. *Anal Biochem*, 150, 76.

16. Plikaytis, B.D. et al. 1994. Determination of parallelism and nonparallelism in bioassay dilution curves. *J Clin Microbiol*, 32, 2441.

17. Barrette, R.W. et al. 2006. Quantifying specific antibody concentrations by enzyme-linked immunosorbent Assay using slope correction. *Clin Vaccine Immunol*, 13, 802.

18. Livak, K.J. and Schmittgen, T.D. 2001. Analysis of relative gene expression data using real-time quantitative PCR and the 2(-Delta Delta C(T)). *Methods*, 25, 402.

19. Zhu, L. et al. 2009. Proximity ligation measurement of the complex between prostate specific antigen and alpha1-protease inhibitor. *Clin Chem*, 55, 1665.

20. Schlingemann, J. et al. 2010. Novel means of viral antigen identification: Improved detection of avian influenza viruses by proximity ligation. *J Virol Methods*, 163, 116.

21. Swartzman, E. et al. 2010. Expanding applications of protein analysis using proximity ligation and qPCR. *Methods*, 50, S23.

22. Pascal, L.E. 2008. Correlation of mRNA and protein levels: Cell type-specific gene expression of cluster designation antigens in the prostate. *BMC Genomics*, 9, 243.

23. Chen, G. 2002. Discordant protein and mRNA expression in lung adenocarcinomas. *Mol Cell Proteomics*, 1, 304.

24. Xu, N. 2009. MicroRNA-145 regulates OCT4, SOX2, and KLF4 and represses pluripotency in human embryonic stem cells. *Cell*, 137, 647.

25. Baek, D. et al. 2008. The Impact of microRNAs on protein output. *Nature*, 455, 64.

26. Chang, S.T. et al. 2009. Identification of a biomarker panel using a multiplex proximity ligation assay improves accuracy of pancreatic cancer diagnosis. *J Transl Med*, 11, 105.

27. Lundberg, M. et al. 2011. Multiplexed homogenous proximity ligation assays for high throughput protein biomarker research in serological material. *Mol Cell Proteomics*, 10, 4978.

28. Pai, S.S. and Ellington, A.D. 2009. Using RNA aptamers and the proximity ligation assay for the detection of cell surface antigens. *Methods Mol Biol*, 504, 385.

29. Bjarnegard, M. et al. 2004. Endothelium-specific ablation of PDGFB leads to pericyte loss and glomerular, cardiac and placental abnormalities. *Development*, 131, 1847.

30. Gustafsdottir, S.M. et al. 2006. Detection of individual microbial pathogens by proximity ligation. *Clin Chem*, 52, 1152.

31. Gustafsdottir, S.M. et al. 2007. *In vitro* analysis of DNA-protein interactions by proximity ligation. *Proc Natl Acad Sci USA*, 104, 3067.

32. Gustafsdottir, S.M. et al. 2008. Use of proximity ligation to screen for inhibitors of interactions between vascular endothelial growth factor A and its receptors. *Clin Chem*, 54, 1218.

28 High-Resolution Melt Analysis

Einar S. Berg and Tania Nolan

CONTENTS

28.1 INTRODUCTION

Upon heating, double-stranded DNA (dsDNA) melts into single-stranded random coils. The DNA unfolding, which occurs within a narrow temperature range, can be monitored in real time by use of generic fluorescent dyes which have high affinity to dsDNA.[1] These specialized dyes have low fluorescence when they are free in solution and a large fluorescence emission upon binding to dsDNA.[2] A characteristic DNA melting curve can be generated by plotting the decrease in fluorescence as the temperature of the sample is increased, such that the duplex dissociation temperature is reached and the dye is released (Figure 28.1a).

The DNA melting temperature (T_m) is defined as the midpoint of the temperature range where DNA melts, that is, the temperature at which half of the dsDNA has melted into single-stranded random coils. By taking the negative derivative of the diminishing fluorescence with respect to temperature ($-dF/dT$ vs. T) the DNA melting curve is converted to melting peak displaying the T_m of the DNA clearly (Figure 28.1b). During the duplex dissociation process, the hydrogen bonds between the paired bases in the dsDNA are disrupted. Each G–C base pair has three hydrogen bonds and therefore gives rise to a more stable duplex and higher melting temperature than an A–T base pair, which has only two hydrogen bonds. In general, a DNA fragment with more bases and/or with a higher GC content and thus more hydrogen bonds melts at a higher temperature than a smaller or a more AT-rich fragment. To demonstrate this effect, the melting temperature of four DNA fragments of 43 bp, 53 bp, 73 bp, and 93 bp were compared. These were derived from pUC 19 sequence and modified *in silico* to contain a G/A/T/C single-nucleotide polymorphism (SNP) which was flanked by G/A/T/C combinations to take the nearest-neighbor interactions into account (Figure 28.2).

The web-based uMelt[SM] application developed by the University of Utah was used to predict the T_m of each of the fragments.[3] This is one of several programs that can be used to calculate the stability of the nearest-neighbor base interaction at each temperature step in the melt process. The dsDNA-specific fluorescent reporter dyes, such as SYBR® Green I dye, enable detection and differentiation of the various products generated in a PCR.[1] Running low resolution (or conventional) melt analysis where the temperature is raised relatively fast is straightforward for most real-time polymerase chain reaction (qPCR) instruments. Low-resolution melt analysis can be used to distinguish primer–dimers from longer PCR products and can be very useful during the initial optimization of the reaction conditions for the most specific generation of a PCR product. Notably, since the

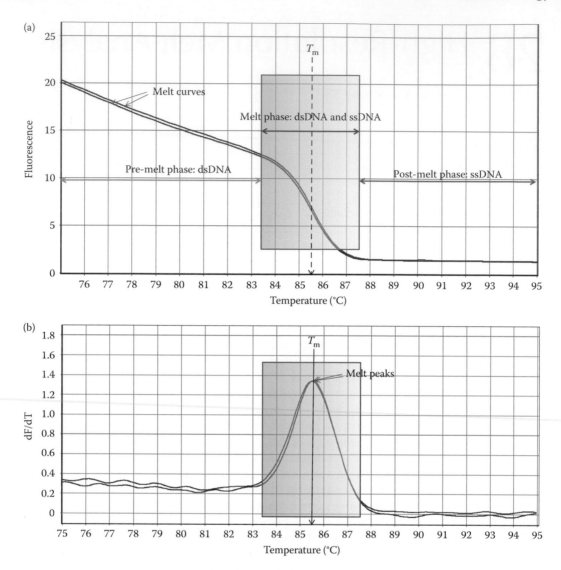

FIGURE 28.1 Melting of DNA. (a) When the temperature of the sample is increased through the melt phase, the double-strand DNA dissociates. Simultaneously the fluorescent dye is released with decrease in the fluorescence emission. (b) Melt peak curve displaying the DNA melting temperature, T_m.

post-PCR melting is a closed tube analysis any exposure of the environment to the PCR products is eliminated. Therefore, in contrast to traditional DNA separation techniques, post-PCR melt analysis is done without carry-over contamination risk.[4]

When the DNA fragments have small length differences, base substitutions, or other minor sequence changes, it is critical to consider that the neighboring bases and the interaction between the bases have a small but significant effect on the DNA melting temperature.[5] This effect can be observed, for instance, by genotyping of the four SNP classes found in the human genome. It has been claimed that high-resolution melt (HRM) instruments are necessary for detection and characterization of minor sequence variation such as SNPs, point mutations, and so on. To demonstrate the need for high-resolution capability of the melt instrument, a pairwise comparison was performed for the computer simulated T_m's for each of the 1024 SNP variants described in Figure 28.2 and plotted in Figure 28.3.

(a)   4 × 1024 artificial SNP sequence combinations based on pUC DNA

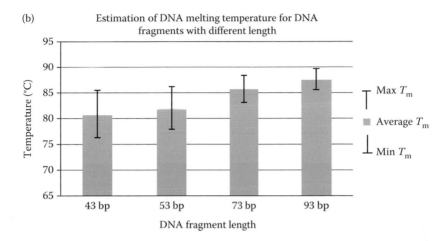

-XX—N_1—N_2—N_3—YY-

(b)

FIGURE 28.2 Prediction of the DNA melting. (a). *In silico* inserted sequence combinations of nearest neighbor nucleotides (N_1 and N_3) flanking a G/A/T/C single nucleotide polymorphism (N_2) in the core of a pUC DNA sequence. (b). The average and the range of the calculated T_m of the 1024 sequence combinations shown for DNA fragments with increasing length.

SNP class I, which has a frequency of 66%, consists of A/G (A–G or G–A) and C/T (C–T or T–C) transitions. SNP class II, which has 18% frequency, consists of C/A and G/T transversions. SNP class III and class IV have G/C- and A/T transversions and frequencies of 9% and 7%, respectively.[b] The detection of SNPs of class III (G/C) and class IV (A/T) relies upon a higher resolution of smaller temperature difference than the SNP of class I (C/T and G/A) and class II (C/A and G/T). There seems to be no practical difference in the span of the T_m shift between class I and class II SNPS, as well as between class III and class IV SNPs. In general, when the size of the DNA fragment is increased, the T_m shift between the alleles shrinks, which makes them more difficult to differentiate. As expected, the computer-simulated data gives no T_m shift for C/G and A/T SNPs that are flanked by nearest-neighbor base symmetry (e.g., –A(G/C)T–). Thus in theory, these special variants cannot be sorted directly by melting regardless of the resolution power of the HRM instrument or which dsDNA fluorescent dye is being used. Indirectly, the typing of these SNPs can yet be performed by spiking the samples with the alternative homozygote sequence to create artificial heterozygote samples, which are then identified as such by the melting curve profile, can be used to perform the typing of these SNPs.

28.2 HIGH-RESOLUTION MELT INSTRUMENTS

Currently, most real-time thermal cyclers have DNA melting analysis capability. Sixteen instruments have been compared with a focus on their SNP resolution power using DNA melting of a 110 bp β-globin gene fragment with an A/T polymorphism.[8,9] A few of the instruments included in the studies were designed specifically for HRM analysis, that is, having more accurate temperature control and ability to acquire more data points during the melting. Ranking of the instruments based on the standard deviation and error rates given in these studies is shown in Figure 28.4.

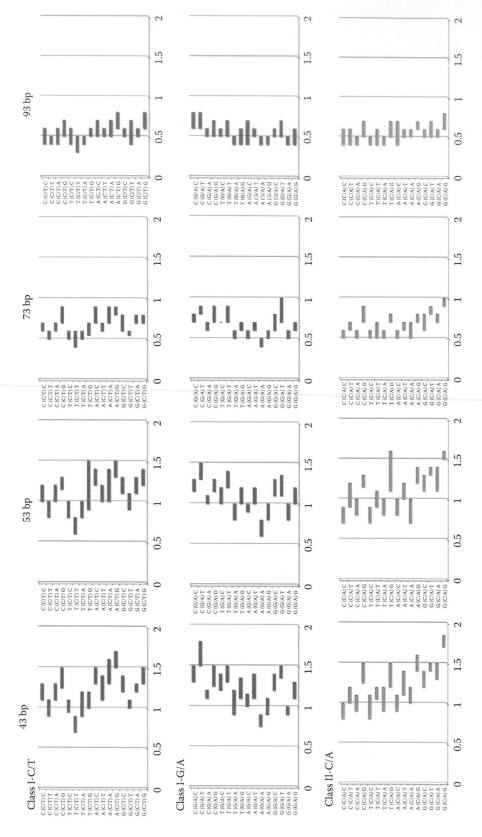

FIGURE 28.3 T_m difference spans within the SNP classes estimated for all possible nearest neighbor nucleotide combinations using artificial pUC DNA fragments with increasing length. X-axis; the calculated temperature difference [°C]. Middle grey bars; SNP class I, light grey bars; SNP class II, dark grey bars; SNP class III and SNP class IV. SNP class III and IV gives typically smaller shift than SNP class I and II.

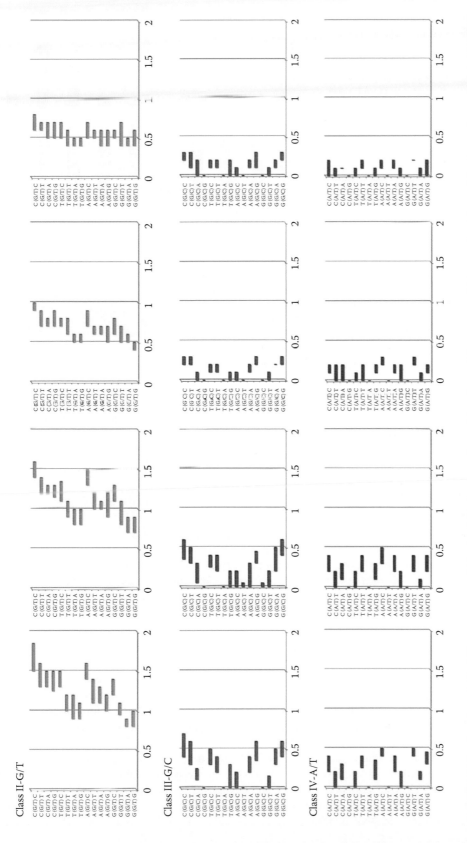

FIGURE 28.3 (continued) T_m difference spans within the SNP classes estimated for all possible nearest neighbor nucleotide combinations using artificial pUC DNA fragments with increasing length. X-axis; the calculated temperature difference [°C]. Middle grey bars; SNP class I, light grey bars; SNP class II, dark grey bars; SNP class III and SNP class IV gives typically smaller shift than SNP class I and II.

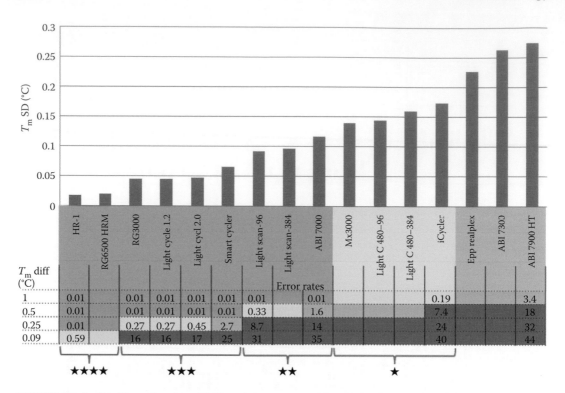

The table below accompanies the figure:

T_m diff (°C)	HR-1	RG6500 HRM	RG3000	Light cycle 1.2	Light cycl 2.0	Smart cycler	Light scan-96	Light scan-384	ABI 7000	Mx3000	Light C 480–96	Light C 480–384	iCycler	Epp realplex	ABI 7300	ABI 7900 HT
1	0.01		0.01	0.01	0.01	0.01	0.01		0.01				0.19			3.4
0.5	0.01		0.01	0.01	0.01	0.01	0.33		1.6				7.4			18
0.25	0.01		0.27	0.27	0.45	2.7	8.7		14				24			32
0.09	0.59		16	16	17	25	31		35				40			44

(Error rates)

★★★★   ★★★   ★★   ★

FIGURE 28.4   Ranking of various DNA melt instruments. The shade of grey color clustering in the table is based on consideration of the reported standard deviation together with the error rates found by repeated melting of a 110 bp DNA fragment. To the left are the instruments which might resolve small. The star classification is derived from the apparent grouping of the resolution capability.

The relative small T_m shifts measured by SNP genotyping are influenced by both the length and base composition of the DNA fragments. The overall span between minimum and maximum T_m shift for each of the SNP class combinations for the 4×1024 *in silico* analyzed SNP fragments are plotted in Figure 28.5.

The data illustrate that HRM capacity is necessary. As a general rule, shorter PCR products display HRM differences better than longer products. The instruments classified with one or zero stars in Figure 28.4 have insufficient precision for genotyping of any SNP containing fragment with size above 73 bp. The instruments classified with three and four stars enable discrimination and genotyping of all class I and class II SNPs. However, with SNP class III and IV the likelihood that they may fail increases with the fragment size.

28.3   HIGH-RESOLUTION MELT ANALYSIS

In principle, HRM runs in the same way as classical low-resolution melting, except that more melt data points are acquired as the temperature is raised. The instruments with dedicated HRM functions, such as Rotor-Gene Q HRM, emit higher-intensity fluorescent light than standard qPCR machines. A typical qPCR followed by both classical melt and HRM peak analysis is shown in Figure 28.6.

In this experiment, the template DNA in the PCR was diluted through a 6 log series of four 73-base-long oligonucleotides, each with a G, A, T, or C in a central position for mimicking SNP homozygote situations. The sequences of the T and C variants were identical to an SNP upstream the IL28 gene (rs12979860). Using EvaGreen® as dsDNA-specific dye and the Corbett's Rotor-Gene 6500 HRM instrument, the DNA melting analysis revealed peak curves from the HRM and normal

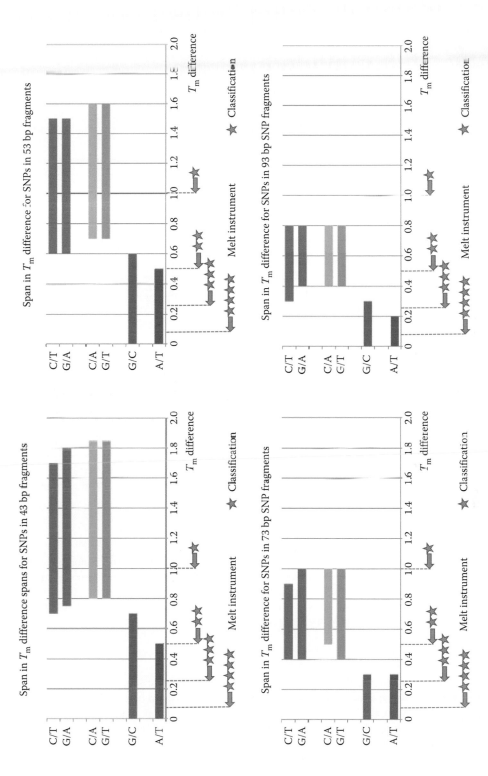

FIGURE 28.5 The resolution power for the star-classified melt instruments by genotyping of SNPs in fragments with different length. Increased fragment length gives smaller temperature differences which excludes melts instruments with fewer stars, that is, with too low resolution power. SNP class I (C/T or G/A), SNP class II (C/A or G/T), SNP class III (G/C), SNP class IV (A/T).

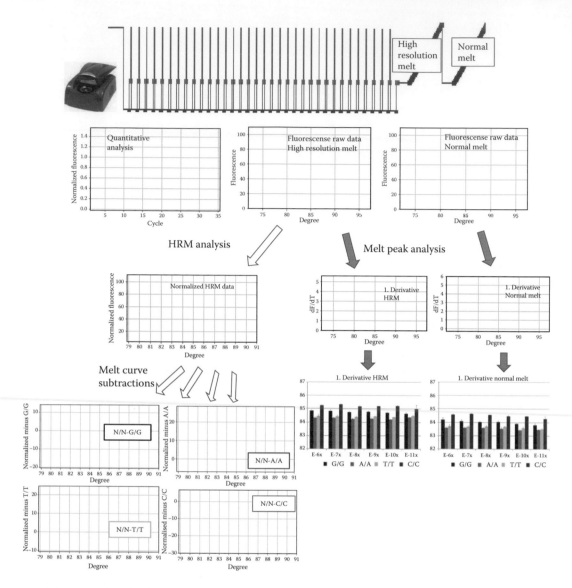

FIGURE 28.6   (See color insert.) The progress of qPCR and melt analysis in Corbett's Rotor-Gene 6500 HRM instrument. PCR input was 10× dilution series of homozygous DNA. Initially, the PCR reveals quantitative analysis data. Subsequently, after the melting, the raw melt data are processed in melt peak analysis to show the T_ms of the samples. Plotting of these results in bar diagrams can easily be done. Alternatively, the raw HRM data can be subjected to special HRM analysis algorithm which normalizes each of the melt curves and then may subtract a predefined reference curve from the sample curves. The resulting melt difference curves are especially beneficial by the genotyping of heterozygote samples.

melting that were very similar. Furthermore, the experiment revealed the same T_ms for each of the four alleles regardless of whether they were amplified from a concentrated or highly diluted sample. By use of the special HRM analysis software provided with the instrument a more detailed analysis of the high-resolution melting data could be performed. The HRM normalization processing is based on the pre- and post-melting fluorescence measurements, allowing the inherent fluorescence variance between the samples to be eliminated. The HRM difference curve is created by subtracting each point of a user-defined standard curve from the sample curve. The resulting graph highlights the curve differences and allows automatic classification of the SNP variants.[10]

28.4 PERFORMANCE OF FOUR HIGH-RESOLUTION MELT DYES

Ethidium bromide was initially used as the stain for qPCR products.[11] The dye has a large Stoke's shift, that is, with short wavelength excitation maxima (300 and 360 nm) and substantial longer wavelength emission maximum (590 nm), which is not compatible with the narrow light sources/detectors in the HRM instruments. Even so, by creating a new channel with special excitation/emission wavelengths of 470/610 nm, ethidium bromide can be used for regular melt analysis, but not for HRM, in Rotorgene thermocyclers. The novel and brighter dyes such as SYBR Green I,[1] EvaGreen,[12,13] LCGreen® Plus,[10] and SYTO®9 [14] are now the preferred dsDNA stains for both HRM and regular melt analysis. These dyes have narrow Stoke's shift with excitation/emission wavelengths similar to fluorescein. Ideally any dye used for HRM should not bind preferentially to pyrimidines or purines, change the T_m of the amplicon, or inhibit the PCR. None of the dyes listed above satisfy all of these requirements. For instance, all of them are inhibitory to the PCR at higher concentrations. Hence the right sub-inhibitory working concentration of the dye must be determined empirically prior to use. At the same time, the dye concentration must be balanced since the use of too little will diminish the target fluorescence signal into the background level. It has been noted that variation of the reaction conditions, that is, ionic strength and cation concentration as well as use of reagents such as formamide, DMSO, can result in the stabilizing/destabilizing of hydrogen bonds (see Chapter 4). The temperature transition rate may also have an effect on the T_m of an amplicon. Furthermore, increasing the dye concentration increases the T_m of the DNA. Therefore, the reaction conditions during the PCR and subsequent melt analysis must be standardized.[1] Normally the T_m of a DNA fragment increases with the nucleic acid concentration. [15] But since different PCR samples tend to plateau at the same product concentration, the T_m of a certain amplicon will be the same.[16] This was demonstrated in Figure 28.6 where both very high- and low-input DNA copy number of the samples were tested. It been alleged that SYBR Green I is not suited for HRM analysis since this dye is not capable of saturating the amplicon at the low concentration needed for running noninhibitory PCR.[17] Experiments aimed to compare the DNA saturating ability of the four minor groove binding dyes revealed no striking difference between the dyes (Figure 28.7).

SYTO9 is exceptionally low-inhibitory and can thus be used in relative high concentrations.[14] The diminishing fluorescence signal demonstrated in Figure 28.7 at high SYTO9 concentration might be explained by self-quenching caused by unbound dye molecules. This phenomenon has been observed by testing of higher concentrations of SYBR Green I dye.[15] As presented in the green areas in Figure 28.7, it seems common for all of the dyes to give the best genotyping of SNP homozygotes at concentrations well below the saturation concentration, that is, at where the curves level. Several studies have reported that SYBR Green I molecules, in contrast to LCGreen Plus, may redistribute from shorter and/or more unstable fragments during the melting, to longer/more stable fragments and thereby display a favorable detection of these.[10,18] It has been postulated that a major disadvantage by SYBR Green I dye is its inability to saturate the DNA in melt analysis.[7,10,19] Since relative low concentrations of all of the dyes gives the best genotyping results, the dyes' saturating/nonsaturating ability seems not relevant. Further work is needed to study the kinetics of the binding of the dyes to DNA.

28.5 SNP GENOTYPING WITH HRM

To demonstrate the SNP genotyping ability of SYBR Green I, LCGreen Plus, EvaGreen, and SYTO9, the dyes were titrated to their optimal concentration for genotyping in Rotor-Gene 6500 HRM instrument. The results of this experiment are shown in Figure 28.8.

Amplifying each of oligonucleotides with G/A/T/C SNP sequence separately (as in Figure 28.6) gave homozygous PCR products. Heterozygote PCR products were made by mixing different pairs of the oligonucleotides prior to the amplification. The comparison of the four dyes

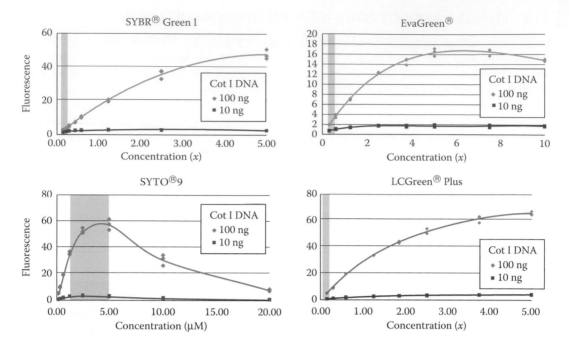

FIGURE 28.7   (See color insert.) The saturating/nonsaturating ability of four dsDNA-specific fluorescent dyes. At extremely high dye concentration all dyes saturated the Cot I DNA. At lower dye concentrations, which would be compatible with noninhibitory PCR, only SYTO9 seems to be able to saturate a typical PCR product at melting. The window for the optimum dye concentration shown as green bars, narrows further when the SNP homozygote and heterozygote genotyping abilities of the dyes are considered.

showed no difference in their ability for detection of the homozygotes. However, especially for SYBR Green I the detection of the heterozygotes was difficult from the melt peak analysis. The difference curve analysis using the least stable A/A homozygote as reference made the interpretation of the data easier. Notably, the analysis revealed that SYBR Green I did not enable separated detection of the G/C heterozygote from the homozygote samples. EvaGreen gave difference curve analysis results which could be classified as tolerable with discriminated detection of heterozygotes from homozygotes. SYTO9 and LCGreen Plus gave the most robust genotyping results. The heterozygote samples could be clearly distinguished from the homozygotes, also by the melt peak analysis. Using these two dyes in low-resolution melt analysis with slow temperature transition rate, as by HRM (data not shown) gave nearly the same melt peak curves. Thus, SYTO9 and LCGreen Plus seem to be applicable for SNP genotyping also in non-HRM instruments for amplicons with smaller sizes.

28.6   DETECTION OF MUTANTS IN WILD-TYPE POPULATIONS WITH HRM

Both traditional Sanger sequencing and the Pyrosequencing technology can be used for quantitative detection of bacteria- and virus mutants in wild-type populations as well as for quantitative detection of mutations associated with cancer diagnostics/treatment.[20–22] In contrast to HRM analysis, the sequencing techniques require additional and open treatments of the PCR product which involve carry-over contamination risk. The sequencing techniques have an analytical sensitivity of 5–25% for detection of minor sequence variants in populations of wild-type sequences. The ability of HRM to detect mutants in wild-type populations is demonstrated in Figure 28.9.

The PCR target was the YMDD motif with the catalytic site of the hepatitis B virus reverse transcriptase where lamivudine resistance may occur, that is, methionine 204 to isoleucine

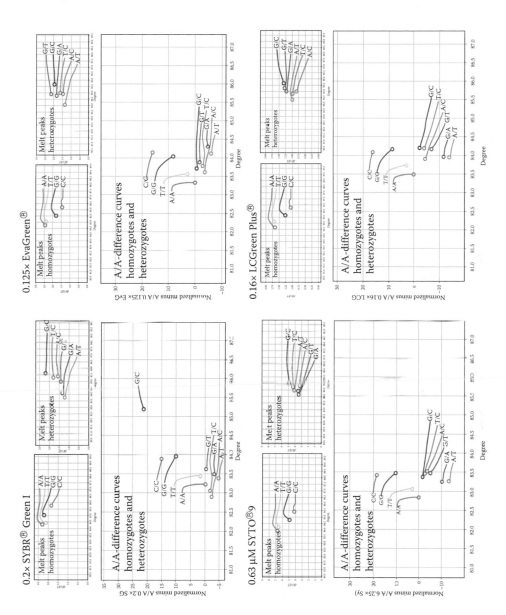

FIGURE 28.8 (**See color insert.**) Comparison of the SNP homozygote and heterozygote genotyping ability of SYBR Green I, LCGreen Plus, EvaGreen, and SYTO9. The homozygote and heterozygote melt peak analyses are shown separately for visualization purposes. By the preparation of the difference curves the homozygous A/A-melt curve was used as reference since this was the least stable homozygote genotype, and thereby separated the heterozygotes from the homozygotes most efficiently.

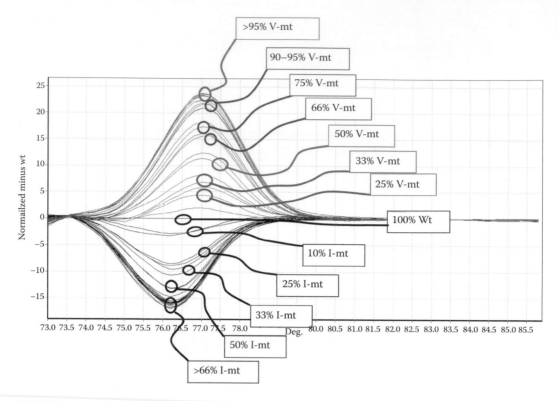

FIGURE 28.9   HRM difference curve analysis of mixed mutant/wild-type virus populations. Melt analysis for detection of hepatitis B virus lamivudine resistance, that is, wild-type methionine (reference curve) to iso-leucine (lower curves) or valine (upper curves) mutants can be an alternative to traditional DNA sequencing. Dependent on the mutation/wild-type sequence context the HRM analysis seems to enable semi-quantitative detection of ratios of wild-type/mutant population comparable with DNA sequencing techniques.

(rtM204I) or valine (rtM204V). The figure demonstrates that the post-PCR HRM analysis can measure the wild-type/mutant ratios at least as efficiently as the sequencing techniques. Advantageously, the HRM analysis is a closed-tube format that can easily be repeated several times. If necessary, the DNA can be subjected to, for example, Pyrosequencing for confirmation of the HRM results.

28.7   CONCLUSION

Successful HRM is dependent on meticulous optimization of PCR conditions as the presence of nonspecific bands and primer dimers can significantly reduce HRM performance. Subsequent stan-dardization of the reaction conditions is important for comparison of the HRM data to enable reli-able genotyping. The use of computer programs can facilitate the design of qPCR and HRM assays. As a general rule, shorter PCR products have lower T_ms and display HRM differences better than longer products. The resolution power of the melt instrument is important, especially for genotyping of SNP class III and IV. Melting of homozygous PCR products gives profiles with a single peak with a characteristic T_m. Heterozygous PCR products consist of both homo- and heteroduplexes and thus have more complex melt profile. Such samples are most reliably interpreted by the HRM difference curve analysis. SYTO9 appears to be an exceptional dsDNA-specific dye that gives robust melt data besides being the least inhibitory to PCR. In general, since the DNA remains intact after HRM, the PCR product can be remelted, or retested by, for example, pyrosequencing.

REFERENCES

1. Ririe, K.M., R.P. Rasmussen RP, and C.T Wittwer. 1997. Product differentiation by analysis of DNA melting curves during the polymerase chain reaction. *Anal Biochem* 245:154–160.
2. Vitzthum, F., G. Geiger, H. Bisswanger, H. Brunner, and J.A. Bernhagen J. 1999. A quantitative fluorescence-based microplate assay for the determination of double-stranded DNA using SYBR® Green I and a standard ultraviolet transilluminator gel imaging system. *Anal Biochem* 276:59–64.
3. Dwight, Z., R. Palais, and C.T. Wittwer. 2011. uMELT: Prediction of high-resolution melting curves and dynamic melting profiles of PCR products in a rich web application. *Bioinformatics* 27:1019–1020.
4. Reed, G.H., and C.T. Wittwer. 2004. Sensitivity and specificity of single-nucleotide polymorphism scanning by high-resolution melting analysis. *Clin Chem* 50:1748–1754.
5. SantaLucia, J. Jr. 1998. A unified view of polymer, dumb bell, and oligonucleotide DNA nearest-neighbor thermodynamics. *Proc Natl Acad Sci USA* 95:1460–1465.
6. Zhao, Z., and E. Boerwinkle. 2002. Neighboring-nucleotide effects on single nucleotide polymorphisms: A study of 2.6 million polymorphisms across the human genome. *Genome Res* 12:1679–1686.
7. Liew, M., R. Pryor, R. Palais, C. Meadows, M. Erali, E. Lyon, and C. Wittwer. 2004. Genotyping of single-nucleotide polymorphisms by high-resolution melting of small amplicons. *Clin Chem* 50:1156–1164.
8. Herrmann, M.G., J.D. Durtschi, L.K. Bromley, C.T. Wittwer, and K.V. Voelkerding. 2006. Amplicon DNA melting analysis for mutation scanning and genotyping: Cross-platform comparison of instruments and dyes. *Clin Chem* 52:494–503.
9. Herrmann, M.G., J.D. Durtschi, C.T. Wittwer, and K.V. Voelkerding. 2007. Expanded instrument comparison of amplicon DNA melting analysis for mutation scanning and genotyping. *Clin Chem* 53:1544–1548.
10. Wittwer, C.T., G.H. Reed, C.N. Gundry, J.G. Vandersteen, and R.J. Pryor. 2003. High-resolution genotyping by amplicon melting analysis using LC Green. *Clin Chem* 49:853–860.
11. Higuchi, R., C. Fockler, G. Dollinger, and R. Watson. 1993. Kinetic PCR analysis: Real-time monitoring of DNA amplification reactions. *Biotechnology* 11:1026–1030.
12. Wang, W., K. Chen, and C. Xu. 2006. DNA quantification using Eva Green and a real-time PCR instrument. *Anal Biochem* 356:303–305.
13. Mao, F., W.Y. Leung, and X. Xin. 2007. Characterization of Eva Green and the implication of its physicochemical properties for qPCR applications. *BMC Biotechnol* 7:76–92.
14. Monis, P.T., S. Giglio, and C.P. Saint CP. 2005. Comparison of SYTO9 and SYBR Green I for real-time polymerase chain reaction and investigation of the effect of dye concentration on amplification and DNA melting curve analysis. *Anal Biochem* 340:24–34.
15. Lipsky, R.H., C.M. Mazzanti, J.G. Rudolph, K. Xu, G. Vyas, D. Bozak, M.Q. Radel, and D. Goldman. 2001. DNA melting analysis for detection of single nucleotide polymorphisms. *Clin Chem* 47:635–644.
16. Gundry, C.N., J.G. Vandersteen, G.H. Reed, R.J. Pryor, J. Chen, and C.T. Wittwer. 2003. Amplicon melting analysis with labeled primers: A closed-tube method for differentiating homozygotes and heterozygotes. *Clin Chem* 49:396–406.
17. Wittwer, C.T., M.G. Herrmann, A.A. Moss, and R.P. Rasmussen. 1997. Continous fluorescence monitoring of rapid cycle DNA amplification. *Biotechniques* 22:130–131, 134–138.
18. Barrett, J., P. Kayser, S. Billingham, N. Simpson, W. Chong, and M. Stevens. 2007. High resolution melt analysis with non-saturating dyes. www.gene-quantification.de/qpcr2007/P064-qPCR-2007.pdf.
19. Reed, G.H., J.O. Kent, and C.T. Wittwer. 2007. High-resolution DNA melting analysis for simple and efficient molecular diagnostics. *Pharmacogenomics* 8:597–608.
20. Clarke, S.C. 2005. Pyrosequencing: Nucleotide sequencing technology with bacterial genotyping applications. *Expert Rev MolDiagn* 5:947–953.
21. Lindström, A, J. Odeberg, and J. Albert. 2004. Pyrosequencing for detection of lamivudine-resistant hepatitis B virus. *J Clin Microbiol* 42:4788–4795.
22. Shen, H., Y. Yuan, H.G. Hu, X. Zhong, X.X. Ye, M.D. Li, W.J. Fang, and S. Zheng. 2011. Clinical significance of K-ras and BRAF mutations in Chinese colorectal cancer patients. *World J Gastroenterol* 17:809–816.

Index